CHEMICAL SENSORS

COMPREHENSIVE SENSOR TECHNOLOGIES

VOLUME 6: CHEMICAL SENSORS APPLICATIONS

CHEMICAL SENSORS

COMPREHENSIVE SENSOR TECHNOLOGIES

VOLUME 6: CHEMICAL SENSORS APPLICATIONS

EDITED BY

GHENADII KOROTCENKOV

GWANGJU INSTITUTE OF SCIENCE AND TECHNOLOGY
GWANGJU, REPUBLIC OF KOREA

MOMENTUM PRESS

MOMENTUM PRESS, LLC, NEW YORK

Chemical Sensors: Comprehensive Sensor Technologies. Volume 6: Chemical Sensors Applications

First published by Momentum Press®, LLC
222 East 46th Street, New York, NY 10017
www.momentumpress.net

ISBN-13: 978-1-60650-239-6 (hard back, case bound)
ISBN-10: 1-60650-239-5 (hard back, case bound)
ISBN-13: 978-1-60650-241-9 (e-book)
ISBN-10: 1-60650-241-7 (e-book)
DOI forthcoming

Cover design by Jonathan Pennell
Interior design by Derryfield Publishing, LLC

10 9 8 7 6 5 4 3 2 1

Printed in Canada

Contents

Preface to Chemical Sensors: Comprehensive Sensor Technologies

In spite of their century-long history, chemical sensors appeared on the commercial market only 50 years ago. In recent years, however, the field of chemical sensors has broadened and expanded greatly. At present, chemical sensors are being used in medicine, agriculture, industry, transport, environmental control, and other fields. However, the process of developing new sensors as well as improving older types of chemical sensors is still ongoing. New technologies and the toughening of ecological standards require more sensitive instruments with faster response times, better selectivity, and improved stability. The second half of this six-volume series on chemical sensors, devoted to *comprehensive sensor technologies,* describes these developments and the new processes and applications. These volumes are intended to be a primary source for both fundamental and practical information about where sensor technologies are now and where they are headed for the future. We are sure that Volumes 4–6 in this series will be a useful addition to the first three volumes, on *fundamentals of sensing materials,* in which various sensing materials that can be used in chemical sensors are discussed in detail. Analysis of chemical sensor design, fabrication, and functioning requires other approaches to description in comparison with materials science problems, and therefore we decided that consideration of materials and devices should be carried out separately. From our point of view, dividing the series into two parts as we have done results in more logical narration and more utility for readers who are interested in different aspects of chemical sensor design.

In this series we provide readers with a thorough understanding of the concepts behind chemical sensors, presenting the information necessary to develop such sensors, covering all aspects including fundamental theories, fabrication, functionalization, characterization, and real-world applications, so as to enable them to pursue their research and development requirements. Therefore, we hope that this series will help readers understand the present status of chemical sensors and will also act as an introduction, which may encourage further study, as well as an estimate of the roles that chemical sensors may play in the future.

Chemical Sensors: Comprehensive Sensor Technologies is a three-volume series, comprising Volumes 4, 5, and 6 in our series, *Chemical Sensors.* Volume 4 deals with solid-state devices, Volume 5 with electrochemical and optical sensors, and Volume 6 with applications of chemical sensors. The chapters included in the volumes consist of review and overview papers written by experts in the field. The authors of each of the chapters were chosen very carefully and are all well known throughout the world in their fields of study. Therefore, these books provide an up-to-date account of the present status of chemical sensors, from fundamental science and processing to applications.

Specifically, Volume 4 includes descriptions of solid-state sensors such as conductometric or resistive gas sensors, Schottky-, FET-, capacitance-, and pyroelectric-type chemical sensors. Pellistors, mass-sensitive, and acoustic wave sensors are described as well. Integrated chemical sensors are also discussed in Volume 4. Volume 5 provides information related to electrochemical and optical sensors. Fundamentals of operation, methods of fabrication, and operating characteristics of electrochemical gas sensors, solid electrolyte–based gas sensors, ion-selective electrodes, CHEMFETs, and different types of optical, fiber optical, and chemoluminescence chemical sensors are discussed. Volume 6 is dedicated to detailed examination of opportunities for applications of chemical sensors in various areas of our lives, including medicine, industry, environmental control, agriculture, and transportation. It is the editor's wish that theses volume will provide the reader with a detailed understanding of the many applications of chemical sensors in both today's world and that of the future. In these chapters one can also find descriptions of architecture and fundamentals of "electronic noses" and "electronic tongues," principles of wireless chemical sensor design, and possibilities for remote chemical sensing for atmospheric monitoring.

In this three-volume series, the authors present sensors that utilize various sensing materials and phenomena. The terminology and concepts associated with sensors are presented, including some of the relevant physical and chemical phenomena applied in the sensor signal transduction system. As is well known, chemical sensing is multidisciplinary by nature. The role of sensing materials in such phenomena is also detailed.

We need to note that the number of disciplines involved in the research and design of chemical sensors has increased dramatically. New knowledge and approaches are needed to achieve miniaturization, lower power consumption, and the ability to operate in complex environments for more selective, sensitive, and rapid determination of chemical and biological species. Compact analytical systems that have a sensor as one of the system components are becoming more important than individual sensors. Thus, in addition to traditional sensor approaches, a variety of new themes have been introduced to achieve the attractive goal of analyzing chemical species on the micro and nano scales. Therefore, throughout these books, numerous strategies for the fabrication and characterization of sensing materials and sensing structures which are employed in sensing applications are provided, and current approaches for chemical sensing are described.

This series can be utilized as a text for researchers and engineers as well as graduate students who are either entering the field for the first time, or who are already conducting research in these areas but are willing to extend their knowledge of the field of chemical sensors. We hope that these volumes will also be of interest to undergraduate students in chemical engineering, electronics, environmental control, and medicine. These books have been written in a way that final-year and graduate university students in the fields of chemistry, physics, electronics, biology, biotechnology, mechanics, and bioengineering can easily comprehend. We believe that practicing engineers or project managers which would like to use chemical sensors but don't know how to do so, and how to select optimal chemical sensors for specific applications, also will find useful information.

It is necessary here to comment briefly on the coverage of the literature. During our work on this series we tried to cover the field more or less completely. However, we need to acknowledge that an appreciable number of relevant papers may remain unknown to the authors. Regarding these, the editors and contributing authors express regret, not only to the authors of such works, but also to the readers of our books.

Finally, we wish to thank all those who participated in the preparation of this series, including the contributing authors and copyright owners in Europe, the United States, Asia, and the rest of the world. We also wish to express our gratitude to the staff of Momentum Press, and in particular Joel Stein, for his kind assistance in bringing these volumes to fruition.

Ghenadii Korotcenkov

Preface to *Volume 6:* *Chemical Sensors Applications*

The market for chemical sensors continues to grow at a rapid rate, reflecting the wide range of possibilities for improving technological processes in industry and agriculture as well as living conditions that can be enhanced by the use of chemical sensors. The military, medicine, air/space, and security markets also continue to drive research and development in this area. At present it is hard to imagine an area where chemical sensors would be useless. On the contrary, we note that every day new areas arise in which new analytical instrumentation with modern functional opportunities is urgently needed.

It is necessary to admit, however, that the development of commercially successful chemical sensors is expensive and technically difficult, and also requires sophisticated market knowledge. This volume provides that knowledge. Just a list of some of the topics discussed—such as biomedical applications, water quality monitoring, food quality monitoring, environmental control, household applications, automotive applications, industrial processes, fire alarm systems, agricultural applications, law enforcement applications, air/space applications, military applications, etc.—provides a clear idea of the extensive analysis presented here, and how broad a perspective of the emerging sensor industry may be acquired.

Proper attention in this book is given to consideration of the principles and construction of "electronic noses" and "electronic tongues." As is becoming widely recognized, the ability to design such devices represents a huge achievement by designers of chemical sensors.

Wireless sensors and remote chemical sensing, which are also discussed in this volume, constitute another trend in chemical sensors design, focused on building systems for global and local environmental monitoring. Nowadays, as our ecosystem is changing so rapidly and human-caused disasters occur more often, establishing such monitoring has become a priority.

Researchers, development engineers, and students can use this volume as a reference source during their work and study according to their interests. We hope that this book and the series as a whole will help all those concerned with sensor research, development, and application to succeed in this promising field. The volume contains comparisons and assessments of the various types of sensors with respect to their practical applications. Therefore, the users of chemical sensors will also benefit.

We believe that readers of this book will be able to do the following:

- Estimate the area of possible applications of chemical sensors.
- Know the advantages and disadvantages of various different types of chemical sensors.

- Understand the special requirements for sensors used in specific applications.
- Estimate the reliability of chemical sensors.
- Determine which sensor types are most appropriate for a given application.

Ghenadii Korotcenkov

About the Editor

Ghenadii Korotcenkov received his Ph.D. in Physics and Technology of Semiconductor Materials and Devices in 1976, and his Habilitate Degree (Dr.Sci.) in Physics and Mathematics of Semiconductors and Dielectrics in 1990. For a long time he was a leader of the scientific Gas Sensor Group and manager of various national and international scientific and engineering projects carried out in the Laboratory of Micro- and Optoelectronics, Technical University of Moldova. Currently, he is a research professor at Gwangju Institute of Science and Technology, Gwangju, Republic of Korea.

Specialists from the former Soviet Union know G. Korotcenkov's research results in the study of Schottky barriers, MOS structures, native oxides, and photoreceivers based on Group III–V compounds very well. His current research interests include materials science and surface science, focused on metal oxides and solid-state gas sensor design. He is the author of eight books and special publications, 11 review papers, 10 book chapters, and more than 180 peer-reviewed articles. He holds 18 patents. He has presented more than 200 reports at national and international conferences. His articles are cited more than 150 times per year. His research activities have been honored by the Award of the Supreme Council of Science and Advanced Technology of the Republic of Moldova (2004), The Prize of the Presidents of Academies of Sciences of Ukraine, Belarus and Moldova (2003), the Senior Research Excellence Award of Technical University of Moldova (2001, 2003, 2005), a Fellowship from the International Research Exchange Board (1998), and the National Youth Prize of the Republic of Moldova (1980), among others.

CONTRIBUTORS

Beongki Cho (Chapter 7)
Department of Material Science and Engineering
and
Department of Nanobio Materials and Electronics
Gwangju Institute of Science and Technology
Gwangju 500-712, Republic of Korea

Zhuang Dafang (Chapter 4)
Data Center for Resources and Environmental Sciences
State Key Lab for Resources and Environmental Information Systems
Institute of Geographical Sciences and Natural Resources Research
Chinese Academy of Sciences
Beijing 100101, China

Saverio De Vito (Chapter 3)
Renewable Energy and Environment Advanced Physical Technologies Department
ENEA—Italian Institute for New Technologies
1-80055 Portici (NA), Italy

Jiang Dong (Chapter 4)
Data Center for Resources and Environmental Sciences
State Key Lab for Resources and Environmental Information Systems
Institute of Geographical Sciences and Natural Resources Research
Chinese Academy of Sciences
Beijing 100101, China

Alessandra Flammini (Chapter 3)
Department of Electronics for Automation
University of Brescia
38-25123 Brescia, Italy

Gábor Harsányi (Chapters 5 and 6)
Department of Electronics Technology
Budapest University of Technology and Economics
1521 Budapest, Hungary

Ghenadii Korotcenkov (Chapters 1 and 7)
Department of Material Science and Engineering
Gwangju Institute of Science and Technology
Gwangju 500-712, Republic of Korea
and
Technical University of Moldova
Chisinau 2001, Republic of Moldova

Joseph R. Stetter (Chapter 1)
KWJ Engineering, Inc.
Newark, California 94560, USA

Fredrik Winquist (Chapter 2)
The Swedish Sensor Centre and the Division of Applied Physics
Department of Physics and Measurement Technology
Linköping University
SE-581 83 Linköping, Sweden

Huang Yaohuan (Chapter 4)
Data Center for Resources and Environmental Sciences
State Key Lab for Resources and Environmental Information Systems
Institute of Geographical Sciences and Natural Resources Research
Chinese Academy of Sciences
Beijing 100101, China

CHAPTER 1

Chemical Gas Mixture Analysis and the Electronic Nose

Current Status, Future Trends

G. Korotcenkov
J. R. Stetter

1. INTRODUCTION

Today's chemical sensors do not have high selectivity, making it difficult to determine the nature of gas molecules in the atmosphere. At the same time this task in many cases is a determining one, in that identification of molecules can provide information necessary for identification of the source of either toxic gases or odor.

The sense of smell is one of the most important senses for human beings, often controlling behavior. For example, the perception of volatile compounds by the human nose is of great importance in evaluating the quality of foods and cosmetics, and thus can enrich our daily lives. Sensory analysis by a panel of experts is a costly process for industries, because it requires trained people who can work for only relatively short periods of time. Therefore, scientists have long been thinking about elaboration of an electronic analog of the human nose—an electronic nose or e-nose—that would not be subjective in evaluation and would have a long life.

The task of odor identification is a complicated one. Odors can be complex mixtures of many hundreds of chemical species, and often even subtle changes in the relative amounts of even one of these species can be detected as a change in the odor. The concentration of those odorant molecules may be as low as a few parts per trillion. We are learning more about the process of olfaction and about odorant molecules, many of which are apparently low-molecular-weight molecules that are partly hydrophobic and sometimes have one polar group, frequently containing an oxygen, nitrogen, or sulfur moiety (Strike et al. 1999).

2. MODERN METHODS OF GAS RECOGNITION

There are many methods for analysis of gases that provide both qualitative and quantitative results. These methods involve obtaining and using multidimensional data. In this section we survey common techniques that have, at least at one time, been called electronic noses. Descriptions of those methods can be found in books and reviews such as those of Kaplan and Braham (1998), Pearce et al. (2003), and Rock et al. (2008).

2.1. GAS CHROMATOGRAPHY

Gas chromatography (GC) is an old and well-accepted technique for the measurement and identification of gases. Simple descriptions (Kaplan and Braham 1998) include a common type of gas chromatographic equipment consisting of a small capillary tube with an interior diameter of about 25–250 μm and a length of 1–30 m (see Figure 1.1).

The inside wall of the hollow tube (the column) is coated with a thin polymeric film, of from one-tenth to a few micrometers thick. As odorants are transported down the tube by the mobile phase (gas), the film, or stationary phase, interacts with the mixture, selectively impeding the progress of each of the components of the mixture toward the outlet. The interaction of the analytes and film depends on the physical and chemical properties of the gas, such as the boiling point, the polarity, hydrogen bonding, polarizability, etc., and the affinity of each single substance for the stationary-phase film material.

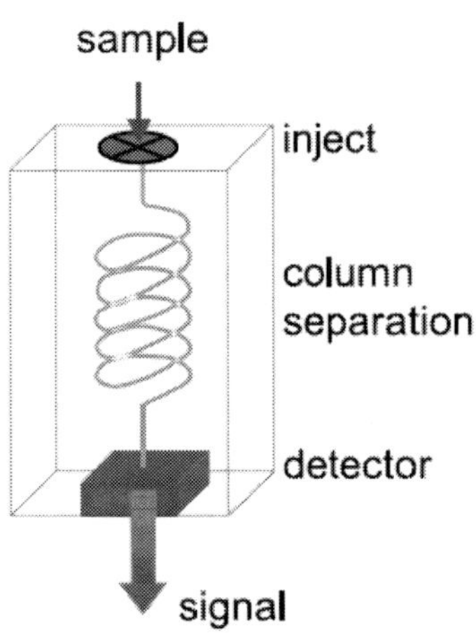

Figure 1.1. Essential parts of a micro-gas chromatograph, which can be built on a single substrate and can be called a micro total analytical system (μTAS).

Therefore, the transport time of each of the various molecular constituents of the mixture vary with their vapor pressure and/or solubility, causing them to emerge from the column outlet separately. One can think of the process as an auto race in which all the contestants begin at the same time in a group, but stretch out and arrive in single file at the finish line. The partitioning behavior in gas/solid and gas/liquid chromatography has been well studied for many film materials and many gas mixtures and can be found in basic textbooks on analytical chemistry (Strobel and Heineman 1989; Willard et al. 1988; Skoog et al. 2007).

While there are many choices for detectors for the eluting components at the end of the column, a gas sensor can often be used to generate a peak-shaped current response with the emergence of each of the constituent gases. The height and the area of the peaks is a measure of the amount of each of the odorant constituents in the carrier gas. While the most common sensor for hydrocarbons is a flame ionization detector, in which a flame heats a small gap between two wires such that each constituent gas produces an electrical current as it passes through the flame, many detectors have been used, both simple ones such as amperometric gas sensors (Stetter 1986) and complex ones such as mass spectrometers (Strobel and Heineman 1989).

The GC method is a batch process and not continuous but, while typically taking some minutes to accomplish, can be completed in less than a second for some mixtures (Eiceman et al. 2006). For any selected chromatographic method, the batch analysis can be optimized by selection of optimum operational parameters including the carrier gas flow rate, an increase of the temperature-program heating rates, a reduction of the column length, and/or a reduction of the thickness of the stationary phase (Rock et al. 2008). Depending on the sample, it is important to avoid using all possibilities at once, because this always results in a decrease of resolution, sample capacity, or both. It is also important to note that these optimizations increase the demands on the detector technology used in terms of sensitivity, speed, and dead volume.

Recent developments in GC instrumentation, using the latest technologies for fabrication of the chromatographic column, have considerably shortened the time needed for analysis .

2.2. MASS SPECTROMETRY

Mass spectrometry (MS) is another method which can be used for identification of a gas or gas mixture (Richardson 2000; Schalley 2001; Mano and Gato 2003; Vinaixa et al. 2004). The first instruments of an e-nose type that were based on MS were developed at the end of the 1990s (Dittmann et al. 1998). Some researchers do not consider the MS system used in this manner as an e-nose (Mielle et al. 2000) and prefer to leave the MS detector in its classical role in analytical science, while others report these instruments as "mass sensors" or "new-generation electronic noses" (Peres et al. 2003). It is a bit of semantics, but MS has been used, like GC, with pattern recognition for identification of gas mixtures in a parallel fashion to the sensor array–based e-nose.

The mass spectrometer contains an inlet, ionization chamber, accelerator, mass filter, and ion detector in a vacuum and represents a process for separation and subsequent detection of molecules or atoms. In MS, each constituent compound molecule is ionized, typically by electron beam (electron ionization) or through interaction with reagent ions (chemical ionization), and the energy absorbed by

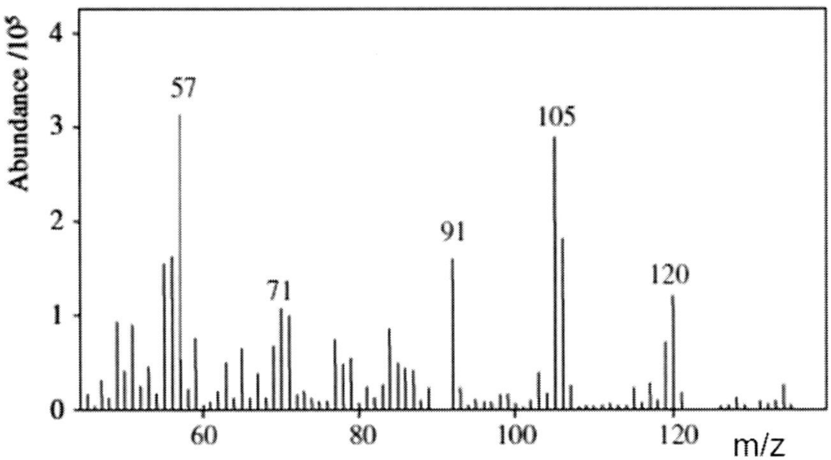

Figure 1.2. Average mass spectrum of a gasoline sample. (Reprinted with permission from Perez Pavon et al. 2006. Copyright 2006 Elsevier.)

the molecule in effect breaks it into fragments (Peres et al. 2000; Dittmann and Nitz 2000). In general, the total abundance of the mass fragments and the discriminating power of the mass spectra increase with the ionization energy level.

However, the energy level that yields the best signal-to-noise ratios depends on the material being studied (Begnaud and Berdagu 2002). Ionized fragments are injected into a mass filter that separates them according to their mass-to-charge ratio (*m*/*z*) (see Figure 1.2), yielding a quantitative abundance measurement of each ion of different mass. The mass separation takes place within an electric and/or magnetic field, and nowadays a variety of mass analyzers are well established. There are many types of mass analyzers, such as double-focusing magnetic and electric sensors, quadrouple mass filters, quadrupole ion traps, time-of-flight mass spectrometers, and Fourier-transform ion cyclotron resonance mass spectrometers, all capable of performing these measurement. The choice of mass analyzer is dictated largely by the goal of the research, based on the required mass range, mass accuracy, resolution, sensitivity, cost, and dynamic range (Mano and Gato 2003). Finally, the ions are collected at the electron multiplier, and the current is measured. The spectrum resulting from simultaneous ionization and fragmentation of the mixture of molecules introduced constitutes a 'fingerprint" that is characteristic of the original sample being analyzed. Exploitation of this spectral information allows determination of the composition of the sample to some extent (Peres et al. 2003; Perez Pavon et al. 2006).

The mass spectrum obtained results from the simultaneous ionization and fragmentation of all the volatile compounds present in the sample. The selection of particular fragment ions may be based on knowledge of the headspace composition—analytical chemistry can provide the necessary tools to characterize the samples—or on the results of mathematical feature extraction. Thus, information about what kinds of molecules are responsible for the differences between samples can be obtained from the ion fragmentation patterns. In some cases, the mass spectrum that results can be used to identify the original molecular composition. These results can then be compared directly with results obtained using conventional GC-MS instruments and more formal analytical methods for validation. The sensitivity and the selectivity of such an instrument can sometimes be improved using MS in the selected ion monitoring (SIM) mode.

One example of a mass spectrometer used as an electronic nose is shown in Figure 1.3. This instrument is simply a direct-injection quadrupole mass spectrometer; it is a conventional analytical instrument, but with reconfigured software to make the output look like an e-nose system. The performance of this type of e-nose is attractive because the technology is mature, the limit of detection is down to the parts-per-billion level, and there is practically no interference from ambient conditions, e.g., humidity, or airborne pollutants such as CO.

There is no doubt that MS-based systems can have considerable advantages over simple sensor arrays because of their inherent direct measurement of a molecular property (the mass-to-charge ratio of a molecule), the response to a broad range of species of interest, and the high sensitivity of ion-based analytical techniques. Advantages of MS over conventional sensor arrays include fewer problems of poisoning, less chance of profile masking by some major constituents of the sample (e.g., ethanol), no strong influence of moisture, and no nonlinearity of signals (Perez Pavon et al. 2006). Of course, the system cost and complexity can be substantially greater than that of a simple chemical sensor array. However, MS is also inherently a very fast technique, and all the compounds can be introduced into the mass spectrometer rapidly and simultaneously, with no chromatographic separation necessary.

Despite the great potential of the MS-based e-noses in the field of food analysis, there are still some aspects of the technique that need to be improved. Signal instability is one of the problems of MS-based e-noses that has not yet been solved. It causes the mass spectrum of a sample to change when the same sample is analyzed some days later. Many factors contribute to signal instability, and the cause may lie in the MS itself (e.g., gradual fouling of the ion source, vacuum instability, aging of the ion multiplier, or change of a filament). This instability is a common problem in the analytical application of the e-nose technique and requires an additional set of samples (i.e., standards) to be analyzed at regular intervals.

Figure 1.3. The Gerstel ChemSensor 4440 manufactured by Gerstel GmbH & Co. KG with participation of Agilent Technologies is a type of quasi-electronic nose based on an automated headspace sampler and an MS. (Uploaded from www.gerstel.de.).

While this approach to analysis (comparison of the unknown to a standard) is the most acceptable for e-noses today, management of the reference samples (selection, physical and chemical stability, and storage), especially in food analysis or other complex tasks, often presents a major problem.

2.3. GC-MS METHOD

The combination of gas chromatography followed by mass spectrometry has become the gold standard for compositional analysis of complex mixtures and is constantly being improved (Lavine 1992; Kaplan and Braham 1998; Aishima 2004; Berna et al. 2008; Capelli et al. 2008). The powerful analytical separation of constituents in time followed by mass spectroscopy of each eluant from the column provides the most data for pattern-recognition analysis. In brief, the GC column separates an odorant mixture into its molecular constituents, and the MS generates a mass spectrum for each peak. The spectra are typically compared to spectra in a large chemical database to help identify the specific chemical compound in each constituent peak. The spectrum of labeled peaks, together with the concentration information and the elution time from the gas chromatographic column, is then used to distinguish the odorant mixture in the original samples. The mass spectrum, with its fragmentation patterns, provides unique fingerprints for discriminating the molecular identity in a sample and between samples. Applications of these instruments to date have included measurement of volatile food components, identification of "off" fragrances, and detection of rancidity in edible oils.

2.4. ION MOBILITY SPECTROMETRY

A major limitation to the field use of mass spectrometry is the need for a vacuum pump, as there are as yet no simple, small, low-cost vacuum pumps for MS. Ion mobility spectrometry (IMS) is an ionization technique with many similarities to MS but operates at ambient pressure and therefore has no need for a vacuum pump. The operating principle of IMS is similar to that of MS in that there is a sample inlet, ionization chamber, time-of-flight (TOF) drift region in which the ions are separated, and an ion detector (Figure 1.4). The major difference in IMS is that, the ions are not separated by the difference in the mass/charge ratio, but rather on the basis of their different ion mobilities in the drift gas (Vakhrushev et al. 2008). Ion mobility depends not just on mass/charge but also on the reduced mass, charge, and the different collision cross sections as determined by the ion's size and shape, all of which directly influence the separation observed. During the process of sampling, ionization, separation, and detection, there are many collisions between the ions and the drift gas (often dry air) molecules, because the measurement is performed at ambient pressure (Creaser et al. 2004).

While the most common agent for ionization is a radioactive alpha emitter such as ^{63}Ni or ^{241}Am, there have recently been many developments relating to nonradioactive, low-cost and low-power ionization sources. Typically, a series of ion–molecule reactions takes place in the ionization source, and often a sample molecule with high proton affinity reacts in air with a trace amount of H_2O and undergoes a proton transfer reaction to a positively charged ion by collision with the ionized water (the ionized water is called the reactant ion and is always present in the source, to chemically ionize the target analyte). By

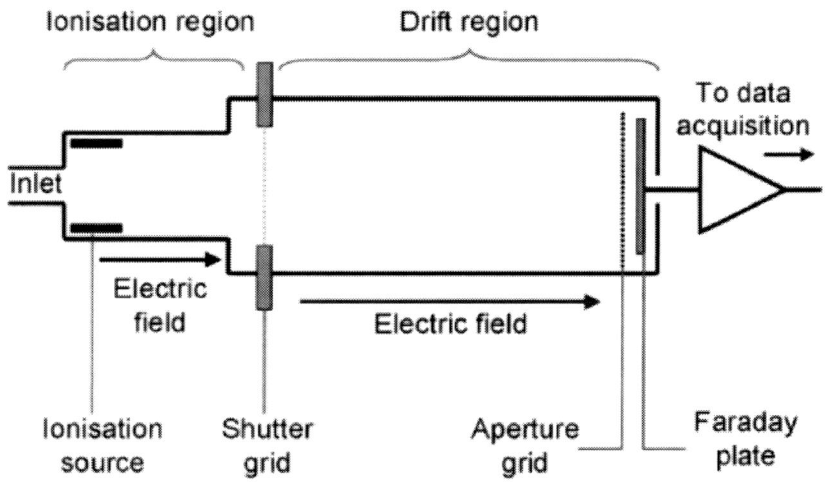

Figure 1.4. Schematic diagram of an ion mobility spectrometer. Ions are generated in an ionization region by any number of methods from plasma and electrospray to a ^{63}Ni source. An ion shutter or electronic method pulses the ions into the drift tube region, where they are accelerated by a uniform weak electric field toward a detector. Their progress is impeded by a number of collisions with the drift gas and each other. Larger ions with greater collision cross sections experience more collisions. Therefore, the separation of ions of differing shape and size becomes possible. (Reprinted with permission from Creaser et al. 2004. Copyright 2004 The Royal Society of Chemistry.)

doping the drift gas with NH_3 vapor, acetone, chlorinated solvents, or other chemicals, the nature of the reactant ion is changed and then the selectivity in the ion source can be modified.

Injection of the ions into the drift region can be controlled electronically such that, using differing polarities for the drift region, either positive or negative ions can be analyzed. Substances with electron-capturing capabilities, such as halogenated compounds, can be detected in the negative-ion mode, while positive ions are analyzed in the positive-ion mode (the detector is at a negative potential relative to the ion source). Another often-used ionization alternative, for compounds with sufficiently low ionization potential, is UV photoionization. This is appropriate for selective measurements of molecules with ionization potentials of less than 11.4 eV.

After ionization of the air sample, the ions are pulsed through a shutter into a drift tube, which is isolated from atmospheric air. The drift tube has a uniform and relatively weak electric field of about 200 V/cm, which accelerates the ions along the drift region. The movement is hindered by collisions, and the ion typically reaches its "ion mobility" or maximum drift velocity in the field until the ions reach the detector at the end. There have been some new methods of injecting the ions to simplify the shutter grid cell design, but in general the drift takes up to about 200 ms.

When the ions approach the detector a shutter grid is used to avoid unwanted charging effects. When the ions finally strike the detector, a current is generated and is measured along with the time of flight. For a manageable and calibrated component amount, this gives information about the identity and concentration of the analyte. When the sample composition is complex, however, there can be many ion–ion interactions, ion–drift gas interactions, or overlapping peaks, complicating interpretation of IMS spectra and lowering resolution. In many of these cases, the classical electronic nose approach using data-evaluation algorithms (Massart et al. 1988) can be applied to gain a maximum of information from the confounding measurements.

Compared to mass spectrometry, the virtual sensor array (many signal channels produced by the IM spectra over time) is not interpretable by discrete mass/charge relationship as in MS, but by the signal integration over definable time intervals to produce patterns that are related to the chemistry of the complex separation produced in the IMS. In this way, the IMS data reflects the sample chemistry and the IMS is operated like the e-nose devices.

2.5. NEAR-INFRARED SPECTROSCOPY

Infrared (IR) spectroscopy can also be used with pattern recognition and a sampling system and therefore can be considered as an electronic nose (MacDonalds and Prebbleis 1993; Blanco and Villarroya 2002; Van Kempen et al. 2002; Cozzolino et al. 2003; Armenta et al. 2006; Rock et al. 2008). The near-infrared (NIR) spectral region is generally defined as the range between 800 and 2500 nm. In this range, molecular vibrations and higher energy levels are excited (see Figure 1.5). Through characteristic absorption bands, the type of chemical bonds can be determined, and pure chemicals can be identified by their unique fingerprint spectra. The spectra of mixtures is evaluated by classical electronic nose algorithms.

Spectral features in the NIR region are overtones and combinations of vibrations observed in the mid-infrared (IR) region and are therefore much less intense (10- to 1000-fold), broader, and more overlapping than the parent absorptions. The consequently greater difficulty of assigning NIR absorptions to structural features may be a disadvantage, but this is compensated by other aspects: Unique combination

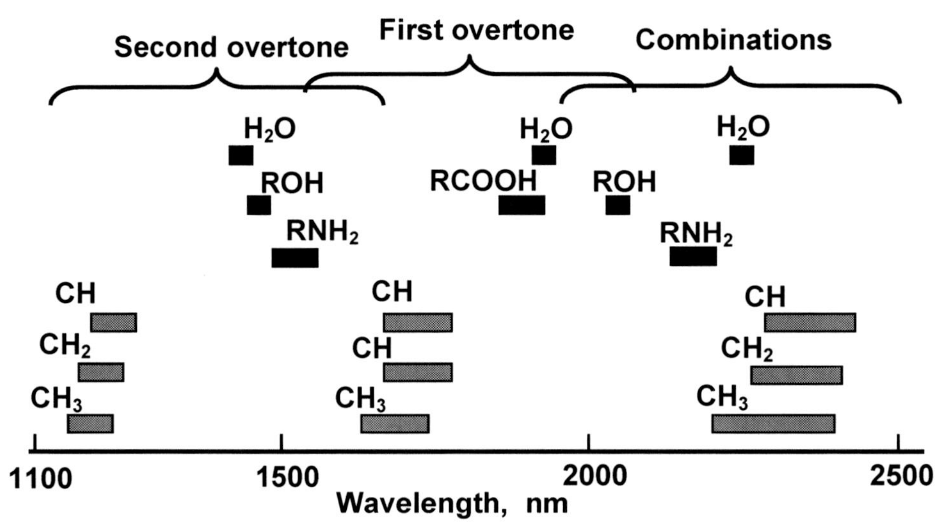

Figure 1.5. Overtones and combinations in the NIR spectral region. (Reprinted with permission from MacDonalds and Prebbleis 1993. Copyright 1993 Elsevier.)

bands offer information that is not available in the mid-IR, the reduced intensities allow direct measurements on undiluted samples, and measurements of, or in the presence of, water are readily performed (MacDonalds and Prebbleis 1993). A more detailed description and analysis of the advantages and disadvantages of NIR spectroscopy are given by Blanco and Villarroya (2002) and Anderson et al. (2005).

For the detection of substances in the gas phase, two affordable methods for use in portable devices are known. In photo-acoustic infrared spectroscopy, a modulation of the intensity of an IR source causes a temperature variation and a resulting expansion and contraction of the gas that is measured at audible frequencies with a microphone. Alternatively, the absorbed energy of a narrow-bandpass infrared beam is measured in filter-based IR spectroscopy. Commercially available nondispersive infrared (NDIR) instruments (e.g., MIRAN SapphIRe from Thermo Scientific) are used mostly for absolute measurements of concentration, either for detection of a single species which has a unique absorbance wavelength or by analysis at multiple wavelengths for a known gas mixture. However, when the constituents of the gas mixture are unknown, these instruments can be combined with pattern recognition and also used as an electronic nose.

In spite of the confirmed feasibility (Van Kempen et al. 2002), the infrared-based e-nose has not become popular, and commercially available devices such as the MIRAN SapphIRe from Thermo Scientific can rather be considered as portable analytic tools and not electronic noses.

2.6. OTHER METHODS OF GAS IDENTIFICATION AND QUANTIFICATION

Other analytical methods can also be used for gas recognition. For example, pulse spectroscopy, a technique patented by Bloodhound Sensors, uses a single solid-state sensor which interacts directly with volatile compounds to give a spectroscopic trace similar to fingerprint mass spectrometry (Chandler et al. 1997). The method has the advantages that it is extremely rapid, with full response times of the order of seconds, and sample processing is minimal. It is also amenable to miniaturization, potentially enabling instrumentation to become hand-held in the future while providing the discriminating power of mass spectrometry.

3. THE ELECTRONIC NOSE: APPROACHES AND ACHIEVEMENTS

3.1. A BRIEF HISTORY OF THE ELECTRONIC NOSE

In spite of the incontestable advantages of the methods so far considered, instrumental methods of determining odors such as gas chromatography and mass spectrometry are expensive and require trained personnel. Moreover, chromatography is a batch process and can take a relatively long time (minutes to hours) in spite of recent improvements in millisecond-timescale GC. Further, the size, cost, and performance of these analytical instruments still does not allow for a low-cost, portable electronic nose.

Therefore, there is still a need for devices for rapid, inexpensive analysis of volatile organic compounds (VOCs) which does not require specialist technicians. Ideally, field devices should be portable, allowing direct online or offsite measurements. As analytical instruments, these systems must be designed for long-term use with high repeatability (the ability to obtain the same pattern for a sample on the same array over short intervals of time) and reproducibility (the ability of different sensor batches or different instruments to produce the same pattern for the same sample) (Schaller et al. 1998). Of course, they also should be capable of continuous measurements in various environments.

The first successful attempts to elaborate such devices were made in the 1980s and 1990s. In particular, in 1982, Persaud and Dodd first reported experiments with three- and four-sensor arrays in an effort to imitate human olfaction. At the same time, a sensor array instrument complete with a pattern-recognition algorithm and an automatic sampling system was developed by Stetter et al. (1985b) (see Figure 1.6). This device, patented by Argonne National Laboratory (ANL) scientists, won an IR-100 award and awards for the inventors from the U.S. Department of Energy.

The term *electronic nose* or *e-nose* was first suggested by Julian Gardner of Warwick University in 1988 (Gardner, 1988) and came into popular use after a 1989 NATO conference on the subject. Since then, development of sensor array–based instruments has been actively pursued in Asia, North America, and Europe. Using instruments based on sensor arrays, the low selectivity of many types of chemical sensors can sometimes be resolved, though at the cost of using an array of reversible but only semiselective detection layers with different chemical properties. An e-nose has been defined as "an instrument which

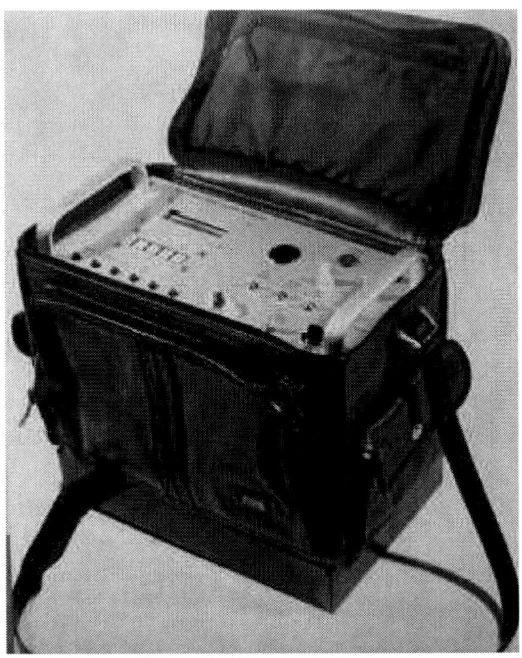

Figure 1.6. View of the fully automated portable Chemical Parameter Spectrometer sensor array (CPS-100 sniffer) designed at Argonne National Laboratory and demonstrated during 1980–1983. The sensor array included four sensors operated in four modes. (Reproduced with permission from the *Electrochemistry Encyclopedia*, http://electrochem.cwru.edu/ed/encycl.)

comprises an array of electronic chemical sensors with partial specificity and an appropriate pattern recognition system, capable of recognizing simple or complex odours" (Gardner and Bartlett 1994). There are also other, similar definitions in the literature (Stetter and Penrose 2001). The sensor arrays feed responses (usually transduced to electronic impulses) into specialized computer software which produces multidimensional response profiles that can be thought of as a digital "smellprint" or "fingerprint" that represents the chemical complexity of the gas mixture.

In the first e-nose device designed by Persaud and Dodd (1982), they used three different metal oxide gas sensors and identified among 20 odorous substances, including essential oils and pure volatile chemicals by the steady-state signals of these sensors. In the work of Stetter et al. (1985), the device identified more than 100 different chemicals. Pattern classification was performed by visual comparison of ratios of sensor responses by Persaud and Dodd (1982), while statistical pattern recognition was coupled to the e-nose in the ANL device (Zaromb and Stetter 1984; Stetter et al. 1986). Focusing primarily on the sensor aspect of the problem, the initial research of Persaud and Dodd only explored the use of metal oxide devices while the ANL device consisted of thermal catalytic sensors and electrochemical sensors to form a heterogenous sensor array. Later work at Warwick University explored the use of conducting polymers, while the work at ANL explored additional sensors and time-dependent responses (Stetter 1986; Zaromb et al. 1986; Stetter and Otagawa 1987). In applications, the ultimate selectivity of chemical sensors is achieved by using multiple sensors in an array, by using different sensors in the array, by modulating the sensors (i.e., using different operating protocols), and by the application of pattern-recognition techniques to interpret responses obtained from the sensor array. Using sensor arrays, we begin to appreciate the analytical power of a collection of sensors which are exposed to the same or nearly the same sample, and their responses are interpreted together.

Research on chemical sensor arrays/electronic noses has progressed in Europe, North America, and Asia since the early 1980s, and there has been a series of international conferences on the subject (see, e.g., Stetter and Penrose 2001b, 2002).

The research on sensor arrays and the resulting instruments that were introduced were made possible by many advances in science and technology, including improved sensors and improved microprocessors, and in pattern recognition and chemometrics that were being developed in analytical chemistry primarily for understanding structure–activity relationships. Early reports of "sniffers" and research on odorants include those of Wilkens and Hatman (1964), who reported on redox reactions of odorants at an electrode; Buck et al. (1995), who discussed the modulation of conductivity by odorants; and Dravieks and Trotter (1965), who reported the modulation of contact potential by odorants.

There have been attempts to obtain multidimensional sensor signals from a single sensor such that one can make an electronic nose using a single sensor. These approaches use either a dynamic working mode for the sensor or simultaneous measurement of several parameters by one sensor. This allows multiple signals characterizing the state of the gas sensing material. For example, in various works (Stetter and Otagawa 1987; Stetter et al. 1989, 1990, 1991; Vaihinger et al. 1991; Nakata et al. 2002), a sinusoidal temperature perturbation was applied to a semiconductor gas sensor and the resulting conductance of the sensor was analyzed. It was found that the sensor response changes in a characteristic way depending on the composition of the gaseous mixtures and can be characterized by fast Fourier transformation (FFT). Others (Takada 1998; Heilig et al. 1999) showed that combined utilization of the temperature drop and the resistance changes of a single SnO_2- or In_2O_3-based sensor provided a new possibility for gas identification and determination of gas concentration. Recently, the modulation of microelectromechanical systems (MEMS) HMOX sensors has resulted in an array of micromachined sensors (Taylor and Semancik 2002; Wheeler et al. 2007). Temperature-programmed sensing for gas recognition has also been proposed (Meier et al. 2007).

None of these approaches, however, found appropriate expansion and commercialization because of the lower reliability of the results obtained when analyzing complicated gas mixtures. Therefore, in this chapter we focus our attention only on electronic noses elaborated on the basis of sensor arrays.

3.2. DEFINITION OF AN ELECTRONIC NOSE

As we mentioned before Gardner and Bartlett (1994) defined the electronic nose as "an instrument, which comprises an array of electronic chemical sensors with partial specificity and an appropriate pattern-recognition system, capable of recognizing simple or complex odours." This seems very far from the human nose, and according to Mielle et al. (1995, 1996), such an analytical system is "obviously electronic but not a nose." In fact, the only aspect in common with our odor-sensing organ is its function. Like the mammalian nose, it detects gases by means of sensors which send signals to a recognition organ—that is, to the brain or to a computer. The operating principle, the number of sensors, as well as the sensitivity and selectivity, are, however, very different (Bartlett et al. 1997). This is why some scientists prefer to call this instrument by other names, for example, "flavour sensor," "aroma sensor" (Mielle 1996), "odour sensing system" (Gardner and Bartlett 1994), "multi-sensor array technology" (Shiers 1995; Mielle and Marquis 2000), or "chemical parameter spectrometer" (Stetter et al. 1984). Recent attempts to more closely mimic the mammalian (dog, rat, mouse) nose have been reviewed (Stetter 2010), and the use of mammalian odor receptors in sensors is showing promising utility (Persaud et al. 2005).

3.3. THE ELECTRONIC NOSE: PRINCIPLES OF OPERATION FOR DETECTION OF ODORS

There are two types of odors: simple odors that tend to arise from single light, polar compounds, such as citral (lemon) or benzaldehyde (bitter almond); and complex odors that comprise hundreds or thousands of different compounds, such as the headspace of coffee or tea. Electronic noses are designed to be able to detect both these odor types (Varanda and Gardner 1999). The general principle of the e-nose technique is that the volatile compounds in the headspace of the sample are introduced into a detection system, where the volatile compounds interact with an array of gas sensors. Then, for each sample analyzed, the detector generates a set of signals that contains information about the volatile composition of the sample. Fundamental to the artificial nose is the idea that each sensor in the array has different sensitivity (Craven et al. 1996; Gardner 2004). Selectivity is maximized when each gas sensor responds more selectively to a certain chemical parameter or chemical quality (such as the polarity, polarizability, or electrochemical redox potential) of chemical compounds such that the array is able to image in "chemical" space. This can lead to multiple responses on multiple sensors in the array for compounds that have some polar character as well as some polarizability or hydrogen-bonding character; compounds with differing amounts of these chemical properties produce unique fingerprints, and the sensors show cross-reactivity or partial selectivity. For example, Odorant No. 1 may produce a high response in one sensor and lower responses in others, whereas Odorant No. 2 might produce high readings for sensor 2 and lower on other sensors. So, each sensor in the array has a unique response profile to different odorants under test. This means that the pattern of response across the sensors is distinct for different odorants. In this way, a small number of gas sensors (usually 3–32) can respond to a variety of different complex odors and there is no need to have one specific sensor for each individual compound. This ability to distinguish chemical identity is a unique property of the sensor array over an individual sensor signal and allows the system to identify an unknown odor from the pattern of sensor responses. When several samples are analyzed, a data matrix is generated that can constitute a "library" of known responses. The matrix of data can be subsequently treated with chemometric techniques so that the unknown samples can be compared to the library on the basis of their multiple responses (i.e., chemical composition), and classified. This approach allows analysis not just of intensity or quantity but a classification of the odor based on its origin, variety, purity, ripeness, aging, or any other complex property of interest. This technique sometimes does not provide information on the amounts of the individual aroma compounds; rather, it makes a global, qualitative estimation of the aroma profile and can sometimes offer an indication of odor intensity based on the intensity of the odor fingerprint. In this respect, it is similar to human olfactory perception. So, the output of the electronic nose may be the identity of the odorant, an estimate of the concentration of the odorant, or the characteristic properties of the odor as might be perceived by a human. When statistical techniques are used, it should be borne in mind that results will be reliable only if a significant number of samples of the different categories of the property of interest are analyzed and kept in the library for comparison. Figure 1.7 shows a simple schematic layout of an electronic nose and the main elements of the artificial olfactory system.

3.4. SIZES AND FEATURES OF SENSOR ARRAYS

Discussions of an electronic nose should include addressing the issue of how many sensors should be included in an array. The most direct way to improve the data would seem to be to increase the number of sensors of the same type, e.g., use 20 or 40 sensors instead of four. The theoretical sensitivity of an optical array of identical redundant sensors has been shown to increase as the square root of the number of sensors, as expected from simple Gaussian error theory (Dickinson et al. 1999). This means that there is a clear benefit to more sensors if they are completely and perfectly redundant. However, this approach meets rapidly diminishing returns (Stetter et al. 1986). Combining all types of available sensors in the many different modes of operation can lead to the incredible diversity that can be produced with sensor arrays. However, it has been found that too many sensors that are not exactly the same but that are also not independently responsive or orthogonal can add noise but no new signal information, and thus can actually reduce the differentiating power of an array (Stetter et al. 1986; Mitrovics et al. 1998). Moreover, an increase in the number of sensors in an array sharply increases the time necessary for processing data obtained from the sensor array. The potential information content of an array of signals is quite large even for a small sensor array (Gopel, 1998; Zellers et al. 1995, 1998). For example, for electrical fires, it was found that one can tell the type of fire as well as the stage of the fire

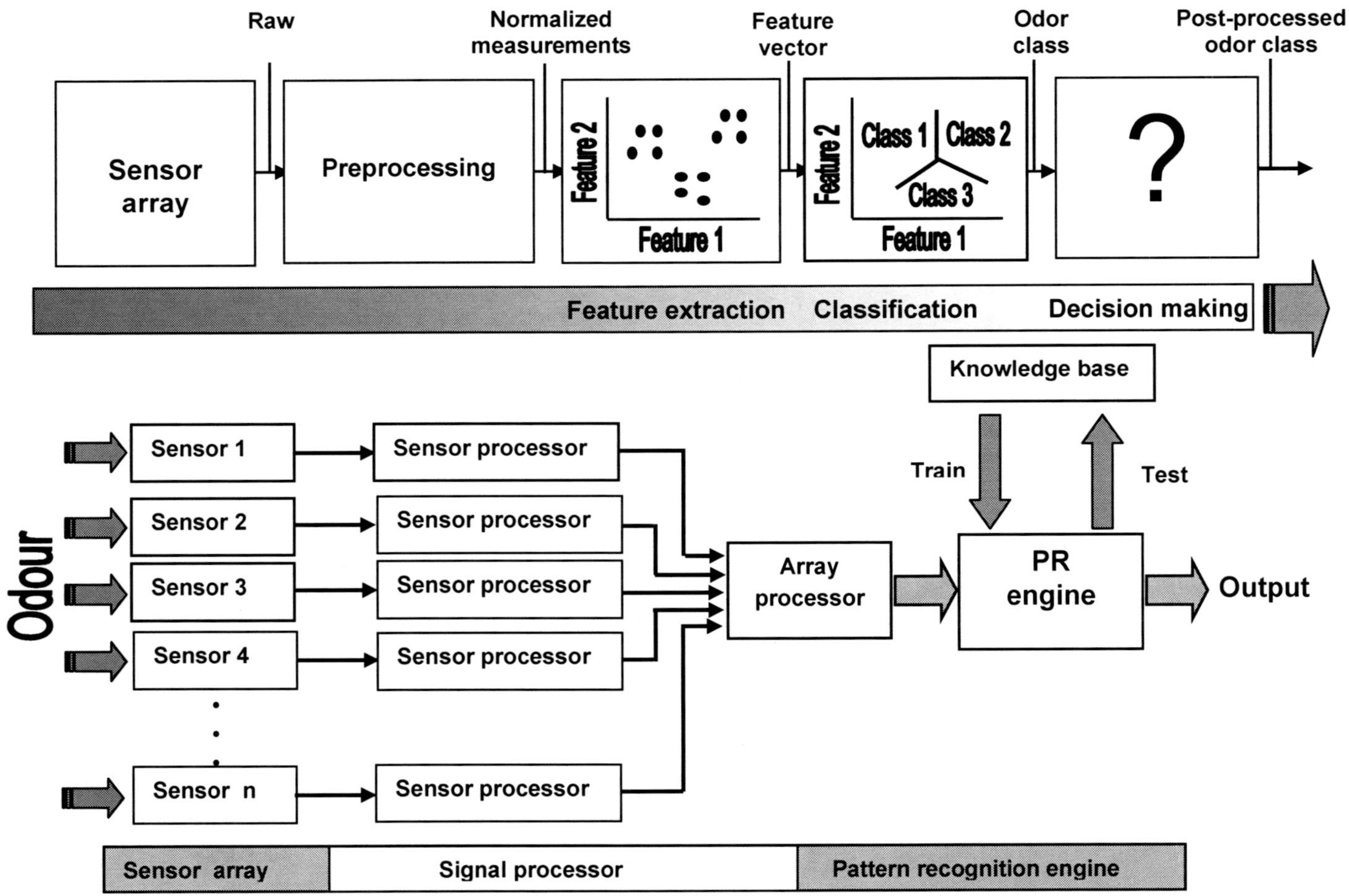

Figure 1.7. Diagram illustrating the functioning of electronic nose. (Adapted with permission from Gardner and Bartlett 1994. Copyright 1994 Elsevier; and Kaplan and Braham 1998. Copyright 1998 IEEE.)

using a subset of three sensors (one electrochemical and two heated metal oxide) in a single array (Ni et al. 2008). Therefore, it is possible that in many cases smaller arrays can have more capabilities than large arrays: All arrays are not created equal. The best arrays are designed for the intended purpose and separate the molecules or molecular mixture by their chemical compositional differences in terms of the sensory responses. The better method of selecting sensors for a sensor-based electronic nose is not to use as many different sensors as are available, but to select them with an eye to the desired application and with knowledge of the requirements for the analytical data. That is the only way to ensure that the substances which are to be detected are causing the signal (Rock et al. 2008) and that the signals can be differentiated from each other and from the background matrix variations. There is an optimum number of sensors for each application, although that number and the choice of sensors may differ greatly from one application to another. For example, a statistical approach illustrates that a small array of six sensors is all that is required to differentiate up to 100 different patterns/compounds (Zaromb and Stetter 1984). In real e-nose instruments the number of elements in sensor arrays is varied from 4 to 32, but for any instrumentation, the simpler the design, the lower will be the cost and the higher the reliability. Finally, research has shown that it is also necessary to take into account the mutual influence of sensors on each other. For example, it is often necessary to specify the order of the sensors in the array as well as the kinds of sensors. This is because some sensors act on the sample and change it. An example of this is metal oxide sensors, which efficiently oxidize the sample and change the response of other, downstream sensors or combustible gas sensors that consume a portion of the sample signal (Stetter et al. 1985a).

4. SIGNAL PROCESSING IN ELECTRONIC NOSES

After the sensor array, the second most important component of an electronic nose is the pattern classifier (Shaffer et al. 1999; Jurs et al. 2000). This is a means of extracting information from the collected sensor array patterns by comparing or associating them. In practice, this means associating a pattern from an unknown sample with a set of patterns from known standards to determine the closest match, for the purpose of identification.

According to Shaffer et al. (1999), an ideal pattern recognition algorithm should have the following six qualities:

1. *High accuracy.* For application of the chemical sensor system to field measurements, the pattern recognition algorithm must accurately classify new sensor signals (i.e., it should have a low false-alarm rate and, ideally, no missed detections). For military applications, such as the detection of toxic chemical vapors, classification rates of >90% accuracy are necessary.
2. *Fast.* For real-time analysis, the pattern recognition algorithm must be able to produce a classification quickly. Thus, algorithms that are computationally intense may not be appropriate for this application.
3. *Simple to train.* The classification rules used by the pattern recognition algorithm must be learned quickly. For many applications, the database of training patterns will need to be updated periodically, thus requiring the algorithm to "relearn" its classification rules. This procedure must be performed as simply and quickly as possible.

4. *Low memory requirements.* For field-portable sensor systems, the pattern recognition algorithm will be embedded in a microcontroller that usually has limited memory resources. Thus, pattern recognition algorithms with large memory requirements may not be appropriate for this application.
5. *Robust to outliers.* For chemical sensor array applications in uncontrolled environments, the pattern recognition algorithm must be able to reduce the potential for false alarms by being able to differentiate between sensor signals it was trained on and those on which it was not. The underlying assumption is that the algorithm has been trained to recognize all the important compounds. Thus, any new or ambiguous sensor signal (e.g., a chemical vapor which it has not been trained to recognize) should be ignored.
6. *Produce a measure of uncertainty.* For many applications of chemical sensor array systems, the pattern recognition algorithm must be able to produce a statistical measure concerning the certainty of the classification. For sensor applications, such as toxic vapor monitoring, such a measure will aid in reducing the occurrence of false alarms by requiring that the sensor system be >80% or 90% certain of a classification decision before a warning is given or an alarm sounded.

4.1. PATTERN RECOGNITION SOFTWARE

There are many tools available for the analysis of data from an array of chemical sensors (Jurs et al. 2000). Earlier, the vast majority of systems reported for electronic noses (including virtually all of the commercially available instruments) were based on steady-state measurements. Sensors were allowed to reach equilibrium and, after a predetermined time, the signal intensities from each sensor were measured. The resulting response vectors were time-independent and represented the absolute change in sensor signal with a measured odor (Dickinson et al. 1998). Early reports of using time-dependent responses (Zaromb et al. 1986) illustrated that inclusion of slow and fast response categories for sensor responses improved the ability to identify different molecular composition of chemical samples.

Recently, the use of temporal data has begun to be explored for odor discrimination (White et al. 1996). This approach more closely parallels the presentation of odors to olfactory cells by the process of sniffing. Saunders et al. (1995) observed that the shapes of the time-dependent responses from piezoelectric crystal sensors were less variable than the maximum frequency changes induced by odorants. In addition to the potential for improving reproducibility, the use of temporal data has been shown to enhance analyte identification and quantification.

The dynamic sensor signal depends on several physical parameters, such as the speed of the flow that carries the odor from the source to the sensor array, the chemical nature of the odor itself, the diffusion and reaction of the odor within the active sensing material, and the ambient conditions such as pressure, temperature, and humidity (Tan et al. 2001). Sophisticated software is necessary to be able to handle such complex information. In general, such software is based on powerful algorithms for pattern recognition. Research into pattern classifiers has been perhaps the most aggressively pursued aspect of electronic nose research. A large body of previous research on pattern classification and recognition has been used in the quest to extract the maximum information from the chemical data produced by the sensors (Pearce et al. 2003; Pardo et al. 2005; Snopok and Kruglenko 2005; Di Natale et al. 2006).

The first step in data analysis is to preprocess the signals generated by the multidimensional sensors (Martı et al. 2005). This process transforms the data into their most appropriate form and enhances the features within the data that are useful in the subsequent steps. The preprocessing steps include normalization, baseline correction, and noise reduction or variable weighing, among others. Many of these techniques have been used in MS and GC for many years. Afterwards, preprocessed data are analyzed by various chemometric techniques (Gardner and Bartlett 1992; Holmberg 1997; Schaller et al. 1998; Keller 1999; Pearce et al. 2003; Barbri et al. 2007), some of which are available in statistical software packages that are available separately or are included in instruments that produce multidimensional data arrays.

4.1.1. Statistical Analysis Techniques

According to Schaller et al. (1998), commercially available techniques fall into three main categories:

1. Graphical analyses: bar chart, profile, polar, and offset polar plots
2. Multivariate analyses: principal components analysis, canonical discriminant analysis, feature weighting, and cluster analysis
3. Network analyses: artificial neural network and radial basis function

The choice of method depends on available data and the type of result that is required.

4.1.1.1. GRAPHICAL ANALYSIS

Graphical analysis is the simplest form of data treatment that may be used with an electronic nose. This option is suitable when visually comparing samples to a single specified reference. However, when several references are used, analysis becomes more complicated and an alternative approach may be necessary.

4.1.1.2. MULTIVARIATE ANALYSIS

Multivariate data analysis generally involves a process for data reduction; it reduces high dimensionality in a multivariate data set or problem in which variables are partly correlated (e.g., sensors with overlapping sensitivities), allowing the information to be displayed in a smaller dimension (typically two or three) (Gardner and Bartlett 1992; Fukunaga et al. 1995). Statistical methods of multivariate analyses used in electronic noses for pattern recognition include principal components analysis (PCA), canonical discriminant analysis (CDA), discriminant factorial analysis (DFA), feature weighting analysis (FW), and cluster analysis (CA) (Schaller et al. 1998).

Principal component analysis is commonly used to analyze the responses of an artificial nose to both simple and complex odors (Dickinson et al. 1998). In this method, response vectors are broken down into their individual components and expressed as linear combinations of orthogonal vectors (Gardner 1991). By convention, the components are numbered in decreasing order of importance. In practice, the majority of variance (and thus information for discrimination) is generally contained in the first two or three orthogonal vectors, owing to the high degree of co-linearity usually found in electronic nose data. Decomposing complex responses into their principal components provides a useful way to display sensory data because it allows a multicomponent dataset to be displayed as a two- or three-dimensional chart (Figure 1.8). Cluster analysis provides a convenient way to isolate clusters or groups of response vectors by calculating the dissimilarity between each vector (dissimilarity being determined by representing each response as a point in multidimensional space and calculating the Euclidean distance between them). Discriminant factorial analysis is a multivariate technique which determines a set of variables which best discriminates one group of objects from another. The first instrumental electronic

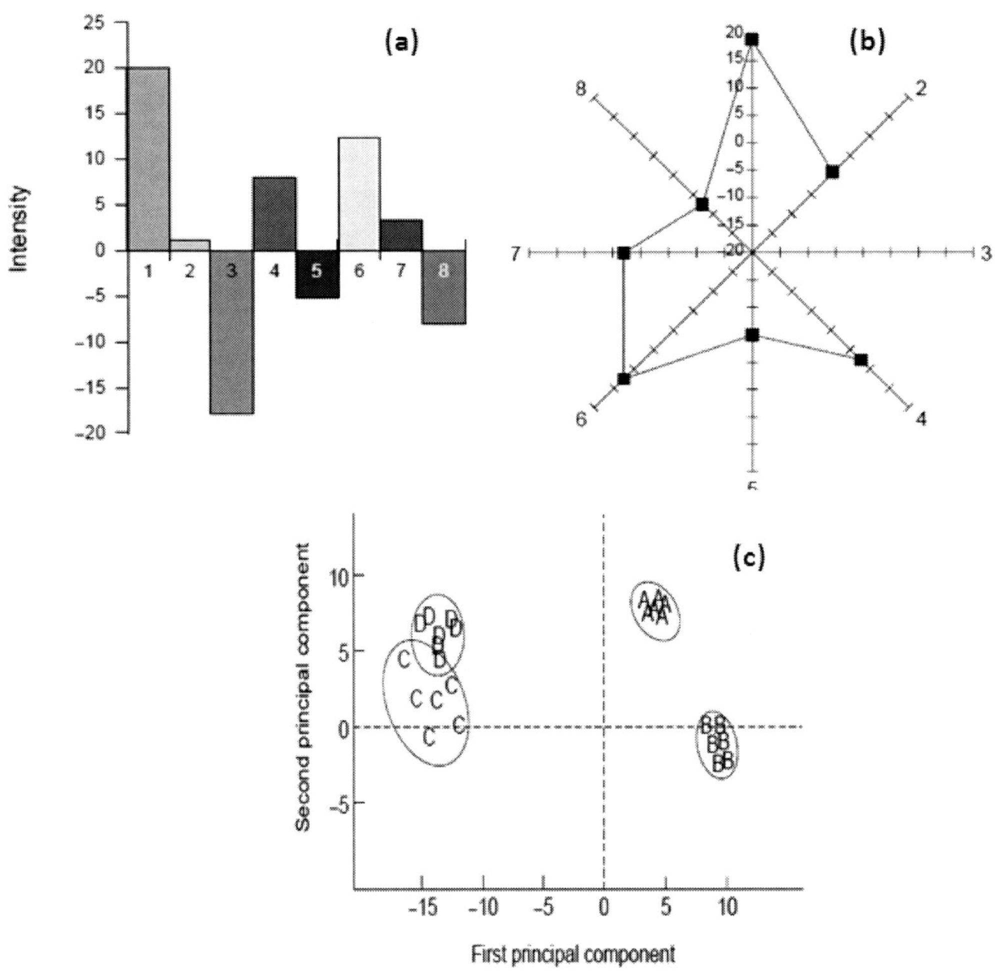

Figure 1.8. Examples of responses from eight-sensor array: (a) a bar graph; (b) a polar plot; (c) results of PCA analysis. (Reprinted with permission from Dickenson et al. 1998. Copyright 1998 Elsevier.)

nose complete with pattern recognition used a K-nearest neighbor (KNN) algorithm to compare unknown vectors determined by the instrument to libraries of known compounds (Stetter et al. 1985b).

In addition to the techniques mentioned above, one can also use multivariate pattern recognition techniques, such as linear discriminant analysis (LDA), soft independent modeling of class analogy (SIMCA), and partial least squares (PLS) regression to analyz output signals from an electronic nose. In particular, PLS regression, currently the most powerful multivariate calibration technique, has been recognized as an indispensable regression technique among analysts working in spectroscopy, chromatography, and sensory sciences (Martens and Naes 1989; Aishima 2004).

4.1.1.3. ARTIFICIAL NEURAL NETWORK

For situations in which quantification or multicomponent mixture classification is desired, a more powerful, nonlinear prediction tool is required. The use of artificial neural networks (ANNs) is such an alternative for analyzing nonlinear e-nose data. A neural network consists of a set of interconnected processing algorithms functioning in parallel. On a very simplified and abstract level, ANNs are based on the cognitive process of the human brain. Mathematical functions, or neurones, link together to build a network, which mimics the human nervous system (Newman 1991). A weight is randomly assigned to each neurone and then adjusted by means of an iterative or "learning" process, for example, error backpropagation, until the desired outputs are obtained. The resulting set of weights and functions is then saved as a "neural network." There are many variations of this approach, with a three-layer feed-forward network being the most widely employed architecture for handling e-nose data (Dickinson et al. 1998).

Unlike other pattern recognition methods, a neural network is a dynamic, self-adapting system that can modify its response to external forces using previous experience, offering a more flexible and, due to the parallelism, faster method of analysis. In addition, it may more closely mimic mammalian neurone processing of odor stimuli (Newman 1991; Persaud and Pelosi 1992). A well-trained ANN is very efficient in comparing unknown samples to a number of known references. A direct comparison of statistical and ANN techniques carried out by Findlay et al. (1993) for the detection of biological activity in grain illustrates that the ANN can be more robust and tolerant to noise and errors than statistical pattern recognition.

More recently, algorithms have been developed to process signals from sensor arrays dynamically (Tan et al. 2001). These dynamic processing techniques include traditional parametric and nonparametric methods as used traditionally in the field of system identification, as well as linear filters, time-series neural networks. In addition to being able to cope with the dynamic nature of the response of the sensor array, such dynamic signal processing techniques are also particularly suited to handle factors such as sensor drift.

4.1.1.4. SUPERVISED/UNSUPERVISED

The pattern recognition methods of multivariate analysis can be divided into unsupervised or untrained techniques, and supervised or trained techniques. Unsupervised learning methods are generally used in exploratory data analysis, because they attempt to identify a gas mixture without prior information about the nature of the samples. These techniques, which include PCA, CA, and multidimensional scaling, are useful when no example of different sample groups is available, or when hidden relationships between samples or variables are suspected (Gardner and Bartlett 1992). Conversely, supervised learning techniques classify an odor by developing a mathematical model relating training data, i.e., samples with known properties, to a set of given descriptors. Test samples are then evaluated against a knowledge base, and predicted class membership is deduced. These methods enable the system to reduce parameters other than volatile, for example, temperature and humidity, and train a system to look only at particular combinations of sensors to measure a given odor.

ANNs as well as statistical techniques can be divided into supervised and unsupervised approaches. Supervised algorithms used in electronic noses include back-propagation-trained feed-forward networks (Rumelhart et al. 1986), learning vector quantizers, and Fuzzy ARTMAPs (Carpenter et al. 1992). Fuzzy ARTMAP is a self-organizing and self-stabilizing supervised classifier that shows generally superior performance in training compared with multilayer perception (MLP) (Carpenter et al. 1992, 1995). This method needs a minimum of known data to train the system correctly. If the number of available data is not sufficient, an erratic result will be obtained. An unsupervised ANN algorithm does not require predetermined odor classes for training and essentially performs clustering of the data into similar groups based on the measured attributes or features that serve as inputs to the algorithm. Unsupervised ANNs used in electronic noses include self-organizing maps (SOMs) (Kohonen 1989) and adaptive resonance theory networks. Generally, electronic nose data are best processed using trained techniques such as CDA, FW, ANN, or RBF (Schaller et al. 1998).

4.1.1.5. LINEAR/NONLINEAR

The above multivariate analyses are all linear pattern recognition methods in which a model is calculated using linear combinations of input data (Holmberg 1997). Therefore, when the instrumental responses recorded are linear, as in MS-based e-noses, all the aforementioned statistical methods give very good results. However, in gas sensor–based e-noses, instrumental responses are essentially nonlinear. In such cases, input data must be transformed into linear responses before applying the chemometric techniques described above. For example, the use of preprocessing algorithms, such as averaging, linearization, or normalization, can improve the performance of these analytical techniques (Gardner and Bartlett 1992; Persaud and Pelosi 1992; Gopel 1995). Such an approach works well if a low concentration of volatiles ensures an approximately linear response.

When high concentrations of volatiles are measured, a nonlinear pattern recognition technique, such as ANN or RBF, may be more appropriate. Nonlinear models usually need more parameters, since some of them are used to describe the shape of the nonlinearity (i.e., they have more input data than linear models). The main advantage of these methods is flexibility, i.e., the ability to adjust to more complex data variations. However, caution is necessary when choosing model flexibility; this can be achieved by selecting the number of parameters. If too many parameters are taken into account, the calculated model will be overly flexible, fitting to all relevant data variations and unwanted sensor noise. The best method to avoid an over-fitted model is to use training data to build a nonlinear model, and validation data to test this model (cross-validation) (Holmberg 1997).

Table 1.1. Qualitative comparison of pattern recognition algorithms

Pattern recognition algorithm	Speed of operation	Simple and fast to train	Memory requirements	Outlier rejection simple	Statistical measure of uncertainty
NN	Slow	Yes	High	Yes	Possible
MLDA	Fast	Yes	Low	Yes	Yes
BLDA	Fast	Yes	Low	Sometimes	Yes
SIMCA	Fast	Sometimes	Low	Yes	Yes
BP-ANN	Fast	No	Low	Sometimes	Possible
PNN	Slow	Yes	High	Yes	Yes
LVQ	Fast	Sometimes	Low	Yes	Possible

NN, nearest-neighbor pattern recognition algorithm; MLDA, Mahalanobis linear discriminant analysis; BLDA, Bayesian linear discriminant analysis; SIMCA, soft independent modeling of class analogy; BP-ANN, back-propagation artificial neural networks; PNN, probabilistic neural networks; LVQ, learning vector quantization.

Source: Reprinted with permission from Shaffer et al. 1999. Copyright 1999 Elsevier.

4.2. COMPARISON OF CHEMICAL SENSOR ARRAY PATTERN RECOGNITION ALGORITHMS

Unfortunately, no pattern recognition algorithm is able to fully meet each of these requirements. Results of comparison study of chemical sensor array pattern recognition algorithms carried out by Shaffer et al. (1999) are presented in Table 1.1. Comparisons were made based on five qualitative criteria (speed, training difficulty, memory requirements, robustness to outliers, and the ability to produce a measure of uncertainty) and one quantitative criterion (classification accuracy). For these purposes four sample datasets, involving simulated data and polymer-coated surface acoustic wave chemical sensor array data, were used.

Based on this analysis, Shaffer et al. (1999) made the following conclusions. Among the seven algorithms in this study and the four datasets, the neural network–based algorithms (LVQ, PNN, and BP-ANN) had the highest classification accuracies. This conclusion will not surprise anyone in the chemical sensor community. When considering the qualitative criteria, the LVQ and PNN approaches fared well compared to BP-ANN due to their simpler training methods. Shaffer et al. (1999) therefore recommendd the LVQ for most applications of chemical sensor arrays and the PNN for special cases where a confidence measure and fast training are critical, while speed and memory requirements are not.

4.3. SHORT VIEW OF THE PROCESS OF SOFTWARE DESIGN FOR PORTABLE DEVICES

As discussed earlier, the usual definition of an electronic nose states that the instrument includes non-specific sensors plus a pattern recognition system. When designing the intelligent processing and smart operation components of an electronic nose, several approaches, ranging from powerful desktop systems to portable systems with limited computational resources, may be considered, depending on the required instrument size and final application. According to Perera et al. (2002), when designing embedded pattern analysis software, it is important to keep in mind that the complexity of training (or estimation) algorithms tends to be much higher than the complexity of the algorithms in the operation phase. A typical example is principal component regression, in which simple matrix manipulation is needed in the operation phase, but more complex routines such as singular value decomposition (SVD) may be needed during model estimation. For platforms with limited computing power, some algorithms should be trained offline, usually in a host computer, and parameters delivered to the system using appropriate digital communications. For more powerful platforms, the same instrument will be able to adapt its data processing scheme to the problem of interest. If some learning engine has to be implemented in a portable device, computational resources are an issue to be taken into account in the system design, since there will be limitations in speed and often in memory.

In developing the e-nose software, some knowledge must also be considered as to the hardware solution (i.e., if dynamic memory is available) (Perera et al. 2002). In any case, to maximize software portability between different platforms, we advocate for a modular system that can be customized at compilation time. This will allow the solution to become portable among different architectures in a range of instrument designs for an e-nose manufacturer.

Perera et al. (2002) believe that the use of embedded technology provides several interesting benefits: availability of an abstraction layer for signal acquisition and control via an operating system, high-level programming of the signal processing algorithms, large data storage in solid-state disks, commercial off-the-shelf hardware for Internet connectivity, I^2C and CAN buses, serial ports, hardware for interfacing various types of displays, etc. Furthermore, an impressive trend toward reducing the cost and size of these systems can be observed in the marketplace.

4.4. ELECTRONIC NOSE CALIBRATION

Any analytical system made of sensors requires calibration. In the case of an e-nose, this can be response magnitude and relative magnitude for an array of chemical responses. Therefore, in order to develop the ability to recognize each chemical, the e-nose must be trained as well as calibrated. This task is very important for proper operation and successful application. In the training of an e-nose, the instrument is exposed repeatedly to varying concentrations of known compounds in varied order, and the resulting patterns of signals across the sensor array are stored and used in a computer algorithm. When consistent circulation of vapors through the e-nose for the vapors of the various calibration substances is complete, a library of responses is formed that is kept in the memory of the processor in the e-nose. This database of labeled signatures is used to train the pattern recognition system. The goal of this training process is to configure the recognition system to produce unique classifications of each chemical so that an automated rapid identification can be implemented in applications. Identification is made through comparison of the response of sample gas to responses obtained with authentic known substances, preprogrammed into the library of responses. Close matches to the library compounds is used to indicate the presence of specific substances or mixtures of molecules. The e-nose can also be trained to recognize partial patterns caused by substances for which it has not been specifically trained—enough to identify the chemical's functional groups. The most important limitation is that the e-nose does not often provide the molecular identity or molecular analysis for the odor. However, it is a powerful tool for comparative analysis and can identify subtle differences in extremely complex molecular mixtures.

5. ELECTRONIC NOSE FABRICATION

5.1. ELECTRONIC NOSE CONFIGURATION

Each e-nose can be divided into three or four basic modules. According to Tan et al. (2001), e-noses are comprised of four basic elements: a sampling system, which includes pumps, air conditioner, and flow controller; an array of sensors; an electronic data acquisition and control system; and pattern recognition software. At the same time, Marti et al. (2005) believe that an e-nose is a union of three elements: the sample-handling system; the detection system; and, last but by no means least, the data analysis system. In any case, one has to have hardware and software to accomplish the sampling, analysis, and reporting of the appropriate analytical information.

5.2. SENSOR ARRAYS

5.2.1. Sensors Acceptable for Application in Electronic Noses

The ideal sensors to be integrated in an electronic nose should fulfill the following criteria (Demarne and Sanjine 1992; Mari and Barbi 1992; Bartlett et al. 1993; Hodgins 1995, 1997; Schaller et al. 1998; Gardner 2004): high sensitivity toward chemical compounds, that is, similar to that of the human nose; low sensitivity toward humidity and temperature; medium selectivity, i.e., they must respond to different compounds present in the headspace of the sample; high stability; high reproducibility and reliability; robustness and durability; easy calibration; easily processable data output; and small dimensions. Fast response is also often desirable and, if a portable e-nose is wanted, the ideal sensors also should have low power requirements. This is a tall order, and sensors can fall short of this mark. The best sensor for the e-nose in this regard may be the mass spectrometer, but then the ideals of small size, low cost, and low maintenance are not realized easily. The most common sensors used in e-noses come from those used for industrial purposes and therefore have the many advantages of commercial practicality. Even if the sensors respond rapidly, like a MS, analysis times of e-nose systems are generally influenced more by the sampling method utilized than by the sensor response time. For example, some static sampling techniques require over 1 h for a sample reading, while some dynamic systems using similar sensors require less than 20 s. An average of around 5–10 min is common (James et al. 2005).

As we noted earlier, for correct detection of a wide range of chemicals and complex mixtures, chemical sensors in the sensor array must have different sensitivities to the target analyte gases and to the matrix gases. Therefore, a sensor array should include a combination of several sensors of different basic type or of sensors of the same type but with different response characteristics. Sensor arrays that have many of the same class of sensor but different types are said to be homogeneous arrays, and those that contain different classes of sensors and different types in each class are said to be heterogeneous arrays. However, it is necessary to note that the majority of electronic noses, commercial and otherwise, use sensors of the same class (homogeneous array), but generally of different types within a class.

There are several different sensor transducer principles that allow us to divide sensors into classes, including electrochemical (e.g., amperometric and potentiometric), thermal (catalytic combustible gas sensors, calorimetric), mechanical (surface acoustic wave, quartz crystal microbalance, cantilever), optical (fiber optic evanescent wave, absorption), and electronic (chemiresistor, chemicapacitor). And there are many types of each class of sensor and many variations even within a type of sensor platform. This leads to a wide variety of sensors and tuning possibilities for operation in an array (James et al. 2005).

Some authors use a simpler approach to classification (Mielle 1996). They divide all sensors for e-nose application into two broad groups:

1. "Hot sensors" (mainly the semiconducting metal oxide gas sensors)
2. "Cold sensors" (conducting polymers, bulk acoustic wave, and surface acoustic wave sensors).

Comparative performance of sensors from these two groups of currently available sensors is summarized in Table 1.2.

The heated metal oxide (HMOX) sensor is one of the most commonly used sensors in the field of electronic noses. For these sensors, different sensitive layers of different morphology with different doping, different production processes, different electrodes, different filter layers, and different operating

Table 1.2. Comparative properties of "cold" and "hot" sensors

Property	"Cold" sensors	"Hot" sensors
Power consumption	Low (a few μW)	High (50–800 mW)
Response time	Slow (20–40 s)	Fast (0.5 s to a few s)
Selectivity	Good (SAW and BAW sensors can be enanitioselective)	Poor (different for each type; can be adjusted by temperature modulation)
Carrier gas	May be inert (avoiding oxidation of sample)	Synthetic air (must contain O_2)
Availability	Commercially available (poor reproducibility)	Commercially available (over 70 types; many providers)
Lifetime	Not guaranteed to be more than 12 months (typically 18 months)	More than 5 years in normal use
Price	Expensive	Low for single sensors; medium for arrays

SAW, surface acoustic wave; BAW, bulk acoustic wave.

Source: Adapted from Mielle 1996.

temperatures all lead to different sensor responses. For example, Figaro Engineering, Inc., lists more than 25 tin oxide sensors in its product line, making a large variety of arrays possible using just this one type of chemiresistor sensor. The same diversity is found for other transducer principles, whether surface or bulk acoustic wave (SAW, BAW) sensors, metal oxide field-effect transistors (MOSFETs), electrochemical sensors, or conducting polymer (CP) sensors.

The particular choice of sensors depends on the individual application (Albert et al. 2000). Advice on choice of sensors is generally provided by the manufacturer. The selection of chemical sensor types to be utilized is of utmost importance if e-nose classifications are to be useful for a wide range of applications or for a specific problem endpoint determination. Most manufacturers seek highly selective sensors, but often, in the case of an electronic nose, every compound present in the gaseous phase should be detected and therefore needs to find a response on at least one sensor. If a new compound is added to a mixture, at least one sensor must detect this addition in order for the sensor array to have an effect beyond simple dilution.

5.2.2. Sensor Array Fabrication

Any sensor that responds reversibly to a chemical in the gas or vapor phase may be suitable for use in an electronic nose instrument. This means that many principles of detection can be used for fabrication of sensor arrays for electronic noses. However, in all cases, the goal is to create an array of differentially sensitive sensing elements. One can find general information about sensor arrays in several good reviews (Albert et al. 2000; Stetter and Penrose 2001; James et al. 2005; Rock et al. 2008).

In spite of the presence of many classes of sensors, the commercial e-nose is dominated by only a few types of sensors, including HMOX, MOSFETs, conducting organic polymers (CPs), piezoelectric crystals (SAW or BAW), amperometric gas sensors (AGS), and electrochemical sensors. Such sensors can be divided into two main classes: hot (HMOX, MOSFETs) and cold (CP, SAW, BAW, and AGS). The former operate at high temperatures and are considered to be less sensitive to moisture with less carryover from one measurement to another, but there is high variability in the reactivity and quality of these sensors.

Other sensor types, such as fiber optic (Dickinson et al. 1996; Sutter and Jups 1997), electrochemical (Mari and Barbi 1992), and calorimetric (Gall 1993; Lerchner et al. 2000), also can be used and may be integrated into electronic noses. Whatever type of sensor is used, its limitations as well as advantages will be reflected in the resulting instrumentation. It is important to consider such instrumental performance parameters as sensitivity, detection limit, selectivity, transient response time, reproducibility, stability, and lifetime before choosing sensors, and to remember that these qualities are, basically, properties of the sensor's chemically sensitive detection layer (Albert et al. 2000; Snopok and Kruglenko 2002).

As can be seen in Figure 1.9, there are two main approaches to design of sensors arrays. Examples of sensors arrays fabricated using the indicated approaches are shown in Figure 1.10. Of course, the second approach is the more progressive one.

Another very interesting approach to gas sensor microarray fabrication was developed at the Forschungszentrum Karlsruhe (Ehrmann et al. 2000). Figure 1.11 shows that, contrary to conventional macroarrays and also other gas sensor microsystems, a single monolithic metal oxide film alone forms

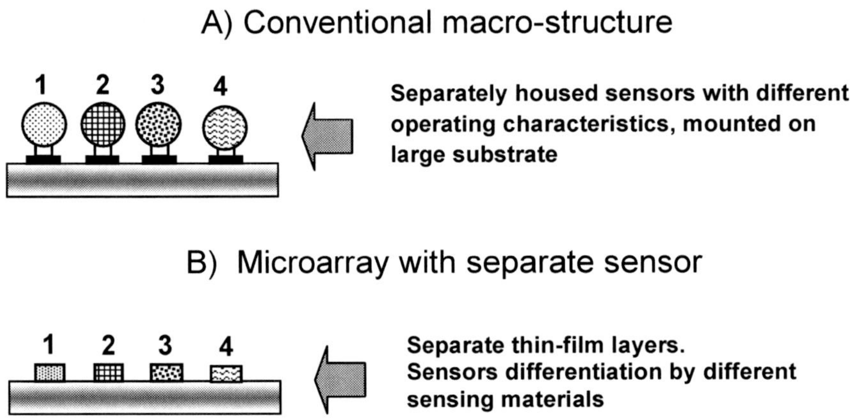

Figure 1.9. Main approaches used for sensor array fabrication.

the basis of the whole array. This film is separated into sensor segments by parallel electrode strips to measure the electrical conductivity of the segments. A simple gradient technique differentiates the gas detection selectivity of the individual sensor segments. The thickness of an ultrathin gas-permeable membrane layer deposited on top of the metal oxide film varies across the array. Additionally, a controlled temperature gradient is maintained across the array by four platinum heating meanders located on the rear side of the chip. As both thickness and temperature have a gas-dependent influence on the diffusion through the membrane and the temperature influence on the gas reaction at the metal oxide interface also depends on the nature of the gases, gas detection selectivity is gradually modified from sensor segment to sensor segment. Therefore, the exposures to single gases or gas ensembles (such as

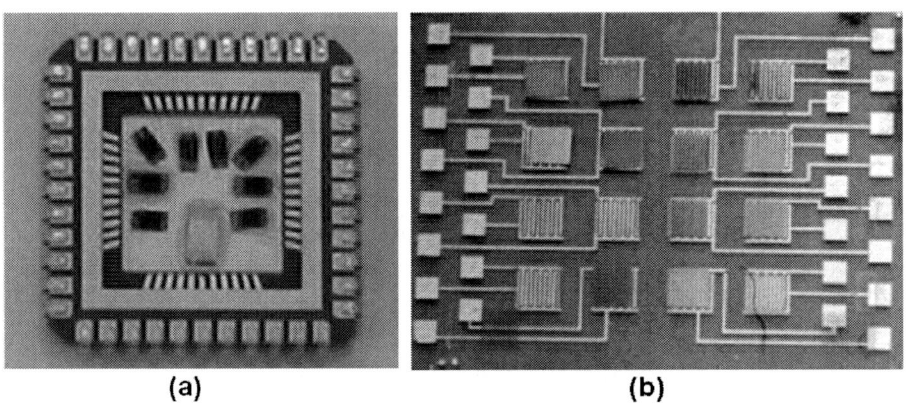

Figure 1.10. Examples of sensor array realization using approaches indicated in Figure 1.9. (a) Conducting polymer sensor array developed using approach A. (Reprinted with permission from Persaud et al. 2005. Copyright 2005 Oxford University Press.) (b) Microfabricated sensor array developed using approach B. Doped SnO_2 films were deposited by pulsed laser deposition. T_{oper} = 400°C. (Reprinted with permission from Aronova et al. 2003. Copyright 2003 American Institute of Physics.)

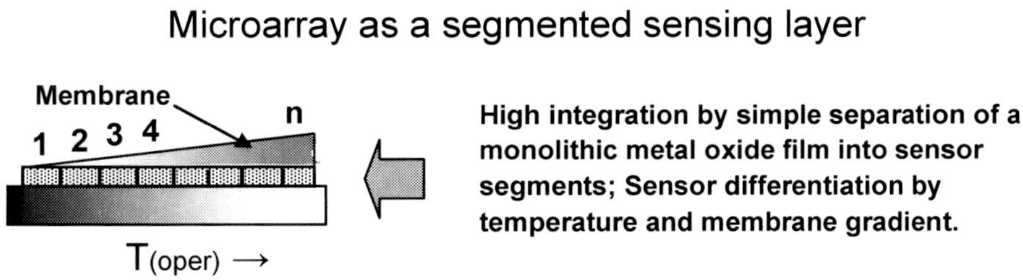

Figure 1.11. Diagram illustrating approach to microarray fabrication proposed by Ehrmann et al. (2000).

odors) cause characteristic conductivity patterns to occur at the gradient microarray. An example of this approach realization is shown in Figure 1.12.

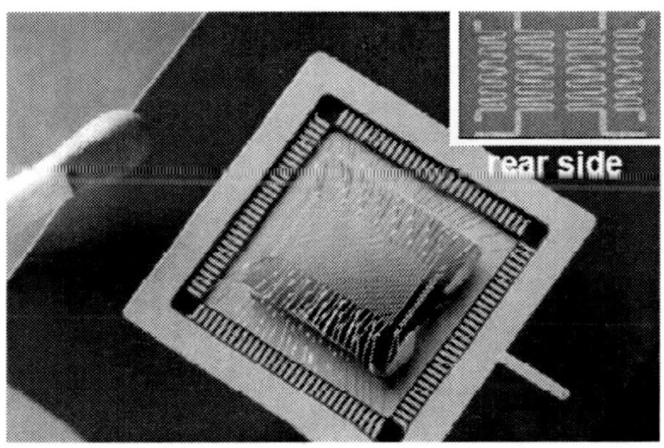

Figure 1.12. Micrograph of sensor microarray chip mounted in its housing. The chip contains four meander-shaped heating elements and a nanometer-thin SiO_2 membrane on top of the metal oxide layer with a thickness gradient. Two thermoresistor meander beside the metal oxide film serve as temperature sensors to feed the heating control. Electrical contacts are provided by gold wire bonds. (Reprinted with permission from Ehrmann et al. 2000. Copyright 2000 Elsevier.)

5.2.2.1. CONDUCTING POLYMERS

One of the early commercial e-nose systems used CP-based sensors, and they have been popular in much sensor array research. CP sensors have been found to respond to a wide range of organic vapors and operate at ambient temperatures with sensitivities of 0.1–100 ppm (De Wit et al. 1998). Compared with the metal oxides, organic polymers are much more diverse and can impart a wide variety of response characteristics to sensors. In the case of conducting polymers, the molecular interaction capabilities of a polymer can be selectively modified by incorporating different counter-ions during polymer preparation or by attaching functional groups to the polymer backbone. Another advantage of conducting polymers is that they can operate at room temperature (Albert et al. 2000). Examples of polymers used for design sensor arrays are given in Table 1.3.

Table 1.3. Examples of polymers used for sensor array design

Reference		
Breer et al. 1996	Kim et al. 2005; Ha et al. 2005	Snopok and Kruglenko 2005
Electrochemical polymerization; array of chemiresistive sensors	Carbon–polymer composites; silicon-based micromachined array of chemiresistive gas sensors	Polymer films; array of QCM-based sensors
1. Polypyrrole(PPy)-butane sulfonic acid 2. PPy-PSA 3. PPy-hexanesulfonic acid 4. PPy-heptane sulfonic acid 5. PPy-octane sulfonic acid 6. PPy-decanesulfonic acid 7. PPy-*p*-toluenesulfonic acid sodium salt 8. PPy-*p*-toluenesulfonic acid monohydrate 9. PPy-tetraethylammonium toluenesulfonate 10. Ployaniline-$NaHSO_4$ 11. Poly(3-methylthiophene-tetraethylammonium tetrafluoroborate	1. Poly(stylene-co-allyl alcohol) 2. Poly(4-methylstyrene) 3. Poly(caprolactone) 4. Polystyrene–black–polyisoprene–black–polystyrene 5. Cellulose acetate 6. Poly(stylene-co-methyl methacrylate) 7. Poly(styrene-co-butadiene) 8. Poly(vinyl stearate) 9. Hydroxypropyl cellulose 10. Poly(butadiene) 11. Poly(vinyl butyral)-co-vinyl alcohol-co-vinyl acetate 12. Ethyl cellulose 13. Poly(vinyl acetate) 14. Poly(4-vinyl pyridine)	1. *t*-Bu-Calix-[6]arene 2. *t*-Bu-Calix-[8]arene 3. Pentacene 4. Tetracene 5. Phtalocyanine 6. Dibenzotetraazaannylene 7. Tetramethyl-dibenzotetraazaannulene

Arrays of conductive polymer sensors are typically prepared electrochemically (Slater et al. 1993; Hodgins 1995b; Gardner and Bartlett 1995; Partridge et al. 1996). The electrochemical deposition of conductive polymers is reasonably controllable because films can be electrodeposited onto metalized areas and the film thickness can be varied by monitoring the total charge passed during the deposition process. The polymerization is usually carried out in a three-electrode configuration in which the working electrode is the sensor substrate. The electrodes are typically interdigitated electrodes or are a pair of metal leads separated by approximately 10–50 μm.

The shortcomings of polymer-based technology are often long response times (20–140 s), inherent time- and temperature-dependent drift, poor batch-to-batch reproducibility, and a relatively high cost of sensor fabrication (Dickinson et al. 1998). The effects of humidity and sensor drift due to oxidation of the polymers over time are other disadvantages of polymers for sensor array fabrication (James et al. 2005), since they lead to instability of the sensory patterns.

Microfabrication of large arrays of CPs with microscale dimensions are possible using interdigitated microelectrodes (Imisides et al. 1996; Stulik et al. 1994) created by photolithography (10-mm gap widths are not uncommon). And CP sensor arrays coupled with application-specific integrated circuits (ASIC) have been reported by Hatfield and co-workers (Hatfield et al. 1994; Neaves and Hatfield 1995) and by Gardner and co-workers (Cole et al. 2003; Garcia-Guzman et al. 2003).

Additionally, CP arrays in e-noses have been used for monitoring the quality of foods and beverages. Conducting polymer–based e-nose technology also has been reported to offer a rapid, reproducible, and objective method for sewage odor assessment. Researchers have utilized electronic noses to monitor air quality in relation to the assessment of malodor in agriculture. The CP electronic nose has been successfully applied to several medical and veterinary science applications as well (Albert et al. 2000; Gardner, 2004).

5.2.2.2. METAL OXIDE SEMICONDUCTOR SENSING TECHNOLOGY

Metal oxide semiconductors (MOS) have been used as gas-sensing elements since some of the earliest e-noses. Metal oxide sensors have fairly good sensitivity, particularly for polar analytes such as ethanol. Sensor selectivity can be shifted to different classes of compounds to some degree, either by changing the operating temperature of the sensors or by modifying the films by incorporating different amounts of noble-metal catalysts during the fabrication process (see Volume 4, Chapter 3, in this series). One drawback to metal oxide array fabrication is the necessity of incorporating a heating element to operate the array at high temperatures. Therefore, relatively high power levels are needed to run the sensors at elevated temperatures, and this is considered to be one of the primary drawbacks of these sensor systems. However, during the last decade, this problem was resolved. For example, AirSense Analytics (Schwerin, Germany) has introduced a multisensor element chip which allows considerable reduction in both the size and overall power consumption of a sensor array. Metal oxides also tend to be susceptible to "poisoning" by irreversible binding of sulfur-containing compounds and weak acids within a sample, as in cheese and vinegar (Dickinson et al. 1998). Despite these drawbacks, metal oxide sensors continue to be the most commonly used materials in the gas-sensing field (Albert et al. 2000; Romain et al. 2000; Mielle and Marquis 2001; Rock et al. 2008).

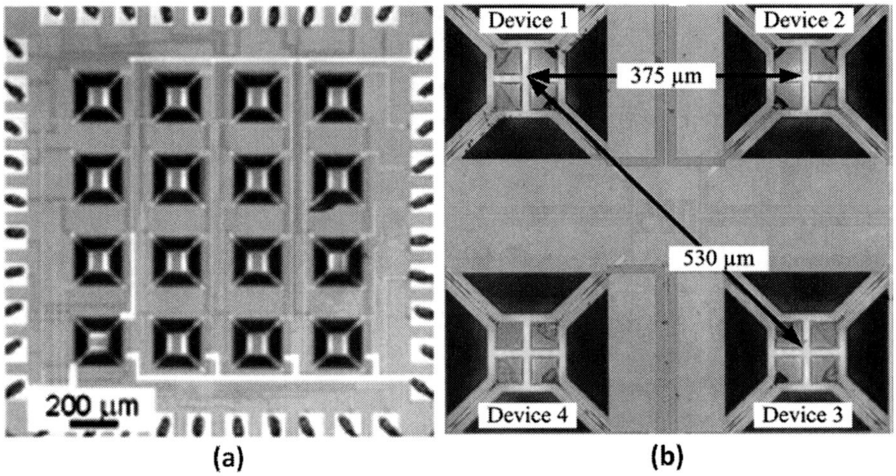

Figure 1.13. Optical micrograph of a 16-element micro-hotplate array (a) and a four-element suspended microhotplate array (b) with surface electrical contacts. [(a) Reprinted with permission from Taylor and Semancik 2002. Copyright 2002 American Chemical Society. (b) Reprinted with permission from Wheeler et al. 2007. Copyright 2007 American Chemical Society.)

The main advantage of MOX sensors is their potential for microminiaturization, which allows much diminished sizes and power consumption of devices. In this direction, considerable success was attained at NIST (USA) (Taylor and Semancik 2002; Wheeler et al. 2007; Meier et al. 2007). The electronic nose elaborated at NIST consists of eight types of sensors in the form of metal oxide film, deposited at the surfaces of 16 microheaters (Figure 1.13a) with two copies of each material. Precise heating control of individual elements, shown in Figure 1.13b, allows using them as a set of "virtual" sensors, working at various temperatures from 150 to 500°C, which increases the number of "virtual" sensors up to 5,600. The combination of sensitive films and the ability to change temperatures allows the device to act as an analytical equivalent of a nose with sensory neurons (Piggee 2008).

Metal oxide sensor arrays have been used to detect a variety of different analytes under varying conditions. Three major groups of analytes for these arrays are toxic gases, volatile organic compounds (VOCs), and food-related species (Albert et al. 2000).

5.2.2.3. ELECTROCHEMICAL SENSORS

Electrochemical sensors can also be used for electronic nose design (Stetter et al. 1986, 2001; Mari and Barbi 1992; Stetter and Penrose 1993). However, disadvantages such as the absence of sensors sensitive to specific gases, the lifetime of sensors, and difficulties related to integration and microminiaturization of electrochemical sensors limit their application in commercialized devices. These sensors are used mainly in the design of hybrid sensor arrays, or for designing an electronic nose aimed for a special application.

Figure 1.14. Sensor array consisting of eight CHEMFETs. (Reprinted with permission from Polk et al. 1999. Copyright 1999 Wiley.)

5.2.2.4. MOSFET ARRAYS

There have been some attempts to design sensor arrays based on MOSFET sensors (Sundgren et al. 1990, 1991, 1992; Sommer et al. 1995; Albert et al. 2008). Arrays have been designed with from 2 to 12 MOSFET sensors. The actual design of the sensor array can vary significantly, depending on the application. Platinum, palladium, and iridium are the three most common catalytic metals used as gate materials in MOSFET sensor arrays. Palladium makes a good hydrogen sensor, while platinum and iridium have sensitivities for analytes such as ammonia and ethanol.

As demonstrated by Polk et al. (1999), the CHEMFET sensor array also can be used for electronic nose design. A view of a CHEMFET-based sensor array is shown in Figure 1.14. A sensor array consisting of eight CHEMFETs was designed for simultaneous potentiometric and impedance sensing in the gas phase. The thickness of the transparent epoxy mask was 100 μm.

However, due to low sensitivity to many gases, this type of sensor is used mainly for hybrid sensor arrays. MOSFET sensors have acceptable sensitivity only to hydrogen, ammonia, and ethanol. Much of the work with MOSFET cross-reactive arrays has been performed at the Linkoping Institute of Technology in Sweden.

5.2.2.5. SAW SENSORS

The first application of pattern recognition methods to data from an array of quartz crystal microbalance (QCM) and surface acoustic wave (SAW) devices was in 1986 (Carey and Kowalski 1986; Ballantine et al. 1986). SAW sensors with polymeric covering find wide application in electronic nose elaboration today (Albert et al. 2000). Such distinctions as high sensitivity, the ability to work at room temperature, and low power consumption make sensors of this type very attractive for some application. Both the advantages and disadvantages of e-noses based on SAW sensor arrays are conditioned by the use of polymers.

5.2.2.6. OPTICAL SENSOR SYSTEMS

The use of optical sensors for chemical sensing is widespread in many areas (see Volume 5, Chapter 6, in this series). In comparison to other sensor types, there is less information on the use of optical sensors in e-nose applications, but since the development of an optical e-nose system by the Walt group in 1996 (Dickinson et al. 1996; White et al. 1996), there has been a significant increase in utilization (Johnson et al. 1997; Sutter and Jups 1997; Bakken et al. 2001; Cho et al. 2001). Optical sensor systems resemble most closely classical sensor array systems, because the dimension of data output can be precisely defined and adapted (Dickinson et al. 1996; Suslick 2004; Suslick et al. 2004). Instead of having transduction principles based on electrical changes in resistance, potential, current, or frequency, the modulation of light properties is measured. In general, optical instruments are more complex but offer a variety of different measuring possibilities. The assortment of applicable technologies is high and ranges from diverse light sources over optical fibers to detectors such as photodiodes and CCD and CMOS cameras (Filippini and Lundstrom 2002). Therefore, various operating modes have been developed and deployed, using changes in absorbance, fluorescence, optical layer thickness, and polarization.

For example, Walt's group at Tufts University has constructed an optical system that interrogates the sensors at an (excitation) wavelength of 500 nm, and measures time and amplitude changes in fluorescence (emission) at 610 nm (Walt et al. 1998). They have explored the optical behavior of polymer beads with different surface chemistries, coated with a solvatochromic fluorescent dye (Dickinson et al. 1999). A solvatochromic fluorescent dye (one that is highly sensitive to the polarity of its local environment) is immobilized in different organic polymers to produce an array of diverse sensors (White et al. 1996). Changes in polarity of the dye's surroundings induce characteristic shifts in the fluorescence emission spectrum, which can be monitored either at a single or at multiple wavelengths. In addition, the polymers used in this approach undergo a characteristic swelling as volatile compounds partition into the polymer matrix.

A large number of different bead types are mixed together. They are attached to the ends of optical fibers, where each may be individually interrogated with a light pulse. The response of each sensor to absorbed vapors is thus based on both the mechanical swelling of the polymer layer and the spectral shifting of the entrapped dye. Because of the method in which the beads and fibers are assembled, the assignment of "sensors" (the sensitive polymer beads) to data channels (individual fibers) is random. A neural network is used to create an association between each sample type and the pattern of responses. In a real sense, this device resembles the structure and ontogeny of the mammalian olfactory sense more than many other realizations of electronic nose technology. This approach has led to the development of sensors that are fast (100 ms to 3 s response time), small (total array diameter 350 mm to 2 mm) (Dickinson et al. 1997), simple to fabricate, inexpensive, and can be made with a highly diverse set of coatings. The approach has been given the trade name of BeadArray and is being commercialized by Illumina, Inc. (San Diego, CA).

Thus an array of fiber sensors can have wide-ranging sensitivities, a feature not easily obtainable with other sensor types. As with other types, differential measurements can also be used to remove common-mode noise effects. In their disfavor are the complexity of the instrumentation control system, which adds to fabrication costs, and their limited lifetime due to photobleaching. The fluorescent dyes are slowly consumed by the sensing process, the way sunlight bleaches fabric dyes (Kaplan and Braham 1998).

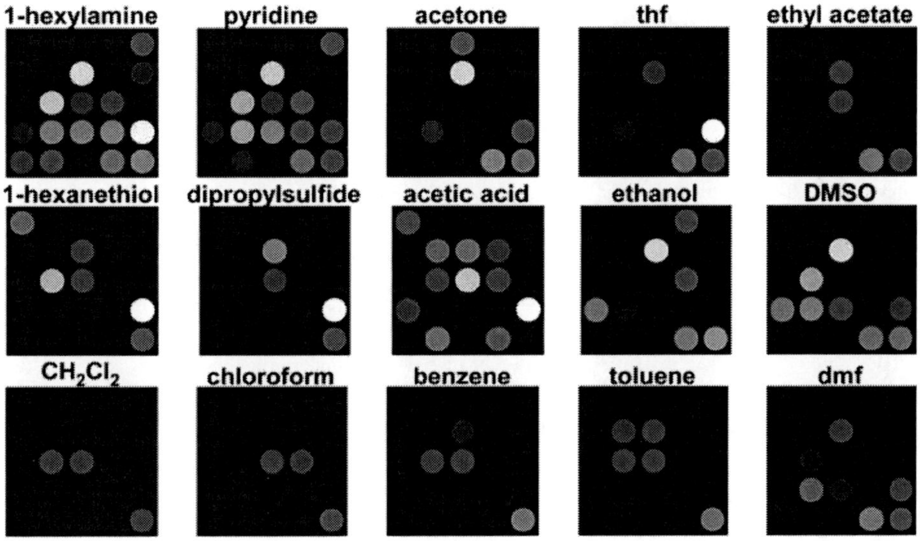

Figure 1.15. Colorimetric array responses for 18 common volatile organic compounds at saturation vapor pressure at 300 K. (Reprinted with permission from Suslick et al. 2004. Copyright 2004 MRS.)

Another unique variation on the electronic nose is the colorimetric sensor array recently proposed by Rakow and Suslick (Rakow and Suslick 2000; Suslick et al. 2004) for identifying solvent vapors. These authors noted that most odorous compounds have at least some Lewis base activity, and would bind to the central atom of a tetraphenylmetalloporphyrin, changing its color in a unique way. The authors spot a series of different metalloporphyrins onto a silica substrate in a stereotyped pattern and expose it to various solvent vapors. Each vapor produces a unique pattern of colors (see Figure 1.15).

Such an "electronic nose" is not necessarily electronic, and does not even require an instrument. The authors, however, have proposed a CCD camera to reduce the array colors to digital form for processing by automatic means. The selectivity of the device is determined by the relative affinity of the volatile for the polymer. The lifetimes of these sensors, however, are presently limited by photobleaching processes. The optical format also requires the use of relatively sophisticated instrumentation, such as CCD cameras and precision optical components (Dickinson et al. 1998).

5.2.2.7. OTHER SENSORS

The thermal sensor subtype of pellistors or calorimetric sensors have not been widely employed for e-nose systems, but some examples have been demonstrated (Gall 1993; Lerchner et al. 2000). There are also communications reporting gas sensor arrays for e-noses using cantilever-based sensors (Thundat et al. 1995) and capacitance-type sensors (Lonergan et al. 1996). For example, Dickert et al. have used an array of interdigital capacitors constructed from noble-metal electrodes to quantify solvent vapors based on the dielectric changes of different materials upon vapor incorporation (Dickert and Keppler 1995).

5.2.2.8. HYBRID SENSOR ARRAYS

Each chemical sensor type has its own advantages for sensor array technologies and its own intrinsic problems. One solution to this problem is the use of more than one type of sensor in the array system; these systems are known as a hybrid e-noses (Holmberg et al. 1995; Ulmer et al. 1997; Mitrovics et al. 1998; Pardo et al. 2004; James et al. 2005). Gopel and co-workers (Gopel 1994; Mitrovics et al. 1998) made the observation that an e-nose that uses chemical sensors of several classes, i.e., sensors whose chemical principles are different, gives data that are more effective when comparing samples than simply increasing the number of sensors of a single class. Moreover, in large arrays of different sensors of one class, sensors that may not contribute information will always contribute noise (Zaromb and Stetter 1984). Experiments have shown that sensors with chemically independent responses are valuable and make the array more versatile and able to distinguish more analyte differences. Sensors of different classes generally provide more chemically different or chemically independent responses than sensors of the same class with small variations in their formula or structure (Mitrovics et al. 1998). Therefore, expanding the sensor array to make it heterogeneous in sensor class can lead to increased chemical independence of the sensor responses (Stetter et al. 1987). The use of different sensor classes, which respond to different physical or chemical properties of the analytes, allows larger numbers of sensors to be used, while still contributing information to the dataset.

The major advantage with this type of e-nose is that it incorporates the advantages inherent in the different transducer technologies, also giving a choice for which chemical sensors should be used. The choice of chemical sensor will be dependent on the sample to be tested. The most common utilization of this system involves MOS and MOSFET sensors, but commercial systems are available that can use MOS, MOSFET, QCM, and AGS in various combinations. Electrochemical sensors also can be included in hybrid sensor arrays (Stetter et al. 2000). In particular, an e-nose designed by Stetter et al. (2000) utilized a sensor array formed by amperometric gas sensors (AGS), metal oxide (MOX) sensors, and piezoelectric quartz microbalance (QMB) sensors.

At present we have information that hybrid sensor arrays have been designed using MOS, PC, electrochemical, MOSFET, and SAW sensors (Albert et al. 2008). Commercial electronic noses made by Alpha MOS (Toulouse), EEV Ltd. (Chelmsford, UK), Nordic Sensor Technologies AB (Linköping, Sweden), RST Rostock Raumfahrt und Umweltschutz GmbH (Rostock, Germany), and Lennartz Electronic GmbH (Tübingen, Germany) have used sensors of mixed type as well (Nagle et al. 1998).

5.3. SAMPLING SYSTEM

The sampling system is usually sufficiently important that it is listed as the third essential component of an electronic nose (Mielle 1996). The role of the sampling system is to collect and convey the volatile sample to the sensors, then to restore previous conditions by means of a cleaning procedure. A good sampling system should both deliver a vapor sample to a sensor array in a reproducible way, and reduce sample-to-sample variation that might result from differences in humidity, temperature, or concentration. In addition, the purpose of the sampling system is to preprocess the sample in any way that increases the quality of the output data. This means that in order to obtain measurements with optimum stability and reproducibility as well as high-amplitude signals and fast sensor response, the sampling device has to be designed in such a way that all factors capable of influencing sensor responses are optimized and kept under control, so that the only variable parameter is odor composition (Falcitelli et al. 2002). However, we have to note that not much research has been carried out on optimized sampling systems, even though some authors have demonstrated dramatic increases in performance of sensor arrays (Stetter et al. 2000). Some sampling systems also provide for automated measurement of a series of samples, although this is not a defining requirement.

The most obvious use of a sampling system is to concentrate the vapors, in order to improve the sensitivity of the sensors (Schaller et al. 2000; Persaud et al. 2005). The purge-and-trap and solid-phase microextraction (SPME) techniques have already been used to improve the selectivity of the sensors. This latter method was successfully employed as a filter for ethanol (Aishima 1991; Privatet et al. 1998), that is, the ethanol contained in the samples was not adsorbed by the porous polymer material, and therefore was not delivered to the sensors. Consequently, the sensors were not blinded by the ethanol content and could response to other components. This approach was applied by Grate et al. (1993) in a four-SAW sensor device which was optimized for chemical warfare agents. It was later commercialized for hydrocarbon measurement by H. Wohltjen at Microsensor Systems, Inc. A short tube of organic sorbent (Tenax) was used to absorb vapors from the air. These were desorbed using a heater and passed through a short GC column to the sensor array. Similar preconcentrator approaches have been used by other workers as well (Strathmann et al. 2001).

Concerning the length of analysis time, it is important to state that sample throughput is related not only to acquisition, but also to the sampling time, headspace equilibration time, and recovery or cleaning time (Mielle and Marquis 2000). For e-nose and MS-based systems, the main part of the analysis time is spent in equilibration. However, the total analysis time may be reduced by using a pipeline mode for the sampling: A new sample equilibrates during the analysis of the current sample. Recovery time for a sensor array system is related to the sensor type and to the nature of the compounds adsorbed. It can last up to 40 min for some low-volatility compounds (Wunsche 1995). The use of an autosampler reduces the acquisition and mainly the recovery time, even at relatively low concentrations, by reducing the amount of volatile reaching the measurement cell. But this also dramatically reduces the sensor response (Lacoste 1999). However, it was shown that, through optimization of the sampling system and measurement cell, the cycle time could be reduced to less than 10 min, retaining high measurement accuracy and low noise level, over a large concentration range (Mielle and Marquis 1999, 2000). The small overshoot was due to a small unbalanced flow rate between the baseline and the sampling period.

Samples often contain substances that are common to all, and although the sensors are dominated by them, they do not contribute to discrimination. In bacterial cultures, for example, the common substance is water. In beer and wine, water and alcohol will be present in all samples in much larger quantity than any other constituent. McEntegart et al. (2000) used Nafion tubing and anhydrous sodium sulfate to remove selectively water, alcohol, and some other hydrophilic vapors from samples. Although sensor signals, on average, were reduced by a factor of 10 or more, the removal of the dominant constituents greatly improved selectivity. Another type of sampling system (Stetter et al. 2000) allowed TNT adsorbed to silica sand to be detected and discriminated from structurally similar compounds by vaporization of a sample from a tiny beaker, using a hot platinum filament. A second filament was located downstream to combust the sample to electrochemically reactive compounds, probably nitrogen oxides and carbon monoxide.

The measurement chamber also plays an important role in the sampling process (Lezzi et al. 2001; Falcitelli et al. 2002). Nonadsorbent and inert materials have to be selected with care to avoid memory effects. The choice of the material has a remarkable effect on the chamber projecting, as unsuitable mechanical properties can hamper refined machining and complex design. Glass and fused silica, for example, are mainly processed as craft work, while stainless steel is very hard to machine on a small scale using common techniques. Polytetrafluoroethylene (PTFE) is soft enough to be machined easily, but, for this same reason, its mechanical properties make it difficult to craft into small components. Moreover, it is microporous and so acts as a polymer trap, so attention has to be paid during use to avoid contaminations and memory effects. All these materials can be treated with aggressive solvents and dried in an oven at high temperatures, even if in the case of Teflon it is suggested not to exceed 200°C.

The volume of the chamber has to be properly dimensioned for any range of gas flow rates, in order to obtain a homogeneous flow with a low-speed gradient with no recirculating zones or stagnant regions. It is also important that all the sensors are exposed at the same time to the same odor concentration, so that their performances will not be related to their position inside the chamber. Several useful advances related to optimization of measurement chamber have been described by Falcitelli et al. (2002).

6. BENEFITS OF ELECTRONIC NOSES

Like the mammalian olfactory system, the electronic nose uses a holistic approach to distinguish aromas. It does not separate or attempt to identify the individual chemicals responsible for the pattern. The electronic nose works equally well on pure compounds or on undefined samples such as flavors, aromas, and other complex odors. Conventional reductionist analytical methods often become less reliable as sample complexity increases. Some work on grain odors, for example, was preceded by a 3-year effort by the U.S. Department of Agricultre (USDA) to use GC-MS to distinguish grain quality. The author of the final report on this study stated, after analysis of more than 300 samples, that "no relationship between the chemical composition and the odor could be found" (Weinberg 1986). Other authors have had similar problems correlating the results of detailed chemical analysis with organoleptic responses (Schaller et al. 1998; Zhou et al. 1999).

Another problem of human testing panels backed by gas chromatography and mass spectrometry is connected with fact that those methods are time-consuming, expensive, and seldom performed in real time in the field (Kaplan and Braham 1998; Zou and Zhao 2008). Electronic noses have the ability to perform sensory-based analyses such as for odor/aroma and taste rapidly and in a cost-efficient manner. This new technology offers the possibility of arranging for continuous real-time monitoring of odor at specific sites in the field over hours, days, weeks, or even months. An electronic device can also circumvent many other problems associated with the use of human panels. Individual variability, adaptation (becoming less sensitive during prolonged exposure), fatigue, infections, mental state, subjectivity, and exposure to hazardous compounds all come to mind (Kaplan and Braham 1998).

Not only are the electronic systems more rapid and objective than human testing panels, they are much more amenable to routine use, for example, in production QC applications. Since they correlate well with human sensory panels, the instruments can be described as a "consumer-based instrumental test." Electronic noses also correlate well with more classical analytical methods such as HPLC and GC/GC-MS instruments which measure a specific analyte or group of analytes (Tan et al. 2001). However, e-noses are much faster. For example, HPLC analysis of caffeine levels in a carbonated soft drink generally takes about 15 min. An e-tongue will give the result in approximately 3 min.

High sensitivity is other advantage of e-noses. Electronic noses are often even more sensitive than the human nose (see Table 1.4). There is some evidence that sensors differentiate aromas on the basis of relatively few compounds, and in the future a relationship between specific chemicals and a single

Table 1.4. Detection threshold levels of human olfactory systems and electronic noses

Volatile compound	Reported human threshold (ppm)	Electronic nose threshold (ppm)	Type of electronic nose	Reference
Ethyl acetate (a)	7–17 (b)	5–25	Fox 3000 (12 MOS)	Stetter et al. 2000
Butyric acid (a)	0.4–10 (b)	<1	Fox 3000 (12 MOS)	Stetter et al. 2000
Diacetyl (a)	(4–15) × 10^{-3} (b)	(50–100) × 10^{-3}	Fox 3000 (12 MOS)	Stetter et al. 2000
n-Hexanal (a)	(10–50) × 10^{-3}	(10–50) × 10^{-3}	Fox 3000 (12 MOS)	Stetter et al. 2000
Methional (a)	(2–50) × 10^{-3}	(10–50) × 10^{-3}	Fox 3000 (12 MOS)	Stetter et al. 2000
Furanol (a)	(20–40) × 10^{-6} (b)	(50–100) × 10^{-6}	Fox 3000 (12 MOS)	Stetter et al. 2000
n-Nonane (c)	0.2–7	<0.2	20 CP composite	Doleman and Lewis 2001
n-Octane (c)	3–9	0.6	20 CP composite	Doleman and Lewis 2001
n-Heptane (c)	7–13	<2	20 CP composite	Doleman and Lewis 2001
n-Hexane (c)	13–30	<10	20 CP composite	Doleman and Lewis 2001
n-Pentane (c)	20–50	40	20 CP composite	Doleman and Lewis 2001
1-Pentanol (c)	0.13–1.3	<0.06	20 CP composite	Doleman and Lewis 2001
1-Butanol (c)	0.2–1.3	0.3	20 CP composite	Doleman and Lewis 2001
1-Butanol (d)	0.7	–	Aromascan (32 CP)	Hudon et al. 2000
1-Butanol (d)		–	Fox 3000 (12 MOS)	Hudon et al. 2000
1-Butanol (d)		+	6 Taguchi (SnO_2)	Hudon et al. 2000
1-Propanol (c)	0.9–1.9	1.3	20 CP composite	Doleman and Lewis 2001
Ethanol (c)	5–500	2	20 CP composite	Doleman and Lewis 2001
Methanol (c)	13–600	3	20 CP composite	Doleman and Lewis 2001
Acetone (d)	141	–	Aromascan (32 CP)	Hudon et al. 2000
Acetone (d)		+	Fox 3000 (12 MOS)	Hudon et al. 2000
Acetone (d)		+	6 Taguchi (SnO_2)	Hudon et al. 2000
Ethanethiol (d)	0.1 × 10^{-3}	–	Aromascan (32 CP)	Hudon et al. 2000
Ethanethiol (d)		–	Fox 3000 (12 MOS)	Hudon et al. 2000
Ethanethiol (d)		–	6 Taguchi (SnO_2)	Hudon et al. 2000

(+), Detected at the same concentration as submitted to human noses; (–), not detected (when response <3× background noise) at the same concentration as submitted to human noses; (a), concentration in water; (b), orthonasal analysis; (c), concentration in air; (d), concentration in vapor in equilibrium with a liquid phase at 22.5–25°C.

Source: Reprinted with permission from Ampuero and Bosset 2003. Copyright 2003 Elsevier.

flavor attribute may be achievable. Also, the possibility exists to differentiate between "top" and "middle" notes of aroma. In addition, the "odor/aroma fingerprint" could then be stored in a database in a way analogous to the memorization of olfaction perception in the human brain.

7. REASONS LIMITING OPTIMAL OPERATION OF ELECTRONIC NOSES

Limitations to utilizing the full potential of electronic noses include loss of sensitivity in the presence of high concentrations of a single component such as alcohol; the inability to provide absolute calibration; the relatively short lifetimes of some sensors; the necessity for considerable method development work for each specific application; and an inability to obtain quantitative data for aroma differences (Harper 2001). Electronic noses are intended to detect odorous compounds, but it is not always clear that discriminations are based on odorous rather than nonodorous, and possibly incidental, components of the headspace.

The drift of output parameters is another problem for e-noses (Zubritsky 2009). All e-noses are subject to drift, but the problem has received the most attention in the resistive sensor instruments. In particular, drift falls into two categories: sensor drift, which is due to the aging or degradation of individual sensors; and system drift, which encompasses all sensors. Some e-noses are designed to monitor both kinds of drift so that users know when sensors need to be replaced. In some instruments, individual sensors can be replaced, but in others, the entire array must be replaced. In all cases (except optical sensors), the e-nose must be retrained afterward. However, it is necessary to take into account that in some cases the "drift" is a consequence of the sensor environment instability. For example, if you put a metal oxide–based e-nose into a room where CO and ozone levels fluctuate dramatically, the variations may register on the instrument. Some people consider these variations as drift, but, strictly speaking, they are not drift.

Related to drift is the issue of an e-nose's sensitivity to humidity. Humidity is a known problem, especially with polymer-based sensors, and care needs to be taken in their use (James et al. 2005). For example, Kaplan and Braham (1998) showed that an e-nose lost sensitivity in the presence of water vapor. This complicates the detection of volatile odors in the presence of high levels of water vapor. Research found that e-noses based on hydrophilic conducting polymers are the most prone to this problems. Therefore, when humidity levels are not compensated for or controlled, it is possible that the humidity is responsible for some successful discriminations. Therefore, minimizing the impact of humidity is a key task for any e-nose designer. Two companies, at least, are addressing this problem. Some instruments are capable of choosing either an ambient air reference (when the humidity is varying) or a filter to remove humidity (when the level is relatively constant). For example, Bloodhound Sensors is developing systems based on discotic liquid crystals, which have very low response to water vapor. Alternative sensor types, described as water-insensitive chemoresistors, are being developed by Marconi Advanced Technologies.

Reproducibility is another common and important problem with e-noses (Zubritsky 2009). One reason that it is difficult to get consistent responses is that sensor materials age or become "poisoned" by some chemicals. In fact, the lifetimes of sensor materials can range from several months to 1 or 2 years, depending on the underlying chemistry and the sensor's environment. The second reason is variations in sampling. To compensate for variations in sampling, some e-noses use auto-sampler systems that control the temperature, shake the samples, and perhaps equilibrate the headspaces. Although this yields better reproducibility, these systems can be large and expensive. However, some companies have put a lot of effort recently into miniaturizing sampling systems.

8. MARKETS FOR ELECTRONIC NOSES

As noted earlier, the first commercial electronic nose devices were launched in the 1990s. Among the first were systems from Alpha MOS in 1993, Neotronics and Aromascan in 1994, and Bloodhound Sensors and HKR Sensorsysteme in 1995. Since then, many companies worldwide have started manufacturing electronic nose instruments, using many different detection principles and sensor types, so that many electronic nose systems are available on the market today, including those listed in Table 1.5.

Each detection method has advantages and drawbacks, so companies are increasingly configuring their products to include several measuring principles in one instrument. For example, Alpha MOS is developing instruments (see Figure 1.16) that include fingerprint mass spectroscopy with the more traditional metal oxide sensor arrays. Lennartz (www.lennartz-electronic.de) and Marconi Advanced Technologies (www.marconitech.com) have integrated quartz crystal microbalance sensors with metal oxide semiconductors, which have been used in parallel with GC-MS and human sensory panel sniff tests to monitor "off" odors from food packaging materials. A similar array, but with conducting polymer chemoresistors also incorporated, has been used with some success to monitor amniotic fluid leakage in expectant mothers.

Applications of commercial e-nose devices range from quality control to landmine detection, from security controls to health screening. Most of the machines listed in Table 1.5 are priced from $20,000 to $100,000 (U.S.), but prices should drop in a few years as competition heats up and sensor technology improves. Most models now on the market are fairly large, none-too-portable, lab-type systems (see Figure 1.16), but several manufacturers have said they are working on miniaturized versions. Low-cost miniaturized devices for personal use are already beginning to emerge on the market.

One example of such a miniaturized instrument is shown in Figure 1.17. The Cyranose 320 requires a one-time training session in which the user exposes the sensor to each type of sample that will be encountered during testing, creating a base-group to which all future samples will be compared. The sampling step involves placing the tip of the device near the sample and simply pressing "Run." Air referencing, vapor sampling, sensor measuring, and data processing are all handled automatically and take roughly 1 min to complete. If the sample matches one of the pretrained samples, the unit reports the identity of the sample. If the unit does not recognize the sample, an "Unknown" reading is given. The unit automatically resets itself, and the polymer composite sensors return to their original resistance. This real-time, portable device enables food companies to spot-test raw materials for batch-to-batch consistency, spoilage, or contamination. The Cyranose 320 is also used by chemical and petrochemical companies for quick assessment of the chemical status associated with various industrial processes. For example, profiling a chemical environment in a hazardous materials situation allows emergency crews to accurately select fire retardants, containment strategies, and protective gear. The device weighs 0.91 kg, and response time is less than 1 min.

Table 1.5. Commercially available electronic nose instruments

Manufacturer	Country	Sensor type	Number of sensors	Size	How to contact
Airsense	Ge	MOS	10	Laptop	www.airsense.com
Airsense	Ge	QMS			www.airsense.com
Agilent Analysis GmbH		QMS			www.chem.agilent.com
Draeger		IMS			www.draeger-safety.com
Environics	Fi	IMS			www.environics.fi
Smith Group		IMS			www.smithsdetection.com
Alpha MOS-Multi	Fr	MOS, CP	6–24	Desktop	www.ALPHA-mos.com
Applied Sensor	Se	MOSSFET	2		www.appliedsensor.com
Orhanoleptic Systems		QCM, SAW			
AromaScan PLC	UK	CP	32	Desktop	www.aromascan.com
Bloodhound Sensors Ltd.	UK	CP	14	Laptop	www.bloodhound.com
Cyrano Sciences Inc.	US	CP	32	Palmtop	www.cyranosciences.com
EEV Ltd. Chemical Sensor Systems	UK	CP, MOS, QCM, SAW	8–28	Desktop	www.eev.com
Electronic Sensor Technology Inc.	US	GC, SAW	1	Desktop	www.estcal.com
Hewlett-Packard Co.	US	QMS	—	Desktop	www.chem.external.hp.com
HKR-Sensorsysteme GmbH	Ge	QMS	6	Desktop	home.t-online.de/home/hkrsensor
Agilent Technology	Ge	QMS			www.agilent.com
Lennartz Electronic GmbH	Ge	MOS, QCM	8–40	Desktop	www.lennartz-electronic.de
Microsensor Systems	US	SAW			www.microsensorsystems.com
Nordic Sensor Technologies AB	Se	MOS, QCM	22		www.nordicsensor.com
		IR, MOSFET		Laptop	
Marconi Applied	UK	MOS, QCM, CP			www.marconitech.com
RaeSystemes		Hybrid	5–6		www.raesystems.com
RST Rostock Raumfahrt undImweltschatz GmbH	Ge	MOS, QCM, SAW	6–10	Desktop	www.rst-rostock.de
Sysca AG	Ge	MOS	38		www.sysca-ag.de
Sawtek Inc.	US	SAW	2	Palmtop	www.microsensorsystems.com
Technobiochip		QCM	8		www.technobiochip.com

CP, conducting polymer; MOS, metal oxide semiconductor; IR, infra red; SAW, surface acoustic wave; QCM, quartz crystal microbalance; QMS, quadrupole mass spectrometry; GC, gas chromatography; FET, field effect transistor; IMS, ion mobility spectrometry.

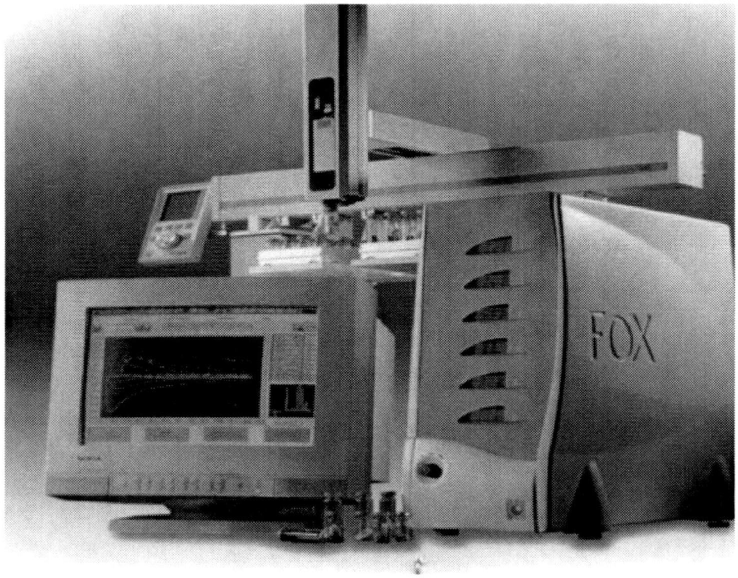

Figure 1.16. Fox electronic nose manufactured by Alpha MOS, France. This unit requires a PC to run the system. (Uploaded from www.alpha-mos.com.)

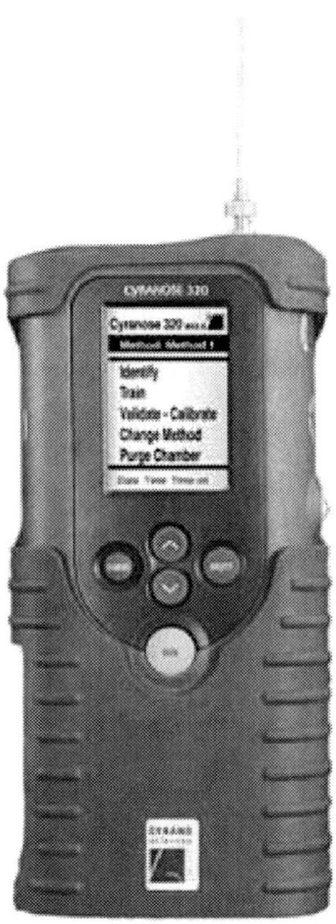

Figure 1.17. View of Cyranose 320 portable electronic nose designed by Cyrano Sciences, Inc., of Pasadena, California, using an ingenious technology licensed from Caltech. Nanocomposite sensor array includes 32 CP-based sensors. (Uploaded from www.smithsdetection.com/eng/Cyranose_320.php.)

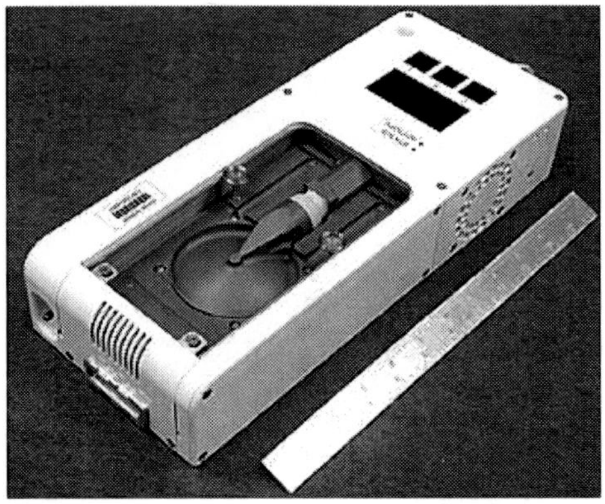

Figure 1.18. Electronic nose installed at ISS in 2008. (Uploaded from enose.jpl.nasa.gov.).

An electronic nose based on a polymer sensor array is shown in Figure 1.18. It has was developed by the Jet Propulsion Laboratory of the California Institute of Technology. Its weight does not exceed 4 kg, and the mass of the sensor unit is only 840 gr. It includes a sensor array consisting from 32 sensors. This e-nose is able to analyze air by itself and then send data to an external computer. Sensitivity of the sensors can be tuned to the parts-per-billion level. Power consumption is only 20 W. In 2008 this device was installed on the International Space Station (ISS). Another version of an electronic nose using a polymer sensor array is shown in Figure 1.19.

A portable electronic nose system based on a laptop PC or PDA has been developed and demonstrated successful differentiation between the complex real samples, i.e., brandy and whiskey. This

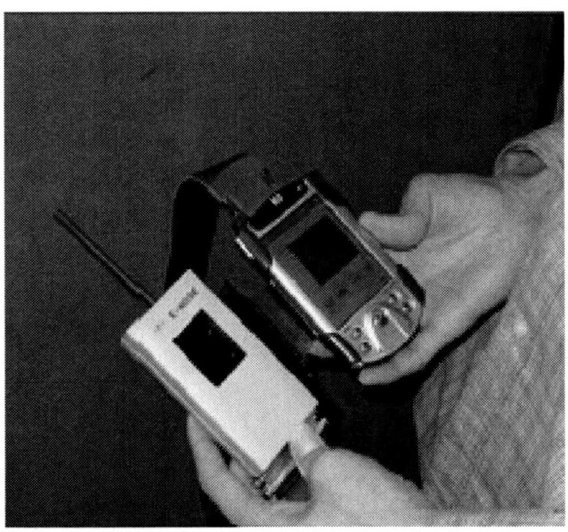

Figure 1.19. Portable electronic nose system based on a carbon black–polymer composite sensor array. (Reprinted with permission from Kim et al. 2005. Copyright 2005 Elsevier.)

system, based on an array of 12 chemiresistive sensors, consists of the hand-held sensing module with vapor delivery components, a single sensor array chip and signal conditioning circuitry, and the personal digital apparatus with the function of processing measured signals, such as data acquisition and pattern recognition. A key part of this e-nose system is the sensor array chip, which is fabricated using Si bulk micromachining and carbon black–polymer composite sensors.

An electronic nose based on a metal oxide sensor array is shown in Figure 1.20. The KAMINA device, developed at the Karlsruhe Research Center (FZK), is comprised of a microarray chip (only some mm^2 in size), fluidic components for gas sampling, and the entire microprocessor-controlled operating electronics in a single unit. Gas sampling can be optionally configured. There are several options for gas guidance (particle filter, tubes of stainless steel, gold-plated stainless steel, or glass). Further, several sampling options are available (ventilator, micropump, or no sampling arrangement) that, where required, allow adjustment of the gas flow. The part of the device below the microarray chip contains the complete controlling measuring electronics as well as the regulated heating supply for the four chip heaters. The operating electronics control each of the four heating meanders individually, to maintain a temperature gradient across the microarray that is responsible for differentiation of the gas selectivity between the individual sensor segments. This temperature gradient is kept constant with the aid of two temperature sensors (Ehrmann et al. 2000). (This approach to sensor array fabrication was discussed in Section 5.2.2.) The gas sensor array consists of 38 or 16 sensor segments. It can be fabricated using Si/SiO_2 or Al_2O_3 substrates and SnO_2 or WO_3 gas sensing layers. A standard serial data interface provides the communication to the controlling PC. A power supply unit for either 230-V or 110-V connection, as well as a 12-V or 24-V inverter, are available. Nearly all gases except for nitrogen and rare gases can be detected, often at very low concentrations, e.g., less than 1 ppm.

An electronic based on an array of electrochemical sensors is shown in Figure 1.21.

Figure 1.20. Operating device of the Karlsruhe Micronose KAMINA with the head cover lifted. (Uploaded from www.sysca-ag.de and www.fzk.de/pmt.)

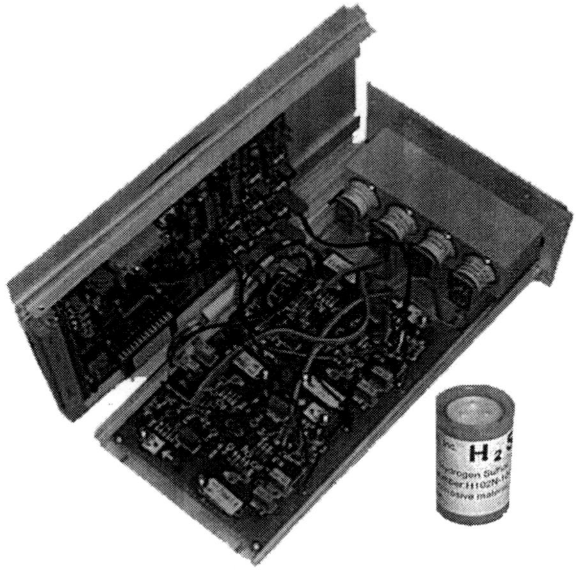

Figure 1.21. Prototype of EC sensor module for the Lennartz electronic MOSES II EN system. Beside it is an EC (amperometric) gas sensor made by TSI, St. Paul, Minnesota. (Reprinted with permission from Stetter et al. 2000. Copyright 2000 Elsevier.)

9. APPLICATIONS OF ELECTRONIC NOSES

Odor is very important in food, beverage, pharmaceuticals, and personal care industries, so one can expect maximal application of electronic nose in these fields. The expectations of consumers regarding the quality of products in these fields are continually increasing as a result of a greater range of choices in the marketplace together with targeted advertising, which emphasizes product quality. Competition for market share and the added emphasis on quality have increased pressure on product development and rigorous quality control to meet consumer expectations. Due to their simplicity, rapidity, and objectivity, electronic noses can be incorporated into instruments aimed for quality control, product matching, origin identification, spoilage detection, and flavor quantification. A particularly potent example is coffee, where the simplest electronic noses can make fine distinctions between blends (Gardner et al. 1992). Vintners have been able to identify wines by provenance as well as vintage (Guadarrama et al. 2000; Marti et al. 2005; Berna et al. 2008).

Just like a human taster and odor controller, e-noses can be used for both quantitative and qualitative applications. Quantitative applications include sensory score correlation and the measurement of the concentration of flavors or fragrances within a food or cosmetic product. Qualitative sensory applications include the determination of the origin and quality of raw materials and consistency of finished products (Tan et al. 2001). Electronic noses can be applied by food manufacturers to such tasks as freshness testing, quality control, and screening of incoming raw materials, not to mention feedback control to optimize bioreactors and minimize batch variation, and monitoring for accidental or intentional contamination or mislabeling of manufactured food products (Kaplan and Braham 1998). Electronic nose systems are well suited to comparing final products to reference standards, even replacing human taste panels in some applications. Testing of food freshness is especially important for the elderly, since the senses of taste and smell dull with advancing age (Gomez et al. 2006) (see Table 1.6). Other applications of the electronic nose technologies include shelf-life measurements and the evaluation of interactions between packaging and product.

In addition to this field, e-noses can be used in other areas such as petroleum qualitative and quantitative analysis, detection of explosives, classification and degradation studies of olive oils, development of a field odor detector for environmental applications, quality control applications in the automotive industry, discrimination between clean and contaminated cows' teats in a milking system, cosmetic raw materials analysis, plus many other important areas such as medicine and space exploration (Baby et al. 2000; Gardner et al. 2000; Morvan et al. 2000; Tan et al. 2001). Electronic noses may be used to analyze breath, sweat, urine, and stools, and possibly diagnose some illnesses as a result; work in this area has been done at companies such as Aromascan, Bloodhound Sensors, and EEV Chemical Sensor Systems (Kaplan and Braham 1998). Thus, an electronic nose could help solve problems in many fields, including health care, environmental monitoring, indoor air quality, safety and security, and the military (Kaplan and Braham 1998; Scorsone et al. 2006).

Table 1.6. Some reported applications of electronic noses in the food industry

Food	Test	Sensors/type	Reference
Seafood (oyster, sardine, etc.)	Freshness	1/MOS[a]	Nanto et al. 1991
Fish (cod, haddock)	Freshness	4/MOS	Olafsson et al. 1992
Fish	Freshness	1/MOS[a]	Egashira et al. 1994
Fish	Freshness	6/MOS	Barbi et al. 2007
Fish (trout)	Freshness	8/EC	Schweizer-Berberich et al. 1994
Grains	Classification	15/mixed	
Ground pork/beef	Discriminate and effect of ageing	15/mixed	Winquist et al. 1993
Boar	Taints in meat	14/MOS	Bourrounet et al. 1995
Sausage	Meats discriminate	6/MOS	Tan et al. 1995
Food flavours (orange, etc.)	Flavor identification	8/BAW	Nakamoto et al. 1993
Wheat	Grade quality	4/4rEC	Stetter et al. 1993
Wheat and cheese	Discriminate and ageing	20/CP	Pisanelli et al. 1994
Tomatoes	Effect of irradiation and stress	7/mixed	Winquist et al. 1995
Cheese	Maturity of cheddar	20/CP	Persaud and Travers 1997
Cheese	Discriminate	8/CP	

[a]Not strictly an electronic nose as defined here, but an odor monitor.

MOS, metal oxide semiconductor; EC, electrochemical; BAW, bulk acoustic wave; CP, conducting polymer.

Source: Adapted with permission from Gardner et al. 2000. Copyright 2000 Elsevier.

10. SUMMARY

In recent years there has been great progress in elaboration of the principles of electronic nose functioning, as well as in commercialization of attained results. Electronic noses are available on the market, and even now allow resolving a wide range of tasks in various areas. However, in spite of the considerable progress, the electronic nose as a commercial device has not reached its full potential. There are still problems with rate of response, stability, and portability. However, the appearance of new materials, elaboration of modern technologies, and development of mathematical methods of information processing offer hope of the appearance in the near future of new devices with much better functional parameters. However, it is hard to expect that multifunctional devices without application limitations will appear in the near future. A single instrument that could be used in every possible application would be overly complicated due to the large number of sensors and time-consuming statistical analysis. Therefore, in our view, the trend toward creating systems for specific applications is well founded.

It is necessary to note that the presence of analytical instruments for gas analysis based on such methods as chromatography and mass spectroscopy does not limit development of the electronic nose market, because devices are designed for different tasks (Mielle and Marquis 2000; Marti et al. 2005). The higher-cost instruments based on traditional analytical methods such as GC or MS analysis, operated by skilled laboratory workers, will always be needed to determine qualitatively or/and quantitatively why one food sample differs from others. The electronic nose is better adapted for such tasks as quick quality tests in various applications, routine work, or in cases where nonodorous or irritant gases need to be detected. There are a number of application areas where the possibilities are extremely good. Using such intelligent sensing systems is a key factor in the modernization of industry and in protection of the environment (Snopok and Kruglenko 2005).

In some cases, where online use and portability are requirements, gas sensors have clear advantages over the GC- and MS-based instruments as well. Portable, application-specific devices could be used for a wide range of tasks (Kaplan and Braham 1998). The reduced need for flexibility would allow more user-friendly operation, so that, for example, environmental field workers, car mechanics, and doctors could find an extra tool in their armories. Similar instruments could also be incorporated into in-process control systems, reducing the time and cost of analyzing samples offline and minimizing waste.

11. ACKNOWLEDGMENTS

G. Korotcenkov is thankful to the Korean BK21 Program for support of his scientific research.

REFERENCES

Aishima T. (1991) Discrimination of liquor aromas by pattern recognition analysis of responses from a gas sensor array. *Anal. Chim. Acta* **243**, 293–300.

Aishima T. (2004) Correlating sensory attributes to gas chromatography–mass spectrometry profiles and e-nose responses using partial least squares regression analysis. *J. Chromatogr. A* **1054**, 39–46.

Albert K.J., Lewis N.S., Schauer C.L., Sotzing G.A., Stitzel S.E., Vaid T.P., and Walt D.R. (2000) Cross-reactive chemical sensor arrays. *Chem. Rev.* **100**, 2595–2626.

Ampuero S. and Bosset J.O. (2003) The electronic nose applied to dairy products: A review. *Sens. Actuators B* **94**, 1–12.

Anderson M.S., Andringa J.M., Carlson R., Conrad P., Hartford W., Shafer M., Soto A., Tsapin A.I, Dybwad J.P., and Wadsworth W. (2005) Fourier transform infrared spectroscopy for Mars science. *Rev. Sci. Instr.* **76**, 034101(1–9).

Armenta S., Coelho N.M.M., Roda R., Garrigues S., and de la Guardia A. (2006) Seafood freshness determination through vapour phase Fourier transform infrared spectroscopy. *Anal. Chim. Acta* **580**, 216–222.

Baby R.E., Cabezas M., and Walsoe de Reca E.N. (2000) Electronic nose: A useful tool for monitoring environmental contamination. *Sens. Actuators B* **69**, 214–218.

Bakken G.A., Kauffman G.W., Jurs P.C., Albert K.J., and Stitzel S.S. (2001) Pattern recognition analysis of optical sensor array data to detect nitroaromatic compound vapors. *Sens. Actuators B* **79**, 1–10.

Ballantine D.S., Rose S.L., Grate J.W., and Wohltjen H. (1986) Correlation of surface acoustic wave device coating responses with solubility properties and chemical structure using pattern recognition. *Anal. Chem.* **58**, 3058–3066.

Barbri N.El., Amari A., Vinaixa M., Bouchikhi B., Correig X., and Llobet E. (2007) Building of a metal oxide gas sensor-based electronic nose to assess the freshness of sardines under cold storage. *Sens. Actuators B* **128**, 235–244.

Bartlett P.N., Blair N., and Gardner J.W. (1993) *Electronic nose. Principles, applications and outlook.* In: ASIC, *15e Colloque*, Montpellier, pp. 478–486.

Bartlett P.N., Elliot J.M., and Gardner J.W. (1997) Electronic noses and their application in the food industry. *Food Technol.* **51**, 44–48.

Begnaud F. and Berdague J.L. (2002) Influence of ionisation energy on the quality of static head space-mass spectrometry spectral fingerprints. *Sens. Actuators B* **81**, 351–358.

Berna A.Z., Trowell S., Cynkar W., and Cozzolino D. (2008) Comparison of metal oxide-based electronic nose and mass spectrometry-based electronic nose for the prediction of red wine spoilage. *J. Agric. Food Chem.* **56,** 3238–3244.

Blanco M. and Villarroya I. (2002) NIR spectroscopy: A rapid-response analytical tool. *Trends Anal. Chem.* **21**, 240–250.

Bourrounet B., Talou T., and Gaset A. (1995) Application of a multigas sensor device in the meat industry for boar-taint detection. *Sens. Actuators B* **26–27**, 250–254.

Breer H., Wanner I., and Strotmann J. (1996) Molecular genetics of mammalian olfaction. *Behav. Genet.* **26**, 209–219.

Buck T., Allen M.F.G., and Dalton M. (1965) Detection of chemical species by surface effects on metals and semiconductors. In: Bregman T. and Dravnieks A. (eds.), *Surface Effects in Detection.* Spartan Books, Washington, DC.

Capelli L., Sironi S., and Del Rosso R. (2008) A comparative and critical evaluation of odour assessment methods on a landfill site. *Atmos. Environ.* **42**, 7050–7058.

Carey W.P. and Kowalski B.R. (1986) Chemical piezoelectric sensor and sensor array characterization. *Anal. Chem.* **58**, 3077–3084.

Carpenter G.A., Grossberg S., Markuzon N., Reynolds J., and Rosen D.B. (1992) Fuzzy ARTMAP: A neural network architecture for incremental supervised learning of analog multidimensional maps. *IEEE Trans. Neural Networks* **3**, 698–713.

Carpenter G., Grossberg S., and Reynolds J. (1995) A fuzzy ARTMAP non parametric probability estimator for non-stationary pattern recognition problems. *IEEE Trans. Neural Networks* **6**, 1330–1336.

Chandler R.C., Clements J., Gibson T.D., Hulbert J.N., Boden N., and Ruck-Keene E. (1997) Solid state spectroscopy with sub-one second sampling time: A rapid technique for complex volatile headspace analysis. In: *Proceedings of the 11th European Conference on Solid State Transducers—Eurosensors XI,* Warsaw, Poland, 21–24 September 1997.

Cho E.J. and Bright F.V. (2001) Optical sensor array and integrated light source. *Anal. Chem.* **73**, 3289–3293.

Cole M., Ulivieri N., Garcia-Guzman J., and Gardner J.W. (2003) Parametric model of a chemoresistor for use in smart sensor design and simulation. *Microelectron. J.* **34**, 865–875.

Cozzolino D., Smyth H.E., and Gishen M. (2003) Feasibility study on the use of visible and near-infrared spectroscopy together with chemometrics to discriminate between commercial white wines of different varietal origins. *J. Agric. Food Chem.* **51**, 7703–7708.

Craven M.A., Gardner J.W., and Bartlett P.N. (1996) Electronic noses—Development and future prospects. *Trends Anal. Chem.* **15**, 486–493.

Creaser C.S., Griffiths J.R., Bramwell C.J., Noreen S., Hill C.A., and Thomas C.L.P. (2004) Ion mobility spectrometry: A review. Part 1. Structural analysis by mobility measurement. *Analyst* **129**, 984–994.

Demarne V. and Sanjine S.R. (1992) Thin film semiconducting metal oxide gas sensors. In: Sberveglieri G. (ed.), *Gas Sensors.* Kluwer, Dordrecht, The Netherlands, 89–116.

De Wit M., Vanneste E., Geise H.J., and Nagels L.J. (1998) Chemiresistive sensors of electrically conducting poly(2,5-thienylene vinylene) and copolymers: their responses to nine organic vapours. *Sens. Actuators B* **50**, 164–172.

Dickert F.L. and Keppler M. (1995) Self-organized phases combined with IDC devices—Switchable materials for solvent vapor detection. *Adv. Mater.* **7**, 1020–1023.

Dickinson T.A., White J., Kauer J.S., and Walt D. (1996) A chemical-detecting system based on a crossreactive optical sensor array. *Nature* **382**, 697–700.

Dickinson T.A., Walt D.R., White J., and Kauer J.S. (1997) Generating sensor diversity through combinatorial polymer synthesis. *Anal. Chem.* **69**, 3413–3418.

Dickinson T.A., White J., Kauer J.S., and Walt D.R. (1998) Current trends in 'artificial-nose' technology. *TIBTECH* **16**, 250–258.

Dickinson T.A., Michael K.L., Kauer J.S., and Walt D.R. (1999) Convergent, self-encoded bead sensor arrays in the design of an artificial nose. *Anal. Chem.* **71**, 2192–2198.

Di Natale C., Martinelli E., Pennazza G., Orsini A., and Santonico M. (2006) Data analysis for chemical sensor analysis. In: Byrnes J. and Ostheimer G. (eds.), *Advances in Sensing with Security Applications.* Springer-Verlag, Berlin, 147–169.

Dittmann B., Nitz S., and Horner G. (1998) Development of a new chemical sensor on a mass spectrometric basis. *Adv. Food Sci.* **20**, 115–120.

Dittmann B. and Nitz S. (2000) Strategies for the development of reliable QA/QC methods when working with mass spectrometry-based chemosensory systems. *Sens. Actuators B* **69**, 253–257.

Doleman B.J. and Lewis N.S. (2001) Comparison of odor detection thresholds and odor discriminabilities of a conducting polymer composite electronic nose versus mammalian olfaction. *Sens. Actuators B* **72**, 41–50.

Dravnieks A. and Trotter P.J. (1965) Polar vapour detection based on thermal modulation of contact potentials. *J. Sci. Instrum.* **42**, 624–627.

Eiceman G.A., Gardea-Torresdey J., Dorman F., Overton Ed., Bhushan A., and Dharmasena H.P. (2006) Gas chromatography. *Anal. Chem.* **78**, 3985–3996.

Falcitelli M., Benassi A., Di Francesco F., Domenici C., Marano L., and Pioggia G. (2002) Fluid dynamic simulation of a measured chamber for electronic noses. *Sens. Actuators B* 85, 166–174.

Filippini D. and Lundstrom I. (2002) Chemical imaging by a computer screen aided scanning light pulse technique. *Appl. Phys. Lett.* **81**, 3891–3893.

Findlay M.W., Penrose W.R., and Stetter J.R. (1993) Quality classification of grain using a sensor array and pattern recognition. *Anal. Chim. Acta* **284**, 1–11.

Fukunaga T., Mori S., Nakabayashi Y., Kanda M., and Ehara K. (1995) Application of flavor sensor to coffee. In: *Proceedings of the ASIC, 16th International Scientific Colloquium on Coffee, 9–14 April, Kyoto, Japan*, 478–486.

Egashira M., Shimizu Y., and Takao Y. (1994) Fish freshness detection by semiconductor gas sensors. In: Kurihara K., Suzuki N., and Ogawa H. (eds.), *Olfaction Taste XI*, Sapporo, Japan, Springer-Verlag, Tokyo, 715–719.

Ehrmann S, Jungst J., Goschnick J., and Everhard D. (2000) Application of a gas sensor microarray to human breath analysis. *Sens. Actuators B* **65**, 247–249.

Gall M. (1993) The Si-planar-pellistor array, a detection unit for combustible gases. *Sens. Actuators B* **16**, 260–264.

Garcia-Guzman J., Ulivieri N., Cole M., and Gardner J.W. (2003) Design and simulation of a smart ratiometric ASIC chip for VOC monitoring. *Sens. Actuators B* **95**, 232–243.

Gardner J.W. (1991) Detection of vapours and odours from a multisensor array using pattern recognition: Part 1. Principal component and cluster analysis. *Sens. Actuators B* **4**, 109–115.

Gardner J.W. (2004) Review of conventional electronic noses and their possible application to the detection of explosives. In: Gardner J.W. and L. Yinou (eds.), *Electronic Noses and Sensors for the Detection of Explosives.* Kluwer, Dordrecht, The Netherlands, 1–28.

Gardner J.W., Shermer H.V., and Tan T.T. (1992) Application of an electronic nose to the discrimination of coffees. *Sens. Actuators B* **6**, 71–75.

Gardner J.W. and Bartlett P.N. (1992) Pattern recognition in odour sensing. In: Gardner J.W. and Bartlett P.N. (eds.), *Sensors and Sensory Systems for an Electronic Nose*. Kluwer, Dordrecht, The Netherlands, 161–179.

Gardner J.W. and Bartlett P.N. (1994) A brief history of electronic noses. *Sens. Actuators B* **18–19**, 211–220.

Gardner J.W. and Bartlett P.N. (1995) Application of conducting polymer technology in Microsystems. *Sens. Actuators A* **51**, 57–66.

Gardner J.W. (1998) Pattern recognition in the Warwick electronic nose. In: *Proceedings of the 8th International Congress of the European Chemoreception Research Organisation*, Coventry, UK.

Gardner J.W. and Bartlett P.N. (1999) *Electronic Noses: Principles and Applications.* Oxford University Press, Oxford, UK.

Gardner J.W., Shin H.W., and Hines E.L. (2000) An electronic nose system to diagnose illness. *Sens. Actuators B* **70**, 19–24.

Gomez A.H., Wang J., Hu G., and Pereira A.G. (2006) Electronic nose technique potential monitoring mandarin maturity. *Sens. Actuators B* **113**, 347–353.

Gopel W. (1994) Chemical imaging: Concepts and vision for electronic noses. *Sens. Actuators B* **18–19**, 1–21.

Gopel W. (1995) Analysis of complex gas mixtures by pattern recognition. In: *Proceedings of the 2nd International Symposium Olfaction and Electronic Nose, Toulouse.*

Göpel W. (1998) Chemical imaging: I. Concepts and visions for electronic and bioelectronic noses. *Sens. Actuators B* **52**, 125–142.

Gouma P. and Sberveglieri G. (2004) Novel materials and applications of electronic noses and tongues. *MRS Bull.* **29**, 697–700.

Grate J.W., Rose-Pehrsson S.L., Venezky D.L., Klusty M., and Wohltjen H. (1993) Smart sensor system for trace organophosphorus and organosulfur vapor detection employing a temperature-controlled array of surface acoustic wave sensors, automated sample preconcentration, and pattern recognition. *Anal. Chem.* **65**, 1868–1881.

Guadarrama A., Fernández J.A., Iñiguez M., Souto J., and de Saja J.A. (2000) Array of conducting polymer sensors for the characterisation of wines. *Anal. Chim. Acta* **411**, 193–200.

Ha S., Kim Y.S., Yang Y., Kim Y.J., Cho S., Yang H., and Kim Y.T. (2005) Integrated and icroheater embedded gas sensor array based on the polymer composites dispensed in micromachined wells. *Sens. Actuators B* **105**, 549–555.

Harper W.J. (2001) The strengths and weaknesses of the electronic nose. *Adv Exp. Med. Biol.* **488**, 59–71.

Hatfield J.V., Neaves P., Hicks P.J., Persaud K., and Travers P. (1994) Towards an integrated electronic nose using conducting polymer sensors. *Sens. Actuators B* **18**, 221–228.

Heilig A., Barsan N., Weimar U., and Gopel W. (1999) Selectivity enhancement of SnO_2 gas sensors: Simultaneous monitoring of resistances and temperatures. *Sens. Actuators B* **58**, 302–309.

Hodgins D. (1995) A sixth sense. *Dairy Ind. Int.* **60**, 32–33.

Hodgins D.M. (1995b) The development of an electronic 'nose' for industrial and environmental applications. *Sens. Actuators B* **27**, 255–258.

Hodgins D. (1997) The electronic nose: Sensor array-based instruments that emulate the human nose. In: Marsili R. (ed.), *Techniques for Analyzing Food Aroma.* Marcel Dekker, New York, 331–371.

Holmberg M., Winquist F., Lundstrom I., Gardner J.W., and Hines E.L. (1995) Identification of paper quality using a hybrid electronic nose. *Sens. Actuators B* **27**, 246–249.

Holmberg M. (1997) Data evaluation for an electronic nose. Doctoral thesis, Department of Physics and Measurement Technology, Linkoping University, Linkoping, Sweden.

Hudon G., Guy C., and Hermia J. (2000) Measurement of odor intensity by an electronic nose. *J. Air Waste Manage. Assoc.* **50**, 1750–1758.

Imisides M.D., John R., and Wallace G.G. (1996) Microsensors based on conducting polymers. *Chemtech* **26**, 19–25.

James D., Scott S.M., Ali Z., and O'Hare W.T. (2005) Chemical sensors for electronic nose systems. *Microchim. Acta* **149**, 1–17.

Johnson S.R., Sutter J.M., Engelhardt H.L., Jurs P.C., White J., Kauer J.S., Dickinson T.A., and Walt D.R. (1997) of multiple analytes using an optical sensor array and pattern recognition neural networks. *Anal. Chem.* **69**, 4641–4648.

Jurs P.C., Bakken G.A., and McClelland H.E. (2000) Computational methods for the analysis of chemical sensor array data from volatile analytes. *Chem. Rev.* **100**(7), 2649–2678.

Kaplan G. and Braham R. (1998) The how and why of electronic noses. *IEEE Spectrum*, September, 22–31.

Keller P.E. (1999) Mimicking biology: Applications of cognitive systems to electronic noses. In: *Proceedings of the 1999 LEEE International Symposium on Intelligent Control/Intelligent Systems and Semiotics*, Cambridge, MA, September 15–17, 1999, 447–451.

Kim Y.S., Ha S.-C., Yang Y., Kim Y.Y., Cho S.M., Yang H., and Kim Y.T. (2005) Portable electronic nose system based on the carbon black–polymer composite sensor array. *Sens. Actuators B* **108**, 285–291.

Kohonen T. (1989) *Self Organization and Associative Memory*, third ed. Springer-Verlag, New York.

Lacoste F. (1999) Is it possible to use E-Nose for organoleptical quality control of virgin olive oil? In: *Proceedings of the User Forum for the Network of Excellence on Artificial Olfactory Sensing 'NOSE'*, June 17th–18th 1999, Ispra, Italy.

Lavine B.K. (1992) Environmental applications of pattern recognition techniques. *Chemometr. Intell. Lab. Syst.* **15**, 219–230.

Lerchner J., Caspary D., and Wolf G. (2000) Calorimetric detection of volatile organic compounds. *Sens. Actuators B* **70**, 57–66.

Lezzi A.M., Beretta G.P., Comini E., Faglia G., Galli G., and Sberveglieru G. (2001) Influence of gaseous species transport on the response of solid state gas sensors within enclosures. *Sens. Actuators B* **78**, 144–150.

Lonergan, M.C., Severin E.J., Doleman B.J., Beaber S.A., Grubbs R.H., and Lewis N.S. (1996) Array-based vapor sensing using chemically sensitive, carbon black–polymer resistors. *Chem. Mater.* **8**, 2298–2312.

MacDonalds B.F. and Prebbleis K.A. (1993) Some applications of near-infrared reflectance analysis in the pharmaceutical industry. *J. Pharmaceut. Biomed. Anal.* **11**, 1077–1085.

Mano N. and Gato J. (2003) Biomedical and biological mass spectrometry. *Anal. Sci.* **19**, 3–14.

Mari C.M. and Barbi G.B. (1992) Electrochemical gas sensors. In: Sberveglieri G. (ed.), *Gas Sensors*. Kluwer, Dordrecht, The Netherlands, 329–364.

Marquis F. and Mielle P. (1996) Le 'nez 'electronique' en quete de credibilite. *Odours and VOC's J., Special Issue*, 40–46.

Martens H. and Naes T. (1989) *Multivariate Calibration*, Wiley, Chichester, UK.

Marti M.P., Boque R., Busto O., and Guasch J. (2005) Electronic noses in the quality control of alcoholic beverages. *Trends Anal. Chem.* **24**, 57–66.

Massart D.L., Vandeginste B.G.M., Deming S.N., Michotte Y., and Kaufman L. (1988) *Chemometrics: A Textbook*. Elsevier, Amsterdam.

Massart D.L., Vandeginste B.G.M., Buydens L.M.C., De Jong S., Lewi P.J., and Smeyers-Verbeke J. (1998) *Handbook of Chemometrics and Qualimetrics*. Elsevier, Amsterdam.

McEntegart C.M., Penrose W.R., Strathmann S., and Stetter J.R. (2000) Detection and discrimination of coliform bacteria with gas sensor arrays. *Sens. Actuators B* **70**, 170–176.

Meier D.C., Evju J.K., Boger Z., Raman B., Benkstein K.D., Martinez C.J., Montgomery C.B., and Semancik S. (2007) The potential for and challenges of detecting chemical hazards with temperature-programmed microsensors. *Sens. Actuators B* **121**, 282–294.

Mielle P, Hivert B., and Mauvais G. (1995) Are gas sensors suitable for on-line monitoring and quantification of volatile compounds? In: Etievant P. and Schreier P. (eds.), *Proceedings of Bioflavour 95*. INRA, Dijon, 81–84.

Mielle P. (1996) 'Electronic noses': Towards the objective instrumental characterization of food aroma. *Trends Food Sci. Technol.* **7**, 432–438.

Mielle P. and Marquis F. (1999) An alternative way to improve the sensitivity of electronic olfactometers—'Electronic Noses'. *Sens. Actuators B* **58**, 526–535.

Mielle P. and Marquis F. (2000) Gas sensors arrays ('Electronic Noses'): A study about the speed-accuracy ratio. *Sens. Actuators B* **68**, 9–16.

Mielle P., Marquis F., and Latrasse C. (2000) Electronic noses: Specify or disappear. *Sens. Actuators B* **69**, 287–294.

Mielle P. and Marquis F. (2001) One-sensor electronic olfactometer for rapid sorting of fresh fruit juices. *Sens. Actuators B* **76**, 470–476.

Mitrovics J., Ulmer H., Weimar U., and Gopel W. (1998) Modular sensor system for gas sensing and odor monitoring: The MOSES concept. *Accounts Chem. Res.* **31**, 307–315.

Morvan M., Talou T., Gaset A., and Beziau J. F. (2000) Electronic-nose systems for control quality applications in automotive industry. *Sens. Actuators B* **69**, 384–388.

Nagle H. T., Gutierrez-Osuna R., and Schiffman S. S. (1998) The how and why of E-noses. *IEEE Spectrum*, September, 22–38.

Nakagawa M., Kawabata S., Nishiyama K., Utsunomiya K., Yamamoto I., Wada T., Yamashita Y., and Yamashita N. (1997) Analysis of constituents in mixed odor vapors using an array of the ç-Al_2O_3 chemiluminescence-based sensors. In: Kunz R.E. (ed.), *Proceedings of the 3rd European Conference on Optical Chemical Sensors and Biosensors: Europt(r)ode III*, 31 April 1996, Zurich, Switzerland. Elsevier, Amsterdam, Lausanne, 76.

Nakamoto T., Fukuda A., and Moriizumi T. (1993) Perfume and flavour identification by odour-sensing system using quartz resonator array and neural network pattern recognition. *Sens. Actuators B* **10**, 85–90.

Nakata S., Hashimoto T., and Okunishi H. (2002) Evaluation of the responses of a semiconductor gas sensor to gaseous mixtures under the application of temperature modulation. *Analyst* **127**, 1642–1648.

Nakata S., Neya K., and Hashimoto T. (2002) A semiconductor gas sensor based on nonlinearity: Utilization of the effect of competition on the sensor responses to gaseous mixtures. *Electroanalysis* **14**, 881–887.

Nanto H., Sokooshi H., Kawai T., and Usuda T. (1991) Freshness detection of sea-foods using ZnO thin film gas sensor. In: *Technical Digest of the 10th Sensor Symposium*, Tokyo, Japan, 195–198.

Neaves P.I. and Hatfield J.V. (1995) A new generation of integrated electronic noses. *Sens. Actuators B* **27**, 223–231.

Newman A.R. (1991) Electronic noses. *Anal. Chem.* **63**, 585A–588A.

Ni M., Stetter J.R., and Buttner W.J. (2008) Orthogonal gas sensor arrays with intelligent algorithms for early warning of electrical fires. *Sens. Actuators B* **130**, 889–899.

Olafsson R., Martinsdottir E., Olafdottir G., Sigfusson S.I., and Gardner J.W. (1992) Monitoring of fish freshness using tin oxide sensors. In: Gardner J.W. and Bartlett P.N. (eds.), *Sensors and Sensory Systems for an Electronic Nose.* Kluwer, Dordrecht, The Netherlands, 257–272.

Pardo M., Kwong L.G., Sberveglieri G., Brubaker K., Schneider J.F., Penrose W.R., and Stetter J.R. (2005) Data analysis for a hybrid sensor array. *Sens. Actuators B* **106**, 136–143.

Partridge A.C., Harris P.D., and Andrews M.K. (1996) High sensitivity conducting polymer sensors. *Analyst* **121**, 1349–1353.

Pearce T.C., Schiffman S.S., Nagle H.T., and Gardner J.W. (eds.) (2003) *Handbook of Machine Olfaction–Electronic Nose Technology.* Wiley-VCH, Weinheim, Germany.

Perera A., Sundic T., Pardo A., and Gutierrez-Osuna R. (2002) A portable electronic nose based on embedded PC technology and GNU/Linux: Hardware, software and applications. *IEEE Sensors J.* **2**(3), 235–246.

Peres C., Begnaud F., Eveleigh L., and Berdague J.L. (2003) Fast characterization of foodstuff by headspace mass spectrometry (HS-MS). *Trends Anal. Chem.* **22**, 858–866.

Perez Pavon J.L., Nogal Sanchez M., Garcıa Pinto C., Laespada E.F., Moreno Cordero B., and Pena A.G. (2006) Strategies for qualitative and quantitative analyses with mass spectrometry-based electronic noses. *Trends Anal. Chem.* **25**, 257–266.

Persaud K. and Dodd G.H. (1982) Analysis of discrimination mechanisms in the mammalian olfactory system using a model nose. *Nature* **299**, 352–355.

Persaud K.C. and Pelosi P. (1992) Sensor arrays using conducting polymers for an artificial nose. In: Gardner J.W. and Bartlett P.N. (eds.), *Sensors and Sensory Systems for an Electronic Nose.* Kluwer, Dordrecht, The Netherlands, 237–256.

Persaud K.C. and Travers P.J. (1997) Arrays of broad specificity films for sensing volatile chemicals. In: Kress-Rogers E. (ed.), *Handbook Biosensors and Electron Noses.* CRC Press, Boca Raton, FL, 563–592.

Persaud K.C., Wareham P., Pisanelli A.M., and Scorsone E. (2005) "Electronic nose"—New condition monitoring devices for environmental applications. *Chem. Senses* **30**(suppl 1), i252–i253.

Piggee C. (2008) What's that smell? *Anal. Chem.* **80**, 8861.

Pisanelli A., Qutob A.A., Travers P., Szyszko S., and Persaud K.C. (1994) Applications of multi-array sensors to food industries. *Life Chem. Rep.* **11**, 303–308.

Polk B.J., Smith A.J., DeWeerth S.P., Zhou Z., Janata J., and Domansky K. (1999) Design of solid state array for simultaneous potentiometric and impedance sensing in gas phase. *Electroanalysis* **11**, 707–711.

Privat E., Roussel S., Grenier P., and Bellon-Maurel V. (1998) Techniques de désalcoolisation en vue de la discrimination de boissons alcoolisées par capteurs darômes. *Sci. Aliments* **18**, 459–470.

Rakow N.A. and Suslick K.S. (2000) A colorimetric sensor array for odor visualization. *Nature* **406**, 710–713.

Richardson S.D. (2000) Environmental mass spectrometry. *Anal. Chem.* **72**, 4477–4496.

Rock F., Barsan N., and Weimar U. (2008) Electronic nose: Current status and future. *Trends Chem. Rev.* **108**, 705–725.

Romain A.C., Nicolas J., Wiertz V., Maternova J., and Andre P. (2000) Use of a simple tin oxide sensor array to identify five malodours collected in the field. *Sens. Actuators B* **62**, 73–79.

Rumelhart D.E., Hinton G.E., and Williams R.J. (1986) Learning internal representations by error propagation. In: Rumelhart D.E. and McClelland J.L. (eds.), *Parallel Distributed Processing: Explorations in the Microstructures of Cognition. Vol. 1: Foundations.* MIT Press, Cambridge, MA, 318–362.

Saunders B.W., Thiel D.V., and Mackay-Sim A. (1995) Response kinetics of chemically-modified quartz piezoelectric crystals during odorant stimulation. *Analyst* **120**, 1013–1018.

Schaller E., Bosset J.O., and Escher F. (1998) Electronic noses and their application to food. *Food Sci. Technol.* **31**, 305–316.

Schaller E., Zenhäusern S., Zesiger T., Bosset J.O., and Escher F. (2000) Use of preconcentration techniques applied to a MS-based "Electronic Nose." *Analysis* **28**, 743–749.

Schalley C.A. (2001) Molecular recognition and supramolecular chemistry in the gas phase. *Mass Spectrom. Rev.* **20**, 253–309.

Schweizer-Berberich M., Vaihinger S., and Gopel W. (1994) Characterisation of food freshness with sensor arrays. *Sens. Actuators B* **18–19**, 282–290.

Scorsone E., Pisanelli A.M., and Persaud K.C. (2006) Development of an electronic nose for fire detection. *Sens. Actuators B* **116**, 55–61.

Shaffer R.E., Rose-Pehrsson S.L., and McGill R.A. (1999) A comparison study of chemical sensor array pattern recognition algorithms. *Anal. Chim. Acta* **384**, 305–317.

Shiers V.P. (1995) Electronic nose technology—Evaluations and developments for the food industry. In: *Food Ingredients Europe: Conference Proceedings, 7–9 November 1995, Frankfurt, Germany*. Miller Freeman, Maarssen, The Netherlands, 198–200.

Skoog D.A., Holler F.J., and Crouch S.R. (2007) *Principles of Instrumental Analysis*. Thomson Brooks, Belmont, CA.

Slater J.M., Paynter J., and Watt E.J. (1993) Multi-layer conducting polymer gas sensor arrays for olfactory sensing. *Analyst* **118**, 379–384.

Snopok B.A. and Kruglenko I.V. (2002) Multisensor systems for chemical analysis: State-of-the-art in electronic nose technology and new trends in machine olfaction. *Thin Solid Films* **418**, 21–41.

Snopok B.A. and Kruglenko I.V. (2005) Nonexponential relaxations in sensor arrays: Forecasting strategy for electronic nose performance. *Sens. Actuators B* **106**, 101–113.

Sommer V., Tobias P. Kohl D., Sundgren H., and Lundstrom I. (1995) Neural networks and abductive networks for chemical sensor signals: A case comparison. *Sens. Actuators B* **28**, 217–222.

Stetter J.R. (1986) Electrochemical sensors, sensor arrays, and computer algorithms. In: Schuetzle D., Hammerle R., and Butler J. (eds.), *Fundamentals and Applications of Chemical Sensors*. ACS Symposium Series, Vol. 309, 299–308. American Chemical Society, Washington, DC.

Stetter J.R. (2010) Experimental methods in chemical sensor and sensor array evaluation and development. In: Ryan M.A., Shevade A.V., Taylor C.J., Homer M.L., Blanco M., and Stetter J.R. (eds.), *Materials and Sensor Arrays—Computational and Experimental Selection Methods*. Springer-Verlag, New York, 3–46.

Stetter J.R., Findlay M.W., Maclay G.J., Zhang J., Vaihinger S., and Gopel W. (1990) Sensor array and catalytic filament for chemical analysis of vapors and mixtures. *Sens. Actuators B* **1**, 43–47.

Stetter J.R., Findlay M.W., Schroeder K.M., Yue C., and Penrose W.R. (1993) Quality classification of grain using a sensor array and pattern recognition. *Anal. Chem. Acta* **284**, 1–11.

Stetter J.R., Jurs P.C., and Rose S.L. (1986) Detection of hazardous gases and vapors: Pattern recognition analysis of data from an electrochemical sensor arrays. *Anal. Chem.* **58**, 860–866.

Stetter J.R., Maclay G.J., and Christesen S.V. (1989) Time-dependent sensor signals for selective detection. *Sens. Actuators* **20**(3), 277–287.

Stetter J.R. and Otagawa T. (1987) A chemical concentration-modulation sensor for the selective detection of airborne chemicals. *Sens. Actuators* **11**(3), 251–264.

Stetter J.R. and Penrose W.R. (2001) Electrochemistry Encyclopedia: The electrochemical nose, http://electrochem.cwru.edu/ed/encycl/art-n01-nose.htm.

Stetter J.R. and Penrose W.R. (2001b) (eds.) *Artificial Chemical Sensing: Proceedings of Eighth International Symposium on Olfaction and the Electronic Nose (ISOEN8 - ISOEN 2001)*, Washington, DC, March 20–24, 2001. ECS, Pennington, NJ.

Stetter J.R. and Penrose W.R. (2001) The electrochemical nose. Department of Biological, Chemical and Physical Sciences, Illinois Institute of Technology, Chicago, IL.

Stetter J.R. and Penrose W.R. (2002) Understanding chemical sensors and chemical sensor arrays (Electronic Noses): Past, present, and future. In: *Sensors Update*, Vol. 10, Chapter 2.3. Wiley-VCH, Weinheim, Germany, 189-229.

Stetter J.R., Penrose W.R., Zaromb S., Nolan M., Christian D., Hampton D., Billings M., Steinke C., and Stull J.O. (1985b) Portable toxic vapor detector and analyzer using electrochemical sensor array. In: Herbst K.S. (ed.), *Analysis Instrumentation: Proceedings of the Annual ISA Analysis Instrumentation Symposium*, Vol. 21. ISA, Research Triangle Park, NC, 163–170.

Stetter J.R., Shi C.-X., and Maclay G.J. (1991) Modulated photoionization detection of hydrazine compounds in mixtures without prior separation. *Anal. Chem.* **63**(17), 1755–1759.

Stetter J.R., Strathmann S, McEntegart C., Decastro M., and Penrose W.R. (2000) New sensor arrays and sampling systems for a modular electronic nose. *Sens. Actuators B* **69**, 410–419.

Stetter J.R., Zaromb S., and Findlay M.W. (1985a) Monitoring of electrochemically inactive compounds by amperometric gas sensors. *Sens. Actuators* **6**, 269288.

Stetter J.R., Zaromb S., and Penrose W.R. (1987) *Sensor Array for Toxic Gas Detection.* U.S. Patent 4670405.

Stetter J.R., Zaromb S., Penrose W.R., Findlay M.W., and Otagawa T. (1984b) Portable device for detecting and identifying hazardous vapors. In: Ludwigson J. (ed.), *Hazardous Material Spills Conference Proceedings, Prevention, Behavior, Control and Cleanup of Spills and Waste Sites*, April 9–12, Rockville, MD, 183–190.

Stetter J.R., Zaromb S., Penrose W.R., and Otagawa T. (1984a) Portable instrument for the detection and identification of air pollutants. In: Extended abstract of paper presented at the 4th National Symposium on Recent Advances in Pollutant Monitoring of Ambient Air and Stationary Sources, U.S. EPA, May 8–10, Raleigh, NC, 1984.

Strathmann S., Hahn S., and Weimar U. (2001) Artificial chemical sensing. In: Stetter J.R. and Penrose W.R. (eds.), *Proceedings of the Eighth International Symposium on Olfaction and the Electronic Nose (ISOEN 2001)*, 26–28 March 2001, Washington, DC. Electrochemical Society, Pennington, NJ, 48–53.

Strike D.J., Meijerink G.H., and Koudelka-Hep M. (1999) Electronic noses—A mini-review. *Fresenius J. Anal. Chem.* **364**, 499–505.

Strobel H.A. and Heineman W.R. (1989) *Chemical Instrumentation: A Systematic Approach*, 3rd ed. Wiley, New York.

Stulik K., Amatore C., Holub K., Marecek V., and Kutner W. (2000) Analytical Chemistry Division Commission on Electroanalytical Chemistry. *Pure Appl. Chem.* **72**, 1483–1492.

Sundgren H., Lundstrom I., Winquist F., Lukkari I., Carlsson R., and Wold S. (1990) Evaluation of a multiple gas mixture with a simple MOSFET gas sensor array and pattern recognition, *Sens. Actuators B* **2**, 115–123.

Sundgren H., Lundstrom I., and Vollmer H. (1992) Chemical sensor arrays and abductive networks. *Sens. Actuators B* **9**, 127–131.

Sundgren H., Winquist F., Lukkari I., and Lundstrom I. (1991) Artificial neural networks and gas sensor arrays: Quantification of individual components in a gas mixture *Meas. Sci. Technol.* **2**, 464–469.

Suslick K.S. (2004) An optoelectronic nose: "Seeing" smells by means of colorimetric sensor arrays. *MRS Bull.* **29**, 720–725.

Suslick K.S., Rakow N.A., and Sen A. (2004) Colorimetric sensor arrays for molecular recognition. *Tetrahedron* **60**, 11133–11138.

Sutter J.M. and Jups P.C. (1997) Neural network classification and quantification of organic vapors based on fluorescence data from a fiber-optic sensor array. *Anal. Chem.* **69**, 856–862.

Takada T. (1998) A new method for gas identification using a single semiconductor sensor. *Sens. Actuators B* **52**, 45–52.

Tan T., Lucas Q., Moy L., Gardner J.W., and Bartlett P.N. (1995) The electronic nose—A new instrument for sensing vapours. *LC-GC Int.* **8**, 218–225.

Tan T.T., Schmitt V.O., Lucas Q., and Isz S. (2001) Electronic noses and electronic tongues. *LabPlus Int.* September/October, 16–19.

Taylor C.J. and Semancik S. (2002) Use of microhotplate arrays as microdeposition substrates for materials exploration. *Chem. Mater.* **14**, 1671–1677.

Thundat T., Chen G.Y., Warmack R.J., Allison D.P., and Wachter E.A. (1995) Vapor detection using resonating microcantilevers. *Anal. Chem.* **67**, 519–521.

Ulmer H., Mitrovics J., Noetzel G., Weimar U., and Gopel W. (1997) Odours and flavours identified with hybrid modular sensor systems. *Sens. Actuators B* **43**, 24–33.

Vaihinger S., Goepel W., and Stetter J.R. (1991) Detection of halogenated and other hydrocarbons in air: Response functions of catalyst/electrochemical sensor systems. *Sens. Actuators B* **4**, 337–343.

Vakhrushev S.Y., Langridge J., Campuzano I., Hughes C., and Peter-Katalinic J. (2008) Ion mobility mass spectrometry analysis of human glycourinome. *Anal. Chem.* **80**, 2506–2513.

Van Kempen T.A., Powers W.J., and Sutton A.L. (2002) Technical note: Fourier transform infrared (FTIR) spectroscopy as an optical nose for predicting odor sensation. *J. Anim. Sci.* **80**, 1524–1527.

Varada V.K. and Gardner J.W. (1999) Smart tongue and nose. In: *Proceedings of SPIE Conference on Smart Electronics and MEMS*, Newport Beach, CA, March 1999. SPIE Vol. **3673**, 67–76.

Vinaixa M., Marn S., Brezmes J., Llobet E., Vilanova X., Correig X., Ramos A., and Sanchis V. (2004) Early detection of fungal growth in bakery products by use of an electronic nose based on mass spectrometry. *J. Agric. Food Chem.* **52**, 6068–6074.

Walt D.R., Dickinson T., White J., Kauer J., Johnson S., Engelhardt H., Sutter J., and Jurs P. (1998) Optical sensor arrays for odor recognition. *Biosens. Bioelectron.* **13**, 697–699.

Weinberg D.S. (1986) Development of an effective method of detecting and identifying foreign odors in grain samples. USDA Report SORI-EAS-86-1208.

Wheeler M.C., Cavicchi R.E., and Semancik S. (2007) Tin oxide microsensor arrays as probes for the oscillatory CO oxidation reaction on supported Platinum. *J. Phys. Chem. C* **111**, 3328–3332.

White J., Kauer J.S., Dickinson T.A., and Walt D.R. (1996) Rapid analyte recognition in a device based on optical sensors and the olfactory system. *Anal. Chem.* **68**, 2191–2202.

Wilkens W.F. and Hatman A.D. (1964) An electronic analog for the olfactory processes. *Ann. NY Acad. Sci.* **116**, 608–612.

Willard H., Merritt L., Dean J., and Settle F. (1988) *Instrumental Methods of Analysis*. Wadsworth, Belmont, CA.

Winquist F., Hornsten E.G., Sundgren H., and Lundstrom I. (1993) Performance of an electronic nose for quality estimation of ground meat. *Meas. Sci. Technol.* **4**, 1493–1500.

Winquist F., Arwin H., Lund E., Forster R., Day C., and Lundstrom I. (1995) Screening of tomatoes by means of an electronic nose. In: *Proceedings of the 8th International Conference Solid-State Sensors Actuators—Eurosensors IX*, June 25–29, Stockholm, Sweden, 691–694.

Wünsche L.F., Vuilleumier C., Keller U., Byfield M.P., May I.P., and Kearney M.J. (1995) Scent characterisation: From human perception to electronic nose. In: *Proceedings of the 13th International Congress of Flavours, Fragrances and Essential Oils,* 15–19 October, Istanbul, Turkey, 295–313.

Zaromb S. and Stetter J.R. (1984) Theoretical basis for identification and measurement of air contaminants using an array of sensors having partly overlapping selectivities. *Sens. Actuators* **6**, 225–243.

Zaromb S., Battin R., Penrose W.R., Stetter J.R., Stamoudis V.C., and Stull J.O. (1986) Extending the capabilities of the portable chemical parameter spectrometer to the identification of up to 100 compounds. In: *Proceedings of the 2nd International Meeting on Chemical Sensors*, Bordeaux, France, July 7–10, 739-742.

Zellers E.T., Batterman S.A., Han M., and Patrash S.J. (1995) Optimal coating selection for the analysis of organic vapor mixtures with polymer-coated surface acoustic wave sensor arrays. *Anal. Chem.* **67**, 1092–1106.

Zellers E.T., Park J., Hsu T., and Groves W.A. (1998) Establishing a limit of recognition for a vapor sensor array. *Anal. Chem.* **70**, 4191–4201.

Zhou M., Robards K., Glennie-Holmes M., and Helliwell S. (1999) Analysis of volatile compounds and their contribution to flavor in cereals. *J. Agric. Food Chem.* **47**, 3946–3953.

Zou X. and Zhao J. (2008) Comparative analyses of apple aroma by a tin-oxide gas sensor array device and GC/MS. *Food Chem.* **107**, 120–128.

Zubritsky E. (2009) Product review: E-noses keep an eye on the future. *Anal. Chem.* **72**(11), 421A–426A.

CHAPTER 2

Electronic Tongues

Approaches and Achievements

F. Winquist

1. INTRODUCTION

1.1. BIOMIMETIC SYSTEMS

To survive, it is vital for living organisms to obtain information concerning environmental changes. This is done by the use of their senses, making them capable of perceiving and analyzing stimuli from the surroundings. Human beings possess five senses: three physical, vision, hearing, and touch; and two chemical, taste and smell. The physical senses respond to changes in physical parameters and the chemical senses respond to changes in the chemical composition of the environment.

The sense of smell, also called the olfactory system, possesses unique qualities. Most living organisms are capable of recognizing a very large number of odors, often with extreme sensitivity. Pioneering research by Linda Buck and Richard Axel (1991), who in 2004 received a Nobel Prize in Medicine "for their discoveries of odorant receptors and the organization of the olfactory system," has led to a detailed description of how the olfactory system operates. Their studies confirmed the opinion of most researchers in the field that there are a large number of receptor cells in the nasal cavity, each affiliated with a limited number of selectivity classes. The introduction of an odor molecule gives rise to a signal pattern, determined by the individual profiles of the selectivity classes. After a first pretreatment in the olfactory bulb, the brain does processing in the olfactory cortex. The olfactory system is optimized to be able to recognize a maximum number of different odors using a minimum number of receptor classes.

The reason for the very large number of odors that can be detected is that although the specificity of each receptor class may be low, the combination of several specificity classes yields a very large information content. Mimicking this information handling has led to new strategies in measurement technology: Instead of measuring specific parameters, it has become advantageous to determine quality properties. This is done by collecting as much information as possible concerning the specific measurement situation. The information is then treated with multivariate data analysis (MVDA), which is a very powerful tool to extract useful information from a large data set. Examples include speech recognition (in analogy with the auditory sense), vibration analysis (in analogy with the sense of feeling), and image analysis (in analogy with the sense of vision). A commercial system for quality assessment of grain, The Grain Analyzer by FOSS, is based on MVDA analyses of NIR spectra.

1.2. THE CONCEPT OF ELECTRONIC TONGUES

The development of biomimetic sensing systems has been inspired by nature. These systems are based on arrays of sensors with partially overlapping selectivities, and the data obtained are treated with MVDA. The first concept of this kind was described in 1982 by Persaud and Dodds (1982), where an array of gas sensors was subjected to various gases or odors and the resulting response patterns were treated with multivariate data analysis. This system simulated the olfactory system and could thus be regarded as an electronic version of the olfactory system, hence the name "electronic nose."

Interest was also raised for sensing in the aqueous phase, since many compounds can only be measured in the aqueous phase, such as ions or compounds that have low vapor pressures. For such liquid-phase sensors, the terms *taste sensor* (TS) or *electronic tongue* (ET) were used, in analogy with the term *electronic nose*. The ETs and TSs are based on the same principles as for the electronic nose. Thus, the sensor array of the electronic tongue produces signals which are not necessarily specific for any particular species. When evaluated with multivariate data analysis, however, the signals can be shown to correlate with certain features or qualities of a sample. Normally, these ETs and TSs are used for qualitative analysis of a liquid, but they can also be used for prediction of concentrations of individual species.

There are two aspects of the human sense of taste. The sensation of basic taste originates from the five basic tastes of the human tongue: sour, salt, bitter, sweet, and "umami," which are generated from different, discrete regions on the tongue containing specific receptors called papillae. The sensation of total taste is the impression obtained when the basic tastes of the tongue are merged with the information from the olfactory receptor cells, originating when aroma from the food in the mouth enters the nasal cavities via the inner passage. This merged sensory experience is the basis for the descriptive taste classification obtained from sensory panels.

The TS system was developed to more specifically determine the sensation of basic taste of a liquid sample. The results obtained are often compared with the human experience of taste. The later-developed ET has a wider application area, since it measures the quality aspect of a sample, which is often independent of the human taste sensation.

Various sensing principles have been used, the most common being the electrochemical methods potentiometry or voltammetry, but also optical methods, impedance spectroscopy, and measurements of mass changes based on surface acoustic waves. For pattern recognition, multivariate data analysis is used.

Two main issues are dealt with using these techniques. One is to search for a structure and correlation in the data, the other is to make a mathematical model from a calibration set of data, which can then be used for prediction of test data (Esbensen 2001).

1.3. A SHORT HISTORY OF THE DEVELOPMENT OF ELECTRONIC TONGUES

The first attempt to specifically mimic the sensation of basic taste was pioneered by the work of Toko and Hayashi, who in 1990 presented the TS (Toko et al. 1990, Hayashi et al. 1990). The system was related to the sense of taste in a similar way as the electronic nose was to olfaction. These first taste sensors were based on ion-sensitive lipid membranes combined with MVDA and developed to respond to the basic tastes of the human tongue.

The term *electronic tongue* was first coined in 1996 at the Eurosensors X conference to describe similar systems used for estimation of general quality aspects of liquid samples. The electronic tongue was based on a number of common ion-selective electrodes and calciogenic ion-selective electrodes, and had been developed as a research collaboration between a Russian group (Legin, Vlasov, and Rudnitskaya) and an Italian group (DiNatale, Davide, and D'Amico) (Legin et al. 1996; Di Natale et al. 1996).

The first electronic tongue based on voltammetry was described in 1997 by Winquist et al. (1997). It was a simple device consisting of two working electrodes of gold and platinum, a reference electrode and a counter-electrode. The sensor setup was connected to a potentiostat controlled by a computer for onset of pulses and to record current generated.

The voltammetric electronic tongue was further developed by chemically modifying the surface of the working electrodes. Similar work was also in progress on chemically modifying the membrane of potentiometric devices.

The development of electronic tongue systems has lately also included the introduction of other measurement techniques, such as mass-sensitive sensors based on surface acoustic waves (SAW) and impedimetric techniques. Further developments include use of flow injection analysis and miniaturization of the technique. In order to improve the recognition capacities of artificial senses, hybrid systems based on fusion of various measurement technologies have also been investigated.

Since the introduction of the TS, the number of publications has grown for each year. In Figure 2.1, the number of yearly publications of electronic tongues and TSs is shown for the years 1990–2008. The numbers of publications start with a modest number of two in the beginning of this period and reach a number of over 85 in 2008.

The concept of the electronic tongue represents a very powerful combination of unselective sensor arrays with pattern recognition algorithms. It has given rise to a large number of applications. Various reviews covering the subject have also been published (Winquist et al. 2004; Vlasov et al. 2005; Ciosek and Wroblewski 2007; Anand et al. 2007; Winquist 2008a). Applications have been found in fields such as the food industry, the clinical and environmental sectors, the paper and pulp industry, and water quality assessments.

One should bear in mind, however, that although the concept of both the electronic nose and the electronic tongue has nature as a model, these systems have little in common with their natural sensory

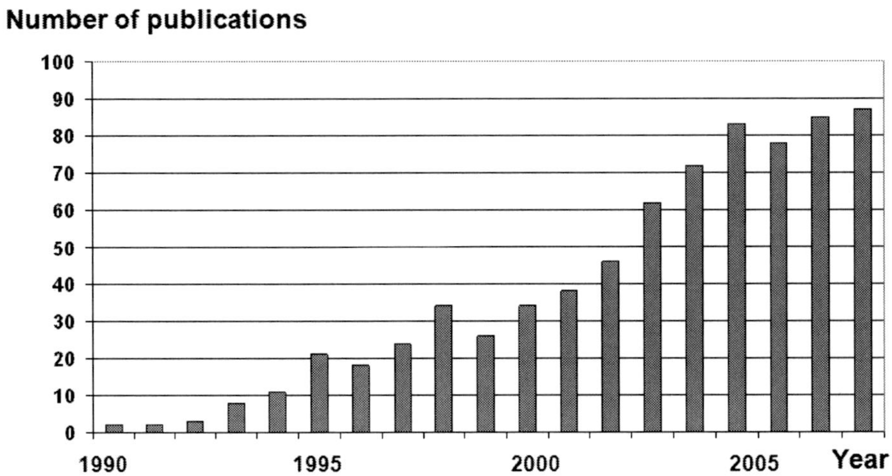

Figure 2.1. The number of yearly publications of electronic tongues, shown for the years 1900–2008.

systems. The terms "electronic nose" and electronic tongue" are thus misleading, but have become well-recognized terms.

2. MEASUREMENT PRINCIPLES

There are several measurement techniques that have potential use for ETs and TSs, although the vast majority described are based on electrochemical techniques such as potentiometry or voltammetry. These techniques are very well established in analytical chemistry. There are a number of textbooks on the subject (Bard and Faulker 1980; Wang 1994). Both techniques use an electrolyte solution and at least two electrodes, where one, the working electrode, responds to the measured specie.

Also other techniques commonly used in liquid sensing have been used, such as optical techniques, impedance spectroscopy, or the detection of changes in mass due to adsorption of measured species.

2.1. POTENTIOMETRIC TECHNIQUES

In potentiometry, a developed potential across a membrane due to the adsorption of charged species is measured. It is a zero-current technique. Due to the adsorption of charged species at the membrane, a potential E will develop, which follows the well-known Nernst relation:

$$E = E_0 + \frac{RT}{nF} \ln a \tag{2.1}$$

where E_0 is a constant for the system given by the standard electrode potentials, R is the gas constant, T is the temperature, n is the number of electrons involved in the reaction, F is Faraday's constant, and a is the activity of the analyte. The potential change has a logarithmic relation to the ionic activity. A 10-fold increase in the activity of a monovalent ion should thus accordingly ideally result in a 59.2-mV change in the membrane potential at room temperature.

2.1.1. Ion-Selective Electrodes

In potentiometric measurements, an ion-selective electrode, a reference electrode, and a potential-measuring device are used. As reference electrode, the silver–silver chloride (Ag/AgCl) electrode is often used.

The ion-selective electrode consists of a reference electrode in which the voltage bridge is replaced with an ion-selective membrane. The ion-selective membranes for conventional electrodes are basically of three types. They may consist of glass (for measurement of pH or some alkali ions), of ionophores embedded in a plastic membrane (for measurement of metal ions), or of insoluble salts (for measurement of halide ions).

Various types of membranes with different recognition profiles have been developed, and ion-selective electrodes are widely used for measurements of many ionic compounds. The most important ion-selective electrode is the pH electrode, but also other electrodes for calcium, sodium, potassium, nitrate, nitrite, and sulfate have found widespread use.

Ion-selective electrodes have many advantages, such as high sensitivity, a well-known measurement principle, low cost, ease of fabrication, and high selectivity. Some disadvantages are temperature sensitivity and that these electrodes can only measure charged species.

The first potentiometric device described in the literature was a pH-sensitive ion-selective electrode (ISE) (Haber and Klemensiewicz 1909). This is still the most widely used sensor for liquids. A number of devices followed, and research on improving the selectivity of ISEs has led to a number of useful and important sensors for a number of ions. The number of selective ISEs is limited. While research continues on selectivity, it was recognized already in 1985 that one way of increasing the analytical applications of these devices was to use arrays of less selective ISEs and to treat the response pattern with signal-processing algorithms. Thus, a first multisystem for liquid analysis was developed (Otto and Thomas 1985), based on potentiometric sensors combined with multivariate data analysis.

2.1.2. Ion-Sensitive Field-Effect Transistors

Ion-selective field-effect transistors (ISFETs) were developed in the early 1970s by Bergweld (1970). A schematic diagram of an ISFET is shown in Figure 2.2. The current between the drain and source (I_{DS}) depends on the charge density at the semiconductor surface, which is controlled by the gate potential. The ion-selective membrane is integrated directly with the transistor, and the potential across the membrane is determined by ions interacting on it. Attractive features of ISFETs are small size, ability to be integrated directly with microelectronics for signal processing, and, if mass fabricated, low cost. Electronic tongues based on the ISFET technique thus have a large future potential of being inexpensive and miniaturized.

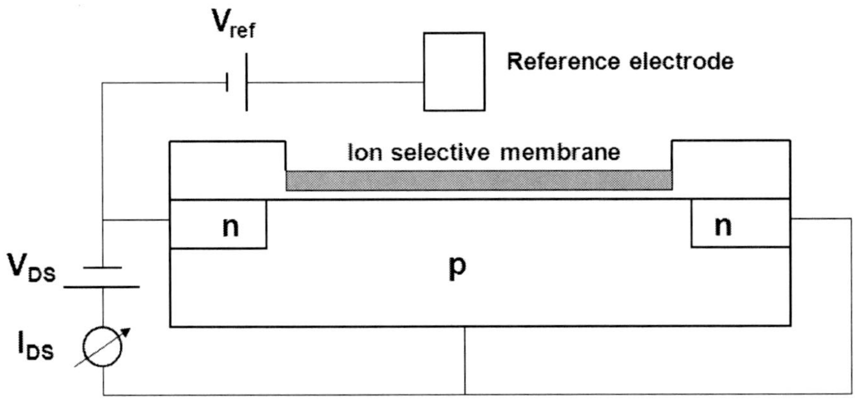

Figure 2.2. Schematic diagram of an ion-selective field-effect transistor (ISFET).

2.1.3. Light-Addressable Potentiometric Sensors

A different approach to a potentiometric system that provides for contactless sensing over a surface has also been developed. It is based on light-addressable potentiometric sensors (LAPS), in which the surface potential of a semiconductor structure is measured locally, thus providing a convenient way to analyze a multifunctional surface (Kanai et al. 1994, Sasaki et al. 1995; George et al. 2000). The semiconductor surface acts as the working electrode in a test solution also containing a reference and a counter-electrode. A light-emitting diode is positioned on the back side of the device and, by illuminating certain regions on the semiconductor, a surface potential map can be generated.

2.2. VOLTAMMETRIC TECHNIQUES

In classical amperometry, a potential is applied to a working electrode and the resulting current obtained when redox active species are reduced or oxidized on the electrode surface is measured. In voltammetry, the potential is swept between different values and the current is subsequently measured. In a two-electrode setup, the current between the working electrode and a reference electrode is measured. In a three-electrode setup, the potential between the reference electrode and the working electrode is set, and the resulting current between the working electrode and a counter-electrode is measured. The current is then a measure of redox-active compounds at a given potential.

A sharp charging current occurs upon onset of a potential pulse when charged species in the solution rush to the working electrode surface and form a so-called Helmholtz layer. The current exponentially decays until reaching an equilibrium determined by the redox current. Upon removal of the potential pulse, a similar current response in the opposite direction occurs. In classical amperometry, measurements are performed when the redox equilibrium current is reached.

The charging current i_c is in principle similar to the charging current of a capacitor in series with a resistor, following an equation of the form:

$$i_c = ER_s e^{(-t/R_s B)} \qquad (2.2)$$

where R_s is the resistance of the circuit (which is equivivalent with the resistance of the solution), E is the applied potential, t is the time, and B is an electrode-related constant.

The redox current i_r that is generated from the reduction or oxidation of a target analyte behaves in a similar way: Initially, a large current pulse is generated when compounds close to the electrode surface are oxidized or reduced, but it decays with time until it reaches the level of the equilibrated redox current. This is illustrated in Figure 2.3, which shows a series of declining voltage pulses and the resulting current pulses. The charging and discharging current pulses are clearly shown at the onset and offset of each voltage pulse.

At equilibrium in unstirred solutions, the redox current is due only to diffusion limitations. If the solution is stirred, redox-active compounds will be physically transported to the electrode surface, thus generating a larger redox current. Due to turbulence, this response is often noisy. If laminar conditions can be reached, stable signals will be obtained. For a planar electrode the current follows the Cottrell equation:

$$i_R = \frac{nFADC}{(\pi Dt)^{1/2}} \qquad (2.3)$$

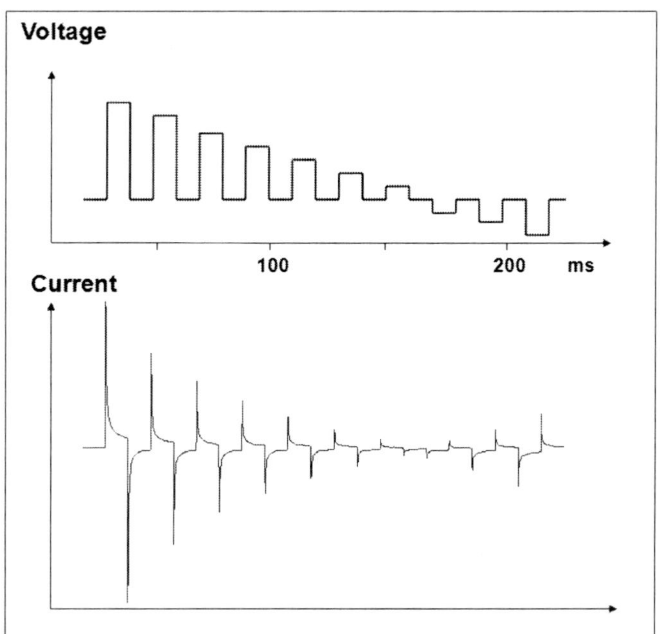

Figure 2.3. Top: A series of declining voltage pulses. Bottom: The resulting current pulses. The charging and decharging of the current pulses are clearly shown at the onset and offset of each voltage pulse.

where A is the area of the working electrode, D is the diffusion constant, and C is the concentration of analyte. At constant concentration, the equation is simplified to

$$i_r = DK_1\left(\frac{1}{t}\right)^{1/2} \tag{2.4}$$

where D is the diffusion constant and K_1 is a constant.

Use of voltammetry as the sensing principle in ETs has several advantages. It is generally acknowledged to be a sensitive, versatile, simple, and robust technique. Various analytical methods can be used, such as cyclic, stripping, and pulse voltammetry, giving various aspects of information.

Drawbacks of the technique are that charged (non–redox-active) compounds cannot be measured, and for very complex solutions, the voltammogram may be difficult to interpret. Furthermore, there is no strategy to optimize the type of applied potential steps. Measurement of pH, however, can be performed. This is because the potential position of the oxidation peak for species adsorbed on the metal working electrode surface is often pH-dependent.

Electronic tongues measurements normally involve comparisons of samples or monitoring a process flow over time. For these situations the values of the absolute voltages applied to the working electrode with respect to a reference electrode are in principle not necessary, as is the case with potentiometric measurements. It is necessary that the surface of the counter-electrode (serving as a pseudo-reference electrode) is sufficiently larger than the surface of the working electrode, resulting in current generation at only the working electrode. This has opened up the possibility of developing voltammetric electronic tongue systems that omit the reference electrode. These systems can be made extremely simple and robust, and in principle maintenance-free, which is of considerable importance for industrial applications.

Voltammetric data generated from measurements of single parameters are normally easy to interpret. In complex solutions containing many redox-active compounds, the signals generated are complex and overlapping, making them very difficult to interpret. Nevertheless, each measurement situation is unique and generates a specific "fingerprint" for a solution, which can be interpreted using multivariate data analysis. The advantages of combining multivariate analysis with voltammetry were recognized early, and the pair has had widespread use in the field (Brown and Bear 1993; Simons and van der Lindenl 1995). Examples include analysis of metal ions (Diaz-Cruz 1995; Menditeta et al. 1996).

2.2.1. Working Electrodes Used in Voltammetric Electronic Tongues

The redox reactions that take place at the working electrode are given mainly by the applied potential, but other factors also contribute, such as catalytic activity or overpotential phenomena. This means that different types of working electrodes will deliver different information about the analyte solution.

Commonly used working electrodes are the catalytic noble metals platinum, gold, rhodium, and palladium, but electrodes of silver, copper, and titanium have also been used. Other examples include working electrodes of conducting polymers or carbon paste modified with phtalocyanines or perylene imide derivatives.

2.2.2. Applied Voltage Techniques

Pulsed voltammetry is often used for voltammetric electronic tongues (VET), since studies of transient responses can be performed when the Helmholz layer is formed. This gives information concerning diffusion coefficients of charged species. Techniques include large-amplitude pulse voltammetry (LAPV) and small-amplitude pulse voltammetry (SAPV), but staircase voltammetry has also been used. In LAPV, potential pulses are applied to a working electrode with a zero potential in between. In SAPV, a slow continuous DC scan is applied to the electrode. Small-amplitude voltage pulses are superimposed at intervals, which results in small perturbations of the surface concentration of adsorbed species. Classical cyclic voltammetry is also often used for electronic tongues, in which the potential is swept back and forward between two values. Another classical technique is differential pulse voltammetry (DPV), in which current is sampled twice, once immediately after an applied potential and at the end of the pulse; the difference between these two values will give a peak-shaped output.

Stripping voltammetry applied to voltammetric ET measurements of heavy metals can be done with high sensitivity. At the onset of a negative potential, metal ions in the solution will be reduced and will accumulate on the electrode. At the onset of a gradually increasing potential, the metals will be stripped out one by one from the metal electrode into the solution, depending on the redox potential of each metal.

To ensure good resolution during VET measurements, the number of current variables stored is normally very large. This results in many of the variables being redundant and having a low level of information. Although the techniques of principal component analysis (PCA) and projection to latent structures (PLS) enable treatment of very large amounts of data, there is nonetheless an incentive in many cases to diminish the amount collected.

2.3. OTHER TECHNIQUES

Surface acoustic wave (SAW) has also been used as a sensing device in electronic tongues. SAW is an acoustic wave traveling along the surface of a material exhibiting elasticity, with an amplitude that typically decays exponentially with depth into the substrate. It uses an input interdigital transducer to convert an electrical signal to a SAW that propagates along a sensing surface, which then is reconverted to an electrical signal via an output interdigital transducer. A schematic picture of a SAW device is shown in Figure 2.4. Adsorption of chemical species on the sensing surface affects the SAW output signal. A short review covering the use of SAW in ET was recently published (Hossenlopp 2006). Basic device configuration is described, and the relation between frequency shifts and physical properties is presented.

A newly developed technique for ETs is based on electrochemical impedance spectroscopy (Bard and Faulker 1980; Wang 1994). This technique has been used extensively in various research areas. It involves the application of a small perturbation of the potential on a working electrode as a sine wave with different frequencies. From the applied perturbation and its measured response, the magnitude of the impedance and phase shift is determined as a function of the frequency of applied potential. An electrochemical cell behaves in principle like a resistor in series with a capacitor. By changing the frequency, the impedance of the tested solution can be divided into two parts, the real impedance Z_{real} and the

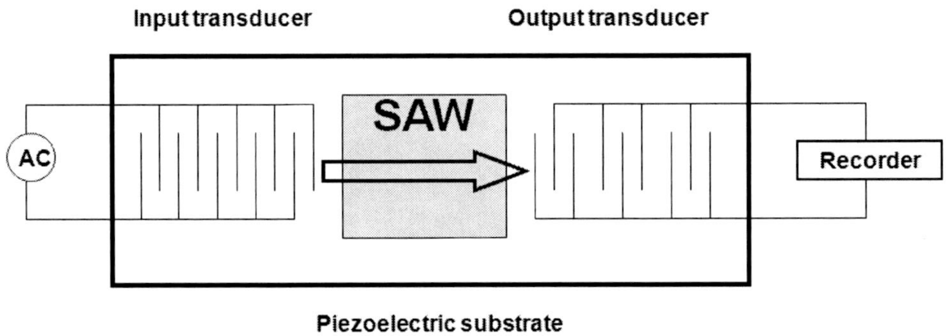

Figure 2.4. Schematic illustration of a SAW device. The SAW is generated by onset of an oscillating voltage applied to the input transducer; it propagates over the sensing surface and is detected by the output transducer.

imaginary impedance Z_{img}. The frequency-related relation between Z_{real} and Z_{img} yields electrochemical information concerning the tested solution. The spectra obtained are, however, often very complex. It is thus advantageous to employ different measurement strategies to enhance the information obtainable.

Optical techniques have also been used for ETs. In principle, they are based on a light source, a chemosensitive layer, and a detector. The chemosensitive layer is normally based on polymeric microspheres with chemically modified surfaces using an indicator that can change optical properties when exposed to a target molecule. Advantages of the technique are that there are a wide range of useful indicators and that many compounds that are difficult to measure electrochemically often can be measured by an optical method. The chemically modified microspheres can be arranged on a small silicon wafer and be illuminated with an appropriate light source, and the resulting color is then detected. The microspheres on the silicon chip form a biomimetic analog to the taste buds, and the system has thus been coined a "taste chip."

3. SIGNAL PROCESSING

The use of signal processing methods has been greatly expanded during the last decade. This is due to the rapid development of computer technology, making it possible to treat vast amounts of data, and the development of various methods for multivariate data analysis or pattern recognition. Two main issues are dealt with using these techniques. One is to search for a structure and correlation in the data. The other is to make a model from a calibration set of data, which is then used to predict test data.

Principal component analysis (PCA) is very useful to correlate data, and projection to latent structures (PLS) is common when making models. Artificial neural nets (ANNs) are also commonly used when making predicting models. For an ANN, the algorithms for learning resemble the way in which the brain learns by altering the strengths in the synaptic junctions between nerve cells. The most often used learning algorithm is the method of back propagation.

These techniques are thoroughly discussed in other chapters of this book.

4. APPLICATIONS

4.1. ELECTRONIC TONGUES BASED ON POTENTIOMETRY

4.1.1. The Taste Sensor

The initial multichannel taste sensor (TS) developed by Toko et al. (1990) is shown schematically in Figure 2.5. It was based on a device consisting of eight electrodes of silver/silver chloride wires inside a potassium chloride solution with an outer membrane with different sensing properties. The membranes were made of ion-sensitive lipid membranes immobilized with the polymer polyvinyl chloride (PVC). Five different lipid analogs were used: n-decyl alcohol, oleic acid, dioctyl phosphate (bis-2-ethylhexyl) hydrogen phosphate, trioctylmethyl ammonium chloride, and oleylamine, together with mixtures of these. The TS was used to study responses from the five basic tastes, NaCl (salt), quinine (bitter), HCl (sour), sucrose (sweet), and monosodium glutamate (umami). The largest responses were obtained from the sour and bitter compounds, thereafter umami and salt, and for sucrose almost no response was obtained. In general, similar substances, such as sour substances like HCl, acetic acid, and citric acid, or salty substances such as NaCl, KCl, and KBr, showed similar response patterns. Nonelectrolytes were not detected, since these compounds had little effect on the membrane potential (Toko 2000).

Since its introduction, the TS system has been further developed and used in a number of different applications, mostly dealing with estimation of the taste of different drinks. In one of these early studies, 33 different brands of beers were analyzed (Toko 1998) both by a sensory panel, expressing the taste in the parameters sharp versus mild and rich versus light, and by the taste sensing system. The responses from the TS were analyzed using PCA, and the first principal component corresponded well to the taste parameters rich versus light taste, and the second principal component corresponded well to the parameters sharp versus mild taste. Mineral water was also analyzed (Toko 2000), and the TS was capable of

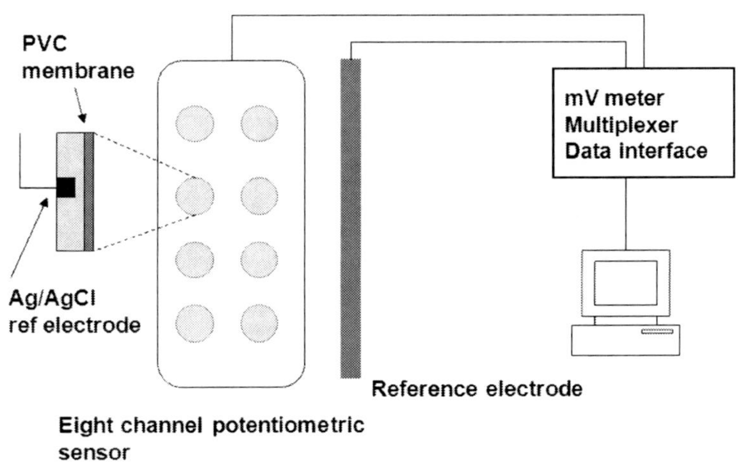

Figure 2.5. Schematic of the first multichannel taste sensor, developed in 1990.

discriminating between different commercial brands as well as determining the hardness of the water. Other applications included estimations of the taste of milk (Yamada et al. 1997) or coffee (Fukuganga et al. 1996) and the development of monitoring systems for water quality (Taniguchi et al. 1999) and fermentation processes for producing soybean paste (Imamura et al. 1996).

A study was performed to investigate the use of a TS system for evaluation of medicinal plants and Chinese medicines with bitter or astringent tastes, and to evaluate possible application of the system for taste and quality control of medical products (Kataoka et al. 2008). The study indicated that different tastes could be distinguished and that the quality of investigated medical plants could be estimated.

The analytical performance of TSs has been further developed to increase the sensitivity and selectivity properties of the membrane. Thus, lipid polymers were also used in a TS for the estimation of sweetness (Habara et al. 2004). By changing the electrode potential of the membranes, the sensitivity for sucrose was increased up to 10 times compared to a conventional TS.

A multichannel TS has been used to evaluate the bitterness of nine different antibiotics, and the results were correlated with the human gustatory sensation (Uchida 2003). The effect of adding sweetener to mask the taste sensation of bitterness of the antibiotic could be estimated with the TS.

4.1.2. Potentiometric Electronic Tongues

The potentiometric electronic tongue (PET), first described in 1996 (Legin et al. 1996; Di Natale et al. 1996), is shown schematically in Figure 2.6. It consisted of potentiometric sensor arrays of two general kinds, conventional ones for pH, sodium, and potassium, and specially designed ones, based on chalcogenide vitreous materials. This sensor array included altogether 20 potentiometric sensors and was used

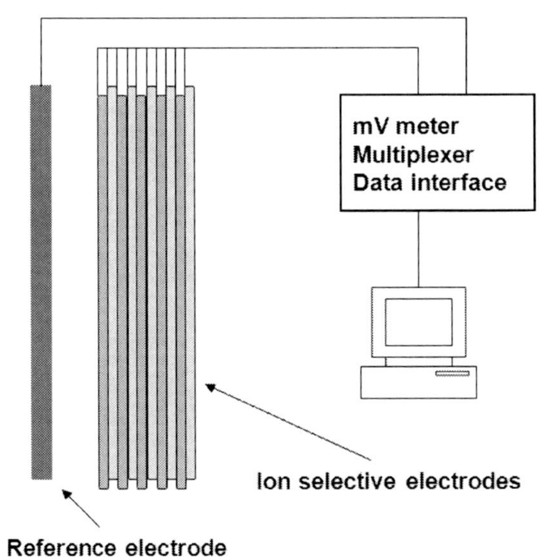

Figure 2.6. Schematic of the first potentiometric electronic tongue, developed in 1996.

for the classification of different kinds of drinks such as soft drinks, juices, tea, and beers. PCA was used as multivariate analytical tool and yielded good separation between samples. With this PET the fouling process of orange juice during storage could also be followed, and a model for storage-time prediction was made using an ANN. The monitoring of polluted river water was also described (Di Natale et al. 1997). River water was artificially polluted with Cu, Cd, Fe, Cr, and Zn, representing a "common" pollution profile, and analyzed with the PET. Other techniques of multivariate data analysis were investigated, such as multiple linear regression, projection to latent structure (PLS), nonlinear least squares, back-propagation ANN, and a self-organizing maps.

Similar PETs have also been used for qualitative analysis of mineral water and wine (Legin et al. 1999a) and some biological liquids (Legin et al. 1999b). A PET based on 34 potentiometric individual sensing units was applied to 56 samples of Italian red wines (Legin et al. 2003). All samples could be classified, and the system was also able to predict human sensory scores. Honey samples from different regions in Portugal have also been distinguished (Dias et al. 2008). Milk samples from both mastitic and healthy gland originating from two different farms have been classified with high accuracy (Mottram et al. 2007). Urinary tract dysfunction has been correlated with both a PET and creatinine levels (Lvova et al. 2009).

An ET based on six metallic potentiometric sensors was developed and used for classification of different types of vinegar and fruit juices (Lvova et al. 2006a). The metals Cu, Sn, Al, brass, and stainless steel were used in the sensor array. The responses obtained were very complex and influenced by several components. The results obtained were unclear, but it was postulated that this sensor type could have a large future potential due to its simplicity and low cost.

An electronic tongue using polymeric membranes doped with porphyrin was used for the analysis of alcoholic beverages (Lvova et al. 2006b), and PVC membranes doped with metallo-porphyrins were used for classification of white wines (Verrelli et al. 2007a). The system could also discriminate between an artificial wine and various wines from different vineyards produced in different years. It could also determine the acetic acid content in wine. Silicon-based potentiometric sensors were developed using conventional Si fabrication techniques (Verrelli et al. 2007b), which allowed for implementation of different structural layers and metal films.

4.1.3. Disposable Systems

Disposable systems have also been developed. Disposable potentiometric TSs were made from conductive films of polypyrrole and polyaniline, PVC and aromatic polyurethane matrices doped with various components, combined with a potentiometric glucose biosensor (Lvova et al. 2003). It was used for the classification of coffee, black tea, and different sorts of green teas. A disposable TS was designed using screen printing techniques (Gotoh 2004). Five reference electrodes (Ag/AgCl) were covered by lipid films. Each lipid film was sensitive to each taste of the five basic tastes. The TS could classify tastes of a large number of various beverages. Disposable PETs were made using a five-sensor array, screen-printed on a polymeric substrate (Guiterrez et al. 2008a). The sensors showed good reproducibility and long time stability, and were used for the simultaneous detection of ammonia, potassium, and sodium in natural surface waters.

4.1.4. Light-Addressable Potentiometric Sensors

Light-addressable potentiometric sensors (LAPS) have been used mainly for the detection of the basic tastes. In one setup (Kanai et al. 1994), five lipid membranes (oleic acid, lecithin, cholesterol, phosphtidyl ethanolamine, and dioctyl phosphate) having different areas were deposited on a semiconductor surface. The sensing area was divided into nine different regions with varying composition and thickness of lipid layers. The sensor system could separate the basic taste substances. The method has also been further developed (Sasaki 1995; Murakami 1998; George 2000).

4.2. ELECTRONIC TONGUES BASED ON VOLTAMMETRY

4.2.1. The Voltammetric Electronic Tongue

The first voltammetric electronic tongue (VET) was described in 1997 (Winquist et al. 1997) and consisted of two working electrodes of platinum and gold, a reference electrode, and a counter-electrode. It could discriminate between various fruit juices and follow the aging process of orange juice stored at room temperature. It was further developed by increasing the number of working electrodes. A common configuration is shown schematically in Figure 2.7. There are five working electrodes (gold, iridium, palladium, platinum, and rhodium) embedded in a dental material with a reference electrode placed in the center of the probe. The stainless steel housing serves as a counter-electrode. Different types of pulsed voltammetry can be used with this sort of probe, such as LAPV, SAPV, and staircase, and the responses generated are treated with PCA, PLS, or ANN.

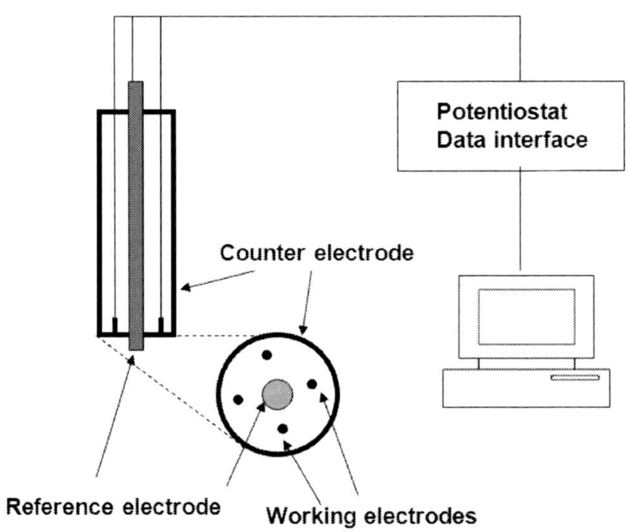

Figure 2.7. Schematic of the first voltammetric electronic tongue, developed in 1997.

This VET was used successfully in a large number of applications. Various types of teas could be separated (Ivarsson et al. 2001a) as well as different detergents (Ivarsson et al. 2001b). A process monitoring application has been described for drinking water production (Krantz-Rülcker et al. 2001). Other applications have been to follow microbial activity (Söderström et al. 2003a, 2003b, 2003c). The deterioration of milk due to microbial growth has been followed and correlated with colony-forming units (CFUs) (Winquist et al. 1998). These studies make it clear that the VET can be used to make detailed analysis of microbial growth, classify various microbial samples, recognize different mold species growing in different media, and predict their incubation time.

In Figure 2.8, three versions of this VET are shown. From left to the right, a miniaturized, a standard, and an industrial VET are shown. The industrial VET is equipped with a reference electrode in the middle.

VET has been further developed, both by incorporating new types of working electrodes and by using new techniques for applying potentials. Thus, multifrequency, large-amplitude pulse voltammetry (MLAPV) has been applied to a voltammetric electronic tongue (Tian et al. 2007) consisting of working electrodes made of platinum, gold, titanium, nickel, and palladium. The device had a Ag/AgCl reference electrode and a platinum electrode as counter-electrode. The waveform of MLAPV consisted of three individual frequency segments, 1 Hz, 10 Hz, and 100 Hz. The electronic tongue could successfully discriminate six different Chinese spirits and seven Longjing teas at different frequency segments.

Using stripping technique on a gold electrode, cadmium in extracted soil samples has been determined down to a concentration of 1 μg/L, corresponding to 9 nM (Winquist et al. 2009). An automatic measurement system for flow injection analysis (FIA), equipped with a self-polishing unit, was described, which has been developed for measurements of cadmium in samples of soils and flour.

Sensor arrays consisting of two types of electrodes, one type based on conducting polymers, the other on carbon paste doped with phthalocyanine, were used successfully to discern fraudulent red

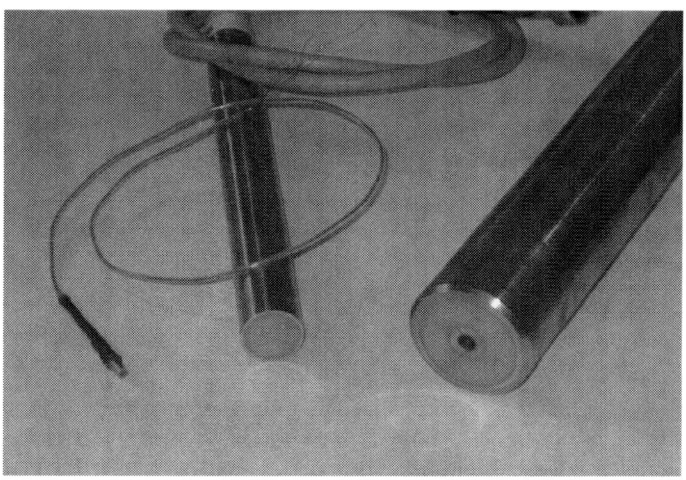

Figure 2.8. Three versions of the VET. Left: A miniaturized VET. Middle: A standard VET. Right: An industrial VET, equipped with a reference electrode in the middle.

wines and to discriminate between red wines aged in oak barrels or matured in steel tanks in contact with oak wood chips (Apertei et al. 2007). The results obtained correlated well with conventional analytical methods. By analyzing voltammograms from a copper electrode in strong alkaline solution (Wu et al. 2005), a classification method for Chinese yellow wines has been achieved.

Voltammetric electrodes based on phtalocyanines have been used for the classification of basic tastes (Arrieta 2003). The electrodes were prepared using either Langmuir-Blodgett (LB) techniques or carbon paste fabrication and were evaluated by voltammetry for response characteristics of the five basic tastes. Both sensor types could classify the tastes in a similar way, but the peaks obtained for the LB films were better resolved compared with the carbon paste–based sensor. A VET, consisting of two conventional electrodes of Au and Pt and an electrode modified with a conducting polymer, were tested for discrimination of fruit juices (Martina et al. 2007). The modified electrode showed the most discriminating ability, and needed less cleaning between measurements. VETs have also been developed for the determination of the basic taste bitter. The working electrodes were made of conducting polymers and carbon paste with phtalocyanine complexes (Apeteri et al. 2004). The VET was tested by evaluating the discriminating capability between $MgCl_2$, quinine, and four phenolic compounds. Voltammograms obtained showed clear differences between redox processes from the tested compounds. VET using carbon paste electrodes modified with three rare-earth bisphthalocyanines and three perylenes have also been studied. Using square voltammetry, different response patterns were obtained for a variety of different compounds, such as acids, bases, ionic solutions, and antioxidants (Rodriguez-Mendez et al. 2008). A VET using a sensor array consisting of polypyrrole sensors doped with different counter-ions, carbon paste electrodes modified with metallophthalocyanines, and carbon paste electrodes modified with perylene imide derivates have been developed (Parra et al. 2006). It was used to follow the aging process of red wine in oak barrels.

VETs have also found an application area in quantification of tastes and masking the effect of sweetener for pharmaceutical purposes (Legin et al. 2004) These qualities of a drug are important for the acceptance of oral intake, especially for children. In one study, a VET was applied to analyze a variety of 41 individual drugs in terms of modality, demonstrating that it could distinguish between different substances having the same basic taste. It could also detect different nuances of basic tastes, and levels of sweetener in a way that was consistent with their masking effect of bitter substances.

A general problem using sensors in the aqueous phase is fouling, which often occurs when measuring in complex media such as milk or various industrial process waters. The conditions in such media often lead to signal drift and loss of sensitivity. The electrodes must thus be cleaned and calibrated regularly, which costs time and money. One way to surmount this problem is to use self-polishing electrodes. Self-polishing VET have been developed where electrode surfaces are continuously polished at regular time intervals (Olsson et al. 2006) These systems have shown improved performance compared with normal VETs (Winquist et al. 2009). For industrial use, self-cleaning electronic tongues possess a considerable advantage in terms of maintenance requirements.

4.2.2. Industrial Applications

Many industrial applications of VETs make use of device properties. The diary industry, for example, requires nonbreakable, hygienic probes for process monitoring. In one application the washing and rinsing cycles in a diary plant (Winquist et al. 2005) could be followed, both inside the stainless steel tubing transporting milk as well as in the heat exchangers. Low fabrication cost makes VETs suitable for consumer products. The VET has been used in household appliances such as dishwashers and washing machines to distinguish among different standardized dirt types both during washing with high levels of detergents (Olsson et al. 2008) and to follow the rinsing process in a washing machine (Ivarsson et al. 2005).

The ability to operate in demanding environments has made possible the monitoring of microbial growth in water from the paper and pulp industry where the detection limit was found to be 2 bacteria/100 mL (Johnsson et al. 2003). VETs have also been used for evaluation of pulp samples from the paper industry with respect to five reference parameters: pH, conductivity, chemical oxygen demand (COD), cationic demand, and zeta potential (Carlsson et al. 2001).

Quality estimations of water samples from a drinking-water purification plant have been performed (Krantz-Rülcker et al. 2001; Winquist et al. 2004, 2008). Water from three parallel slow sand filters as well as incoming raw water in a drinking-water production plant could easily be discriminated from PCA plots on signals from a VET. It was observed during the study that water from one of the sand filters deviated from the others during a three-week period. This sand filter was later taken out of production due to malfunction.

4.2.3. Variable Reduction

The shape of the current responses for LAPV follows Eqs. (2) and (4), making possible calculation of constants that express the response of the exponential current decay. Constants fitting Eq. (2) have accordingly been calculated for classification of different teas. The responses were treated with PCA, and better separation was obtained compared with using original data (Holmin et al. 2001; Artursson et al. 2002).

Multivariate approaches can also be used for variable reduction. In hierarchical PCA, PCA or PLS scores are used as descriptors for further multivariate data analysis. Thus a large data set can be divided into meaningful blocks, followed by a PCA to extract score values (Holmin et al. 2004). These are in turn combined into a new data set and PCA is performed again. Another strategy is to extract relevant variables from a larger data set by the use of PCA. In one experiment, a data set of 1200 variables was reduced to 20, and these were used for classification using an ANN (Winquist et al. 1998).

Data compression of voltammograms obtained from amperometric sensors has also been achieved by employing discrete wavelet transform followed by artificial neural networks (Moreno-Baron et al. 2005, 2006).

Other variable treatment procedures have been based on wavelet neural networks to resolve overlapping signals (Guiterrez et al. 2008b). This method was performed on highly overlapping signals from three phenolic compounds that were simultaneously oxidized at the surface of an epoxy-graphite transducer.

4.3. MINIATURIZED SYSTEMS

Miniaturized VETs have also been developed. In one configuration the VET consisted of three types of working electrode (gold, platinum, and rhodium) inserted in a platinum tube with a size of 2 mm in diameter and 4 mm in length (Winquist et al. 2008b). Trace amounts of cadmium and lead (in the micromolar range) were detected in 2-µl samples. The saliva composition of a volunteer during exercising could also be followed by placing the miniaturized ET under the real tongue of the volunteer.

A miniaturized PET with an integrated reference electrode has been developed which is capable of discriminating different samples of milk (Ciosek and Wroblewski 2008). It was based on an array of microelectrodes fabricated from epoxy-glass laminate and integrated with PVC membranes containing different additives. A miniature TS chip based on lipid/polymer membranes has been described, with integrated working and reference electrodes (Etoh et al. 2008). The system was connected to a laptop computer for signal processing, thus making a portable TS system.

A miniaturized PET was made by depositing PVC membranes, doped with metalloporphyrin complexes, onto miniaturized potentiometric sensors. It was used for classification of different wine samples (Verrelli et al. 2007b).

4.4. FLOW INJECTION ANALYSIS

Flow injection analysis (FIA) is used in many analytical applications (Ruzicka and Hansen 1975) to improve the accuracy of the analyses. A reference solution is continuously pumped through a sensor cell and test samples are injected into the flow. Since relative measurements are performed, these systems are less influenced by sensor baseline drift compared to the absolute response values obtained by a probe. Furthermore, calibration samples can be injected within measurement series, and the system is well adapted for automating. It can also be noted that most electronic nose systems are based on a gas-phase FIA technique, mainly to compensate for sensor drift. The approach of combining flow injection analysis with multisensor systems and data evaluation using multivariate data analysis appears very advantageous. A flow injection system based on chalcogenide glass electrodes for the determination of the heavy metals Pb, Cr, Cu, and Cd has been developed (Mortensen et al. 2000). The flow-through technique has also been applied to a TS system for food quality control (Iwakura et al. 2003).

VETs have been incorporated into FIA systems. In one early system, repeated analyses of standard solutions of H_2O_2, KCl, $CuNO_3$, $K_4[Fe(CN)_6]$, and NaCl were performed, showing that electrode drift was considerably decreased. The system was later used for classification of different apple juices (Winquist et al. 2002). In a later study, it was used for characterization and prediction of conductivity, pH, and chemical oxygen demand (COD) of waste water from a paper mill (Gutes et al. 2006).

In a sequential injection analysis (SIA) system based on three working electrodes of platinum, gold, and epoxy-graphite disks, the compounds ascorbic acid, uric acid, and paracetamol could be discriminated (Gutes et al. 2007).

A PET based on a sensor array of miniaturized solid-state potentiometric electrodes in a flow-through system were used for analysis of beverages (Ciosek and Wroblewski 2006a). It was tested for classification of various beverages such as milk, juice, and beer. Partial selective microelectrodes were introduced in a flow-through system for analyses of different brands of beer (Ciosek and Wroblewski 2006a). Samples of the same type of beer, but from different manufacturers, could be discriminated.

An integrated multisensor consisting of six ISFET devices was incorporated into a FIA system for analysis of grape juice and wine samples. Results obtained showed that samples could be distinguished according to both grape variety and vintage year of the wine samples (Moreno et al. 2008).

Pulse transient responses obtained when introducing samples in a flow stream leading to an ET also contains useful information. In a SIA system, pulse sample transient responses of an eight-sensor array were analyzed using fast Fourier transform. Three coefficients per sensor were applied in an ANN. By injection of a mixture of four different ions (Ca^{2+}, Mg^{2+}, K^{+}, and Na^{+}), a large improved classification capability for this technique was found compared to conventional ET techniques (Calvo et al. 2008).

SIA technique has also been applied to an integrated multisensor chip consisting of an array of ISFETs (Ipatov et al. 2008). A custom-made flow cell was designed for analysis of the mineral water components K^{+}, Na^{+}, and Cl^{-}.

4.5. HYBRID SYSTEMS

When collecting information from various sensors in a given application, the quality of the information is increased if different sensor technologies are combined. This was first demonstrated by the combination of electronic noses and electronic tongues (Winquist et al. 1999).

Hybrid ETs have also been descrbed. In a first application, the measurement techniques potentiometry, voltammetry, and conductivity were combined for classification of different types of fermented milk (Winquist et al. 2000). The composition of the microorganisms of the different fermentations was also reflected in a PCA plot.

In another hybrid system (Men et al. 2005), a multiple light-addressable potentiometric sensor (MLAPS) was combined with two voltammetric electrodes. Stripping voltammetry was applied to the voltammetric electrodes for general detection of heavy metals, and a MLAPS based on chalcogenide thin film was used for the simultaneous detection of Fe(III) and Cr(VI). This system has been used for detection of heavy metals in wastewater or seawater.

Different brands of mineral water were classified (Moreno et al. 2006) using a hybrid system consisting of six ISFETs, an interdigitated platinum electrode for measurement of conductivity, a redox potential electrode, and a temperature sensor.

A VET and an EN have been combined for the classification of wines originating from different areas (Buratti et al. 2004). Sensory descriptors of Italian red wines of different origins could be predicted using a similar combination. The results obtained showed the possibility of using these combined techniques to describe and predict a large part of the sensorial information.

An ET and an EN were combined for verifying geographic origin and uniqueness of virgin olive oils (Cosio et al. 2006). Other parameters were measured, such as free acidity, peroxide value, UV index, and phenol content by use of standard methods. Using ANN yielded satisfactory classifications.

4.6. PERFORMANCE OF ELECTRONIC TONGUES COMPARED WITH OTHER ANALYTICAL TECHNIQUES

A potentiometric electronic tongue, Fourier-transform infrared spectroscopy (FTIR), and high-performance liquid chromatography (HPLC) were used for measurements on five apple varieties (Rudnitskaya et al.

2006). HPLC was used to measure organic acids used for calibration of the PET and the FTIR. Different approaches to data processing were performed. In this study, the combination of a PET with an attenuated total reflectance Fourier-transform IR spectroscopy (ATR-FTIR) were also evaluated as a novel and rapid method for taste research. The PET consisted of 27 potentiometric sensors, and the ATR-FTIR was used to determine sugar and organic acid in four tomato cultivars. HPLC was used as a reference method. Data obtained were treated with PCA, canonical discriminate analysis, and PLS.

The results from a VET were compared with results obtained with a HPLC equipped with an electrochemical detector (Söderström et al. 2005a). Results from a VET have been compared with those from a PET concerning studies of bacterial growth; both systems were capable of separating different bacterial strains (Söderström et al. 2005b).

The commercial ET system ASTREE has been compared with a conventional PET (Beullens et al. 2008). The ASTREE system was developed by the company Alpha M.O.S. (Alfa MOS, Toulouse, France), and consisted of a set of seven sensors that are commercially available. The PET was developed at the University of Saint Petersburg and consisted of 18 potentiometric sensors. Six Belgian tomato cultivars were classified by both systems and a trained sensory panel. Content of sugar, organic acids, and mineral was also determined by classical analytical systems. Both ET systems could classify tomato cultivars based on their taste profile as judged by the sensory panel. The performance of a conventional PET was compared with a method based on gas chromatography–mass spectrometry (GC-MS), when analyzing onions and shallots (Auger et al. 2005). Both methods were able to separate all tested samples, and it was concluded that the two techniques were compatible and in some cases complementary.

4.7. SURFACE ACOUSTIC WAVES, IMPEDANCE MEASUREMENTS, AND OPTICAL TECHNIQUES

An ET based on a dual shear horizontal SAW has been described. It was microfabricated on LiTaO3 as dual-line sensor, operated at 60 MHz, and placed in a Teflon housing (Sehra et al. 2004). Synthetic samples with the four basic tastes sour, sweet, salt, and bitter were injected, and the ET correctly classified all samples.

A SAW microsensor with a polymer microfluidic cell as a part in an EN–ET system were used for identification of liquids (Jacesko et al. 2005). Due to the high electromechanical coupling of the substrate, it could detect differences in electrical properties and thereby identify various liquids, such as strawberry Hi-C, teriyaki sauce, and water of different pH.

Nanostructured films of conducting polymer were used as sensing units (Riul et al. 2003) for electrochemical impedance spectroscopy (EIS). Using the frequency range 1–10 MHz allowed detection of trace amounts of taste substances. Brands of a variety of commercial beverages such as mineral waters and wines could be distinguished.

Carbon nanotube polymer composite sensors have been developed for an impedimetric ET (Pioggia et al. 2008). It was tested using five different tastes.

EIS was applied to an electrode with an ionophore membrane, and the data obtained were processed using an ANN (Cortina-Puig et al. 2007). The SIA technique was applied to introduce various concentrations of the ions K^+, Na^+ and NH_4^+. The ANN could both classify and predict the concentration of the ions. Commercial fertilizers were also classified.

A micromachined fluidic structure for the introduction of liquid samples into a chip-based sensor based on an array of polymeric microbeads has been developed (Young-Soo et al. 2004, 2005). It consisted of micromachined storage cavities combined with a covering glass layer that contained microbeads and fluidic channels. Receptors that could change their optical properties upon exposure to test molecules were chemically attached to the microbeads. Spectral data were acquired for each of the individual microbeads using a charged-coupled device (CCD), allowing for real-time analysis of liquid samples.

An important part of the structure was the passive fluid introduction system, driven only by capillary force, making it compact and well suited for the "lab-on-a-chip" concept. Results showed that the system might be useful in a micro-total analysis system (μ-TAS) and in biomedical applications.

This novel lab-on-a-chip system has also proven valuable for the measurement of the biomarker C-reactive protein (CRP), which is an acute-phase reactant and a well-accepted indicator of inflammation (Christodoulides et al. 2005). Measurements were performed on saliva, and the sensitivity of the microchip system was found to be 10 pg/ml of CRP with 1000-fold dilution of saliva sample. The very high sensitivity of the system was due to increased signal-to-noise ratio, resulting from the higher bead surface area available. One conclusion from the study was that chronic periodontal disease may be associated with higher levels of salivary CRP.

5. SOME PRACTICAL CONSIDERATIONS

As has been shown in this chapter, the number of possible applications for the ET is very large, including determination of single compounds as well as estimation of taste and quality properties of a sample. The most appropriate technique used for ET for a given application will be dependent on the measurement conditions. In general, if it is expected that ions are most important, then a PET should be used. If it is expected that redox-active compounds are important, then a VET should be used. Thus, for the quality estimation of mineral water, a PET is appropriate. To follow a biotechnical process, a VET is probably the best choice. In most measurement situations, however, information is obtained from ionic compounds as well as from redox-active compounds. It will thus be very difficult to foresee which kind of technique will work best. In such cases, an old-fashioned trial-and-error approach is called for. It should also be remembered that the best results may be obtained when techniques are combined.

In the multivariate approach, a large data set of sensor signals is collected, and if it is large enough, a correlation with almost any sample property can be found. This can lead to nonsensical conclusions, and it is thus important that calibration and prediction data sets are carefully chosen. If a prediction model needs to be very sophisticated in order to ensure correct predictions, there is a large risk that the prediction algorithm is unstable. This will result in small changes in the sensor signal causing disproportionately large changes in the prediction. The sensor signals thus need to be very stable and free from drift, which in real life is very difficult to obtain. On the other hand, if sensor signals are stable, prediction algorithms can be kept simple, leading to a very robust measurement system.

Signal drift is a problem affecting all sensor systems. There are various reasons for drift in ET signals. General reasons are temperature changes and sensor surface fouling. Specific reasons for the PET may be leakage of recognition elements from the membrane and slow equilibration between the sensor surface and the surroundings. Specific reasons for the VET may be oxide-layer formation or reorganization of the active surface on the working electrode. For some electrode materials (e.g., gold), poisoning due to exposure to sulfur-containing compounds may occur.

Several solutions to signal drift problems have been introduced. The taste sensing system of the Anirutsi company (as described in Section 6) washes and calibrates the electrodes between measurements, which is an often-used solution to drift problems in analytical chemistry. For the VET, a polishing step can be used to regenerate the sensing surface between measurements (See Section 4.2.1.).

6. COMMERCIALIZATION

The taste sensing system originally developed by Toko et al. (1990) was commercialized in 1992 by Anirutsi (Taste Sensing System SA401, Anritsu Corp., Japan). The system consists of an eight-channel multisensor placed on a robot arm, controlled by a computer. The tested samples are placed in sample holders that also contain cleaning and reference solutions. A measurement sequence starts with a cleaning step in which the multisensor is introduced to the cleaning solution, thereafter into the sample solution, and the cycle repeats. The multisensor is introduced to the reference solution at certain intervals for calibration purposes.

The voltammetric electronic tongue, as originally developed by Winquist et al., was commercialized in 2001 by SensET AB (Linköping, Sweden). This company now offers a continuous measurement system for many important industrial parameters such as chemical oxygen demand (COD), total carbon content (TOC), and sulfite. The VET is equipped with a self polishing system, making it possible to measure in dirty and crude solutions, and is in principle maintenance-free. Main customers have been the paper and pulp industry, and for the COD application. These VETs have been in use for several years, giving correct predictions each fifth minute.

Alpha MOS (Toulouse, France), which earlier commercialized the electronic nose, has recently commercialized the ET. Their system, called ASTREE, consists of an array of ISFETs and is designed mainly for the pharmaceutical industries. The application area has been mainly to estimate the bitterness of drugs and the masking effect of sweetener.

7. CONCLUSIONS

Since the introduction of the TS in 1990, followed by the potentiometric electronic tongue in 1995 and the voltammetric electronic tongue in 1997, the interest in ETs has steadily increased. The technique has been shown to be quite versatile, and a large number of applications have been developed. Applications have been reported in food processing, clinical analysis, environmental monitoring, paper manufacturing, and water quality assessments.

Although the technique is well established, there still exists room for further development. Interesting areas include miniaturization, use of microelectrodes, development of new electrode materials, and hand-held disposable systems.

8. ACKNOWLEDGMENTS

The Swedish Sensor Centre (S-SENCE) was supported by grants from the Swedish Governmental Agency for Innovation Systems (VINNOVA) during 1995–2006. At present, S-SENCE is running a drinking-water surveillance project in collaboration with Linköping municipality, Tekniska Verken i Linköping AB, SAAB Security, and Adixen Scandinavia, with funding from the Swedish Civil Contingencies Agency (MSB), the Swedish Defence Materiel Administration (FMV), and The Swedish Governmental Agency for Innovation Systems (VINNOVA).

The Swedish Farmers' Foundation for Agricultural Research, the VL foundation, and S-SENCE at Linköping University are also gratefully acknowledged for financial support.

9. NOMENCLATURE

ANN	Artificial neural net
ATR-FTIR	Attenuated total reflectance Fourier-transform infrared spectroscopy
DPV	Differential pulse voltammetry
EIS	Electronic impedance spectroscopy
EN	Electronic nose
ET	Electronic tongue
FIA	Flow injection analysis
FTIR	Fourier-transform infrared spectroscopy
GC	Gas chromatography
HPLC	High-performance liquid chromatography
ISE	Ion-selective electrode
ISFET	Ion-selective field-effect transistor
LAPS	Light-addressable potentiometric sensor
LAPV	Large-amplitude pulse voltammetry
LB	Langmuir-Blodgett
NIR	Near-infrared spectroscopy
MLAPV	Multifrequency large-amplitude pulse voltammetry
MLAPS	Multiple light-addressable potentiometric sensor
MVDA	Multivariate data analysis
PCA	Principal components analysis
PET	Potentiometric electronic tongue
PLS	Projection to latent structures
PVC	Polyvinyl chloride
SAPV	Small-amplitude pulse voltammetry
SAW	Surface acoustic wave
SIA	Sequential injection analysis
TS	Taste sensor
VET	Voltammetric electronic tongue

REFERENCES

Anand V., Kataria M., Kukkar V., Saharan V., and Choudhury P.K. (2007) The latest trends in the taste assessment of pharmaceuticals. *Drug Discov. Today* **12**, 257–265.

Apetrei C., Apetrei I., Nevares I., Del Alamo M., Parra V., Rodriguez-Mendez M.L., and De Saja J.A. (2007) Using an e-tongue based on voltammetric electrodes to discriminate among red wines aged in oak barrels or aged using alternative methods. *Electrochim. Acta* **52**, 2588–2592.

Apetrei C., Rodriguez-Mendez M.L., Parra V., Gutierrez F., and de Saja J.A. (2004) Array of voltammetric sensors for the discrimination of bitter solutions. *Sens. Actuators B* **103**, 145–152.

Arrieta A., Rodriguez-Mendez M L., and de Saja J A. (2003) Langmuir-Blodgett film and carbon paste electrodes based on phthalocyanines as sensing units for taste. *Sens. Actuators B* **95**, 357–365.

Artursson T., Spångeus P., and Holmberg M. (2002) Variable reduction on electronic tongue data. *Anal. Chim. Acta* **452**, 255–261.

Auger J., Arnault I., Legin A., Rudnitskaya A., Seleznev B., Sparfel G., and Dore C. (2005) Comparison of gas chromatography-mass spectrometry and electronic tongue analysis for the classification of onions and shallots. *Int. J. Environ. Anal. Chem.* **85**, 971–980.

Bard A.J. and Faulkner L.R. (1980) *Electrochemical Methods—Fundamentals and Applications.* John Wiley, New York.

Bergveld P. (1970) Development of an ion-sensitive solid state device for neurophysiological measurements. *IEEE Trans. Biomed. Eng.* **17**(1), 70–71.

Beullens K., Kirsanov D., Irudayaraj J., Rudnitskaya A., Legin A., Nicolai B., and Lammertyn J. (2006) The electronic tongue and ATR-FTIR for rapid detection of sugars and acids in tomatoes. *Sens. Actuators B* **116**, 107–115.

Beullens K., Meszaros P., Vermeir S., Kirsanov D., Legin A., Buysens S., Cap N., Nicolai B., and Lammertyn J. (2008) Analysis of tomato taste using two types of electronic tongues. *Sens. Actuators B* **131**, 10–17.

Brown S. and Bear R. (1993) Chemometric techniques in electrochemistry: A critical review. *Crit. Rev. Anal. Chem.* **24**, 99–131.

Buck L. and Axel R. (1991) A novel multigene family may encode odorant receptors: A molecular basis for odor recognition. *Cell* **65**(1), 175–187.

Buratti S., Benedetti S., Scampicchio M., and Pangerod E.C. (2004) Characterization and classification of Italian Barbera wines by using an electronic nose and an amperometric electronic tongue. *Anal. Chim. Acta* **525**, 133–139.

Carlsson A., Krantz-Rülcker C., and Winquist F. (2001) An electronic tongue as a tool for wet-end monitoring. *Nord. Pulp. Pap. Res. J.* **16**, 319–326.

Calvo D., Duran A., and del Valle M. (2008) Use of pulse transient response as input information for an automated SIA electronic tongue. *Sens. Actuators B* **131**, 77–84.

Christodoulides N., Mohanty S., Miller C.S., Langub M.C., Floriano P.N., Dharshan P., Ali M.F., Bernard B., Romanovicz D., Anslyn E., Fox P.C., and McDevitt J.T. (2005) Application of microchip assay system for the measurement of C-reactive protein in human saliva. *Lab. Chip* **5**, 261–269.

Ciosek P., Brzozka Z., and Wroblewski W. (2006a) Electronic tongue for flow-through analysis of beverages. *Sens. Actuators B* **118**, 454–460.

Ciosek P. and Wroblewski W. (2006b) The recognition of beer with flow-through sensor array based on miniaturized solid-state electrodes. *Talanta* **69**(5), 1156–1161.

Ciosek P. and Wroblewski W. (2007) Sensor arrays for liquid sensing—Electronic tongue systems. *Analyst* **132**(10), 963–978.

Ciosek P. and Wroblewski W. (2008) Miniaturized electronic tongue with an integrated reference microelectrode for the recognition of milk samples. *Talanta* **76**(3), 548–556.

Cortina-Puig M., Munoz-Berbel X., Alonso-Lomillo A., Munoz-Pascual F., and del Valle M. (2007) EIS multianalyte sensing with an automated SIA system—An electronic tongue employing the impedimetric signal. *Talanta* **72**(2), 774–779.

Cosio M.S., Ballabio D., Benedetti S., and Gigliotti C. (2006) Geographical origin and authentication of extra virgin olive oils by an electronic nose in combination with artificial neural networks. *Anal. Chim. Acta* **567**(2), 202–210.

Dias L., Peres A., Vilas-Boas M., Rocha, M., Estevinho L., and Machado A. (2008) An electronic tongue for honey classification. *Microchim. Acta* **163**, 97–102. Diaz-Cruz J.M., Tauler R., Grabaric B., Esteban M., and Casassas E. (1995) Application of multivariate curve resolution to voltammetric data. Part 1. Study of Zn(II) complexation with some polyelectrolytes. *J. Electroanal. Chem.* **393**, 7–16.

Diaz-Cruz J.M., Tauler R., Grabaric B., Esteban M., and Casassas E. (1995) Application of multivariate curve resolution to voltammetric data. Part 1. Study of Zn(II) complexation with some polyelectrolytes. *J. Electroanal. Chem.* **393**, 7–16.

Di Natale C., Davide F., D'Amico A., Legin A., Rudinitskaya A., Selezenev B.L., and Vlasov Y. (1996) Applications of an electronic tongue to the environmental control. In: *Technical Digest of 10th European Conference Eurosensors X*, Leuven, Belgium, 1345–1348.

Di Natale C., Macagnano A., Davide F., D'Amico A., Legin A., Vlasov Y., Rudinitskaya A., and Selezenev B.L. (1997) Multicomponent analysis on polluted water by means of an electronic tongue. *Sens. Actuators B* **44**, 423–428.

Esbensen K. (2001) *Multivariate Analysis in Practice*, 5th ed. Camo ASA/CAMO Process AS, Norway.

Etoh S., Feng L., Nakashi K., Hayashi K., Ishii A., and Toko K. (2008) Taste sensor chip for portable taste sensor system. *Sensor Mater.* **20**(4), 151–160.

Fukunaga T., Toko T., Mori S., Nakabayashi Y., and Kanda M. (1996) Quantification of taste of coffee using sensor with global selectivity. *Sens. Mater.* **8**(1), 47–56.

George M., Parak W., and Gaub H. (2000) Highly integrated surface potential sensors. *Sens. Actuators B* **69**, 266–275.

Gotoh M. (2004) Disposable taste biosensor. *Chem. Sens.* **20**(Suppl. B), 486–487.

Gutes A., Cespedes F., del Valle M., Louthander D., Krantz-Rülcker C,. and Winquist F. (2006) A flow injection voltammetric electronic tongue applied to paper mill industrial waters. *Sens. Actuators B* **115**, 395–401.

Gutes A., Calvo D., Céspedes F., and del Valle M. (2007) Automatic sequential injection analysis electronic tongue with integrated reference electrode for the determination of ascorbic acid, uric acid and paracetamol. *Microchim. Acta* **157**(1–2), 1–6.

Gutierrez M., Moo V., Alegret S., Leija L., Hernandez P., Munoz R., and del Valle M. (2008a) Electronic tongue for the determination of alkaline ions using a screen-printed potentiometric sensor array. *Microchim. Acta* **163**(1–2), 81–88.

Gutierrez J.M., Gutes A., Cespedes. F., del Valle M., and Munoz R. (2008b) Wavelet neural networks to resolve the overlapping signal in the voltammetric determination of phenolic compounds. *Talanta* **76**(2), 373–381.

Habara M., Ikezaki H., and Toko K. (2004) Study of sweet taste evaluation using taste sensor with lipid/polymer membranes. *Biosens. Bioelectron.* **19**(12), 1559–1563.

Haber F. and Klemensiewicz Z. (1909) Ueber elektrische Phasengrenzkräfte. *Z. Phys. Chem.* **64**, 385–392.

Hayashi K., Yamanaka M., Toko K., and Yamafuji K. (1990) Multichannel taste sensor using lipid membranes. *Sens. Actuators B* **2**, 205–213.

Holmin S., Spångeus P., Krantz-Rülcker C., and Winquist F. (2001) Compression of electronic tongue data based on voltammetry—A comparative study. *Sens. Actuators B* **76**, 455–501.

Holmin S., Krantz-Rulcker C., and Winquist F. (2004) Multivariate optimisation of electrochemically pre-treated electrodes used in a voltammetric electronic tongue. *Anal. Chim. Acta* **519**(1), 39–46.

Hossenlopp J. (2006) Applications of acoustic wave devices for sensing in liquid environments. *Appl. Spectrosc. Rev.* **41**(2), 151–164.

Imamura T., Toko K., Yanagisawa S., and Kume T. (1996) Monitoring of fermentation process of *miso* (soybean paste) using multichannel taste sensor. *Sens. Actuators B* **37**, 179–185.

Ipatov A., Abramova N., Bratov A., and Dominguez C. (2008) Integrated multisensor chip with sequential injection technique as a base for "electronic tongue" devices. *Sens. Actuators B* **131**, 48–52.

Ivarsson P., Holmin S., Höjer N.-E., Krantz-Rülcker C., Lundström I., and Winquist F (2001a) Discrimination of tea by means of a voltammetric electronic tongue and different applied potential modes. *Sens. Actuators B* **76**, 449–454.

Ivarsson P., Kikkawa Y., Winquist F., Krantz-Rülcker C., Höjer N.-E., Hayashi K., Toko K., and Lundström I. (2001b) Comparison of a voltammetric electronic tongue and a lipid membrane taste sensor. *Anal. Chim. Acta* **449**, 59–65.

Ivarsson P., Johansson M., Höjer N.-E., Krantz-Rülcker C., Winquist F., and Lundström I. (2005) Supervision of rinses in a washing machine by a voltammetric electronic tongue *Sens. Actuators B* **108**, 851–858.

Iwakura M., Isomoto Y., Inoue T., and Toko K. (2003) Development of a flow-type taste sensor. *Sens. Mater.* **15**(7), 371–380.

Jacesko S., Abraham J., Taeksoo J., Varadan V., Cole M., and Gardner J. (2005) Investigations on an electronic tongue with polymer microfluidic cell for liquid sensing and identification, *Smart Mater. Struct.* **14**, 1101–1016.

Johnson H., Karlsson O., Winquist F., Krantz-Rülcker C., and Ekedahl L.-G. (2003) Predicting microbial growth in pulp using an electronic tongue. *Nord. Pulp. Pap. Res. J.* **18**, 134–141.

Kanai Y., Shimizu M., Uchida H., Nakahara H., Zhou C.G., Maekawa H., and Katsube T. (1994) Integrated taste sensor using surface photovoltage technique. *Sens. Actuators B* **20**, 175–179.

Kataoka M., Tokuyama E., Miyanaga Y., and Uchida T. (2008) The taste sensory evaluation of medicinal plants and Chinese medicines. *Int. J. Pharm.* **351**(1–2), 36–44.

Krantz-Rülcker C., Stenberg M., Winquist F., and Lundström I. (2001) Electronic tongues for environmental monitoring based on sensor arrays and pattern recognition: A review. *Anal. Chim. Acta* **426**, 226–232.

Legin A., Rudinitskaya A., Vlasov Y., Di Natale C., Davide F., and D'Amico A. (1996) Tasting of beverages using an electronic tongue based on potentiometric sensor array. In: *Technical Digest of Eurosensors X,* Leuven, Belgium, 427–430.

Legin A., Rudinitskaya A., Vlasov Y., Di Natale C., Mazzone E., and D'Amico A. (1999a) Application of electronic tongue for quantitative analysis of mineral water and wine. *Electroanalysis* **11**(10–11), 814–820.

Legin A., Smirnova A., Rudinitskaya A., Lvova L., Suglobova E., and Vlasov Y. (1999b) Chemical sensor array for multicomponent analysis of biological liquids. *Anal. Chim. Acta* **385**, 131–135.

Legin A., Rudnitskaya A., Lvova L., Vlasov Y., Di Natale C., and D'Amico A. (2003) Evaluation of Italian wine by the electronic tongue: Recognition, quantitative analysis and correlation with human sensory perception. *Anal. Chim. Acta* **484**, 33–44.

Legin A., Rudnitskaya A., Clapham D., Seleznev B., Lord K., and Vlasov Y. (2004) Electronic tongue for pharmaceutical analytics: Quantification of tastes and masking effects. *Anal. Bioanal. Chem.* **380**(1), 36–45.

Lvova L., Legin A., Vlasov Y., Cha G.S., and Nam H. (2003) Multicomponent analysis of Korean green tea by means of disposable all-solid-state potentiometric electronic tongue microsystem. *Sens. Actuators B* **95**, 391–399.

Lvova L., Martinelli E., Mazzone E., Pede A., Paolesse R., Di Natale C., and D'Amico A. (2006) Electronic tongue based on an array of metallic potentiometric sensors. *Talanta* **70**(4), 833–839.

Lvova L., Paolesse R., Di Natale C., and D'Amico A. (2006) Detection of alcohols in beverages: An application of porphyrin-based electronic tongue. *Sens. Actuators B* **118**, 439–447.

Lvova L., Martinelli E., Dini F., Bergamini A., Paolesse R., Di Natale C., and D'Amico A. (2009) Clinical analysis of human urine by means of potentiometric electronic tongue. *Talanta* **77**, 1097–1104.

Martina V., Ionescu K., Pigani L., Terzi F., Ulrici A., Zanardi C., and Seeber, R. (2007a) Development of an electronic tongue based on a PEDOT-modified voltammetric sensor. *Anal. Bioanal. Chem.* **387**, 2101–2110.

Martina V., Ionescu K., Pigani L., Terzi F., Ulrici A, Zanardi C., and Seeber R. (2007b) Development of an electronic tongue based on a PEDOT-modified voltammetric sensor. *Anal. Bioanal. Chem.* **387**, 2110–2115.

Men H., Zou S., Li Y., Wang Y., Ye X., and Wang P. (2005) A novel electronic tongue combined MLAPS with stripping voltammetry for environmental detection. *Sens. Actuators B* **110**, 350–355.

Menditeta J., Diaz-Cruz M.S., Tauler R., and Esteban M. (1996) Application of multivariate curve resolution to voltammetric data. Part 2. Study of metal-binding properties of the peptides. *Anal. Biochem.* **240**, 134–141.

Moreno-Baron L., Cartas R., Merkoci A., Alegret S., Gutierrez J., Leija L., Hernandez P., Muñoz R., and del Valle M. (2005) Data Compression for a voltammetric electronic tongue modeled with artificial neural networks. *Anal. Lett.* **38**, 2189–2193.

Moreno-Baron L., Cartas R., Merkoci A., Alegret S., del Valle M., Leija L., Hernandez P.R., and Muñoz R. (2006) Application of the wavelet transform coupled with artificial neural networks for quantification purposes in a voltammetric electronic tongue. *Sens. Actuators B* **113**, 499–503.

Moreno L., Merlos A., Abramova N., Jimenez C., and Bratov A. (2006b) Multi-sensor array used as an " electronic tongue" for mineral water analysis. *Sens. Actuators B* **116**, 130–134.

Moreno i Codinachs L., Kloock J., Schoening M., Baldi A., Ipatov A., Bratov A., and Jimenez-Jorquera C. (2008) Electronic integrated multisensor tongue applied to grape juice and wine analysis. *Analyst* **133**, 1440–1448.

Mortensen J., Legin A., Ipatov A., Rudnitskaya A., Vlasov Y., and Hjuler K. (2000) A flow injection system based on chalcogenide glass sensors for the determination of heavy metals. *Anal. Chim. Acta* **403**, 273–277.

Mottram T., Rudnitskaya A., Legin A., Fitzpatrick J., and Eckersall D. (2007) Evaluation of a novel chemical sensor system to detect clinical mastitis in bovine milk. *Biosens. Bioelectron.* **22**(11), 2689–2693.

Murakami Y., Kikuchi T., Yamamura A., Sakaguchi T., Yokoyama K., Ito Y., Takiue M., Uchida H., Katsube T., and Tamiya E. (1998) An organic pollution sensor based on surface photovoltage. *Sens. Actuators B* **53**, 163–172

Olsson J., Winquist F., and Lundstrom I. (2006) A self polishing electronic tongue. *Sens. Actuators B* **118**, 461–465.

Olsson J., Ivarsson P., and Winquist F. (2008) Determination of detergents in washing machine wastewater with a voltammetric electronic tongue. *Talanta* **76**, 91–95.

Otto M. and Thomas J.D.R. (1985) Model studies on multiple channel analysis of free magnesium, calcium, sodium and potassium at physiological concentration levels with ISE. *Anal. Chem.* **57**, 2647–2651.

Parra V., Arrieta A., Fernandez-Escudero J., Iniguez M., De Saja J.A., and Rodriguez-Mendez M.L. (2006) Monitoring of the ageing of red wines in oak barrels by means of an hybrid electronic tongue. *Anal. Chim. Acta* **563**(1–2), 229–237.

Persaud K. and Dodd G. (1982) Analysis of discrimination mechanisms in the mammalian olfactory system using a model nose. *Nature* **299**, 352–355.

Pioggia G., Di Francesco F., Ferro M., Sorrentino F., Salvo P., and Ahluwalia A. (2008) Characterization of a carbon nanotube polymer composite sensor for an impedimetric electronic tongue. *Microchim. Acta* **163**(1–2), 57–62.

Rodriguez-Mendez M.L., Parra V., Apetrei C., Villanueva S., Gay M., Prieto N., Martinez J., and de Saja, J. (2008) An electronic tongue based on voltammetric electrodes modified with materials showing complementary electroactive properties. Applications. *Microchim. Acta* **163**(1–2), 23–31.

Rudnitskaya A., Kirsanov D., Legin A., Beullens K., Lammertyn J., Nicolai B., and Irudayaraj J. (2006) Analysis of apples varieties—Comparison of electronic tongue with different analytical techniques. *Sens. Actuators B* **116**, 23–28.

Riul A., Malmegrim R., Fonseca F., and Mattoso L. (2003) An artificial taste sensor based on conducting polymers. *Biosens. Bioelectron.* **18**(11), 1365–1369.

Ruzicka A. and Hansen E. (1975) Flow injection analyses, I. New concepts of fast continuous flow analysis. *Anal. Chim. Acta* **78**, 145–157.

Sasaki Y, Kanai Y., Uchida H., and Katsube T. (1995) Highly sensitive taste sensor with a new differential LAPS method. *Sens. Actuators B* **24–25**, 819–822.

Sehra G., Cole M., and Gardner J.W. (2004) Miniature taste sensing system based on dual SH-SAW sensor device: an electronic tongue. *Sens. Actuators B* **103**, 233–239.

SensET AB, Linköping, Sweden, http://www.senset.se.

Simons J., Bos M., and van der Linden W.E. (1995) Data processing for amperometric signals. *Analyst* **120**, 1009–1012.

Söderström C., Boren H., Winquist F., and Krantz-Rülcker C. (2003a) Use of an electronic tongue to analyze mold growth in liquid media. *Int. J. Food Microbiol.* **83**, 253–260.

Söderström C., Winquist F., and Krantz-Rulcker C. (2003b) Recognition of six microbial species with an electronic tongue. *Sens. Actuators B* **89**, 248–255.

Söderström C., Boren H., Winquist F., and Krantz-Rülcker C (2003c) Use of an electronic tongue to analyze mold growth in liquid media. *Int. J. Food Microbiol.* **83**, 253–260.

Söderstrom C., Rudnitskaya A., Legin A., and Krantz-Rulcker C. (2005a) Differentiation of four Aspergillus species and one Zygosaccharomyces with two electronic tongues based on different measurement techniques. *J. Biotech.* **119**(3), 300–308.

Söderstrom C., Boren H., and Krantz-Rulcker C. (2005b) Use of an electronic tongue and HPLC with electrochemical detection to differentiate molds in culture media. *Int. J. Food Microbiol.* **97**(3), 247–257.

Taniguchi A., Naito Y., Maeda N., Sato Y., and Ikezaki H. (1999) Development of a monitoring system for water quality using a taste sensor. *Sens. Mater.* **11**(7), 437–446.

Taste Sensing System SA401, Anritsu Corp., Japan, http://www.anritsu.com.

The Astree Liquid & Taste Analyzer, Alfa MOS, Toulouse, France, http://www.alpha-mos.com.

Tian S., Deng S., and Chen Z.X. (2007) Multifrequency large amplitude pulse voltammetry: A novel electrochemical method for electronic tongue. *Sens. Actuators B* **123**, 1049–1053.

Toko K., Hayashi K., Yamanaka M., and Yamafuji K. (1990) Multichannel taste sensor with lipid membranes. In: *Technical Digest of 9th Sensor Symposium,* Institute of Electrical Engineers of Japan, Tokyo, Japan, 193–196.

Toko K. (1998) Electronic tongue. *Biosens. Bioelectron.* **13**, 701–709.

Toko K. (2000) *Biomimetic Sensor Technology*. Cambridge University Press, Cambridge, UK.

Uchida T., Tanigake A., Miyanaga Y., Matsuyama K., Kunitomo M., Kobayashi Y., Ikezaki H., and Taniguchi A. (2003) Evaluation of the bitterness of antibiotics using a taste sensor. *J. Pharm. Pharmacol.* **55**(11), 1479–1485.

Verrelli G., Lvova L., Paolesse R., di Natale C., and D'Amico A. (2007a) Metalloporphyrin-based electronic tongue: An application for the analysis of Italian white wines. *Sensors* **7**(11), 2750–2762.

Verrelli G., Francioso L., Paolesse R., Siciliano P., Di Natale C., D'Amico A., and Logrieco A. (2007b) Development of silicon-based potentiometric sensors: Towards a miniaturized electronic tongue. *Sens. Actuators B* **123**, 191–197.

Vlasov Y., Rudniskaya A., Di Natale C., and D'Amico A. (2005) Nonspecific sensor arrays for chemical analysis of liquids (IUPAC Technical Report). *Pure. Appl. Chem.* **77**, 1965–1983.

Wang J. (1994) *Analytical Electrochemistry*. Wiley-VCH, New York.

Winquist F., Wide P., and Lundström I. (1997) An electronic tongue based on voltammetry. *Anal. Chim. Acta* **357**, 21–31.

Winquist, F., Krantz-Rülcker C., Wide P., and Lundström I. (1998) Monitoring of milk freshness by an electronic tongue based on voltammetry *Meas. Sci. Technol.* **9**, 1937–1946.

Winquist F., Lundstrom I., and Wide P. (1999) The combination of an electronic tongue and an electronic nose. *Sens. Actuators B* **58**, 512–517.

Winquist F., Holmin S., Krantz-Rülcker C., Wide P., and Lundström I. (2000) A hybrid electronic tongue. *Anal. Chim. Acta* **406**, 157–162.

Winquist F., Rydberg E., Holmin S., Krantz-Rülcker C., and Lundström I. (2002) Flow injection analysis applied to a voltammetric electronic tongue. *Anal. Chim. Acta* **471**, 172–179.

Winquist F, Krantz-Rülcker C., and Lundström I. (2004c) Electronic tongues. *MRS Bull.* **29**,726–732.

Winquist F., Bjorklund R., Krantz-Rülcker C., Lundström I., Östergren K., and Skoglund T. (2005) An electronic tongue in the dairy industry. *Sens. Actuators B* **112**, 299–309.

Winquist F. (2008a) Electronic tongues. *Microchim. Acta* **163**, 3–10.

Winquist F., Krantz-Rülcker C., and Lundström I. (2008b) A miniaturized voltammetric electronic tongue. *Anal. Lett.* **41**(5), 917–924.

Winquist F., Krantz-Rülcker C., Olsson T., and Jonsson A. (2009) Measurement of cadmium in soil extracts using multivariate data analysis and electrochemical sensors. *Precis. Agric.* **10**, 231–246.

Wu J., Liu J., Fu M., Li G., and Lou Z. (2005) Classification of Chinese yellow wines by chemometric analysis of cyclic voltammogram of copper electrode. *Sensors* **5**, 529–536.

Yamada H., MizotaY., Toko K., and Doi T. (1997) Highly sensitive discrimination of taste of milk with homogenization treatment using a taste sensor. *Mater. Sci. Eng. C* **5**, 41–45.

Young-Soo S., Goodey A., Anslyn E., McDevitt J.T., Shear J.B., and Neikirk D.P. (2004) A micromachined fluidic structure for capillary-based sample introduction into a microbead array chemical sensor. *Sensor Lett.* **2**, 69–72.

Young-Soo S., Goodey A., Anslyn E., McDevitt J.T., Shear J.B., and Neikirk D.P. (2005) A microbead array chemical sensor using capillary-based sample introduction: Toward the development of an "electronic tongue." *Biosen. Bioelectron.* **21**, 303–312.

CHAPTER 3

WIRELESS CHEMICAL SENSORS

A. Flammini
S. De Vito

1. WIRELESS SENSORS: AN INTRODUCTION

Chemical sensors are currently used in several application fields, from home automation to industrial process control, health, and agriculture. Usually, chemical sensors are connected to an acquisition system. This extracts information from the output signal of chemical sensors and elaborates such data together with the information coming from other sensors (e.g., temperature sensors, humidity sensors) and also using "*a priori*" knowledge. This is the architecture of a typical electronic nose, in which the acquisition system has the additional task of providing the needed operating conditions. This centralized architecture (Figure 3.1a) can be network-enabled through the acquisition system and can be realized in a low-cost fashion (Branzila et al. 2007).

A modular approach can improve the adaptability of a centralized architecture (Depari et al. 2007), but a true, versatile, flexible, and scalable solution can be obtained only if the centralized architecture can be replaced by a distributed one (Figure 3.1b), with great advantages in terms of signal-to-noise ratio, significance of the measurement, reliability, and scalability.

Some of the main advantages are the following.

- In a distributed architecture, every sensor has its own "intelligence"; this way the sensing element is close to the elaborating unit, improving the signal-to-noise ratio.
- Several sensors of the same type can be placed in different locations, each with its own characteristics in terms of environmental effects (air flow, temperature, humidity, different gas

concentration, etc.). Data fusion methods can be used to merge information coming from all these equivalent sensors in order to provide more precise, localized, and reliable information.

- Each smart chemical sensor comprising the distributed architecture has communication capability, and its information is available for more than one client. This way, the sensor information can be used for multiple aims. Considering, for instance, a home environment, data from a CO_2 sensor can be used not only for safety purposes, but for better management of fan coils as well (Chung et al. 2006).
- The network can adapt itself to a variable number of chemical sensors, improving reliability. If a sensor fails, the network can estimate its response on the basis of previous behavior and the response of the closest sensors (Pan et al. 2007a). Also, sensor calibration can be performed by temporarily adding other sensors and a specific elaboration system that, thanks to data fusion techniques, can recalibrate the considered sensor.
- Sensor redundancy is easier to implement, providing diagnostic facilities and improving sensor metrological characteristics. Furthermore, a distributed architecture means that every sensor has computing capabilities; it can provide diagnostic services and elaborate strategies, for instance, to improve the stability of its metrological characteristics.

For all these reasons, a distributed architecture is better than a centralized one and, as a consequence, a "smart" chemical sensor is preferable with respect to a "raw" sensor; such "smart" behavior can be obtained with either wired or wireless communication [IEEE1451.0-2007, IEEE Standard for a Smart Transducer Interface for Sensors and Actuators—Common Functions, Communication Protocols, and Transducer Electronic Data Sheet (TEDS) Formats].

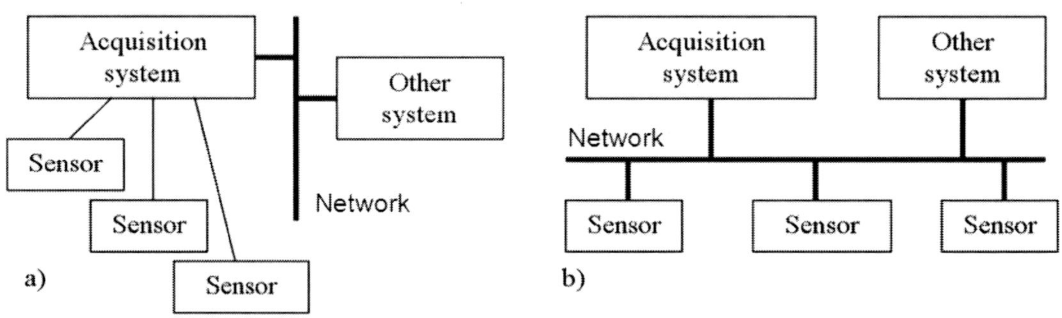

Figure 3.1. Chemical sensors in (a) a centralized architecture and (b) a distributed one.

1.1. SMART SENSORS: FROM WIRED TO WIRELESS TECHNOLOGIES

A block diagram of a "smart" chemical sensor is shown in Figure 3.2. It is composed of five sections: the power supply, one or more sensing/actuating elements, electronic circuits, an elaborating unit (typically a low-cost microcontroller), and a communication interface (typically a receiving and transmitting device, i.e., a transceiver).

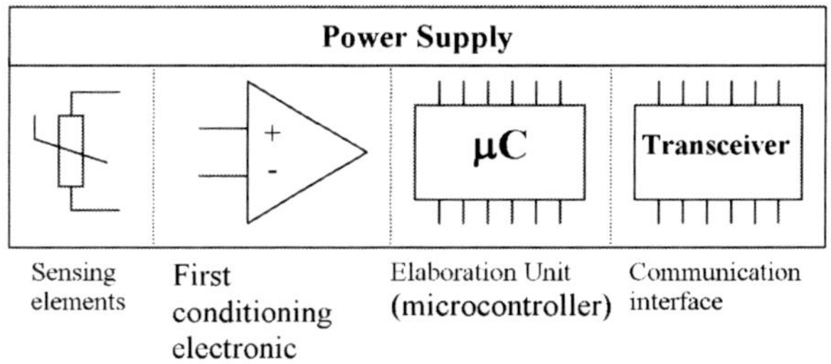

Figure 3.2. Architecture of a smart chemical sensor.

The sensing/actuating element section is composed mainly of the chemical sensor, but it can also include one or more heaters, temperature and/or humidity sensors, etc.

First conditioning electronic circuits are generally provided in order to interface the sensing/actuating elements to the elaboration unit. Low noise is one of the most common requirements of this section, together with a wide measuring range, in order to improve the readout resolution. In fact, the resolution should be independent from the sensor working point, which can be due to the chemical substance but also to sensor realization tolerances and sensor aging effects (Frey et al. 2008; Ferri et al. 2008). The microcontroller is used to capture and elaborate information from the sensing elements, but also to control the sensor temperature by tuning the correct current amount in the heater, or to perform additional functions, such as diagnostics or other digital signal processing techniques. For instance, if a metal oxide (MOX) chemical sensor is considered, sensor information can be retrieved by measuring the sensor resistance, but also sensor capacitance and the rate of sensor resistance can provide information about the correctness of the sensor operating mode and, more generally, about the sensor behavior.

The communication unit is often composed of a single device, the transceiver, that allows data transmission and reception on the selected media (wired or wireless). According to the ISO-OSI model (Hayyes 1988) (see also Section 1.4), the transceiver implements at least the physical interface. Normally it also provides connection to higher layers of the communication stack, such as the data link layer. For instance, the transceiver generally manages the media access control, allowing the microcontroller not to care about the state of occupancy of the communication medium. If a wired communication method is selected, such as USB or Ethernet, or the fieldbuses used in homes, buildings, and industrial automation (e.g., CANbus, ProfiBus, DeviceNet, EIB, LonWorks, ModBus, ModBus [IEC/TR 61158-1 Ed. 2.0, Industrial Communication Networks—Fieldbus Specifications—Part 1: Overview and Guidance for the IEC 61158 and IEC 61784 Series, 14 Nov. 2007], the chemical sensor position cannot be changed as easily.

It should be highlighted that the power supply unit may be the most critical block of a smart chemical sensor, in that it should provide different and very stable power supply voltages and currents to all the units. For instance, the sensing elements and the first conditioning electronic circuits need a very low-noise power supply and ground reference, which may not be easy to implement in the same circuit where a heater must be supplied. The source of power may be the main supply, batteries, or, in some cases, the power can be recovered by the same cable used for communication. A typical example is USB (USB—Universal Serial Bus Rev 2.0 Specification, 27 April 2000) (Depari et al. 2008); a USB port can provide up to 2.5 W (500 mA @ 5 V) on a specific connector pin for direct sensor powering. Ethernet can also provide power to a generic device, according to the more general method known as Power over LAN. The project Power over Ethernet, also known as IEEE 802.3af (IEEE 802.3-2005 LAN/MAN CSMA/CD Access Method), allows power retrieval directly from the data lines (Ethernet 10/100/1000 Mbit/s) or from specific connector pins dedicated to power transmission (Ethernet 10/100 Mbit/s only). Another very interesting method which uses the same cable for sensor powering and communicating is power line communication, also known as power line carrier or mains communication (UPA—Universal Power line Association) (Wall et al. 2005). In this case, data is carried on a conductor that is normally used for electric power transmission by impressing a modulated carrier signal on the wiring system. Different types of power line communications use different frequency bands; broadband communications are receiving special attention, especially in home applications, where chemical sensor networking could have an interesting application field (for more detailed information, see also IEEE P1901 Draft Standard for Broadband over Power Line Networks).

More generally, a smart chemical sensor can be wired or wireless according to different schemes, as depicted in Figure 3.3. Figure 3.3a shows a smart chemical sensor with a wired communication interface, also called a wired chemical sensor; power supply is derived from mains. Figure 3.3b represents a wired chemical sensor with power recovered by the wired interface, directly from the data lines or from

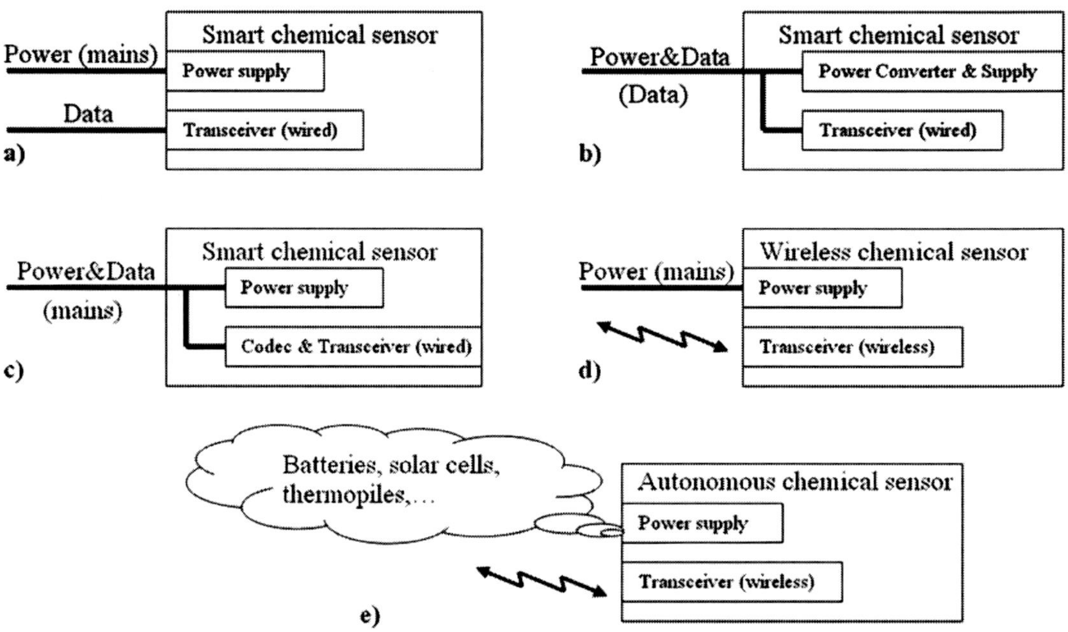

Figure 3.3. Possible schemes for a smart chemical sensor: (a) smart chemical sensor with wired communication interface (wired chemical sensor) and power supply from mains; (b) wired chemical sensor with power recovered by this interface (e.g., USB); (c) wired chemical sensor with data coded in the power interface; (d) smart chemical sensor with wireless interface and wired connection for power supply; (e) autonomous chemical sensor with no wires, the sensor is movable.

specific connector pins dedicated to power transmission (e.g., USB). Figure 3.3c is a block diagram of a wired chemical sensor with data coded in the power interface, according, for instance, to Power Line Communication technologies. In Figure 3.3d the smart chemical sensor has a wireless communication interface, but wires are still necessary for power supply. This is the case for MOX sensors, where the power needed by the heater essentially precludes the use of batteries. The sensor can be moved, but only where there is a power source; this is the situation in industrial, building, and home applications, where power is present almost anywhere. In this case, advantages with respect to the scheme depicted in Figure 3.3c are in a more reliable and fast communication system and a better integration with information and communication technologies such as personal computers, Internet, and Web technologies. A true mobile smart chemical sensor is shown in Figure 3.3e, where a wireless communication interface has been adopted and the sensor is powered without the need for wires, simply using energy stored in batteries or in supercapacitors, powered by solar cells or other forms of energy harvesting. It should be highlighted that a distributed architecture can include both autonomous mobile sensors and wireless sensors, as the only requirement is to share the communication interface protocol.

1.2. WIRELESS SENSOR ARCHITECTURE

Figure 3.4 shows a block diagram of an autonomous sensor, which is a special case of the more general scheme of Figure 3.2 (Bicelli et al. 2005) As stated before, one of the critical requirements of an autonomous sensor is power saving, in order to prolong sensor life. Power saving often guides sensor choice and, for instance, sensors operating at room temperature are usually preferred. More generally, to save power, the sensing elements and the related circuits (interface electronic circuits, microcontroller, transceiver) are turned off when they are not used. This strategy, which is normally used with most kinds of wireless sensors (temperature, humidity, mechanical quantities, etc.), may present some problems with chemical sensors. In fact, for instance, it is well known that MOX chemical sensors require a long warm-up period to ensure reliable and repeatable behavior, thus they should be continuously powered and heated, whereas the interface circuits, the microcontroller, and the transceiver can be turned on only periodically or as needed.

Sensor life depends directly on battery life. A battery has an energy storage or capacity *C* that is normally computed in terms of ampere-hours (A h), supposing the battery to be able to provide a stable and constant voltage for an infinite time. For instance, if a battery has a capacity of 1 A h and the wireless sensor has a mean current consumption $I_{cc,\mathrm{mean}}$ of 100 μA, the battery life should be equal to 10,000 h.

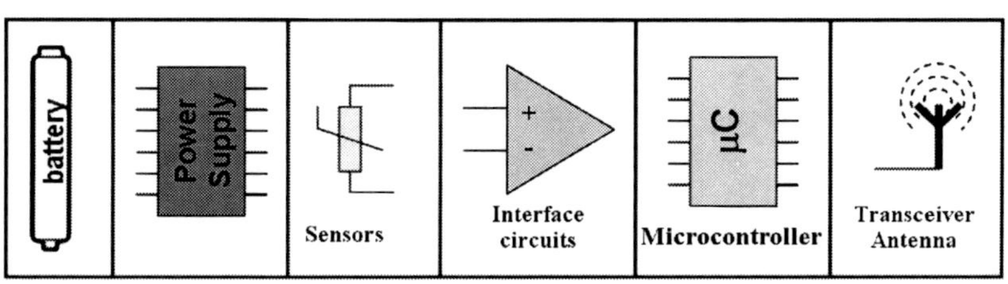

Figure 3.4. Architecture of an autonomous sensor.

Real batteries behave differently, however, and their discharge follows different profiles, according to the discharging mode: burst, continuous, with constant current, continuous with constant power, etc. (Zheng et al. 2001; Matsushima et al. 2006).

Low-cost batteries provide a voltage of the order of 1 V, while electronic circuits normally operate at 3.3 V, so an additional power supply unit is often used. Some batteries (lithium cells) provide an output voltage of about 3 V; if the power and the voltage are suitable to supply the chemical sensor and all the related circuits (heater, interface circuits, etc.), an extra power supply can be avoided and the batteries can be connected directly to the circuits. The power supply unit, which is normally composed of a single-chip low-power step-up converter, is often present also to allow use of the battery toward the end of its life, when it provides very low voltage. This power supply is in fact a sort of power converter to obtain a constant voltage V_{out} of the order of 3 V; particularly, it converts V_{in}, I_{in} from the power source (e.g., batteries) into the values V_{out}, I_{out} needed by the electronic circuits (interface circuits, microcontroller, and transceiver). The power supply may have more sections to provide different values of output voltage; for each section, a voltage ratio $K_V = V_{out}/V_{in}$ and an efficiency $\eta = (V_{out}I_{out})/(V_{in}I_{in})$ can be computed. Starting from these parameters, the capacity C of the battery and an estimate of the mean current consumption $I_{cc,\mathrm{mean}}$ of the wireless sensor, the following equation can be used to roughly estimate sensor life L:

$$L = \frac{\eta C}{K_V I_{cc,\mathrm{mean}}} \tag{3.1}$$

Regarding the conditioning circuits, analog devices must have "rail-to-rail" behavior to exploit the whole signal dynamic and to offer very low bias current. If the interface circuits are turned off when they are not used, as is generally done, it is important that they have fast recovery—that is, they need to reach stable behavior quickly when they are turned on. Therefore, the use of heavy analog low-pass filters should be discouraged, so there is no need to achieve a stable and accurate measurement. Well-tailored digital low-pass filters are normally preferred (Bicelli et al. 2008).

The power consumption of a microcontroller can be reduced by using a low-frequency clock; however, then the microcontroller must be turned on for a long time in order to perform a given task. A smarter design is to dynamically adapt the clock frequency according to CPU load demand. A low-power microcontroller with fast wake-up procedures, data processing, and protocol managing should be preferred.

The transceiver should be turned on only when it is needed, that is, when transmitting or receiving. It should be noted that, for a given frequency band, the transmission power is similar for most of the wireless technologies. The difference is in the low-power modes, that is, power consumption when no transmission or reception occurs, and in the time spent to recover from the low-power to the active mode. Thus, it is clear how a simple communication protocol implies a power consumption decrease. Another important aspect to point out is that reception may consume more power than transmission; in fact, the sensor can turn the transceiver on at the proper time to transmit, but it must turn the transceiver on and wait for the reception (overhearing). For this reason, good synchronization can decrease the sensor power consumption (Cox et al. 2005).

In conclusion, the sensor life, which is the most critical factor in an autonomous battery-powered sensor, can be estimated starting from the mean current consumption $I_{cc,\mathrm{mean}}$ of the wireless sensor.

Supposing that the sensor wakes up periodically (period = T_m) and performs a measurement taking a time T_a and, if necessary, communicates data taking a time T_{RF} with a mean communication period of T, then the mean current consumption $I_{cc,\text{mean}}$ can be computed according to Eq. (3.2), where some assumptions have been made: I_0 is the power consumption of the elements that must be continuously turned on (e.g., a chemical sensor with its heater); the elements that must be turned on during the measuring phase (e.g., microcontroller, interface circuits) have a contribution of I_a during the measurement phase and of I_{as} otherwise; the element (transceiver) that must be turned on during the communication phase has a contribution of I_{RF} during the active phase and of I_{RFs} otherwise.

$$I_{cc,mean} = I_0 + I_{as}\left(1 - \frac{T_a}{T_m}\right) + I_{RFs}\left(1 - \frac{T_{RF}}{T}\right) + I_a\frac{T_a}{T_m} + I_{RF}\frac{T_{RF}}{T} \tag{3.2}$$

Many methods can be implemented to dynamically adapt $I_{cc,\text{mean}}$ to the residual energy, according to adaptive sensing strategies (Jain and Chang 2004; Alippi et al. 2009). An adaptive sensing strategy can increase times T_m and T if the battery voltage is close to its minimum value, or can reduce the number of sensor samples or the measuring accuracy, limiting the required processing time and the length of the packet to be communicated.

1.3. BATTERIES, POWER HARVESTING, AND POWER SCAVENGING

Electrochemical batteries are the most common power source in wireless sensor networks, since they are perhaps the most economic and versatile solution. A low-cost alkaline manganese battery provides about 1.5 V with a capacity of more than 1 A h; this means that this source of energy can continuously supply about 100 μA for more than 1 year. Lithium batteries are a more powerful primary source of energy, but their cost is still high. Rechargeable batteries (such as lithium ion or nickel metal-hydride) can be used, especially if there are external sources of power that can be used for battery recharge. This is called energy harvesting (also known as power harvesting or energy scavenging); it is the process by which energy is derived from external sources (e.g., solar power, thermal energy, wind energy, salinity gradients, or kinetic energy such as vibration), captured, and stored in a capacitor, supercapacitor, or rechargeable battery. Capacitors are used when the application needs to provide short bursts of energy, while batteries are employed when a steady flow of energy is needed. Use of supercapacitors (also called ultracapacitors) is growing thanks to their low cost and high energy density; they are electrochemical double-layer capacitors (EDLCs) with a capacity of thousands of farads. Rechargeable batteries, supercapacitors, or capacitors are secondary power sources, and they need primary power to be charged. The easiest primary power source is the power grid, but this approach may be unsuitable for some applications of autonomous chemical sensors. Fuel cells and, particularly, micro fuel cells are likely to replace batteries thanks to their very high energy density. Fuel cells normally use hydrogen atoms as fuel, but hydrocarbons such as methanol can also be used. Fuel cells are a promising research sector, although many other ways to power wireless sensors are also under investigation.

In the following, some techniques for power transmission or power recovery from the environment (power harvesting or power scavenging) are briefly presented (Roundy et al. 2003; Mateu and Moll 2005).

1.3.1. Radiofrequency (RF) Power Transmission

An available product, in which a sensor network is powered by a low-frequency electromagnetic field in an industrial cell, with distances of the order of a few meters, is described by Scheible et al. (2007). Received power is proportional to transmitted power and depends directly on the square of the wavelength and on the inverse of R^{α}, where R is the distance between the transmitter and the receiver and typically is $2 < \alpha < 5$ (depending on the transmitting medium, reflections, etc.). For these reasons, electromagnetic power recovered by data transmission, normally at 2.4 GHz, can be used only for very short distances (RFID technologies operate at lower frequencies).

1.3.2. Solar Cells

Light is another form of energy that can be quite easily transmitted; for instance, a lamp can be used to power a wireless sensor (Brunelli et al. 2005). As in the case of RF power transmission, in light transmission the received power also depends on the inverse of distance and, for this reason, the quantity of available power can be optimized in an outdoor environment, where the direct powerful sunlight makes negligible distance variations (on a sunny day, the power density is of the order of 100 mW/cm^2, while in an illuminated indoor environment it can vary from 100 μW/cm^2 down to less than 1 μW/cm^2). Solar harvesting by means of solar cells is a quite mature technology, and most wireless sensor networks for outdoor applications are normally powered this way. On the other hand, research teams are investigating solar cells with better efficiency for indoor applications (thin-film amorphous silicon or cadmium telluride cells), where the harvested power is still quite small.

1.3.3. Vibrations

Vibrations and, more generally, mechanical energy can be converted into electrical energy by means of piezoelectric materials (Ottman et al. 2003). Vibrations are present in industrial plants, but also in automobiles, trains, and large commercial buildings. Voltages furnished by a piezoelectric generator are normally very small (millivolts) and should be rectified, requiring additional electronic circuits. For this reason, mechanical energy to be harvested should be relevant and continuously present, as it occurs in industrial plants or regarding human body. Researchers are investigating energy recovery from blood pressure or from human body movements (e.g., walking). Commercial piezoelectric harvesters are available for industrial and automotive applications (www.ambiosystems.com).

1.3.4. Thermoelectric

The temperature difference between two adjacent materials can be used, exploiting the Seebeck effect. Best performance can be achieved using materials with high Seebeck coefficient, high electrical conductivity, and low thermal conductivity, the latter in order to maintain a high thermal gradient. Commercial products are available, producing tens of microwatts at 3 V from a 5°C temperature difference using a 5-cm^2-wide module (www.poweredbythermolife.com). Even in this case, the human body is an attractive source of energy, as its temperature is almost constant and normally with a good gradient with respect to the environment.

1.3.5. Other

Power can be harvested in other ways, such as from radioactive substances, electrochemical reactions, acoustic energy, or from wind, but at the present state of the art, the most popular and efficient way to recover energy for wireless sensor networks is the use of solar cells for outdoor applications. For indoor applications, there are some specific cases where piezoelectric generators or thermopiles or solar cells are advantageous, but RF power transmission and batteries are the most often used solutions. Recent research activities concern harvesting power from the human body, which is a source of thermal energy, mechanical energy, and vibrations. Recently, biofuel cells have been announced which are based on oxidization of blood glucose (Holmes 2007). The growing interest in harvesting power from the human body is because wearable electronic wireless chemical sensors might be employed in this field for health monitoring purposes, together with temperature and blood pressure sensors (Starner and Paradiso 2004).

1.4. COMMUNICATION PROTOCOL STACK

A wireless chemical sensor is usually an autonomous sensor node within a network that is normally managed by a master or coordinator, usually main-powered. In some applications, where heated sensors are employed, the sensor nodes are also main-powered (wireless chemical sensors). The ensemble of rules that specify the behavior of every network device is called the protocol stack. A well-accepted classification of the tasks performed by such a stack is the ISO/OSI model (Hayyes 1988). This model defines seven layers, each of which has dedicated functions.

The lowest layer is the physical one, which covers the physical part of the communication. For instance, it contains all the hardware specification data, such as the modulation scheme and other data. Additional details about this layer are reported in following sections, where some radio technologies are briefly described.

All nodes sharing the same radio domain are called neighbors and are usually identified by a unique address within the whole network. The formation and maintenance of such direct links is a task of the second layer of the ISO/OSI stack, called the data link layer. In particular, the protocol at this layer tries to ensure the reliability of the link, solving packets collision/corruption that may occur in a shared unreliable medium such as air.

The third layer of the ISO/OSI stack is the so-called network layer. A network differs from an ensemble of nodes in that a communication link between two nodes may involve more nodes that cooperate to that link. It should be noted that some simple protocol stacks are defined only at the physical layer and the data link layer, while the network layer has great importance in mesh networks, as described later.

Usually, in wireless sensor networks, upper layers are not implemented, except for the application layer (the seventh layer of the ISO/OSI model), which provides an interface that can be used by the application. For example, it provides access to distributed databases or other high-level services. The application layer depends strongly on what the applications (or the operating system) need, and it is usually designed to meet these requirements.

1.4.1. The Physical Layer

In wireless communications, the cable is replaced by a frequency channel, which is a frequency bandwidth where communication can happen. It is well known that user data must be modulated in order to be transmitted over the air. More precisely, modulation is the process of varying a periodic waveform, usually a high-frequency sinusoid called the carrier, in order to use such a signal to convey a message. The carrier can be amplitude, frequency, or phase (or a combination of all) modulated, referring to which of the parameters of the carrier is changed to carry information. Wireless communications normally adopt digital and quite complex modulations, whose aim is to ensure a fast data rate and good noise immunity—that is, good resistance to interference with other RF sources operating at nearby frequencies. Narrow band radios normally operate on a single center frequency and occupy only the bandwidth required to accommodate their data transmission requirements. On the opposite side, the UWB (Ultra Wide Band) solutions, operating at higher frequencies and with wider frequency channels, are more robust to interference and allow a higher throughput, at the expense of more complicated hardware and smaller area coverage. There are other solutions between these two extremities, such as FHSS (Frequency Hopping Spread Spectrum) and DSSS (Direct Sequence Spread Spectrum). Simply speaking, frequency hopping radios move a narrowband communication over different frequencies in different time instants, following a quasi-random pattern, while direct-sequence radios uses random noise–like signals to modulate a narrowband signal with the effect of spreading the latter over a wider band of frequencies. Another popular technique is OFDM (Orthogonal Frequency Division Multiplexing), which generates a large number of narrow signals adjacent to each other and sharing the overall data payload. In DSSS systems, the user bit stream is grouped into symbols and the symbols are transmitted using sequence codes (PN, Pseudorandom Noise sequences) formed by several binary units called chips. The chip rate, higher than the bit rate, is limited by the width of the frequency channel.

1.4.2. The Data Link Layer

The protocol at the data link layer is responsible for the reliability of the link, identifying nodes with a local address and solving packet collisions or corruptions that may occur. For this reason, error-correction algorithms are used and redundant information is added in the transmitted data. In the unfortunate event of a packet loss, the protocol may deal with it through some contention-resolution algorithm, for example, by resending the message later, after a randomly selected time. Alternatively, the protocol may simply discard the message and leave the retransmission strategy—if any—to the higher layers in the protocol stack. Traditionally, Medium Access Control (MAC) approaches can adapt two kinds of strategies; a fixed scheduling of the accesses or a contention among all nodes that want to use the medium. The former approach is well suited for cyclic data transmission, as is typically required in industrial applications, while the latter is more efficient with sporadic data, as it occurs in home automation. In other words, continuous monitoring of a physical quantity is usually performed by means of a schedule-based MAC protocol, which is inherently collision-free and ensures great efficiency if the synchronization among nodes is ensured; in fact, overhearing is avoided, since receiving nodes (sinks) know beforehand when to expect incoming data. On the contrary, event notifications, such as alarms, are better handled with contention-based protocols, which are extremely flexible regarding the time, location, and amount of data transferred by individual nodes, easily handling the traffic fluctuations. Since the data link layer controls the radio, it has a large impact on the overall energy consumption, and hence the lifetime of a node. A brief summary of MAC protocols for wireless sensor networks is given by Demirkol et al. (2006). In particular, in schedule-based MAC protocols, a central authority (the access point or the network coordinator) regulates access to the medium by broadcasting a schedule specifying when, and for how long, each controlled node may transmit over the shared channel. Since there is no contention overhead, the network cannot collapse under high loads. In addition, nodes get deterministic access to the medium and can provide delay-bounded services. Typical protocols of this class are Time Division Multiple Access (TDMA), Frequency Division Multiple Access (FDMA), and Code Division Multiple Access (CDMA). The TDMA scheme subdivides the time axis into fixed-length super-frames, that is, a repetitive pattern of time slots. These time slots are assigned exclusively to the nodes. TDMA requires tight time synchronization among nodes to avoid superposition of slots. In the FDMA scheme, each user receives individual allocation of one or several frequency bands, or channels. This scheme requires frequency synchronization and relatively narrowband filters, resulting in a more expensive solution than TDMA. Finally, the CDMA technique is a form of spectrum spreading, exploiting the random nature of PN sequences. Such sequences can be easily computed by means of low-complexity circuits (usually shift registers) and, even if they can be reproduced in a deterministic way, behave just like noise. In CDMA systems, unique PN codes, rather than separate RF channels or time slots, are assigned to the devices of the network. Source devices use their own code as a signature to mark their data; generally the user data sequence is simply multiplied by the PN sequence. In this way, a correlator in the receiver, fed with the correct PN code, can extract the original user data sequence, rejecting both noise and other user data. It may be easier to visualize the differences between these methods by imagining a cocktail party where two people wish to converse with each other. With FDMA, everyone in the room must be silent except for the speaker; different speakers can talk simultaneously in different rooms. With TDMA, speakers use a common clock to know when they have the right to talk. In CDMA, the speaker and the listener(s) have agreed beforehand to use a language that no one else at the party understands. Many speakers may talk at a CDMA party, each using a different language, and it is relatively easy for the listener to hear and understand the speaker as long as there are not too many speakers talking at the same time.

For the sake of completeness, the Carrier Sense Multiple Access with Collision Avoidance (CSMA/CA) mechanism is briefly described here, as an example of a contention-based protocol. A node desiring to transmit first of all senses the medium. If the medium is busy (someone else is already transmitting), the transmission will be delayed; otherwise the transmission may occur. This approach is very effective when the medium is not heavily loaded, but collisions may occur. They are minimized using an increasing contention window, in which retransmission are deferred by a randomly chosen amount of time.

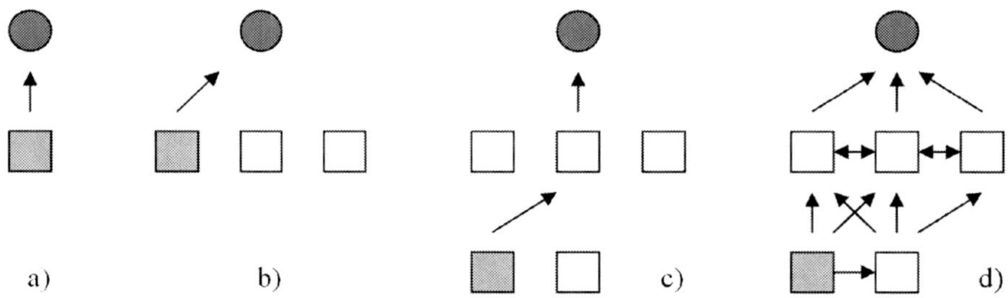

Figure 3.5. Network topologies: (a) peer to peer; (b) star; (c) tree; (d) mesh.

1.4.3. The Network Layer

Almost all communication stacks are described at least at the physical and data link layers of the ISO/OSI stack. A network layer description is required to better define how information travels through more nodes. In a peer-to-peer (Figure 3.5a) or in a star (Figure 3.5b) network, each sensor, indicated with a square node, communicates with only one node, typically the acquisition system that normally performs the function of the network coordinator. If a source node is not able to send data directly to the destination node, it has to rely on the assistance of intermediate nodes (normally other sensors) to forward the information. If the path is unique, as occurs in tree networks (Figure 3.5c), the complexity of the stack is quite similar to the previous ones. Otherwise, as occurs in mesh networks (Figure 3.5d), the path can be chosen according to different strategies:

- To limit the delay time
- To spread the job in a fair manner
- To use the most reliable intermediate nodes
- To avoid decreasing the amount of energy to energy-dependent nodes

The network is called a *multihop* network because more than one hop is needed to reach the destination (called the sink, indicated in dark gray in Figure 3.5) from a source (light gray in Figure 3.5). In such a network, nodes have to decide to which neighbor an incoming packet should be passed; this decision is made by means of routing tables, listing the most appropriate neighbor (called the router) for any given destination. The maintenance and construction of these tables is the main task of the third layer of the ISO/OSI stack, the network layer. Layer 3 defines addresses, which are not related to the addresses on Layer 2 (if they are implemented). The pattern in which nodes (source, sink, and routers) are connected within the network is the topology. A brief discussion of the network topologies applying to wireless sensor networks is outlined below (Akkaya and Younis 2005). A star network is a communication topology in which the network remote nodes can send and/or receive a message only to/from the coordinator—they are not permitted to send messages to each other. Such a network does not have a consistent routing protocol, since it is a single-hop network. The advantage of this type of network for wireless sensor networks is its simplicity and its ability to keep the remote node's power consumption to a minimum. It also allows for low-latency communications between the remote node and the coordinator. The disadvantage of such a network is that the coordinator must be within the radio domain of all the individual nodes and may be as not as robust as other networks, since the coordinator is a single point of failure. On the contrary, a network in which any node can send or receive packets to/from any other node is called a mesh. This network topology has the advantage of redundancy and scalability. If an individual node fails, a remote node still can communicate to any other node in its range, which, in turn, can forward the message to the desired location. In addition, the range of the network is not necessarily limited by the range between single nodes, since the network itself behaves like a giant antenna. The disadvantage of this type of network is in the power consumption, since routers must be always in the receiving state, i.e., in a high power-consumption mode, thus limiting the battery life. The latency is another disadvantage, since the time to deliver the message is related to the number of hops and the delivery time for each hop.

A tree network can be considered as a mesh with a fixed routing table. A hybrid between the star and mesh network provides for a robust and versatile communication network, while maintaining the ability to keep minimum the power consumption of the wireless sensor nodes. In this network topology, the lowest-power sensor nodes implement a reduced set of functionalities—e.g., they are not enabled to forward messages. This allows for minimal power consumption to be maintained. However, other nodes implement all the functionalities, and can forward messages from low-power nodes to other nodes on the network. Generally, nodes with multihop capability require more power, and they are often connected to power mains.

1.5. INTERFERENCE, COEXISTENCE, SECURITY, SYNCHRONIZATION, LOCALIZATION

A wired sensor network can be considered a reliable link, because the medium can be dedicated to this communication. Interference immunity (the ability to continue communicating in the presence of other systems sharing the medium) and coexistence (the ability to share the medium without affecting the performance of both networks) are not true problems, although with Ethernet-based fieldbus (real-time Ethernet) this assumption should be revised (Hillebrand et al. 2007; Ferrari et al. 2008). A node fault in a wired network could cause a crash of the whole network, depending on the fault conditions. Under this point of view, wireless network suffer from problems due to the unknown state of the medium (reflections, attenuation of the transmitted signal, interference with other RF systems, low coexistence among different networks), but a node fault does not affect the network communication and there is no need to check the integrity of the cables. As stated before, the 2.4-GHz bandwidth has numerous widespread standards [WiFi—www.wi-fi.org, Bluetooth—IEEE Standard 802.15.1-2005, Standard for Information Technology—Telecommunications and Information Exchange Between Systems—Local and Metropolitan Area Networks—Specific Requirements, Part 15.1: Wireless Medium Access Control (MAC) and Physical Layer (PHY) Specifications for Wireless Personal Area Networks (WPANs), ZigBEE (ZigBee Specification Document 053474r17, January 2008, www.zigbee.org), and so on]. For this reason, interference and, more recently, coexistence are hot topics, and there are research teams and standardization groups working on it (IEEE 802.15.2-2003, Coexistence of Wireless Personal Area Networks with Other Wireless Devices Operating in Unlicensed Frequency Band) (Ferrari et al. 2008).

Security in wireless sensor networks is also receiving growing attention, and the most attractive technologies in this field, such as ZigBee (Dissanayake et al. 2008) and Wireless Hart (IEC/PAS

62591 Ed. 1.0, Industrial Communication Networks—Fieldbus Specifications—WirelessHART Communication Network and Communication Profile, Publication Date 2009-01-22), support data encryption (Dworkin 2004). Security is strictly connected to coexistence; in fact, one of the problems in a crowded area (e.g., small and close together rooms of a hotel) is that similar networks could interfere due to a nonperfect exclusive setup of each network.

Synchronization among nodes of a wireless sensor network is useful for at least three reasons: It is necessary for proper management of some solutions, such as Time Division Multiplexed Access (TDMA); it provides efficient power management of autonomous nodes; and it may be a requirement for performing distributed measurements. In the first case, that is, with TDMA, all the nodes of the networks should share a common sense of the time, in order to occupy the shared medium in the correct time slot, without interfering with each other. In the second case, if a communication is scheduled between two nodes, the receiving node should turn the transceiver on in advance, to be sure not to lose the message; but if it is not well synchronized with the transmitting node, it must keep the transceiver on for a long time, increasing power consumption. Distributed and synchronized measurements (the third case) might be a particular need of a wireless chemical sensor network for gas flow detection. In fact, if we assume many chemical sensors in a close environment, we need to have a simultaneous measurement in each point of the area, in order to follow the motion of the gas flow. Most stacks for wireless sensor networking provide their own synchronization method, such as as ZigBee or WirelessHART, but there are synchronization methods specifically designed for wireless sensor networks. Some of them are based on the so-called sender–receiver approach, in which the node operating as the clock source and the node representing the clock slave exchange messages and the clock slave can calculate the clock correction starting from the transmitting and receiving times (timestamps) of these messages (Ganeriwal et al. 2003). Other designers use the receiver–receiver method, in which more nodes compare the receiving instants of the same packet in a more fair and democratic manner (Elson et al. 2002). Synchronization is a very important topic in wireless sensor networks, and a lot of research works deals with it (Ageev et al. 2008).

Another important topic in wireless sensor networks is localization. If we propose to spread autonomous sensors for chemical detection in an open and wide environment or a network of mobile autonomous sensors, node localization allows referring a measurement to a particular location. Normally, the localization of a node is done by referring its position with respect to fixed nodes with known position, the so-called anchor nodes. Estimation of the distance with respect to an anchor node can be made by two different approaches: measurement of the received power being known as the transmitted power (Bergamo and Mazzini 2002; Patwari et al. 2003; Elnahrawi et al. 2004), or measurement of the time of flight (time of arrival or round-trip time) of a message (Alsindi et al. 2007; Zhang and Dong 2008). The first approach can be affected by reflections, multipath fading, and the unknown attenuation of the medium, while the second can be well implemented with sharp UWB pulses (Nanotron is a commercial product that employs UWB to provide localization to a wireless sensor network (www.nanotron.com).

1.6. WIRELESS TECHNOLOGIES

Microwave transmissions have moved toward 2.4-GHz frequencies, thanks to the license-free ISM (Industrial, Scientific, and Medical) band radio frequencies available in every region of the world. In fact, the first devices for wireless, low-cost, low-power communications are disappearing. For instance, in Europe the bandwidth at 434 MHz was used by devices employing simple AM/FM (amplitude modulation/frequency modulation) schemes. These devices are suitable for very simple peer-to-peer communications (e.g., radio commands), but they are unsuitable for networks due to the unreliable modulation scheme and the poor performance. In addition, these kinds of devices have been developed in the United States at frequency around 915 MHz. On the other hand, the frequency band between 2.4 GHz and 2.5 GHz is free all over the world, as is the frequency band around 5.8 GHz, allowing improved performance. Another bandwidth shared by all the world is 13.56 MHz, used for radiofrequency identification or RFID (ISO/IEC 15693-1:2000, Identification Cards—Contactless Integrated Circuit(s) Cards—Vicinity Cards—Part 1: Physical Characteristics). Nowadays, the more widespread applications, such as as Bluetooth, WirelessLAN, and ZigBee, exploit the 2.45-GHz band, and the diffusion of these technologies has contributed to lower the prices of transceivers.

Generally speaking, there are a lot of wireless technologies that can be suitable for wireless sensor networks (WSNs), from RFID (ISO15693) to the widespread technologies developed for cellular phones or around the ISM 2.4-GHz band. Early proprietary solutions are disappearing. In fact, whatever the sensing principle, interoperability among different manufacturers can be reached only by means of standard communication solutions. Wireless networks are classified according to their extension: wireless wide area networks (WWANs), used for instance for environmental monitoring (e.g., pollution); wireless local area networks (WLANs), with more limited range, typically of the order of 100 m; and wireless personal area networks (WPANs), with a communication range of the order of 10 m (Bisdikian et al. 2001). Wireless sensor applications cannot ignore RFID technology, although it cannot be considered a suitable technology for sensor networking.

RFID technology is very attractive due to its low cost and low power. The sensor node is a passive element powered by RF energy emitted by a reader. When the reader is close to the passive element, the electromagnetic power can be collected by a suitable antenna and can be used to perform the measurement and transmit the result. Wireless chemical sensors based on passive RFID cannot be assembled in a real network, but are limited to applications where a reader can be moved toward each sensor. On the contrary, active RFID technology, that is, RFID with a battery that is able to elaborate and communicate autonomously, is receiving growing interest, especially for telemetry applications.

Wireless wide area networks use the mobile telecommunication cellular network. In Europe, the most widespread solution is still the Global System for Mobile Communication (GSM) network, even if the coverage of third-generation UMTS systems is greater (Redl et al. 2005). GSM cellular technology uses 200-kHz RF channels, with Gaussian minimum shift keying (GMSK) modulation. These channels are time-division multiplexed to enable up to eight users to access each carrier. Thus, this technology can be considered a hybrid TDMA/FDMA system. Each sensor device should appear to the base station (i.e., the core of a cell) as a mobile station (i.e., the mobile phone). For this reason, it should include an identification number such as the International Mobile Equipment Identity (IMEI), set by the manufacturer and used during registration to check whether the equipment has been reported as stolen. In addition, each piece of terminal equipment embodies a subscriber identity module, which includes a number known as the International Mobile Subscriber Identity (IMSI). Coverage is generally offered on a nationwide level, with the wireless network infrastructure furnished by a wireless service provider. The main shortcoming in using GSM for WSNs is the need to pay a fee to the infrastructure provider, together with the high cost of such devices and their high power consumption (Flammini et al. 2007; Zhao et al. 2008). For these reasons, most applications use different technologies for wireless sensor networks, and only a few nodes have GSM capability.

The term wireless local area network identifies short-range (less than 100 m) packet data communication between base stations, known as access points, and user terminals, even if direct communication between terminals is also possible (infrastructure versus ad-hoc mode). There are several standards for the WLAN air interface, but recent market developments have clarified that IEEE 802.11b (IEEE Standard 802.11b-1999, Supplement to IEEE Standard for Information Technology—Telecommunications and Information Exchange Between Systems—Local and Metropolitan Area Networks—Specific Requirements—Part 11: Wireless LAN Medium Access Control (MAC) and Physical Layer (PHY) Specifications: Higher-Speed Physical Layer Extension in the 2.4 GHz Band) and g are the dominating standards. The term WiFi is often misused as a synonym for WLAN, but in fact it simply refers to an industrial alliance that certifies the interoperability among different IEEE802.11 devices. Since it is a member of the IEEE802 family, IEEE802.11 defines only the physical and data link layers of the ISO/OSI stack. Several amendments to the original proposal have increased the data transfer rate from 1 Mbps to up to 54 Mbps (even if the draft version of IEEE802.11n claims hundreds of Mbps). Typical transmission range is 100 m with a standard antenna. Both frequency hopping and direct sequence spread spectrum modulation schemes are available, and medium access is regulated by the CSMA/CA protocol. The adoption of the ISM band around 2.4 GHz and 5 GHz allows for worldwide availability. Even if the data rates are high enough for wireless sensor applications and it has been shown that ad-hoc networks can be used effectively to implement WSN, the power requirements generally preclude its use in autonomous sensor applications (Ferrari et al. 2006; Rentel et al. 2008).

The term wireless personal area network is used to describe all solutions that allow to coverage of an operating space of about 10–100 m. The forerunner of such systems is Bluetooth, standardized by the IEEE802.15.1 working group. It was originally designed to serve cable replacement applications such as data transfer from personal computers to peripheral devices such as cell phones or personal digital assistants. Bluetooth implements a star network topology supporting up to seven active nodes (slaves) communicating with a single base station (master). It operates in the ISM 2.4-GHz region, exploiting FHSS together with frequency and/or phase modulation. The MAC technique is TDMA, with 0.625-ms-wide time slots. Bluetooth 2.0 specs define modulation changes and additional packet types that allow designers to deliver a peak rate of 3 Mbps (2.1 Mbps of user-available throughput), compared to the original spec that defined a maximum data rate of 1 Mbps (723 kbps real throughput). Even though some wireless sensors based on Bluetooth have been proposed (Zhang et al. 2005), they have not met with wide market acceptance due to limitations of the protocol, including: (1) poor power efficiency due to the absence of low-power modes and the long time needed to rejoin the network when returning from sleep; (2) the low number of active nodes per network (fewer than 8 nodes per Piconet); and (3) a protocol stack that is overly complex in comparison to that required for wireless sensor applications. It should be noted that TDMA access to the medium makes easy the transmission of time-bounded data, e.g., as shown by Ferrari et al. (2005).

Probably the most interesting proposal for WSNs today is IEEE802.15.4 [IEEE Standard 802.15.4-2006, IEEE Standard for Information Technology—Telecommunications and Information Exchange Between Systems—Local and Metropolitan Area Networks—Specific Requirements Part 15.4: Wireless

Medium Access Control (MAC) and Physical Layer (PHY) Specifications for Low-Rate Wireless Personal Area Networks (WPANs)].

1.6.1. IEEE 802.15.4

In the following, IEEE802.15.4 is described, according to the ISO/OSI approach.

1.6.1.1. PHYSICAL AND MAC LAYERS

IEEE802.15.4-PHY offers three operational frequency bands, all within the ISM: 2.4 GHz, 915 MHz, and 868 MHz. There is a single channel between 868 and 868.6 MHz, 10 channels between 902 and 928 MHz, and 16 channels between 2.4 and 2.4835 GHz (the most used). It must also be remembered that, in March 2005, baseline requirements for accurate ranging and localization in IEEE802.15.4 WPAN appeared. Two optional PHYs were added: a Continuous Pulsed UWB Radio (operating in unlicensed UWB spectrum), and a Chirp Spread Spectrum (operating in the same unlicensed 2.4 GHz band of legacy IEEE802.15.4). In March 2007, the resulting IEEE802.15.4a was approved as an amendment to IEEE Standard 802.15.4-2006 (www.ieee802.org/15/pub/TG4a.htm).

The higher data rate of 250 kbps is achieved in the 2.4-GHz band, where the user data stream is spread by means of a 32-chip-long sequence, according to the DSSS scheme. The chip streaming is then modulated by offset quadrature phase-shift keying, with half-sine shaping (OQPSK-HSS). This modulation scheme achieves a high degree of spectral occupancy, while having very little spectral leakage outside the signal band. This strategy results in a system with excellent sensitivity, little interference to other users in the spectrum, and low power implementation.

The IEEE802.15.4-MAC supports beacon-enabled or non–beacon-enabled operational modes. The media access implements a "listen-before-talk" scheme (slotted or unslotted CSMA/CA) flanked by an (optional) TDMA strategy.

In the beacon-enabled mode, beacon telegrams are sent periodically by the coordinator to announce itself, to synchronize devices associated with it, and to describe the so-called superframe structure. The superframe is made up of an active period and, optionally, an inactive period exploited to extend the life of battery-powered devices. The active period, corresponding to the superframe duration (SD), is divided into 16 equal-sized time slots, where data transmission is allowed. Each active period can be further divided into a contention access period (CAP) and an optional contention free period (CFP), composed of guaranteed time slots (GTSs). Each superframe can contain up to 7 GTSs, and each of them may last one or more time slots.

If the superframe structure is used and beacons are defined in the PAN, then slotted CSMA/CA needs to be used in the CAP. If beacons are not used in the PAN or a beacon cannot be located in a beacon-enabled network, the unslotted CSMA/CA algorithm is used. The only difference is in the need for synchronization among the "atomic" unit of time, called the back-off period, and the beacon boundaries. No matter the mode, the core of the algorithm is always the delay of a random number of complete back-off periods in accessing the medium. In this way, collisions are almost avoided.

However, allocating guaranteed time slots in the beacon-enabled modality allows for bandwidth reservation. GTSs implement TDMA and are the only way to ensure determinism, since in all other modalities, data coming from upper layers is stored in buffers and there is no control on the actual transmit instants.

1.6.1.2. NETWORK TOPOLOGIES AND DATA TRANSFER MODEL

An IEEE802.15.4 network consists of two types of devices: full-function devices (FFDs) and reduced-function devices (RFDs). An FFD can implement all kinds of devices (for instance, a PAN coordinator), while an RFD can act only as an end device. An FFD can talk to RFDs and other FFDs, while an RFD can talk only to FFDs. An RFD is intended for very simple applications, where there is no need to send large amounts of data. As a result of these restrictions, an RFD needs only minimal resources and memory capacity. An IEEE802.15.4 network is constituted of at least two devices within a personal operating space communicating on the same physical channel. It contains at least one FFD, acting as the PAN coordinator, and one FFD or RFD. The network may operate in several topologies. In the star topology, all devices communicate only with a single central PAN coordinator. The peer-to-peer topology also has a PAN coordinator, but, additionally, devices can communicate directly with each other. This allows for more complex topologies, such as a mesh topology or a cluster tree. The cluster tree network is a special case of a peer-to-peer network with mostly FFD devices. The PAN coordinator starts the network and forms the first cluster, becoming also its cluster head (CLH). Other devices may then join the network, but if no more devices can be added, the PAN coordinator may instruct another FFD to become the CLH of a new adjacent cluster. As soon as other devices connect, a multicluster network structure is formed. In its simplest form, the cluster tree network consists of only a single cluster, but larger networks may be formed as a mesh of multiple neighboring clusters. The multicluster structure trades increased message latency for an increase in coverage area. Such a peer-to-peer network can clearly be ad-hoc, self-organizing, and self-healing.

Devices are recognized by means of their addresses. The IEEE802.15.4 standard provisions for two types of addresses. All devices have a unique 64-bit extended IEEE address; additionally, once joined to a network, devices can receive 16-bit short addresses from the PAN coordinator during the association phase.

Three types of data transfer transactions exist: (1) from a coordinator to a device; (2) from a device to a coordinator; and (3) between two peer devices. Data transfers are completely controlled by the devices on a PAN, rather than by the coordinator. This provides the energy-conservation feature of the network, since end devices usually have limited power sources (batteries). This means that, when a coordinator wishes to transfer data to a device, it stores the data until the end device requests the data. On the contrary, in a peer-to-peer network, every device can communicate with any other device in its transmission radius. There are two options for this. In the first case, the node will listen constantly and transmit its data using unslotted CSMA-CA. In the second case, the nodes synchronize with each other so that they can save power.

Often, IEEE802.15.4 networks are confused with ZigBee solutions. The ZigBee Alliance is an association of companies that has developed a protocol stack specifying the IEEE 802.15.4 as the physical and MAC layer. In other words, it adds a complete network layer, which allows it to implement star, mesh, and hybrid star mesh networks, and standardizes the application layer, making multivendor compatibility easier.

However, ZigBee is not the only standard proposal exploiting the lower levels of the IEEE802.15.4. Other interesting solutions, especially designed for process control in industrial environments, include WirelessHART and ISA100 (Jianping et al. 2008).

IEEE802.15.4 draws large interest from sensor constructors, thanks to the availability of low-cost transceivers and protocol stacks. Table 3.1 indicates some of these transceivers, together with the available protocol stacks.

2. WIRELESS CHEMICAL SENSING

Chemical signal propagation has very different characteristics from other physical signals. In real environments, diffusion, as a propagation mechanism, is negligible with respect to fluid dynamic effects such as flows or turbulence (Nakamoto et al. 1999). Figure 3.6 shows an example of a common complex diffusion pattern that a chemical sensing platform might find. Furthermore, the presence of moving objects and/or individuals as well as varying conditions in outdoor environments strongly affect the propagation of chemical signals by modifying flow paths, inducing turbulences, and producing interferents. This basically makes a single point of measurement not a viable solution for constructing a consistent "olfactive image" of a sensed environment. Evolution has dealt with this reality by producing and adapting very complex olfactive systems. In some cases, when olfaction is crucial for species where individuals are strongly interconnected, such as bees or wolves, distribution and cooperation are strategies for transforming olfactive data in real knowledge about the environment that in turn means food, protection, or security.

Distribution and cooperation may be very interesting solutions to the intricacies of spatial-temporal chemical propagation patterns and to solid-state sensor issues in real-life chemical sensing application fields. On the other hand, sensor power consumption as well as reliability problems have significantly restricted development. As a results, wireless chemical sensing is now still in an early phase, but advancements in wireless technologies, as described in Section 1, may now be exploited to build working architectures that may foster the development of efficient platforms and increase knowledge about distributed olfaction sensory data fusion. In fact, as shown in Section 1, nowadays wireless sensor network technologies are available to allow widespread use of wireless chemical sensor networks (WCSNs), and the actual limit is in designing a low-power, low-cost chemical sensor with good metrological characteristics to be used as an autonomous sensing node.

2.1. APPLICATION SCENARIOS

Chemical sensors are today embedded in a wide number of measurement instruments used in several application fields, from home and building automation to industrial process control, health, and agriculture. A significant number of these different applicative areas can take advantage of emerging wireless technologies, to enable truly pervasive knowledge about chemicals in the target environment (see

Table 3.1. Transceivers and available protocol stacks

<table>
<tr><th></th><th colspan="4">Features</th></tr>
<tr><th>Manufacturer</th><th>RF Transceiver for IEEE 802.15.4 PHY</th><th>Microcontroller</th><th>ZigBee stack</th><th>Embedded operating system for application layer</th></tr>
<tr><td>Atmel</td><td>• AT86RF212 for 800/900-MHz band
• AT86RF231 for 2.4-GHz band</td><td>• AVR/AVR32 RISC controller with in-system programmable FLASH</td><td>BitCloud</td><td>• FreeRTOS
• TinyOS
• μC/OS-II
• NORTi RTOS</td></tr>
<tr><td rowspan="2">Freescale</td><td>• MC13192 for 2.4-GHz band
• MC13202 for 2.4-GHz band</td><td>• MC9S08GT
• Flexis ultralow-power MC9S08QE128 (8-bit) and MCF51QE128 (32-bit)</td><td rowspan="2">BeeStack</td><td rowspan="2">• FreeRTOS
• μC/OS-II</td></tr>
<tr><td colspan="2">• MC1321x system-in-package (SiP)
• MC1322x platform-in-package (PiP)</td></tr>
<tr><td rowspan="3">TI</td><td>• CC2420 for 2.4-GHz band
• CC2520 for 2.4-GHz band</td><td>• 8051
• MSP430</td><td rowspan="2">Z-Stack</td><td rowspan="2">• FreeRTOS
• TinyOS
• μC/OS-II</td></tr>
<tr><td colspan="2">• CC2430 system-on-chip (SoC)
• CC2431 system-on-chip (SoC)</td></tr>
<tr><td colspan="3">CC2480 ZigBee network processor with Z-Stack + MSP430</td><td></td></tr>
<tr><td>Ember</td><td colspan="2">EM250 system-on-chip (SoC)</td><td>EmberZNet Stack</td><td>• FreeRTOS
• TinyOS</td></tr>
</table>

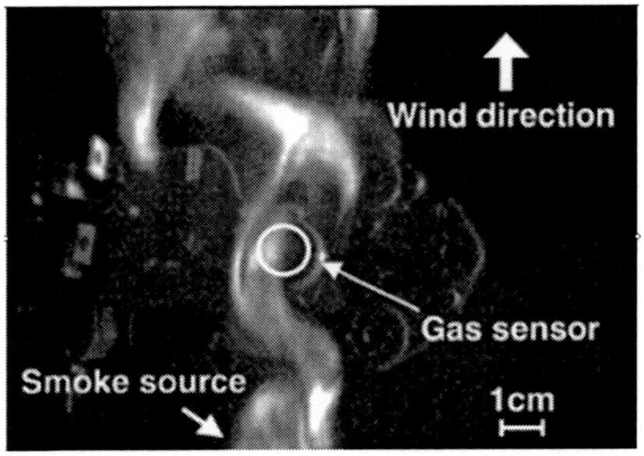

Figure 3.6. Instantaneous smoke image showing gas concentration. (Reprinted with permission from Nakamoto et al. 1999. Copyright 1999 American Chemical Society.)

Figure 3.7). In fact, smart environments often need to gather data from distributed locations by means of a multitude of sensing elements, process the collected data, and react by modifying relevant working parameters. As an example, indoor volatile organic compounds (VOCs), pollution arising from the use of special cleaning products, solvents, furniture glues, smoking, and other vapor sources represent an increasing concern for public safety. The availability of a truly pervasive monitoring system, capable of detectinging and assess indoor pollution levels, could enable a smart conditioning system to perform reaction strategies and mitigate issues for living people. Cooperation among the sensing elements, with regard to sensor fusion, will be crucial for consistent assessment in complex indoor environments (Chunga et al. 2006).

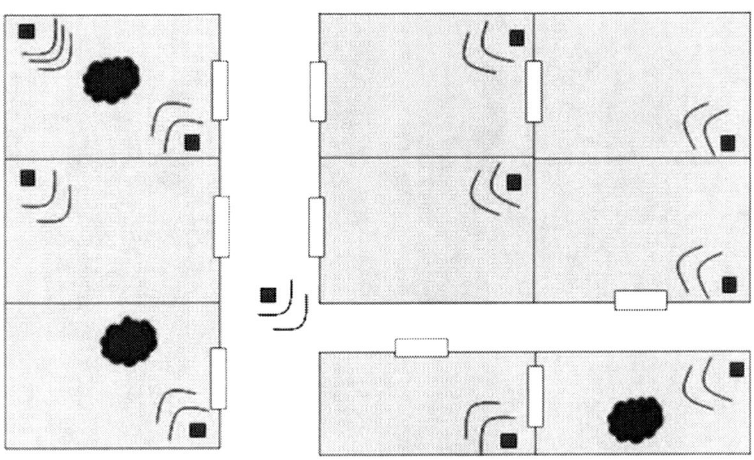

Figure 3.7. Concept of a wireless chemical sensor network deployment in an indoor volatile pollution monitoring scenario.

Similar considerations can be applied to the safety of working people in industrial environments. In fact, wireless chemical sensors are used in industrial applications, especially in food industries, to reveal the presence of contaminants or bacteria (Changa et al. 2007), but their use could be extended to a more general concept of safety improvement.

More generally, distributed and cooperative chemical sensing will allow a significant leap in environmental monitoring. Besides indoor environments, it will also lead to more precise and timely assessment of outdoor city pollution; in fact, a dense number of chemical sensing elements can reconstruct the real pollution pattern without suffering from the fluid dynamic influences of the intricate network of roads and building. These influences actually make classic solutions based on sparse networks and diffusion models practically ineffective.

Landfill gas and off-odor monitoring and control, as well as distributed geochemical monitoring for volcanic or seismic systems situational awareness, or outdoor fire detection, are other envisaged applications that will benefit from the possibility of a network of light, solid-state, wireless detectors.

Safety, on its own, is one of the most interesting fields for the application of distributed chemical sensing technologies. The introduction of dangerous explosive, or toxic gases in complex environments such as docks, airports, military installations, and critical infrastructures is another major concern. A network of very sensitive chemical detectors acting together is one of the foreseen architectures to efficiently detect these targets at the very low concentrations expected in these scenarios.

Leaving safety scenarios, being able to truly understand the running metabolism of valuable cultures by monitoring their specific gaseous markers, such as CO_2 or ethylene, as well as understanding the microclimate they are harvested in, could be a welcome application in modern agriculture scenarios. Another interesting application field for wireless chemical sensors is in health-care monitoring systems. Recently, solutions and devices have been proposed for this purpose (Hsu et al. 2007).

As will be described in more detail in the following sections, a number of experiments are now underway to test different possible architectures, tailoring them to the specific application. The different requirements give birth to several different solutions, as we are still far away from a unified architectural framework.

2.2. CHEMICAL SENSORS TECHNOLOGIES AND WIRELESS CHEMICAL SENSING

The characteristics of an optimal sensor for distributed chemical sensing should include low-power operation, low-cost, long-term reliability, and stability. Additionally, it should also be easy to integrate the sensor with simple signal-conditioning electronic circuits. Depending on the application, the sensor should possibly express good specificity properties and very low detection limits. By far, this description applies more to an ideal device than to a real one; in fact, no current chemical sensor technology seems near to obtaining such results simultaneously. Nonetheless, some of the available technologies can be considered good candidates for use in wireless chemical sensing, especially if tackling particular issues in order to customize the device for the wireless requirements of the specific application.

In particular, the well-known metal oxide (MOX) chemical sensor technology can exploit the relatively stable knowledge about the working principles as well as its diffusion and, last but not least, the availability of a wide range of commercial products, in order to select the appropriate device and driving technology to obtain significant results in this field. Unfortunately, MOX sensors suffer from the high temperatures at which they need to operate, and consequently the power consumption is high. Instead of MOX sensors, polymer-based chemiresistors, resonators, and mass sensors (quartz microbalances, surface acoustic waves) usually operate at room temperature. Although they are not as common as their MOX counterparts, their low power requirements and their relatively simple readout electronics can be quickly turned to huge advantage with respect to other technologies, especially for detection of VOCs.

Electrochemical sensors, have, of course, really interesting power profiles and can be targeted to a significant range of inorganic pollutants and toxic gases, but their cost and low sensitivity may prevent their use in applications where sub-parts-per-million or VOC detection are needed.

For particular applications, where sensitivity is not a primary issue, LED/polymer-based optical sensing can be an extremely interesting solution because of very low-cost, reliability, and low power demand (Sheperd et al. 2007). However, in order to obtain suitable sensitivity in most applications, a laser source should be used, and this may increase significantly both costs and power needs.

Table 3.2 summarizes the power consumption of some commercially available sensor technologies. Most of the currently proposed wireless chemical sensing nodes are in fact based on MOX or polymeric sensors.

Table 3.2. Power consumption of some commercially available sensor technologies

SENSOR TECHNOLOGY	TARGET	POWER CONSUMPTION	EXAMPLE
Thermistor	Temperature	Less than 1 mW	Radionics 813-806
LDR	Light	250 mW	NORP12
LED/polymer	Several VOCs	10–30 mW	N/A
Laser diode		225 mW	SIGMA/635/1
MOX	Methane	280 mW	Figaro TGS2611
	CO	14 mW	Figaro TGS2442
	Organic solvents	660 mW	Figaro TGS822
IR sensor	N/A	6000 mW	KT sensor (PPM 4022H)
Photoionization detector		110 mW	Alphasense PID-AH
Electrochemical	CO_2	0 mW	Alphasense CO2-D1
	PH	50 mW	Global Water 407228

2.2.1. Metal Oxide Sensors for WCSNs

Metal oxide sensors are very common gas-sensing platforms and have been available as commercial devices for decades. Resistive MOX sensors have been widely studied in order to overcome their lack of specificity or to exploit their suitability for developing fully integrated devices. The working principle is basically due to exchange of electrons between the gas and the device surface, leading to a change of the metal oxide layer conductance (Barsan and Weimar 2001). Typically, in order to activate the reaction, significantly high temperatures are needed (>200°C), thus MOX sensors are mostly coupled with micro-hotplates used to increase the temperature of the sensitive layer. A significant number of designs have been proposed, but the need to operate at high temperatures represents a serious issue for their use in wireless chemical sensing applications. In fact, their average power consumption is in the range of 200–800 mW with continuous operation. It is easy to understand that this power consumption is not practically sustainable for any battery-operated wireless node (sometimes called mote), especially if considering warm-up and response times, i.e., the time needed to achieve a steady-state response.

As shown in Section 1, we can define the battery life L [h] of a wireless sensor as the operating time until the provided voltage falls below a defined threshold (cutoff voltage). Battery life can be approximated by using Eq. (3.1), shown earlier, where a battery with capacity C [A h] feeds a power supply with efficiency η and output voltage ratio K_v and the sensor has a mean current consumption of $I_{cc,\mathrm{mean}}$ including the heater. As shown in Eq. (3.2), if the sensor must be continuously powered and heated, the term I_0 is dominant, practically impeding the realization of an autonomous sensor. On the contrary, if the sensor has been designed to operate in a pulsed mode, as for CO sensor TGS2442, the term I_0 can be neglected and Eq. (3.2) can be simplified. A typical sampling duty cycle is designed so that the microcontroller wakes up every T_m s, executes a T_a (active) phase to start up and collect measured quantities, and transmits and receives information by the RF link every T s ($T >> T_m$) taking time T_{RF}.

In this case the average current $I_{cc,\mathrm{mean}}$ is given by

$$I_{cc,\mathrm{mean}} = I_a \frac{T_a}{T_m} + I_{\mathrm{RF}} \frac{T_{\mathrm{RF}}}{T} + I_{\mathrm{sleep}} \frac{T_m \cdot T - T_m \cdot T_{\mathrm{RF}} - T \cdot T_a}{T \cdot T_m} \tag{3.3}$$

where I_a, I_{RF}, and I_{sleep} are the whole-circuit current consumptions in the active, RF, and sleeping phases, respectively.

If we try to predict operating life of a real mote based on a commercial low-power MOX CO sensor (Figaro TGS2442), we have to compute that the active phase foresees a heating subphase ($I_h \sim 250$ mA for 14 ms every 1 s) and a measuring subphase (requiring less than 1 mA for 5 ms every 1 s), each with a $I_{cc,\mathrm{mean}}$ contribution of 3.5 mA and 5 μA, respectively. In this case, considering typical values for the above parameters, i.e., $T_{\mathrm{RF}} = 10$ ms, $T = 60$ s, $T_m = 1$ s, $I_{\mathrm{RF}} = 30$ mA, $I_{\mathrm{sleep}} = 20$ μA, $\eta = 0.9$, $C = 1000$ mA h, and $K_v = 1.4$, the operating battery life does not exceed 1 week.

Nonetheless, MOX sensors have been investigated for their possible use in wireless chemical sensing platforms by numerous authors (e.g., Pan et al. 2007), mostly without assessing mote operating life. However, it is interesting to note that the use of particular low-power thermal profiles in MOX operative heating can significantly extend the network operating life, retaining metrological characteristics at the cost of extending the sampling period, as demonstrated experimentally by Bicelli et al (2007) and supported by simulations obtained for a specifically designed model (Bicelli et al. 2009). Simpler heating schemes have also been proposed which yield significant power consumption reduction, but they may also significantly affect metrological characteristics (Nomura et al. 1998). Micromachining is also emerging as an promising alternative investigation route toward low-power MOX sensor operations (Jaegle et al. 1999; Tsai et al. 2001; Choa et al. 2003).

In the last few years, the increasing demand for less power-greedy chemical sensing products has led to the development of "low-power" commercial sensing platforms whose average power needs can be significantly lowered if they are appropriately driven (as low as 14 mW at 1 sample/second). The above-mentioned product by Figaro Inc., targeted to CO detection, is an excellent example. Although its power consumption may still be too high for a WCSN application, it represents a definite step toward the development of commercial power-aware MOX sensors that might in turn lead to a breakthrough for wireless chemical sensing platforms.

2.2.2. Polymer-Based Chemical Sensors

Polymer-based sensors are small, low-power devices that produce chemical transduction at room temperature and have shown good sensitivity properties to various chemicals while retaining reversibility. Polymer-based chemiresistors are typically based on very simple designs, being built around a polymeric layer that swells upon being exposed to the analyte. In common sensor architectures, a conductive nanoscale powder may be dispersed in a polymeric matrix; upon polymer swelling, conductive paths among interdigitated electrodes may be disrupted, thus increasing device resistance. The extent of swelling and hence the electrical response of the entire device has been reported to be dependent basically on the solubility of the polymer in the particular vapor. As regards their use in WCSN applications, their most appealing characteristic is their capability of operating at low temperatures, with consequent low power consumption. Unfortunately, they may need preconcentration in order to be efficient at very low concentrations, and most of them are significantly affected by relative humidity (RH) levels. Hence, in an operative scenario, the measurements should be calibrated by using the response of a RH sensor. The last few years have seen the first wireless electronic nose prototype based on an array of room-temperature-operating polymeric sensors (De Vito et al. 2008) designed for indoor pollution detection and monitoring.

For about a decade now, polymer/nanocomposite reactivity to chemicals has also been applied to the development of passive resonant sensors that are capable of wireless remote operations in a very simple fashion. They are interrogated by a radio chirp; in response to the radio wave, they resonate at a frequency that is modulated by the sensing-layer impedance, which is, in turn, quenched by the presence of the target chemical at different concentrations (Cai et al. 2001; Ong et al. 2002; Plum et al. 2006; Yoon et al. 2006). This particular architecture, depicted in Figure 3.8, is also very promising for the design of very simple remote-monitoring sensing devices based on the common RFID technology (see Section 2.3.1).

SAW sensors (see Fisheraur 1995 for a review) have also been adapted to work as passive chemical sensors by Reindl et al. (1996). Wang et al. (2006) later targeted a similar design to CO_2 detection. Vadapalli and Misra (2003) designed a custom interface for acoustic wave sensor readout in a sensor network scenario. The interface chip detects the frequency shift as a function of the chemical species absorbed by the sensor and alerts the ad-hoc network controller when the monitored parameter exceeds a specified threshold, based on local processing and measurements.

2.3. WIRELESS CHEMICAL SENSING ARCHITECTURES

Although there has been progress in the field of wireless sensor networks, distributed chemical sensing is still in the earliest phase of development. Clearly, the reasons are not a result of issues in the technologies

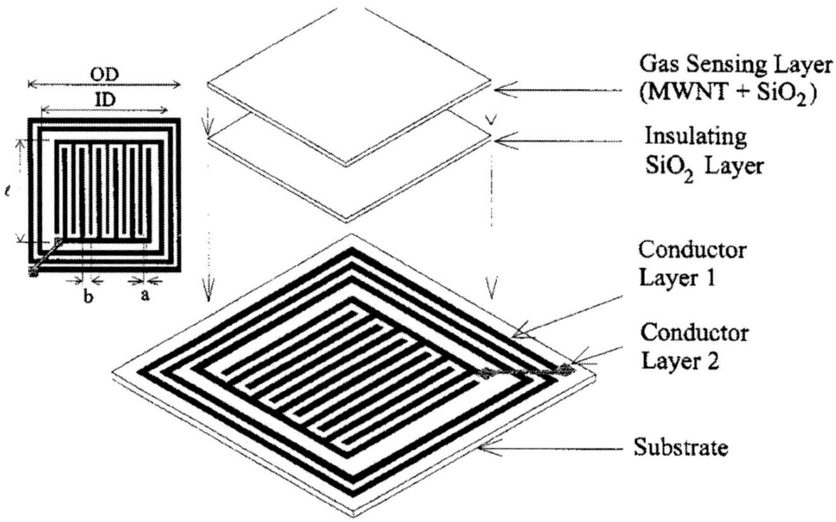

Figure 3.8. Wireless passive carbon nanotube–based resonant gas sensor. (Reprinted with permission from Ong et al. 2002. Copyright 2002 IEEE.)

available in the wireless sensor networks framework; rather, they may be found in the intrinsic limitations of currently available chemical sensing devices, architectures, and technologies. Apart from the issues already discussed in previous sections, a further issue can be seen in the well-known stability and selectivity problems. In fact, low-power solid-state sensors typically have greater stability and selectivity problems and are often unsuitable for real-world applications. Furthermore, most of the current multiple-gas sensing architectures have been designed for fixed, single-unit, in-lab usage, and their dimensions are often relatively cumbersome. They also have limited on-board computing power and, needless to say, have been designed for wired connectivity. We can thus conclude that most of the solid-state gas-sensing instrumentations are simply not available for WCSN use, lacking fundamental features and properties for operating in distributed sensing scenarios, while conventional gas sensor technology seems not to be ready for such applications.

In recent years, a number of novel approaches have been proposed and experimented with; this field is hence rapidly evolving, exploiting the plethora of results obtained by researchers in the wireless technology field. Basically, most of the work done in the direction of distributed chemical sensing can be classified as using three different main architectures, here listed in terms of increased complexity:

1. **RFID-based wireless chemical sensors units.** In this framework, a single wireless chemical sensor is operated as a standalone transducer readable by a simple wireless information channel.
2. **Wireless single-chemical sensor networks.** Prototypes belonging to this class are built by coupling a single chemical sensor to an embedded architecture featuring simple data acquisition, processing, and transmission capabilities. Actually, most of the desired features are available within the shell of a single microcontroller. These wireless chemical sensor nodes can be dispersed in the environment and are capable of simultaneously reporting the chemical status at multiple

points. A significant amount of effort has to be devoted to enabling cooperative reconstruction of the chemical status.

3. **Wireless electronic nose networks.** Here the data-capture chain is connected to a small array of chemical sensors. In this framework a network of wirelessly connected electronic noses exploits the flexibility of the sensor array approach to chemical sensing and advantages of the spatial distribution. Specific algorithms have to be implemented for multilevel sensor fusion.

2.3.1. RFID-Based Wireless Chemical Sensors

In particular applications, sensors are within a data aggregator (data sink) radio range; or, more generally, a simple star topology is required for the monitoring network to build a chemical "image" of the sensed environment. When sub-parts-per-million resolution or high specificity are not a concern, very simple schemes, practically without on-board computation or signal conditioning, can be adopted. In this case, the coupling of passive, battery-less, resonant chemical sensors with RFID-like technologies (see Section 1.6) could be an elegant solution to distributed chemical sensing. Within the above-mentioned constraints, applications such as measurement of toxic industrial chemical concentrations, monitoring of volatile organic compounds, or chemicals in liquids, where inexpensive probes have to be placed remotely and read periodically, could be well supported by this architecture.

In principle, the sensor node could be based on a common RFID coated with a film that is sensitive to the desired chemical compound. When a measurement is needed, the RFID reader sends an inquiry to the sensing tag, which sends the encoded data embedding chemical concentration–related information (www.theengineer.co.uk). Sensing technology relies basically on a passive resonator scheme. Recently, an interesting example of this technology has been developed by Steinberg et al. (2009) for pH measurement. The optical chemical sensing capability of the RFID tag is demonstrated by measuring pH using a thin sol-gel film containing a pH-sensitive indicator dye. The film exhibits a colorimetric response to pH in the range 5.0–8.5 pH units. As shown in Figure 3.9, the optoelectronic interface of the tag comprises a silicon photodiode and two light-emitting diode (LED) sources that measure the optical absorption of the pH-induced color change of the thin film at two discrete wavelengths (Steinberg et al. 2009).

Of course, star topology and hence single-hop radio range pose a significant limit to the complexity of the environment this architecture can efficiently serve. More stringent limitations are cast on the available sensor technologies, as well as by the very limited computational capabilities available on the mote and the simplicity of the signal-conditioning schemes. As mentioned previously, this strongly influences the overall performance, especially in terms of limits of detection and sensitivity. On the other hand, power is needed only at the data sink, and consequently, the sensor life represents the only limit to the entire network operative life.

2.3.2. Wireless Single-Chemical Sensor Networks

Going back to the early 1990s, one of the first contributions toward distributed and cooperative chemical sensing was presented by Yamasaki and Hiranaka (1992). Their approach was based on a wired grid

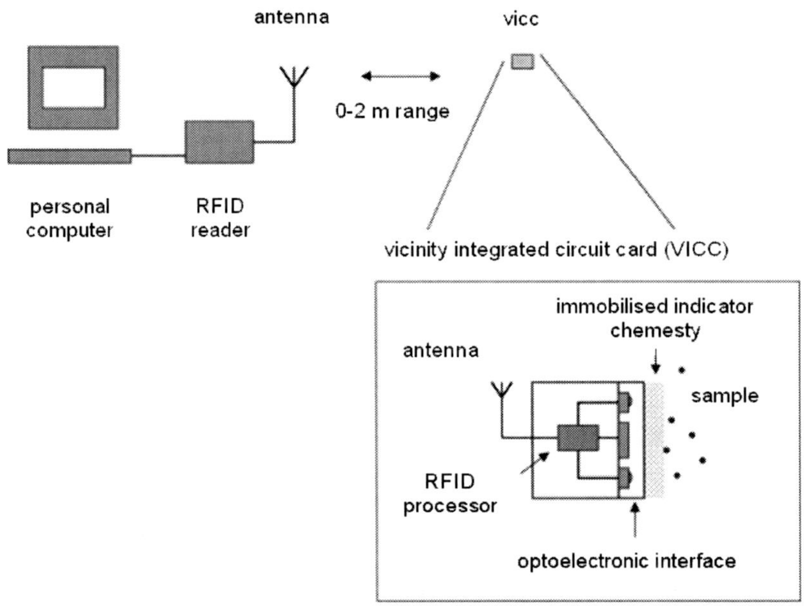

Figure 3.9. Simplified schematic of a pH RFID-based, polymer-mediated optical sensor. The sensor is totally battery-free and operates by exploiting optical quenching by a polymer layer recovering a photodetector diode illuminated by a couple of LEDs. (Reprinted with permission from Steinberg et al. 2009. Copyright 2009 Elsevier.)

of metal oxide sensors: 64 instances of the same MOX sensor were homogeneously spaced by 20 cm on a rectangular grid (see Figure 3.10).

The resistance response of sensors was used for estimating the spatial/temporal distribution of concentrations of several vapors (ethanol, ethyl ether), as well as human body off-odor, shedding a first light

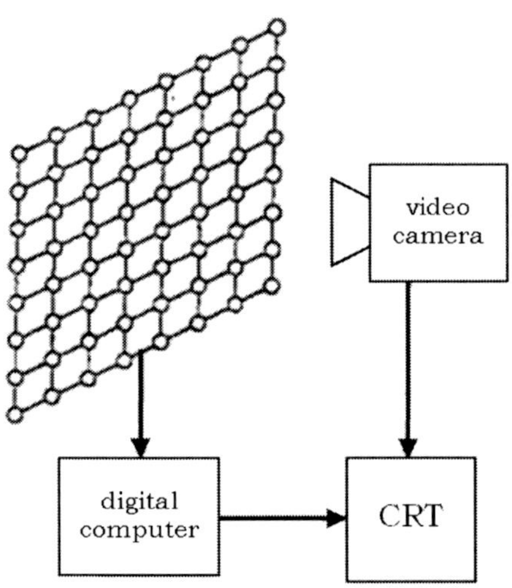

Figure 3.10. Pioneering wired chemical sensor matrix designed by Yamasaki and Hiranaka to gain insight into chemical signal propagation by different emitters. (Reprinted with permission from Yamasaki and Hiranaka 1992. Copyright 1992 Elsevier.)

on the complexities involved in distributed chemical sensing. Many years later, Maruo and co-workers (2003) first engineered and tested a network of chemical sensors for the distributed detection of NO_2 in a city air pollution framework; their work was based on several stations equipped with single optical sensors. Tsujita and co-workers (2005) experimented with a similar setup for NO_2 distributed detection and quantification while providing a first interesting example of a cooperative calibration algorithm. The work focused on the analysis of simultaneous response of their network of solid-state measuring stations while using conventional NO_x pollution analyzer output as a reference (Tsujita et al. 2005). They based their algorithm on the reasonable assumption that, in particular conditions, the concentration at the measuring site can be estimated as very near to zero when the concentration of all the nearest stations is detected as zero. Considering the complicated diffusion patterns generated by complex city air fluid dynamics, these conditions can occur at night, or on stormy days as well as on holidays. Results were very interesting reporting the effective capability to maintain the error to a stable value below 40 ppb using the proposed cooperative auto-calibration algorithm notwithstanding the intrinsic long-term instabilities of MOX sensors. Grimes et al. (2003) proposed a wireless chemical sensing architecture for the detection of hydrogen that was powered by three AA batteries. Interestingly, it could be applied to the monitoring of hydrogen leaks in hydrogen filling stations for fuel cell–powered cars. Based on titania nanotubes, the sensors were heated continuously at 290°C, and nodes were capable of operating for more than 100 h while collecting the resistance of the single sensing element every 2 s. Although the system was deployed in a mesh-shaped network framework, the authors only reported results using a single sensing unit operating in the range 50–10,000 ppm H_2.

A miniaturized low-power wireless monitoring system based on electrochemical analysis was proposed by Yun et al. (2004) for the detection of heavy-metal pollution in water. They reported results for single-unit operation, claiming positive detection for 1 mM cadmium contamination in 10 mM acetate buffer with a very interesting total power consumption, averaged over the duty cycle, below 1 mW. However, detection time and limits were not reported, nor were the duty-cycle parameters, and no computation or calibration was performed on board.

In 2005, Yokosawa et al. (2005) proposed a wireless chemical sensing network for dangerous gases (hydrogen), based on a single FET sensor (see Figure 3.11). However, they only reported results for a single monitoring node. One of the most interesting features of this unit was the capability of partially driving the sensor unit at room temperature. Practically, the single-sensor device is driven at room temperature until the change of the resistance value signals the detection of hydrogen. As soon as the presence of hydrogen is detected, the heating subsystem is started, driving the sensors to 100°C and signaling the output to the base station within 0.3 s from acquisition. In this case, the proposed detection limit was 1 ppm and the response time of sensors was estimated to be less than 2 s, while total detection time was 4 s.

Shepherd et al. (2007) were the first to show the development and operation of a network of battery-operated wireless chemical sensing nodes. Using an ad-hoc measuring chamber, they showed the response of a small network of wireless motes equipped with a single sensible polymer-coated LED/detector designed to measure the color changes inducted by exposure to a moving plume of acetic acid. The entire node, based on mica motes, could be encased in a 35-mm film can (see Figure 3.12). Sensibility and detection limits were not investigated; on the contrary, the authors focused their study on the cooperative response of the networked motes, reporting delays and magnitudes with which the

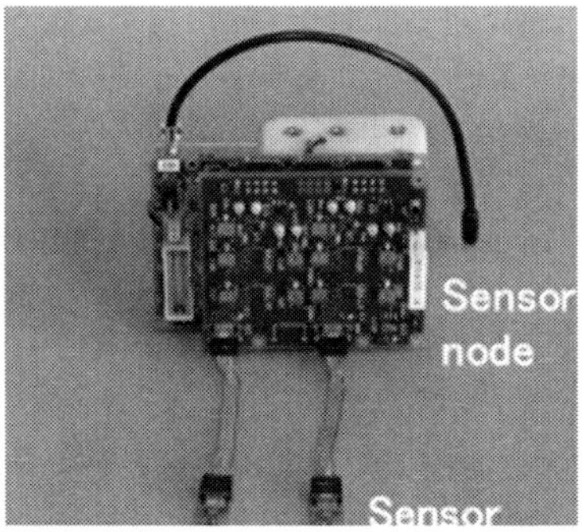

Figure 3.11. Prototype hydrogen-sensing node proposed by Yokosawa et al. for hydrogen leak detection in hydrogen refilling stations. (Reprinted with permission from Yokosawa et al. 2005. Copyright 2005 IEEE.)

motes responded to acetic acid injection within the ad-hoc measuring chamber. In later work, they provided a calibration for the single-sensor system with respect to acetic acid, finding a limit of detection (LOD) as low as 0.03 ppm (Beirne et al. 2008).

Cooperative plume detection and tracking was also the subject of work by Dilum Bandara et al. (2008). In that work, a custom network of five chemical sensing nodes, based on Crossbow TelosB motes and a Decagon ECH2O-TE sensor, was investigated in a lab environment for the detection of plumes released by underground improper disposal of pollutants or accidental chemical spills (Dilum Bandara et al. 2008).

In the same period, Bicelli et al. (2006) proposed a wireless chemical sensor for CO detection based on a well-known and reliable MOX sensor from Figaro Inc. The power dissipation due to the heater was reduced thanks to a suitable thermal profile; in this way the battery life increased from less than 3 weeks to about 1 year. A robust model has been developed for this sensor which has been exploited to simulate sensor behavior when operating at different temperature profiles (Fort et al. 2007). Recently, a suitable thermal profile has been designed to extract measurement information from the dynamic behavior of the sensor resistance while still keeping low the power dissipation due to the heater (Bicelli et al. 2009).

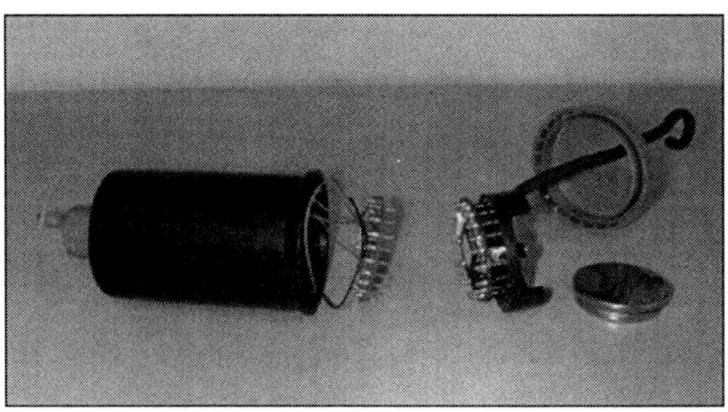

Figure 3.12. Miniaturized wireless chemical sensing mote proposed by Beirne et al. in 2007. The sensor case is based on a conventional film canister. Volatile detection is performed by means of a coupled LED/photodiode system. (Reprinted with permission from Shepherd et al. 2007. Copyright 2007 Elsevier.)

The algorithm readout, in terms of ppm CO, has a measurement updating time of 120 s and seems to be more robust against calibration errors and interfering quantities.

2.3.3. Wireless Electronic Nose Networks

The limitations of the MOX single-sensor approach are well known in the scientific community, and most of the complex classification and quantification tasks on real gas mixtures are now conducted with the use of sensor arrays in electronic nose architectures. It should be clear that, at least where gas mixtures are concerned, distributed chemical sensing can yield significant benefits by using a multisensor approach involving centralized or distributed sensor fusion capabilities. Typical applications might be VOC pollution detection for indoor environments, space, or military security. In any case, the added power requirements and the computational cost due to the adoption of a sensor array make the development of such platforms a definitely more challenging task, and prototype architectures are only starting to be proposed and reported in the scientific literature.

In early 2007, De Vito et al. (2007) proposed one of the first prototypes for a wireless electronic nose architecture designed for networking operations based on commercial WINS nodes running TinyOS, one of the most-praised operating system in the wireless sensor networks field. The prototype was equipped with a small array of polymer sensors with operating capabilities at room temperature for very-low-power operations, making it one of the few sporting true battery operation capability. Polymers used for sensor array fabrication are shown in Table 3.3. The prototype was tested both in alone and in a networked scenario (De Vito et al. 2008, 2009).

In networked scenarios, a small network produced delay and magnitude data results that were essentially compatible with expected fluid dynamics of the mock pollutant; the e-nose located near the gas source was the fastest to detect the incoming plume, showing also a definitely higher magnitude in the response with respect to the node located near the extraction fan. The response pattern, however, was very similar, indicating the possibility of performing analyte discrimination. In a further work, performance tests on each single unit were conducted to establish on-board classification and quantification capabilities. In fact, a sensor fusion component, based on the neural network paradigm, was embedded in the on-board software running on a TinyOS architecture. Two e-noses were equipped with a trained feed-forward neural network and revealed a capability of detecting and distinguishing between two mock pollutants (acetic acid, ethanol) while simultaneously estimating their concentrations in a real-time fashion. Classification capabilities were good, but performance of the concentration estimation

Table 3.3. Chemical sensor arrays

Polymer	Abbreviation	Chemical structure
Poly-(methylmethacrylate)	PMMA	$[CH_2C(CH_3)(CO_2CH_3)]_n$
Poly-(2-hydroxyethylmethacrylate)	PHEMA	$H(NHCH_2CH_2)_nNH_2$
Poly-(styrene)	PS	$(C_6H_{10}O_3)_n$
Poly-(ethylenimine)	Linear PEI	$[CH_2CH(C_6H_5)]_n$

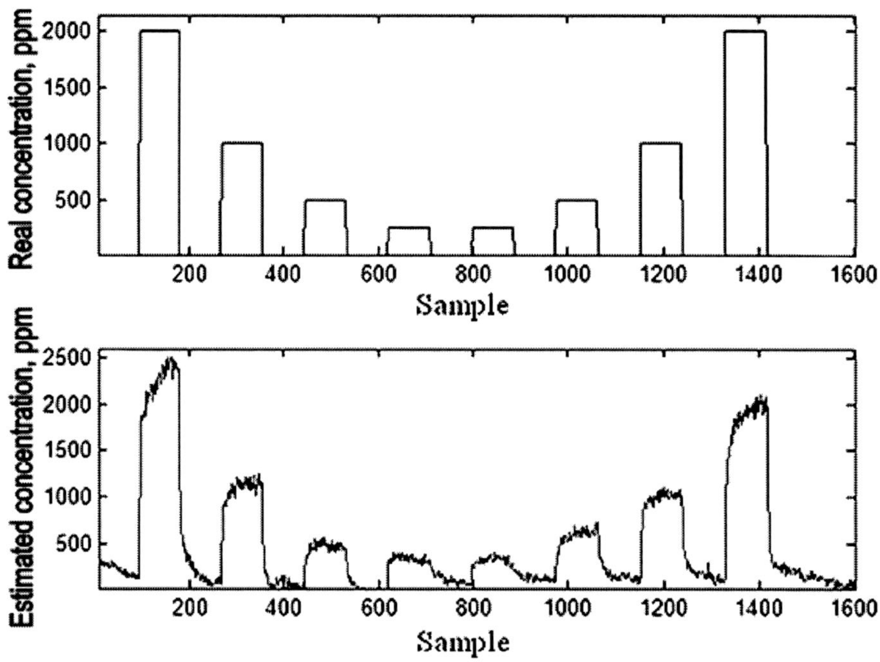

Figure 3.13. Ethanol concentration results as locally estimated by one of the TinyNose motes (De Vito et al. 2009) by mean of an on-board sensor fusion component. Local sensor fusion allows for the implementation of sensor censorship strategies so to avoid uninformative data transmissions. Only detection events related to significant pollutant concentrations trigger data transmission to the data sink. As computation is generally less power-intensive than transmission, the energy balance for sensor censorship is definitely positive.

indicated that the proposed prototype was only capable of rough quantification. In any case, this capability was clearly good enough to support sensor censorship strategies for enhanced power management (see Figures 3.13, 3.14, and 3.15).

In the same year, Pan et al. (2007) proposed a wireless electronic nose architecture devised for outdoor use in a scenario of monitoring livestock farm emissions using commercial MOX sensors. In this case, only results from a single unit were presented. However, the unit, using a data sink based on off-line neural network analysis, proved capable of predicting livestock odor emission as measured by a panelist group. The use of four commercial MOX sensors, however, make the operating life on batteries at least questionable, but the authors did not focus on that aspect.

In the city pollution framework, De Vito and co-workers proposed the use of on-field multivariate calibration of a network of wireless gas multisensor devices (an electronic nose) as opposed to univariate in-lab controlled calibration (De Vito et al. 2008). The work can be compared with the work of Tsujita et al. (2005), in which a single-sensor approach was used. De Vito et al. (2008) provided results computed over a 1-year measurement campaign for a single multisensor device targeted to benzene, NO_x, NO_2, CO, and nonmethanic hydrocarbons, estimating the number of measurement days required to train a supervised regression algorithm to obtain optimal performance. Their conclusions state that

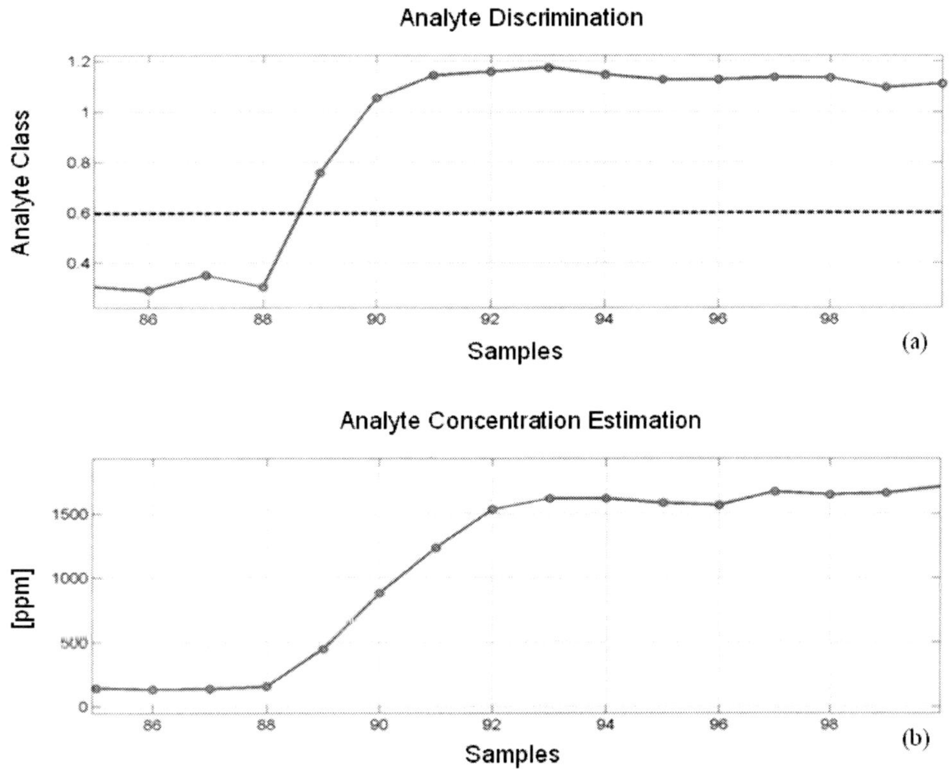

Figure 3.14. Time response of the simple sensor fusion components running on board the TinyNose prototype designed by De Vito and co-workers. The sensor fusion component processes instantaneous response of the sensor array, showing dynamic behavior that is itself strongly affected by the sensor dynamics.

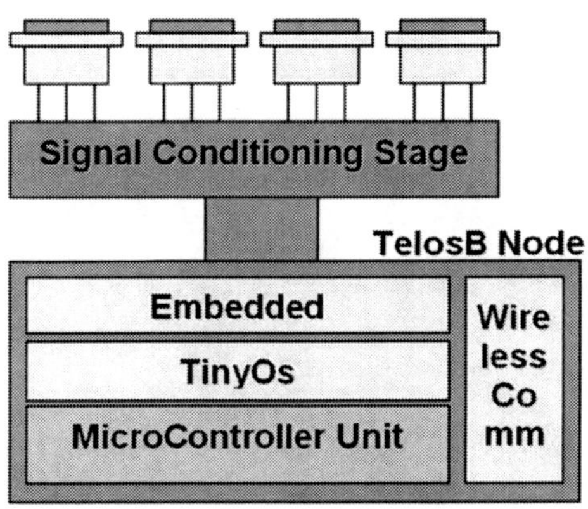

Figure 3.15. Simplified schematic of TinyNose prototype as designed by De Vito and co-workers.

10 days were sufficient to obtain stable performances for more than 6 months. Although the station could transmit the gathered data through a GPRS channel, no wireless operation was shown, and even if the application framework was intended to build a dense network of measuring station, results were reported only for a single installation and the station was not designed for battery operation.

A very recent and interesting application to the security field was proposed by a joint effort of industrial (JLM Innovations) and academic researchers (Warmer et al. 2009) (see Figure 3.16). Safety and security applications are in fact among the most appealing application fields for distributed chemical sensing. These researchers developed a mesh-shaped, four MOX sensor–based, IEEE 802.15.4–enabled wireless electronic nose network for the detection and classification of hazardous (flammable) chemicals in a naval security framework, the use of flammable chemicals by terrorists being among the most feared dangers in naval docks. The single sensing element was designed to react to flammable chemical vapors released by attackers moving in a naval corridor, which persons had to pass in order to enter a restricted area; multiple points of measure were hence a key factor for the feasibility of detection at very low concentrations. A TCP/IP-enabled data sink allowed for the broadcasting of sensed data to a sensor fusion workstation, where data was used for classification purposes. Interestingly, the data coming from chemical sensing units were coupled to laser ranging scanners and video recording systems so as to associate the chemical contamination detection to a single person carrying the volatiles source among a group of moving targets. The authors showed that it was possible to discriminate among different solvents and fire-accelerating fuels while tracking the menace-carrying person in a significant fraction of the test cases. Unfortunately, they also reported that the power consumption of the single unit was still not suitable for long-term battery operation, but the platform seems very promising for a network-powered distributed chemical sensing application.

2.3.4. Wireless Electronic Nose Networks and Data Fusion

Data fusion has been recognized as a focal point for solid-state chemical sensing since the first electronic nose platform was developed. It is easy to foresee an even more crucial role for data fusion in distributed

Figure 3.16. The zSense four MOX sensors based e-nose prototype by JLM Innovations (www.jlm-innovation.de) designed for naval safety applications. It demonstrated capability of detecting and classifying vapors of explosives transported by simulated attackers.

chemical sensing. At this early stage, researchers are only starting to understand how multiple gas-sensing units can cooperate to reconstruct an "olfactive image" of the sensed environment. It is still unclear whether providing redundancy can compensate for a lack of sophisticated calibration procedures, an area that remains a subject to investigate. A few pioneering and preliminary studies (Tsujita et al. 2005; Shepherd et al. 2007; De Vito et al. 2009), as well as a review by Nakamoto and Ishida (2008), underline the importance of spatial/temporal response pattern analysis in order to extract the desired information content. From these studies it can also be noted that local sensor fusion capabilities may be helpful for enhanced power management, allowing one to avoid uninformative data transmission. Focus on the mechanisms with which chemical sensing swarms might possibly provide reliable detection and quantification capabilities was only recently highlighted by authors such as Hayes et al. (2001). In their work, they described and tested a distributed algorithm that allows a group of mobile robots to perform odor source localization using swarm intelligence and showing enhanced performance with respect to a single agent. Hence, in a swarm intelligence scenario, local sensor fusion capabilities might be used for the implementation of local reaction strategies, as well as a first stage in a multiple-stage sensor fusion architecture.

3. ACKNOWLEDGMENTS

The authors would like to acknowledge Alessandro Depari, Emiliano Sisinni, and Paola Di Palma for their contributions to the presentation of this chapter.

REFERENCES

Ageev A., Macii D., and Flammini A. (2008) Towards an adaptive synchronization policy for wireless sensor networks. Precision clock synchronization for measurement, control and communication. In: *Proceedings of IEEE International Symposium on Precision Clock Synchronization for Measurement, Control and Communication, ISPCS 2008*, Sept. 22–26, Ann Arbor, MI, 115–120.

Akkaya K. and Younis M. (2005) A survey on routing protocols for wireless sensor networks. *Ad Hoc Networks* **3**(3), 325–349.

Alexandru C., Branzila M., Cretu M., and Trandabat Al. (2007) A prototype of web-E-nose device to be used in medicine fields. In: *Proceedings of 15th Symposium on Novelties in Electrical Measurements and Instrumentation*, Sept. 19–21, Iasi, Romania, 1–3.

Alsindi N., Li X., and Pahlavan K. (2007) Analysis of time of arrival estimation using wideband measurements of indoor radio propagations. *IEEE Trans. Instrum. Meas.* **56**(5), 1537–1545.

Anastasi C., Francesco M., and Roveri M. (2009) Energy management in wireless sensor networks with energy-hungry sensors. *IEEE Instrum. Meas. Mag.* **12**(2), 16–23.

Bandara H.M.N.D., Jayasumana A.P., Illangasekare T.H., and Han Q. (2008) *A Wireless Sensor Network Based System for Underground Chemical Plume Tracking*, Colorado State University Libraries.

Barsan N. and Weimar U. (2001) Conduction model of metal oxide gas sensors. *J. Electroceram.* **7**(3), 143–167.

Beirne S., Corcoran B., Lau K.T., and Diamond D. (2008) Chemical event tracking using a low-cost wireless chemical sensing network. In: *Proceedings of IEEE Sensors Conference, Sensors 2008*, 26–29 October, Lecce, Italy, 1615–1618.

Bergamo P. and Mazzini G. (2002) Localization in sensor networks with fading and mobility. In: *Proceedings of the 13th IEEE International Symposium on Personal, Indoor and Mobile Radio Communications*, 15–17 September, Pavilhao Atlantico, Portugal, vol. **2**, 750–54.

Bicelli S., Flammini A., Marioli D., Sisinni E., and Taroni A. (2005) Implementation of an energy efficient wireless smart sensor. In: *Proceedings of Eurosensors XIX*, September 11–14, Barcellona, Spain.

Bicelli S., Flammini A., Depari A., Marioli D., Ponzoni A., Sberveglieri G., and Taroni A. (2007) Low-power carbon monoxide MOX sensors for wireless distributed sensor networks. In: *Proceedings of Instrumentation and Measurement Technology Conference, IMTC 2007*, 1–3 May, Warsaw, Poland, 1–5.

Bicelli S., Depari A., Flammini A., Marioli D., Serpelloni M., Sisinni E., and Taroni A. (2008) Sensor signal processing for autonomous wireless sensors. In: *Proceedings of Instrumentation and Measurement Technology IEEE Conference, IMTC 2008*, May 12–15, Victoria, BC, Canada, 1210–1215.

Bicelli S., Depari A., Faglia G., Flammini A., Fort A., Mugnaini M., Ponzoni A., Rocchi S., and Vignoli V. (2009) Model and experimental characterization of the dynamic behavior of low-power carbon monoxide MOX sensors operated with pulsed temperature profiles. *IEEE Trans. Instrum. Meas.* **58**(5), 1324–1332.

Bisdikian C., Bhogwat P., and Golmie N. (2001) Wireless personal area networks. *IEEE Network* **15**(5), 10–11.

Brunelli D., Moser C., Thiele L., and Benini L. (2009) Design of a solar harvesting circuit for battery-less embedded systems. *IEEE Trans. Circuits Syst. I* **56**(11), 2519–2528.

Cai Q.Y., Grimes C.A., and Jain M.K. (2001) A wireless, remote query ammonia sensor. *Sens. Actuators B* **77**, 614–619.

Chang E.Y. and Jain A. (2004) Adaptive sampling for sensor networks. In: *Proceedings of the First International Workshop on Data Management for Sensor Networks, DMSN 2004*, August 30th, Toronto, Canada, 10–16.

Changa C.K., Changa K.S., and Chen C.Y. (2007) A surface acoustic wave sensor modified from a wireless transmitter. *Sens. Actuators B* **125**, 207–213.

Choa J.H., Kima Y.W., Na K.G., and Jeon G.J. (2008) Wireless electronic nose system for real-time quantitative analysis of gas mixtures using micro-gas sensor array and neuro-fuzzy network. *Sens. Actuators B* **134**, 104–111.

Chung W.Y. and Oh S.J. (2006) Remote monitoring system with wireless sensors module for room environment. *Sens. Actuators B* **113**, 64–70.

Cox D., Jovanov E., and Milenkovic A. (2005) Time synchronization for ZigBee networks. In: *Proceedings of the 37th South-Eastern Symposium on System Theory, SSST '05,* March 20–22, Tuskegee, AL, 135–138.

Demirkol I., Ersoy C., and Alagz F. (2006) MAC protocols for wireless sensor networks: A survey. *IEEE Commun. Mag.* **44**(4), 115–121.

Depari A., Falasconi M., Flammini A., Marioli D., Rosa S., Sberveglieri G., and A. Taroni (2007) A new low cost electronic system to manage resistive sensors for gas detection. *IEEE Sensors J.* **7**(7), 1073–1077.

Depari A., Ferri G., De Marcellis A., Flammini A., and Stornelli V. (2008) Novel CMOS fully integrable interface for wide-range resistive sensor arrays with parasitic capacitance estimation. *Sens. Actuators B* **130**, 207–215.

Depari A., Flammini A., Marioli D., and Taroni A. (2008) USB sensor network for industrial applications. *IEEE Trans. Instrum. Meas.* **57**(7), 1344–349.

De Vito S., Massera E., Burrasca G., Di Girolamo A., Della Sala D., and Di Francia G. (2007) Enabling distributed VOC sensing alications: Toward TonyNose, a polymeric wireless e-nose. In: *Proceedings of XII Italian Sensors and Microsystems Conference, AISEM' 2007*, Naples, Italy, 270–277.

De Vito S., Massera E., Burrasca G., Di Girolamo A., Miglietta M., Di Francia G., and Della Sala D. (2008a) TinyNose: Developing a wireless e-nose platform for distributed air quality monitoring applications. In: *Proceedings of IEEE Sensors Conference, Sensors 2008,* Oct. 26–29, Lecce, Italy, 701–n704.

De Vito S., Massera E., Piga M., Martinotto L., and Di Francia G. (2008b) On field calibration of an electronic nose for benzene estimation in an urban pollution scenario. *Sens. Actuators B* **129**, 750–757.

De Vito S., Massera E., Burrasca G., Miglietta M., and Di Francia G. (2009) Power savvy wireless e-nose network using in-network intelligence. In: *Proceedings of 13th International Symposium on Artificial Olfaction and Electronic Noses, ISOEN, Brescia, Italy, AIP Conf. Proc.* **1137**, 211–214.

Diamond D., Coyle S., Scarmagnani S., and Hayes J. (2008) Wireless sensor networks and chemo/biosensing. *Chem. Rev.* **108**, 652–679.

Dissanayake S.D., Karunasekara P.P.C.R., Lakmanaarachchi D.D., Rathnayaka A.J.D., and Samarasinghe A.T.L.K. (2008) Zigbee wireless vehicular identification and authentication system. In: *Proceedings of 4th International Conference on Information and Automation for Sustainability, ICIAFS 2008*, Dec. 12–14, 257–260.

Dworkin M. (2004) *Recommendation for Block Cipher Modes of Operation: The CCM Mode for Authentication and Confidentiality.* National Institute of Standards and Technology. Special Publication 800-38C.

Elnahrawy E., Li X., and Martin R.P. (2004) The limits of localization using signal strength: a comparative study. In: *Proceedings of First Annual IEEE Communications Society Conference on Sensor and Ad Hoc Communications and Networks, IEEE SECON 2004.*, Oct. 4–7, Santa Clara, CA, 406–414.

Elson J., Girod L., and Estrin D. (2002) Fine-grained network time synchronization using reference broadcasts. In: *Proceedings of the 5th Symposium on Operating Systems Design and Implementation, OSDI'2002*, Boston, MA. Also published as UCLA Computer Science Technical Report 020008.

Ferrari P., Flammini A., Marioli D., Rinaldi S., and Taroni A., (2008) Testing coexistence of different RTE protocols in the same network. In: *Proceedings IEEE International Workshop on Factory Communication Systems, WFCS'2008*, Dresden, Germany, May 21–23, 179–187.

Ferrari P., Flammini A., Marioli D., Sisinni E., and Taroni A. (2008) Synchronized wireless sensor networks for coexistence. In: *Proceedings of IEEE International Conference on Emerging Technologies and Factory Automation, ETFA'2008*, Hamburg, Germany, Oct. 15–18, 656–663.

Fischerauer G. (1995) Surface acoustic wave devices. In: Gopel W., Hesse J., Zemel J.N., Meixner H., and Jones R. (eds.), *Sensors. A Comprehensive Survey*, Vol. **8**. VCH, Weinheim, Germany.

Flammini A., Marioli D., Sisinni E., and Taroni A. (2007) Environmental telemonitoring: A flexible GSM-DECT-based solution. *IEEE Trans. Instrum. Meas.* **56**(5), 1688–1693.

Frey U., Graf M., Taschini S., Kirstein K.-U., and Hierlemann A. (2008) Digital systems architecture to accommodate wide range resistance changes of metal-oxide sensors. In: *Proceedings of IEEE Sensors Conference, Sensors'2008,* Oct 26–29, Lecce, Italy, 593–595.

Ganeriwal S., Kumar R., and Srivastava M.B. (2003) Timing-sync protocol for sensor networks. In: *Proceedings of the 1st International Conference on Embedded Networked Sensor Systems (SenSys)*, Los Angeles, CA, 138–149.

Grimes C.A., Ong K.G., Varghese O.K., Yang X., Mor G., Paulose M., Dickey E.C., Ruan C., Pishko M.V., Kendig J.W., and Mason A.J. (2003) A sentinel sensor network for hydrogen sensing. *Sensors 2003*, **3**, 69–82.

Hayes A.T., Martinoli A., Goodman R.M., (2001) Swarm robotic odor localization. In: *Proceedings of International Conference of Intelligent Robots and Systems,* Maui, HI, 1073–1078.

Hayyes J.P. (1988) *Computer Architecture and Organization*, 2nd ed. McGraw-Hill, New York.

Hillebrand J., Rahmani M., Steinbach E., and Bogenberger R. (2007) Coexistence of time-triggered and event-triggered traffic in switched full-duplex ethernet networks. In: *Proceedings of International Symposium on Industrial Embedded Systems, SIES '07,* July 4–6, Lisbon, Portugal, 217–224.

Jaegle M., Wöllenstein J., Meisinger T., Böttner H., Müller G., Becker T., and Bosch-v.Braunmühl C. (1999) Micromachined thin film SnO_2 gas sensors in temperature-pulsed operation mode. *Sens. Actuators B* **57**, 130–134.

Maruo Y.Y., Ogawa S., Ichino T., Murao N., and Uchiyama M. (2003) Measurement of local variations in atmospheric nitrogen dioxide levels in Sapporo, Japan, using a new method with high spatial and high temporal resolution. *Atmos. Environ.* **37**(8), 1065–1074.

Mateu L. and Moll F. (2005) Review of energy harvesting techniques and applications for microelectronics. In: *Proceedings of SPIE—The International Society for Optical Engineering, Vol.* **5837**, *Part I, VLSI Circuits and Systems II*, 359–373.

Matsushima T. and Horie T. (2006) Residual capacity estimation of stationary lithium-ion secondary cells in telecommunications systems using a brief discharge. In: *Proceedings of 28th Annual International Telecommunications Energy Conference, INTELEC '06,* Sept. 10–14, Providence, RI, 1–7.

Nomura T., Fujimori Y., Kitora M., Matsuura Y., and Aso I. (1998) Battery operated semiconductor CO sensor using pulse heating method. *Sens. Actuators B* **52**, 90–95.

Nakamoto T., Ishida H., and Moriizumi T. (1999) A sensing system for odor plumes. *Anal. Chem.* **71**(15), 531A–537A.

Nakamoto T. and Ishida H. (2008) Chemical sensing in spatial/temporal domains. *Chem. Rev.* **108**, 680–704.

Ong K.G., Zeng K., and Grimes C.A. (2002) A wireless, passive carbon nanotube-based gas sensor. *IEEE Sensors J.* **2**(2), 82–88.

Ottman G.K., Hofmann H.F., and Lesieutre G.A. (2003) Optimized piezoelectric energy harvesting circuit using step-down converter in discontinuous conduction mode. *IEEE Trans. Power Electron.* **18**(2), 696–703.

Pan L., Rui L., Shanghong P., Chai Y., and Yang S.X. (2007) A wireless electronic nose network for odours around livestock farms. In: *Proceedings of 14th International Conference on Mechatronics and Machine Vision in Practice, M2VIP 2007,* Dec. 4–6, Xiamen, China, 211–216.

Pan L. and Yang S.X. (2007) A new intelligent electronic nose system for measuring and analysing livestock and poultry farm odours. *Environ. Monit. Assess.* **135**, 399–408.

Patwari N., Hero A.O., Perkins M., Correal N.S., and O'Dea R.J. (2003) Relative location estimation in wireless sensor networks. *IEEE Trans. Signal Process.* **51**(8), 2137–2148.

Plum T.J., Saxena V., and Jessing J.R. (2006) Design of a MEMS capacitive chemical sensor based on polymer swelling. In: *Proceedings of IEEE Workshop of Microelectronics and MicroDevices*, April 14, Boise, ID, 49–50.

Redl S.H., Weber M.K., and Oliphant M.W. (1995) *An Introduction to GSM.* Artech House, Boston.

Rentel C.H. and Kunz T. (2008) A mutual network synchronization method for wireless ad hoc and sensor networks. *IEEE Trans. Mobile Comput.* **7**(5), 633–646.

Roundy S., Wright P.W., and Rabaey J.M. (2003) *Energy Scavenging for Wireless Sensor Networks, with Special Focus on Vibrations.* Kluwer Academic, London.

Scheible G., Dzung D., Endresen J., and Frey J.E. (2007) Unplugged but connected—Design and implementation of a truly wireless real-time sensor/actuator interface. *IEEE Ind. Electron. Mag.* **1**(2), 25–34.

Shepherd R., Beirne S., Lau K.T., Corcoran B., and Diamond D. (2007) Monitoring chemical plumes in an environmental sensing chamber with a wireless chemical sensor network. *Sens. Actuators B* **121**, 142–149.

Song J., Han S., Mok A.K., Chen D., Lucas M., and Nixon M. (2008) WirelessHART: Applying wireless technology in real-time industrial process control. In: *Proceedings of Real-Time and Embedded Technology and Applications IEEE Symposium, RTAS '08.* April 22–24, San Francisco, CA, 377–386.

Starner T. and Paradiso J. (2004) Human generated power for mobile electronics. In: Piguet C. (ed.), *Low-Power Electronics Design.* CRC Press, Boca Raton, FL, 1–35.

Steinberg I.M. and Steinberg M.D. (2009) Radio-frequency tag with optoelectronic interface for distributed wireless chemical and biological sensor applications. *Sens. Actuators B* **138**, 120–125.

Tsai P.P., Chen I.-C., and Ho C.J. (2001) Ultralow power carbon monoxide microsensor by micromachining techniques. *Sens. Actuators B* **76**, 380–387.

Tsujita W., Yoshino A., Ishida H., and Moriizumi T. (2005) Gas sensor network for air-pollution monitoring. *Sens. Actuators B* **110**, 304–311.

Vadapalli Y.R. and Misra D. (2003) An interface circuit for a SAW based chemical sensor in an ad-hoc network. In: *Proceedings of Modelling and Simulation*, Feb. 24–26, Palm Springs, CA, 60–63.

Yamasaki H. and Hiranaka Y. (1992) Multidimensional intelligent sensing system using sensor array. *Sens. Actuators A* **35**, 1–8.

Yokosawa K., Nakano S., Goto Y., and Tsukada K. (2005) Prototype system comprising battery-powered sensor nodes and a wireless network for detecting hydrogen leaks. In: *Proceedings of the 22nd Sensor Symposium on Sensors, Micromachines, and Applied Systems,* Oct. 20–21, Tokyo, Japan, 435–438.

Yoon H., Xie J., Abraham J.K., Baradan V.K., and Ruffin P.V. (2006) Passive wireless sensors using electrical transition of carbon nanotube junctions in polymer matrix. *Smart Mater. Struct.* **15**(1), 14–20.

Yuen S.C.L., Lee J.M.H., Li W.J., and Leong P.H.W. (2007) An AA-sized vibration-based microgenerator for wireless sensors. *IEEE Pervas. Comput.* **6**(1), 64–72.

Yun K.S., Gil J., Kim J., Kim H.J., Kim K., Park D., Kim M., Shin H., Lee K., Kwak J., and Yoon E. (2004) A miniaturized low-power wireless remote environmental monitoring system based on electrochemical analysis. *Sens. Actuators B* **102**, 27–34.

Wall R.W. and King B.A. (2005) Power line carrier communications for center pivot irrigation control. In: *Proceedings of International Symposium on Power Line Communications and Its Applications*, 6–8 April, Moscow, ID, 109–115.

Warmer J, Mitrovics J., Kaul P., and Becher C. (2009) A distributed sensing platform for the detection of evaporated hazardous material released from moving sources. In: *Proceedings of International Symposium on Artificial Olfaction and Electronic Noses, ISOEN' 2009*, Apr. 2009, Brescia, Italy, *AIP Conf. Proc.* **1137**, 215–219.

Zhang X. and Riley G.F. (2005) Energy-aware on-demand scatternet formation and routing for Bluetooth-based wireless sensor networks. *IEEE Commun. Mag.* **43**(7), 126–133.

Zhang R. and Dong X. (2008) A new time of arrival estimation method using UWB dual pulse signals. *IEEE Trans. Wireless Commun.* **7**(6), 2057–2062.

Zhao Y. and Ye Z. (2008) A low cost GSM/GPRS based wireless home security system. *IEEE Trans. Consum. Electron.* **54**(2), 567–572.

Zheng J.P., Jow T.R., and Ding M.S. (2001) Hybrid power sources for pulsed current applications. *IEEE Trans. Aerosp. Electron. Syst.* **37**(1), 288–292.

CHAPTER 4

Remote Chemical Sensing

Application for Atmosphere Monitoring

Dong Jiang
Yaohuan Huang
Dafang Zhuang

1. INTRODUCTION

Accompanying the acceleration of urbanization and industrialization, air pollution has become one of the most serious environmental problems on Earth. It affects not only human health but also the health of ecological systems. The atmosphere protects life on Earth by absorbing ultraviolet solar radiation, warming the surface through heat retention, and reducing temperature extremes between day and night. However, severe loss of stratospheric ozone has been detected in the high latitudes of the Northern Hemisphere as well as over the Antarctic. At the same time, intensification of ultraviolet radiation has been observed. Ultraviolet radiation is known to be a danger to human beings as well as having an effect on agriculture, forests, and water ecosystems.

Global air pollution studies have been an important topic. The observation and collection of reliable data on regional and global air quality has a rather brief history. Routine atmospheric measurements of gas and particle concentrations have been conducted at sites with ground-based instruments (such as atmospherically emitted radiance interferometers), which has severely restricted the area of land that can be monitored. The ground instruments are designed to monitor specific pollutants (e.g., carbon dioxide), and many of these instruments cannot provide an accurate description of the total concentration of all pollutants at a regional level (Mark et al. 2004). Remote sensing satellites have many advantages for monitoring air quality. Satellite observations can provide a complete survey of a region, showing the major sources of pollution and the distribution pattern (Xuemei et al. 2001). Since the early 1970s, remote sensing instrumentation have been developed, fabricated, and operated for remotely measuring several atmospheric parameters. They have proved to be efficient tools for atmospheric monitoring. Many new chemical remote sensing sensors, such as NASA's Aura (EOS-CHEM), have been established in recent years. Trace gas and aerosol instrumentation have been developed and operated to measure ambient concentrations of trace gases and aerosols and the exchange of trace gases with the Earth's surface. Such data and information support the research into the atmospheric energy balance, the hydrological cycle, climate trends, and other aspects of the atmospheric system that are of vital interest to us. Remote sensing is central to this effort because it is the only way we can obtain full spatial and temporal perspective needed to understand atmospheric processes (Michael 1993).

This chapter focuses mainly on the remote sensing sensors onboard satellites. Satellite remote sensing of trace gases and aerosols for air quality applications appeared in the middle of the last century. In the 1970s, the global distributions of H_2O, CH_4, and HNO_3 were obtained from the U.S. GOES meteorological satellite. Lyons et al. presented an image from the GOES satellite showing a large area of haze covering the Midwest United States (Lyons et al. 1976). Fraser et al. used GOES observations to conduct the first retrieval of aerosol optical depth over land and applied it to examine a haze event over the eastern United States (Fraser et al. 1984). After that, the TOMS instrument on board the Nimbus-7 meteorological satellite gathered important information about the O_3 distribution in the troposphere. In recent years, satellite remote sensing of air quality has evolved dramatically. Global observations are now available for a wide range of species, including aerosols, tropospheric O_3, tropospheric NO_2, CO, HCHO, and SO_2 (Randall 2008).

2. TECHNIQUES AND INSTRUMENTS FOR ATMOSPHERE MONITORING

2.1. TECHNIQUES FOR ATMOSPHERE MONITORING

Remote sensing in the most generally accepted meaning refers to instrument-based techniques employed in the acquisition and measurement of spatially organized (most commonly, geographically distributed) data/information on some properties (spectral, spatial, physical) of an array of target points (pixels) within the sensed scene that correspond to features, objects, and materials, doing this by applying one or more recording devices not in physical, intimate contact with the item(s) under surveillance (Xuemei et al. 2001). Suffice to say that remote sensing is a tool for gathering information, usually about the surface of the Earth and the atmosphere. Remote chemical sensing techniques for retrieval of atmosphere parameters fall broadly into three categories.

2.1.1. Optical Remote Sensing

Optical remote sensing makes use of visible, near-infrared, and short-wave infrared sensors to form images of the Earth's surface by detecting the solar radiation reflected from targets on the ground (wavelength < 4 mm). Different materials reflect and absorb differently at different wavelengths. When a remote sensing satellite sends a signal toward the Earth, the signal comes into contact with the atmosphere and is modified by the interaction between the radiation and the atmospheric components. The sensor on the satellite then records the modified signal and determines both the geometric and radiometric changes in the signal. The change in the signal is due to particle absorption and elastic scattering. By adding the scattering and absorption components, and integrating these components along the path between the Earth's surface and the satellite's altitude, the particulate optical thickness can be calculated. Additional corrections may be made to the particulate optical thickness (Chance 2006).

Trace gas remote sensing using solar backscatter takes advantage of attenuation in the intensity of radiation traversing a medium. This attenuation is commonly expressed as Beer's law (Randall 2008):

$$I_{\lambda} = I_{\lambda,0} e^{-\sigma_{\lambda} \Omega_s} \tag{4.1}$$

where I_{λ} is the backscattered intensity observed by a satellite instrument at a specific wavelength λ, $I_{\lambda,0}$ is the backscattered intensity that would be observed in the absence of absorption, σ_{λ} is the absorption cross section of the trace gas, and Ω_s is the trace gas abundance over the atmospheric path length, which is commonly referred to as the slant column. Trace gas retrieval using solar backscatter exploits the spectral variation in σ_{λ} to infer Ω_s, including a spectral fit to determine atmospheric abundance over the radiation path and a radiative transfer calculation to determine the path of radiation through the atmosphere (Bowman et al. 2006; Randall 2008).

2.1.2. Thermal Infrared Remote Sensing

Atmosphere monitoring in the thermal infrared uses spectral variation in absorbed and emitted radiation to infer trace gas abundance (spectral range: 4–50 mm). The upwelling thermal intensity at the top of the atmosphere is the sum of contributions from the surface and the atmosphere. The vertical distribution of a trace gas can be obtained by exploiting the pressure dependence of the trace gas spectral emission lines (Bowman et al. 2006).

Trace gas profiles derived from thermal infrared observations typically have little sensitivity near the surface because infrared instruments depend on thermal contrast, although boundary-layer sensitivity is possible under conditions with high contrast between the skin temperature and the air temperature, and with enhanced boundary-layer concentrations (Randall 2008).

2.1.3. Active Remote Sensing

Active sensors onboard satellites, such as lidar, transmit energy downward and measure the backscatter. The differential absorption lidar technique provides three-dimensional mapping of gas distributions in the atmosphere. Pulses from a tunable laser are transmitted into the atmosphere, and photons, elastically backscattered from aerosols and major constituents, are collected by an optical telescope, giving rise to an electrical transient after detection in a photomultiplier tube. It has been found that NO_x, CO, CO_2, SO_2, O_3, etc., can be monitored using active laser remote sensing techniques (Edner et al. 1992).

Laser radar monitoring of the environment is an application of time-resolved laser spectroscopy. Differential optical absorption as well as laser-induced fluorescence can be used for this type of remote sensing. Apart from providing range-resolved data, the use of an active illumination source provides a more accurate assessment than if just the ambient passive radiation is employed (Andersson 1997).

2.2. SENSORS FOR ATMOSPHERE MONITORING

As mentioned in Section 2.1, satellite remote sensing of the atmosphere falls broadly into three categories, so sensors for atmosphere monitoring can also be classified into three types accordingly, as optical sensors, thermal sensors, and active laser radar sensors.

2.2.1. Optical Sensors

The main optical sensors used for atmosphere monitoring are listed in Table 4.1. The objectives, spectral characters, and main applications are described in detail in the following paragraphs.

Table 4.1. Optical sensors for atmosphere monitoring

Sensors	Platform	Period	Spectral range (μm)	Main applications
TOMS	Nimbus-7	1978–1993	0.31–0.38	O_3, aerosol
	Meteor-3	1996–2005		
	AURA			
GOME-1	ERS-2	1995–2003	0.23–0.79	NO_2, HCHO, SO_2, O_3
GOME-2	MetOP	2006–	0.24–0.79	NO_2, HCHO, SO_2, O_3
MOPITT	Terra	1999–	4.7	CO
MODIS	Terra	1999–	0.4–14.4	Aerosol
	Aqua	2002–		
SCIAMACHY	ENVISAT	2002	0.23–2.3	NO_2, HCHO, SO_2, O_3
OMI	Aura	2004	0.27–0.50	NO_2, HCHO, SO_2, O_3, aerosol
POLDER	PARASOL	2004	0.44–1.0	Aerosol

2.2.1.1. TOTAL OZONE MAPPING SPECTROMETER (TOMS)

The TOMS program (Total Ozone Mapping Spectrometer), specializing in ozone retrieval, began with the launch of TOMS Flight Model #1 on the Nimbus-7 spacecraft on October 24, 1978. The Nimbus-7 TOMS instrument measures backscattered ultraviolet radiance from Earth at wavelength bands centered at 312.5, 317.5, 331.3, 339.9, 360.0, and 380.0 nm. The first four wavelengths are sensitive to ozone; the two longer wavelengths are used for estimating the scene reflectivity necessary for deriving ozone amounts. The TOMS instrument fell silent in May 1993. It was the only source of high-resolution global information about ozone. Meteor-3, a Russian satellite, also carried a TOMS. It stopped operating on December 27, 1994. Earth Probe TOMS, onboard AURA, was launched on July 2, 1996, and continued to present TOMS data until it experienced calibration problems in recent years. The Ozone Monitoring Instrument (OMI) onboard AURA is currently the only NASA spacecraft in orbit that specializes in ozone retrieval (Qiu et al. 2008).

TOMS measures the total solar radiance incident on the satellite andcompares it to the ultraviolet radiation scattered back from the atmosphere (40–400 nm). Total column ozone is inferred from the differential absorption of scattered sunlight in the ultraviolet using the ratio of two wavelengths, 312 and 331 nm, for instance, where one wavelength is strongly absorbed by ozone while the other is weakly absorbed. Because it depends on scattered solar radiation, TOMS does not work at night. Ozone measurements given by TOMS are in Dobson units and give the total ozone in a column (http://toms.gsfc.nasa.gov/index_v8.html).

2.2.1.2. GOME-1 AND GOME-2

GOME, a nadir-scanning ultraviolet and visible spectrometer for global monitoring of atmospheric OZone, was launched onboard ERS-2 in April 1995. It measures solar backscatter with broad spectral coverage (230–790 nm) and moderate resolution (0.2–0.4 nm). A key feature of GOME is its ability to detect other chemically active atmospheric trace gases as well as aerosol distribution (www.esa.int/esaLP/ESAST5VTYWC_LPmetop_0.html). GOME-1 has been measuring ozone (total column and profile), nitrogen dioxide, and other minor trace gases since 1995. An advanced GOME-2 instrument on the METOP satellites will provide the input for the ozone data record in the timeframe 2005–2020, provided by the EUMETSAT Polar System. Tropospheric NO_2, HCHO, SO_2, and tropospheric O_3 can be retrieved from GOME-1 data. GOME-2 observes all the species of GOME-1. Information about the spatial-temporal distribution of tropospheric trace gases has been presented based on GOME-2.

2.2.1.3. MOPITT

The MOPITT instrument onboard NASA's Terra satellite is a nadir-viewing gas correlation radiometer operating in the 4.7-μm band, measuring tropospheric carbon monoxide on the global scale. MOPITT has been operational since March 2000. The MOPITT instrument measures upwelling infrared radiances in absorption bands of both CO and methane using the technique of gas-filter correlation radiometry. Ultimately, MOPITT-retrieved CO profiles are either analyzed directly or are assimilated into models to study the chemistry and dynamics of CO (and other constituents) in the lower atmosphere (www.acd.ucar.edu/mopitt/concepts.shtml). MOPITT retrievals of CO have been thoroughly validated in a variety of geographical settings (Emmons et al. 2007). MOPITT measurements enable scientists to analyze the distribution, transport, sources, and sinks of CO, a trace gas produced by methane oxidation, fossil fuel consumption, and biomass burning.

2.2.1.4. MODERATE-RESOLUTION IMAGING SPECTRORADIOMETER (MODIS)

MODIS (Moderate Resolution Imaging Spectroradiometer) is a key instrument aboard the Terra (EOS AM) and Aqua (EOS PM) satellites. Terra's orbit around the Earth is timed so that it passes from north to south across the equator in the morning, while Aqua passes south to north over the equator in the afternoon. Terra MODIS and Aqua MODIS are viewing the entire Earth's surface every 1–2 days, acquiring data in 36 spectral bands with varying spatial resolution of 250, 500, and 1000 m (http://modis.gsfc.nasa.gov/). The channels span the spectral range from 405 to 14,385 nm, and bandwidth varies from channel to channel. Aerosol retrievals over land from MODIS were described originally by Kaufman et al. (1997). Two independent retrievals are conducted at 470 and 660 nm, and subsequently interpolated to 550 nm. The surface reflectance for the channels at 470 and 660 nm are estimated from measurements at 2.1 mm using empirical relationships (Levy 2007). The products from MODIS for atmosphere monitoring include aerosol products, water vapor products, and atmosphere profile products. The aerosol products from MODIS include aerosol type, aerosol optical thickness, particle size distribution, aerosol mass concentration, optical properties, and radioactive forcing; The water vapor product monitors atmospheric water vapor and precipitable water. The atmosphere profile product monitors profiles of atmospheric temperature and moisture, atmospheric stability, and total ozone burden (http://modis.gsfc.nasa.gov).

2.2.1.5. SCANNING IMAGING ABSORPTION SPECTROMETER FOR ATMOSPHERIC CARTOGRAPHY (SCIAMACHY)

The SCanning Imaging Absorption spectroMeter for Atmospheric CartograpHY (SCIAMACHY) is an imaging spectrometer whose primary mission objective are global measurements of trace gases in the troposphere and in the stratosphere. The solar radiation transmitted, backscattered, and reflected from the atmosphere is recorded at relatively high resolution (0.2–1.5 nm) over the range 240–1700 nm, and in selected regions between 2.0 and 2.4 μm (www.iup.uni-bremen.de/sciamachy). SCIAMACHY has three different viewing geometries, nadir, limb, and sun/moon occultation, which yield total column values as well as distribution profiles in the stratosphere and (in some cases) the troposphere for trace gases and aerosols. The large wavelength range of SCIAMACHY is also ideally suited for the detection of clouds and aerosols.

2.2.1.6. OMI

OMI measurements are one of the four instruments on the Aura platform launched on July 15, 2004. The OMI instrument can distinguish among aerosol types, such as smoke, dust, and sulfates, and measures cloud pressure and coverage, which provide data to derive tropospheric ozone. OMI continues the TOMS record for total ozone and other atmospheric parameters related to ozone chemistry and climate (http://aura.gsfc.nasa.gov/instruments/omi.html). OMI is a nadir-viewing imaging spectrometer that uses two-dimensional CCD detectors to measure the solar radiation backscattered by the Earth's atmosphere and surface over 270–500 nm with a spectral resolution of 0.5 nm (Randall 2008). The OMI instrument employs hyperspectral imaging in a push-broom mode to observe solar backscatter radiation in the visible and ultraviolet. The hyperspectral capabilities improve the accuracy and precision of the total ozone amounts and also allow for accurate radiometric and wavelength self-calibration over the long term.

2.2.1.7. POLARIZATION AND DIRECTIONALITY OF THE EARTH'S REFLECTANCE (POLDER)

POLDER (Polarization and Directionality of the Earth's Reflectance) is settled on the PARASOL, the second microsatellite in the Myriade series. POLDER is designed to improve our knowledge of the radiative and microphysical properties of clouds and aerosols by measuring the directionality and polarization of light reflected by the Earth–atmosphere system (http://smsc.cnes.fr/PARASOL). The POLDER instrument consists of a digital camera with a CCD detector array, wide-field telecentric optics, and a rotating filter wheel enabling measurements in nine spectral channels from blue (0.443 mm) through to near-infrared (1.020 mm) and in several polarization directions. Polarization measurements are performed at 0.490, 0.670, and 0.865 mm. The bandwidth is between 20 and 40 nm, depending on the spectral band (Chance 2006; Randall 2008).

2.2.2. Thermal Sensors

The main thermal sensors for atmosphere monitoring are listed in the Table 4.2. The objectives, spectral characters, and main applications are described in detail in the following paragraphs.

Table 4.2. Thermal sensors for atmosphere monitoring

Sensors	Platform	Period	Spectral range (μm)	Application
TOVS	TIROS	1978–	3.5–15.5	O_3
CLAES	UARS	1991–1993	3.5–12.9	O_3, NO, NO_2, HNO_3, $ClONO_3$
IMG	ADEOS	1996–1997	3.3–16.7	CO, HNO_3
AIRS	EOS	2002–	3.7–15.4	O_3, CO, CH_4
TES	Aura	2004–	2.3-15.3	O_3, CO, CH_4, NO, NO_2, HNO_3
IASI	MetOP	2006–	3.62–15.5	O_3, CO, CH_4

2.2.2.1. TOVS

The TIROS Operational Vertical Sounder (TOVS) aboard NOAA's TIROS series of polar orbiting satellites, launched in 1978, consists of three instruments: the High Resolution Infrared Radiation Sounder (HIRS), the Microwave Sounding Unit (MSU), and the Stratospheric Sounding Unit (SSU). The MSU and SSU have been replaced with improved instruments, the AMSU-A and AMSU-B, on the newer satellites (www.ozonelayer.noaa.gov/action/tovs.htm). TOVS has a band at 9700 nm, an important ozone-absorption band. Like the TOMS data, TOVS gives total ozone column concentrations in

Dobson units, but the quality and accuracy of its data are dependent on cloud conditions. TOVS data are best when collected under cloudless conditions. The NOAA satellite that carries TOVS is a polar orbiter that passes close enough to the poles to give continuous data.

2.2.2.2. CRYOGENIC LIMB ARRAY ETALON SPECTROMETER (CLAES)

The CLAES (Cryogenic Limb Array Etalon Spectrometer) instrument was launched on the Upper Atmosphere Research Satellite (UARS) in September 1991. Instead of measuring reflected radiation by using spectroscopy, it determines the amount of ozone by measuring the radiance emitted at several wavelengths. CLAES makes measurements of thermal emission from the Earth's limb in a number of spectral regions which are then used to derive stratospheric altitude profiles of temperature, pressure, ozone (O_3), water vapor (H_2O), methane (CH_4), nitrous oxide (N_2O), nitrogen oxide (NO), nitrogen dioxide (NO_2), dinitrogen pentoxide (N_2O_5), nitric acid (HNO_3), chlorine nitrate ($ClONO_2$), $CFCl_3$, and CF_2Cl_2. Aerosol extinction coefficients are also calculated for each spectral region (http://badc.nerc.ac.uk/data/claes) Unlike data from TOMS, CLAES can provide data at night, since it measures emitted radiation rather than solar radiance. However, its orbit prevents collection of data in the vicinity of the poles (www.lmsal.com/9130.html).

2.2.2.3. INTERFEROMETRIC MONITOR FOR GREENHOUSE GASES (IMG)

The IMG (Interferometric Monitor for Greenhouse gases) was launched as one of eight sensors boarding the ADEOS satellite (Advanced Earth Observing Satellite) in August 1996. The ADEOS satellite ceased to collect and transmit data in June 1997 due to a power failure in its solar panel. IMG is a Michelson-type Fourier transform spectrometer (FTS) with two mirrors and a beam splitter. The incident radiation received from the Earth is divided by the beam splitter into two paths. One mirror is moved so that the two paths produce an interference pattern when they are recombined. The signal measured by the detector, the interferogram, can be inverse Fourier-transformed to obtain the incident spectrum. The diameter of the entrance aperture for the optics is 10 cm. The scanning mirror is suspended on magnetic bearings and scans a 10-cm-long path in 10 s (www.eorc.jaxa.jp/AtmChem/IMG).

IMG was the first high-resolution nadir infrared tropospheric sounder that allowed simultaneous retrieval of several trace gases. IMG obtained detailed spectra of thermal infrared radiation from the Earth's surface and the atmosphere. Thermal infrared spectra include absorption and emission signatures of many atmospheric gases. IMG's high-resolution spectra give atmospheric concentrations of water vapor and other greenhouse gases, and also temperature profiles.

2.2.2.4. ATMOSPHERIC INFRARED SOUNDER (AIRS)

The Atmospheric Infrared Sounder (AIRS), an advanced sounder containing 2378 infrared channels, four visible/near-infrared channels, and a 13.5-km nadir field of view, aimed at obtaining highly accurate temperature profiles within the atmosphere plus a variety of additional Earth/atmosphere products (http://aqua.nasa.gov/about/instrument_airs.php). AIRS is the highlighted instrument in the AIRS/AMSU-A/HSB triplet centered on measuring accurate temperature and humidity profiles throughout the atmosphere. AIRS measures the Earth's outgoing radiation at 0.4–1.0 μm and at 3.7–15.4 μm with 1 K temperature retrieval accuracy per 1-km layer in the troposphere (Liu 2008).

AIRS uses cutting-edge infrared technology to create three-dimensional maps of air and surface temperature, water vapor, and cloud properties. With 2378 spectral channels, AIRS has a spectral resolution more than 100 times greater than previous infrared sounders and provides more accurate information on the vertical profiles of atmospheric temperature and moisture. AIRS can also measure trace greenhouse gases such as ozone, carbon monoxide, carbon dioxide, and methane (http://airs.jpl.nasa.gov/overview/overview).

2.2.2.5. TROPOSPHERIC EMISSION SPECTROMETER (TES)

The Tropospheric Emission Spectrometer (TES) was launched into sun-synchronous orbit aboard Aura, the third of NASA's Earth Observing System (EOS) spacecraft, in July 2004. The primary objective of TES is to make global, three-dimensional measurements of ozone and other chemical species involved in its formation and destruction.

TES is a Fourier-transform infrared emission spectrometer with high spectral resolution (0.1 cm^{-1}) and coverage over a wide spectral range (650–3050 cm^{-1}) (Randall 2008). TES is a high-resolution imaging infrared Fourier-transform spectrometer that operates in both nadir and limb-sounding modes. TES global survey standard products include profile measurements of ozone, water vapor, carbon monoxide, methane, nitrogen dioxide, and nitric acid for 16 orbits every other day. TES Special Observations are research measurements of targeted locations or regional transects which are used to observe specific phenomena or to support local or aircraft validation campaigns (Beer 2006). Tropospheric O_3 and CO are retrieved with an optimal estimation method. In cloud-free conditions the vertical resolution of the O_3 estimate is about 6 km, with sensitivity to both the lower and upper troposphere but reduced sensitivity in the boundary layer (Worden et al. 2004).

2.2.2.6. INFRARED ATMOSPHERIC SOUNDING INTERFEROMETER (IASI)

IASI (Infrared Atmospheric Sounding Interferometer) is a state-of-the-art, sophisticated sounding instrument that will be used for global measurements of atmospheric temperature and moisture with unprecedented accuracy and spectral resolution to improve weather prediction. The IASI instrument consists of a Fourier-transform spectrometer associated with an imaging system, designed to measure the infrared spectrum emitted by the Earth in the thermal infrared using a nadir geometry. The instrument is providing spectra of high radiometric quality at 0.5 cm^{-1} resolution, from 645 to 2760 cm^{-1} (Randall 2008). This instrument is also destined to provide a wealth of data on various components of the atmosphere to further our understanding of atmospheric processes and the interactions among atmospheric chemistry, climate, and pollution. In addition, the IASI will deliver data on land-surface emissive and sea-surface temperature (in cloud-free conditions) (www.esa.int/esaLP/SEMM36BUQPE_LPmetop_0.html).

2.2.3. Laser Radar Sensors

Laser radar sensors are more complex than optical sensors but provide more accuracy in retrieving atmosphere parameters. A laser radar instrument was launched on the Space Shuttle in September 1994. The methodology of global cloud and aerosol monitoring has been tested and interesting results have been achieved. The main laser radar sensors include CALIOP and GLAS, with abilities of cloud monitoring and aerosol profile retrieving.

2.2.3.1. CLOUD-AEROSOL LIDAR WITH ORTHOGONAL POLARIZATION (CALIOP)

Cloud-Aerosol LIdar with Orthogonal Polarization (CALIOP) is a two-wavelength polarization-sensitive lidar that provides high-resolution vertical profiles of aerosols and clouds as well as their optical and physical properties (Winker 2007). CALIOP utilizes three receiver channels: one measuring the 1064-nm backscatter intensity and two channels measuring orthogonally polarized components of the 532-nm backscattered signal. Dual 14-bit digitizers on each channel provide an effective 22-bit dynamic range. The receiver telescope is 1 m in diameter. A redundant laser transmitter is included in the payload. Cloud and aerosol layers are discriminated using the magnitude and spectral variation of the lidar backscatter (Randall 2008). Aerosol extinction profiles are computed with a vertical resolution of 120–360 m from an extinction-to-backscatter ratio, or lidar ratio. Aerosol layers can be detected with sufficient averaging (Winker 2004).

2.2.3.2. GLAS LIDAR

The GLAS lidar (Geoscience Laser Altimeter System) onboard ICESat (the Ice Cloud and Elevation Satellite) was launched in January 2003. It was the first laser altimeter system onboard a spacecraft. The GLAS lidar was designed to measure ice sheet elevation, but it is being applied to retrieve aerosol profiles. GLAS make unique atmospheric observations, including measuring ice-sheet topography, cloud and atmospheric properties, and the height and thickness of radioactively important cloud layers needed for accurate short-term climate and weather prediction (Spinhirne 2005).

3. APPLICATIONS

3.1. AEROSOL RETRIEVAL

Aerosol thickness is an indicator of the overall pollution of an area. Tropospheric aerosols are important components of the earth–atmosphere–ocean system (Kaufman 2005), affecting climate through three primary mechanisms. First, it causes direct radiative forcing results when radiation is scattered or absorbed by the aerosol itself. Second, indirect radiative forcing results when enhanced concentrations of aerosol particles modify cloud properties, resulting in more cloud drops, albeit smaller in size, that generally increase the albedo of clouds in the Earth's atmosphere. Finally, aerosol particles can have an indirect effect on heterogeneous chemistry, which in turn can influence climate by modifying the concentration of climate-influencing constituents (such as greenhouse gases) (Song et al. 2007).

The relative effects of aerosol optical thickness and single scattering albedo on satellite reflection function measurements is the basis for remote sensing of aerosol optical thickness and single scattering albedo from reflected solar radiation measurements. The maximum sensitivity to aerosol optical thickness occurs over dark surfaces. For surfaces brighter than $A_g = 0.1$, where A_g is the surface reflectance, the sensitivity is much reduced and depends on aerosol absorption. Therefore, measurements over ocean surfaces or dark targets over land are most frequently used to detect aerosol optical thickness from space-based sensors, and a combination of dark and bright surfaces are used to detect aerosol single scattering albedo (Kaufman et al. 1997). Many methods have been used over the past 30 years to monitor aerosol thickness, and atmospheric aerosols can be retrieved by different methods, which can be classified by single- and multiple-channel reflectance, multiangle reflectance, the contrast-reduction method, and polarization (Sifakis, 1998).

Remote sensing of aerosol optical properties from space has, in the past, been accomplished using satellite data not explicitly designed with this application in mind. This has included AVHRR data, whose primary purpose was the determination of sea surface temperature and vegetation index, and TOMS data, whose primary purpose was the derivation of total ozone content (Carlson 1977). Since then, not only aerosols above seas but also aerosols above lands have been studied and tested. Sensors applied for these studies include AVHRR, TOMS, TOVS, SeaWiFS, MERIS, GLI, OMI, etc. An in-depth discussion of those sensors was presented by Michael et al. (1999).

Since the end of the last century, quite a few satellite sensors, such as MODIS, HIRIS, PICASSO (USA–NASA), ILAS (Japan), and POLDER (European Space Agency), have been launched successfully with the goal of monitoring cloud and aerosol at the global scale. Many types of international scientific projects, such as IGAC (International Global Atmospheric Chemistry Project), APEX (Asian Atmospheric Particulate Environment Change Studies), etc., had been conducted for global aerosol monitoring with support of remote chemical sensing methods (Dubovik et al. 2008).

3.2. WATER VAPOR RETRIEVAL

Water vapor is one of the most important and most abundant greenhouse gases in the Earth's atmosphere, keeping the temperature of the Earth's surface above the freezing level. Atmospheric water vapor plays a key role in the hydrological cycle, whose distribution is essential in understanding weather and global climate. The distribution of water vapor varies greatly both in space and time, with values ranging from about 5 cm near the equator to less than one-tenth as much at the poles, which can lead to sudden changes in local weather (www.ae.utexas.edu/courses/ase389p_gps/projects99/whitlock /intro.html). In order to develop accurate weather prediction and global climate models, it is vital to monitor water vapor as accurately as possible. The radiosonde network has long been the primary in situ observing system for monitoring atmospheric water vapor. Radiosondes provide vertical profiles of meteorological

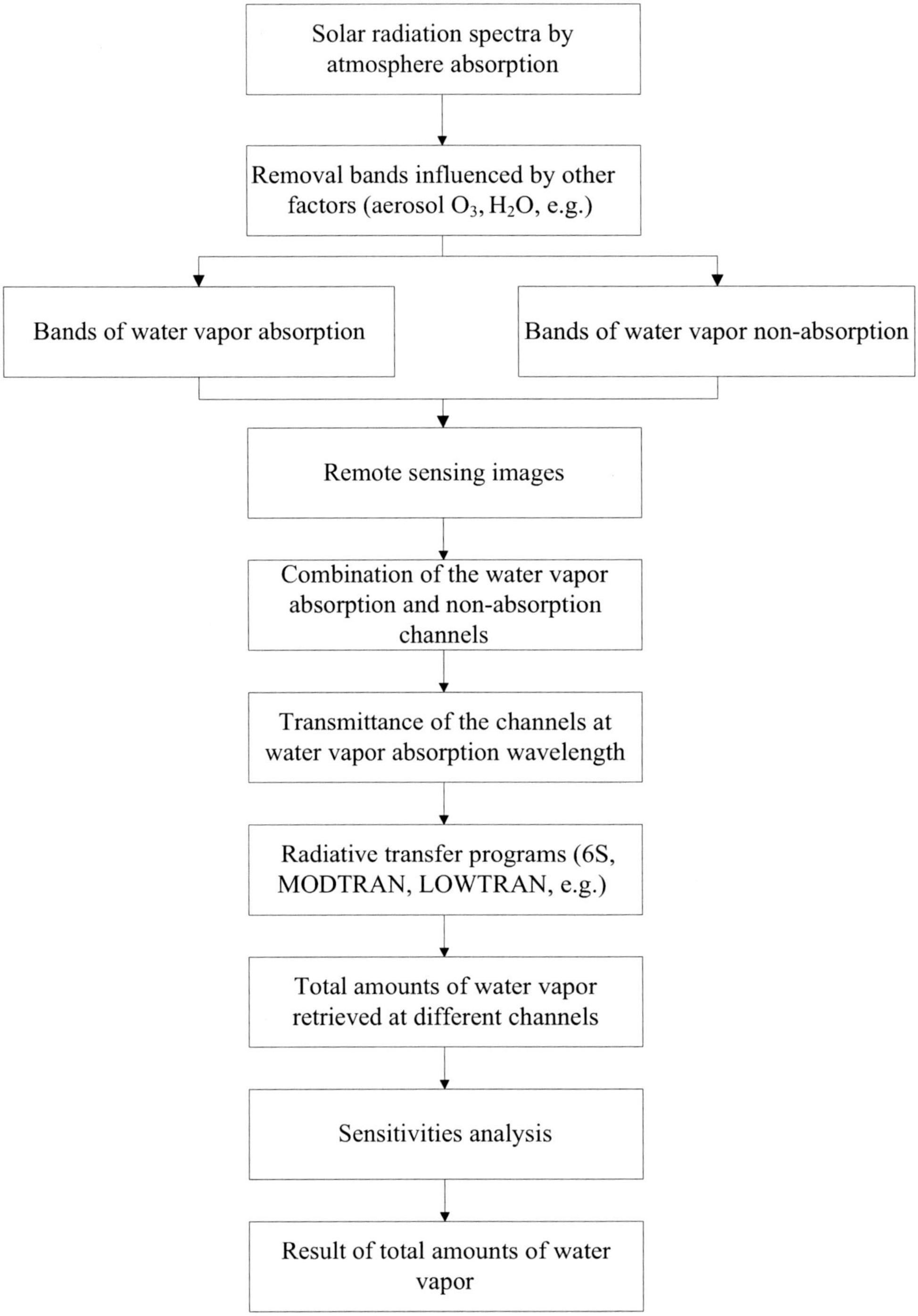

Figure 4.1. Diagram of retrieval of water vapor from MODIS data.

variables such as pressure, temperature, and relative humidity. Sometimes, wind information can be obtained as well.

The Global Positioning System (GPS) is an increasingly operational tool for measuring precipitable water vapor. GPS signals are delayed when propagating through the troposphere. The total tropospheric delay can be divided into a hydrostatic term (ZHD), caused primarily by dry gases in the atmosphere, and a wet term (ZWD), caused by the refractivity due to water vapor (Zhenhong et al. 2008). GPS measurements provide estimates of the total zenith delay (TZD) using mapping functions. If surface air pressure is known with an accuracy of 0.3 hPa or better, ZHD can be estimated. The primary advantage of GPS is that it makes continuous measurements possible. Furthermore, the spatial density of the current Continuous GPS (CGPS) network is much higher than that of the radiosonde network, and its capital and operational costs are much lower than for remote sensing (Zhenhong et al. 2008).

The remote sensing method is based on detecting the absorption by water vapor of the reflected solar radiation after it has transferred down to the surface and back up through the atmosphere. The near-infrared total-column precipitable water is very sensitive to boundary-layer water vapor, since it is derived from attenuation of reflected solar light from the surface. This data product is essential to understanding the hydrological cycle, aerosol properties, aerosol–cloud interactions, energy budget, and climate. Of particular interest is the collection of water vapor data above cirrus cloudiness, which has important applications to climate studies (http://modis-atmos.gsfc.nasa.gov/MOD05_L2/index.html). As a case in point, the technique implemented for the MODIS water vapor retrievals uses ratios of radiance from water vapor–bsorbing channels centered near 0.905, 0.935, and 0.94 μm with atmospheric window channels at 0.865 and 1.24 μm. Both the two-channel and three-channel ratioing techniques are used to retrieve the water vapor for MODIS (http://modis-atmos.gsfc.nasa.gov/MOD05_L2/index.html). Figure 4.1 illustrates the steps in retrieving water vapor from MODIS data.

The output from the Level 2 near-IR water vapor algorithm includes column water vapor amounts on a pixel-by-pixel basis and an associated quality assurance parameter. In addition to the Level 2 near-IR water vapor product, Level 3 (MOD43) gridded products are produced daily, every 8 days, and monthly (Zhenhong et al. 2008).

3.3. ATMOSPHERIC TRACE GASES DETECTING

Oxides of carbon, sulfur, nitrogen, and ozone are serious environmental pollutants produced by the productive and social activities of humans. Hence real-time monitoring and comprehensive control of these pollutants are very important. Atmospheric trace gases data, such as for ozone, are also important for climate research and as an input to numerical weather prediction models. Trace gas measurements will be important for monitoring the long-term effects of global climate change.

3.3.1. Ozone

Since the mid-1950s, total stratospheric ozone amounts have been regularly measured. In the early part of this period, all measurements were made *in situ* by instruments released from the ground. From 1979 until today, a steady decrease in stratospheric ozone has been noted. The decrease has been especially obvious over Antarctica, where an "ozone hole" appears in the spring and disappears in the summer. Each year, this springtime hole covers a larger area than it had the previous year (www.cotf.edu/ete/modules/ozone/ozremote.html).

In 1978 NASA launched the Nimbus-7 satellite, equipped with a Total Ozone Mapping Spectrometer (TOMS). Several other satellites were launched subsequently. Since 1979, several satellites have been equipped with sensors that collect data on ozone, as well as other atmospheric constituents that affect the amounts of ozone present. American, Russian, and Japanese satellites have carried ozone sensors. The satellites that have flown the Total Ozone Mapping Spectrometer (TOMS) have included Meteor-3, Nimbus-7, and, most recently, ADEOS and Earth Probe. The Cryogenic Limb Array Etalon Spectrometer (CLAES) is an instrument that has flown on the Upper Atmosphere Research Satellite (UARS) since 1991. The TIROS-N Operational Vertical Sounder (TOVS) measures radiances.

Ozone retrieval is based on comparison between measured radiances and radiances based on radiative transfer calculations for different amounts of ozone in the atmosphere. Ozone absorbs strongly between 312 and 380 nm, in the ultraviolet region. Comparing what fraction of the incoming radiance in this band is reflected, it is possible to relate this value to the total amount of ozone. NASA maintains a TOMS homepage, with an extensive database and additional information about the project. It is also possible to retrieve the ozone concentration at any point and any time until the previous day (http://toms.gsfc.nasa.gov/index_v8.html).

3.3.2. Nitrogen Dioxide

Nitrogen dioxide (NO_2) is one of the key species in atmospheric chemistry. In the stratosphere, it is involved in catalytic ozone destruction, whereas the photolysis of tropospheric NO_2 results in O_3 formation. In addition, it is indirectly responsible for the atmosphere oxidizing capacity and contributes to radiative forcing of climate (Burrows et al. 1999). However, for a long time, the global distribution of NO_2 could only be analyzed by global chemistry transport models, because ground-based or airborne measurement campaigns were temporally and spatially limited.

Stratospheric NO_2 has been measured by a number of satellite instruments, e.g., LIMS (Limb Infrared Monitor of the Stratosphere), SME (Solar Mesosphere Explorer), SAGE-II/III (Stratospheric Aerosol and Gas Experiment), ISAMS (Improved Stratospheric and Mesospheric Sounder), HALOE (Halogen Occultation Experiment), and POAM (Polar Ozone and Aerosol Measurement). Despite the global coverage of satellite observations, these measurements are characterized by the limited time sampling and high uncertainty in the lower stratosphere (Ionov et al. 2006). The Global Ozone Monitoring Experiment (GOME) in 1995 was the first satellite mission to provide a global picture of atmospheric NO_2 with reasonable spatial and temporal resolution. Unlike previous satellite systems, aiming at individual NO_2 vertical profile measurements, GOME is designed to map the global distribution of the NO_2 vertical column. Since then, similar instruments such as the SCanning Imaging Absorption spectroMeter for Atmospheric CartograpHY (SCIAMACHY), Ozone Monitoring Instrument (OMI), and GOME-2 have been launched into sky. The tropospheric NO_2 maps derived from these instruments have been used to study many scientific applications, such as pollution emissions and pollutant distributions (Hans et al. 1987).

The NO_2 inverse algorithm, the Differential Optical Absorption Spectroscopy (DOAS), from satellite measurements are most popular retrieve methods. The DOAS method determines the NO_2 state column density (SCD) along the light path through the atmosphere based on the Lambert-Beer law. It makes use of a differential absorption signal with respect to an extraterrestrial solar spectrum. The first step of this technique is the removal of aerosol scattering and surface reflecting effects by a low-order polynomial function, then the ring effects have to be considered, and finally the NO_2 SCD is derived based on a spectral fit of NO_2 to a reflectance spectrum. Based on the total SCD, the tropospheric NO_2 SCD is calculated by subtracting the stratospheric NO_2 concentration, and the tropospheric NO_2 SCD is converted to VCD by air mass factor (AMF) (Chen et al. 2009). Kokhanovsky and Rozanov (2009) studied the accuracy of the retrieved NO_2 vertical columns using satellite observations under cloudy conditions using the radiative transfer code SCIATRAN. It was found that the tropospheric nitrogen dioxide columns can be retrieved in the case of thin clouds, if their optical properties and the altitude are retrieved from independent observations. A diagram of the retrieval of NO_2 column concentration information is shown in Figure 4.2.

3.3.3. Other Trace Gases

Most estimates of air quality from satellite observations have focused on ground-level aerosol mass concentration of NO_2 and O_3. However, information on other trace gases (such as CO) concentration is becoming available, in part due to the increasing spatial resolution afforded by more recent instrumentation.

The MOPITT instrument onboard Terra is a nadir-viewing gas correlation radiometer operating in the 4.7-mm band of CO (Drummond and Mand 1996). Satellite retrievals of CO exhibit strong signals from the free troposphere due to broad averaging kernels of current instruments and reduced thermal contrast near the surface. Nonetheless, enhanced signals in CO columns over cities are apparent in long-term averages for SCIAMACHY and MOPITT (Clerbaux et al. 2008). Furthermore, ground-level CO concentrations in regional air quality models are sensitive to boundary conditions, which can be constrained by satellite observations. CO retrievals feature lower tropospheric information in regions with strong thermal contrast such as arid environments (Randall 2008).

Retrieved global SO_2 slant columns from GOME with sufficient accuracy have been very useful to study volcanic plumes and major pollution sources. In 2009 the satellite instruments OCO and GOSAT were launched, which promise a revolutionary improvement in our ability to monitor the greenhouse gases CO_2 and CH_4. HCHO columns and SO_2 columns for both absorption and extinction can be retrieved from OMI (Randall 2008; Chen et al. 2009). To date, significant uncertainties remain in our understanding of the global cycles of trace gases. The large number of measurements and global coverage as provided by satellite instruments could significantly accelerate progress in process understanding, which is urgently needed to understand the atmospheric evolution of trace gases, and prerequisite to climate change prediction (Khokhar et al. 2005).

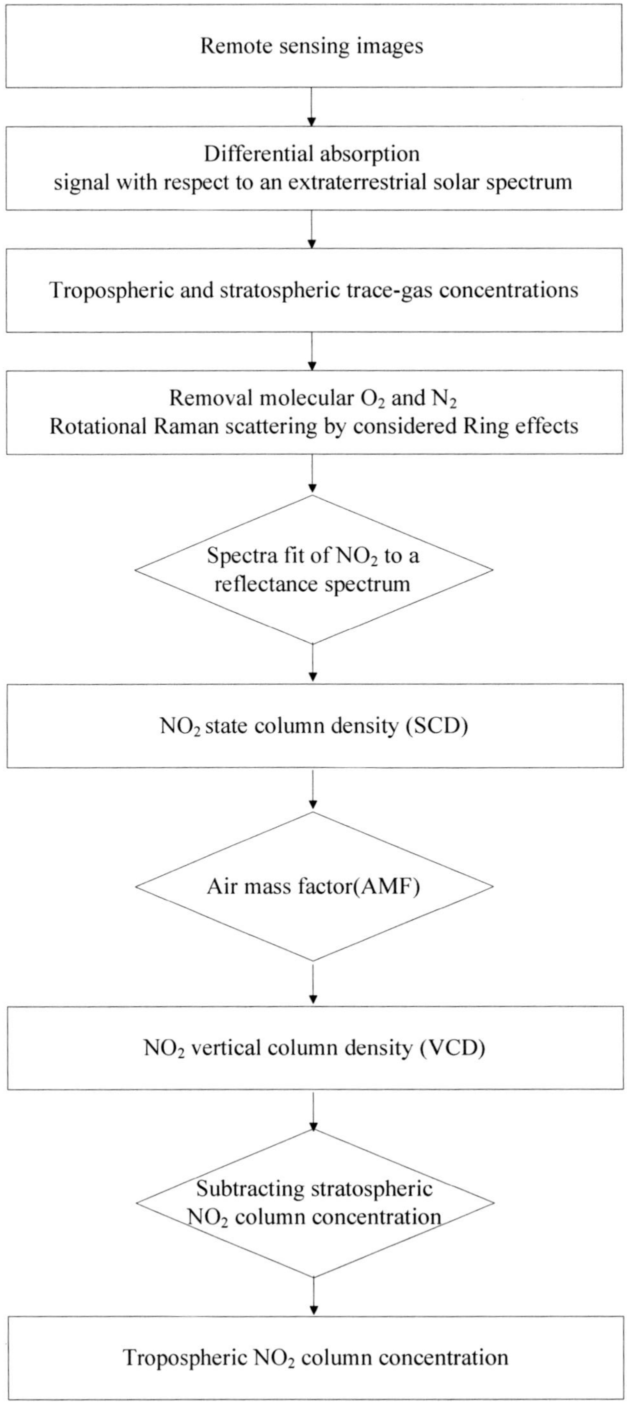

Figure 4.2. Diagram of retrieval of NO_2 column concentration.

4. CONCLUSION

One of the most important ecological issues for our planet is climate change. It is generally agreed that the Earth's climate will modify in response to radiative forcing induced by changes in atmospheric trace gases, cloud cover, cloud type, solar radiation, and tropospheric aerosols (liquid or solid particles suspended in the air). In order to develop conceptual and predictive global climate models, it is vital to monitor these properties. Unfortunately, our knowledge of most climatic parameters is limited, so good climate models exist only for very limited areas of the Earth. For example, atmospheric temperature data, a good index for measuring global warming, are very limited over the ocean. The distribution and sources of greenhouse gases are other major unknowns (http://modis.gsfc.nasa.gov).

Satellite-derived data are essential to obtain global knowledge about these parameters. This requirement led to the development of chemical sensors for monitoring atmosphere quality from space. Remote sensing satellites have many advantages for monitoring air quality. Satellite observations can provide a complete survey of a region, show the major sources of pollution, and the distribution pattern. Since the early 1970s, remote sensing instrumentation had been developed for remotely measuring several atmospheric parameters. Space-based monitoring is the only effective way to assess atmosphere contents distribution on a global basis, and many new chemical remote sensing sensors have been developed in recent years. Trace gas and aerosol instrumentation have been developed and operated to measure ambient concentrations of trace gases and aerosols and the exchange of trace gases with the Earth's surface. Such data and information support researches into the atmospheric energy balance, the hydrological cycle, climate trends, and other aspects of the atmospheric system that are of vital interest to us.

REFERENCES

Andersson M., Edner H., Johansson J., Svanberg S., Wallinder E., and Weibring P. (1997) *Remote Sensing of the Environment Using Laser Radar Techniques*. Springer–Verlag, Berlin, Heidelberg, 257–268.

Beer R. (2006) TES on the aura mission: Scientific objectives, measurements, and analysis overview. *IEEE Trans. Geosci. Remote Sens.* **44**(5), 1102–1105.

Bowman K.W., Rodgers C.D., Kulawik S.S., Worden J., Sarkissian E., Osterman G., Steck T., Ming L., Eldering A., Shephard M., Worden H., Lampel M., Clough S., Brown P., Rinsland C., Gunson M., and Beer R. (2006) Tropospheric emission spectrometer: Retrieval method and error analysis. *IEEE Trans. Geosci. Remote Sens.* **44**, 1297–1307.

Burrows J.P., Richter A., Dehn A., Deters B., Himmelmann S., Voigt S., and Orphal J. (1999) Atmospheric remote-sensing reference data from Gome-2. Temperature-dependent absorption cross sections of O_3 in the 231-794 nm range. *J. Quant. Spectrosc. Radiat. Transfer* **61**(4), 509–517.

Carlson T.N. and Wending P. (1977) Reflected radiance measured by NOAA - 3 AVHRR as a function of optical depth for Saharan dust. *J. Appl. Meteorol.* **16**, 1368–1371.

Chance K. (2006) Spectroscopic measurements of tropospheric composition from satellite measurements in the ultraviolet and visible: Steps toward continuous pollution monitoring from space. In: Perrin A., Ben Sari-Zizi N., and Demaison J. (eds.), *Remote Sensing of the Atmosphere for Environmental Security.* Springer-Verlag, Dordrecht, The Netherlands, 1–25.

Chen L.F., Han D., Tao J.H., and Su L. (2009) Overview of tropospheric NO_2 vertical column density retrieval from space measurement. *J. Remote Sens.* **13**(3), 1–5.

Clerbaux C., Edwards D.P., Deeter M., Emmons L., Lamarque J., Tie X.X., Massie S.T., and Gille J. (2008) Carbon monoxide pollution from cities and urban areas observed by the Terra/MOPITT mission. *Geophys. Res. Lett.* **35**, L03817.

Drummond J.R. and Mand G.S. (1996) The Measurements of Pollution In The Troposphere (MOPITT) instrument: Overall performance and calibration requirements. *J. Atmos. Ocean. Technol.* **13**, 314–320.

Dubovik O., Lapyonok T., Kaufman Y.J., Chin M., Ginoux P., Kahn, R.A., and Sinyuk A. (2008) Retrieving global aerosol sources from satellites using inverse modeling. *Atmos. Chem. Phys.* **8**(2), 209–250.

Edner H., Ragnarson P., Svanberg S., Wallinder E., De Liso A., Ferrara R., and Maserti B.E. (1992) Differential absorption lidar mapping of atmospheric atomic mercury in Italian geothermal fields. *J. Geophys. Res. D* **97**, 3779–3786.

Emmons L.K., Pfister G.G., Edwards D.P., Gille J.C., Sachse G., Blake D., Wofsy S., Gerbig C., Matross D., and Nedelec P. (2007) Measurements of Pollution In The Troposphere (MOPITT) validation exercises during summer 2004 field campaigns over North America. *J. Geophys. Res.* **112**, D12S02.

Fraser R.S., Kaufman Y.J., and Mahoney R.L. (1984) Satellite measurements of aerosol mass and transport. *Atmos. Environ.* **18**, 2577–2584.

Hans E., Kent F., Anders S., Sune S., Leif U., and Wilhelm W. (1987) Mobile remote sensing system for atmospheric monitoring. *Appl. Opt.* **26**(19), 4330–4335.

Ionov D.V., Sinyakov V.P., and Semenov V.K. (2006) Validation of GOME (ERS-2) NO_2 vertical column data with ground-based measurements at Issyk-Kul Kyrgyzstan. *Adv. Space Res.* **37**, 2254–2260.

Kaufman Y.J., Remer L.A., Tanre D., Rong-Rong L., Kleidman R., Mattoo S. Levy R.C., Eck T.F., Holben B.N., Ichoku C., Martins J.V., and Koren I. (2005) A critical examination of the residual cloud contamination and diurnal sampling effects on MODIS estimates of aerosol over ocean. *IEEE Trans. Geosci. Remote Sens.* **43**, 2886–2897.

Kaufman Y.J., Tanré D., Gordon H.R., Nakajima T., Lenoble J., Frouin R., Grassl H., Herman B.M., King M.D., and Teillet P.M. (1997) Passive remote sensing of tropospheric aerosol and atmospheric correction for the aerosol effect. *J. Geophys. Res.* **102**, 16815–16830.

Khokhar M.F., Frankenberg C., Beirle S., Kuhl S., Van Roozendael M., Richter A., Platt U., and Wagner T. (2005) Satellite observations of atmospheric SO_2 from volcanic eruptions during the time period of 1996 to 2002. *Adv. Space Res.* **36**, 879–887.

King M.D., Kaufman Y.J., Tanré D., and Nakajima T. (1999) Remote sensing of tropospheric aerosols from space: Past, present, and future. *Bull. Am. Meteorol. Soc.* **80**(11), 2229–2259.

Kokhanovsky A.A. and Rozanov V.V. (2009) Retrieval of NO_2 vertical columns under cloudy conditions: A sensitivity study based on SCIATRAN calculations. *Atmos. Res.* **93**(4), 695–699.

Levy R.C., Remer L.A., Mattoo S., Vermote E.F., and Kaufman Y.J. (2007) Second generation operational algorithm: Retrieval of aerosol properties over land from inversion of moderate resolution imaging spectroradiometer spectral reflectance. *J. Geophys. Res.* **112**,. D13211.

Liu C.Y., Li J., Weisz E., Schmit T.J., Ackerman S.A., and Huang H.L. (2008) Synergistic use of AIRS and MODIS radiance measurements for atmospheric profiling. *Geophys. Res. Lett.* **35**(21), L21802.

Lyons W.A. and Husar R.B. (1976) SMS/GOES visible images detect a synoptic-scale air pollution episode. *Month. Weath. Rev.* **104**, 1623–1626.

Mark T. and Jonathan L. (2004) Applications of satellite remote sensing to urban air-quality monitoring: Status and potential solutions to Canada. *Environ. Inform. Arch.* **2**, 846–854.

Michael A.J. (1993) *Atmospheric Remote Sensing by Microwave Radiometry.* Wiley-Interscience, New York.

Qiu J.H., Wang P.C, Xia X.A., Duan M.Z., and Zong X.M. (2008) Recent progresses in atmospheric remote sensing researches. *Chinese J. Atmos. Sci.* **32**(4), 841–853.

Randall V.M. (2008) Satellite remote sensing of surface air quality. *Atmos. Environ.* **42**, 7823–7843.

Sifakis N.I., Soulkellis N.A., and Paronis D.K. (1998) Quantitative mapping of air pollution density using earths observations: A new process method and application to an urban area. *Int. J. Remote Sens.* **19**(17), 3289–3300.

Song W. and Zhang L. (2007) A review of remote sensing for aerosol optical depth. *Arid Meteorol.* **25**(3), 77–81.

Spinhirne J.D., Palm S.P., Hart W.D., Hlavka D.L., and Welton E.J. (2005) Cloud and aerosol measurements from GLAS: Overview and initial results. *Geophys. Res. Lett.* **32**, L22S03.

Winker D.M., Hunt W.H., and Hostetler C.A. (2004) Status and performance of the CALIOP Lidar. In: *Proc. SPIE*, **5575**, 8–15.

Winker D.M., Hunt W.H., and McGill M.J. (2007) Initial performance assessment of CALIOP. *Geophys. Res. Lett.* **34**, L19803.

Worden J., Kulawik S.S., Shephard M.W., Clough S.A., Worden H., Bowman K., and Goldman A. (2004) Predicted errors of tropospheric emission spectrometer nadir retrievals from spectral window selection. *J. Geophys. Res.* **109**, D09308.

Xuemei W., Ruru D., and Zhijian H. (2001) Application of remote-sensing technology in atmospheric environmental assessment. *Air Pollution* **IX**, 521–528.

Zhenhong L., Jan-Peter M., Paul C., Hong G., Yang P., Baum B.A., and Heymsfield A.J. (2008) Relationship between ice water content and equivalent radar reflectivity for clouds consisting of nonspherical ice particles. *J. Geophys. Res.* **113**(D20), D20205.

CHAPTER 5

CHEMICAL SENSORS IN OUR LIVES

G. Harsanyi

1. INTRODUCTION

Although materials research and development of sensor devices have been progressing rapidly, practical applications have requirements that are often mismatched to achieved results. Sensors are often applied without clear knowledge of underlying theory. There are thus debates about signal excitation mechanisms, signal conditioning methods, and the interpretation of both practically measurable and theoretically accepted results. On the other hand, many applications have requirements that can be fulfilled using existing methods only with difficulty and by accepting compromises. Although the development of sensor materials and devices needs multidisciplinary knowledge, this is even truer for the applications: Each application field is a "vocation" with very special challenges, both from the application environment and from the requirements on sensor parameters.

The present chapter provides a broad survey of the practical applications of chemical sensors in our lives. In dividing the chapters into sections, we concentrate not on the measured parameter but on the application fields. This chapter is devoted to those areas where everybody may meet sensors: biomedical, food quality, water quality, environmental air monitoring, and household applications.

The biggest competitor of chemical sensors in practice is the conventional analytical chemical instrumentation, with its always-emphasized disadvantages: large instruments and high-cost measurements. Sensors offer the possibility of continuous monitoring using low-cost devices, but the always returning problem is the accuracy and reliability of the measurements.

The following main areas are discussed in this chapter: biomedical sensor applications; sensors for monitoring blood gases and pH; invasive and transcutaneous electrochemical/optical sensors; combined biomedical sensors; ionic compounds in blood and other secretions; pH in gastric acid; enzymatic biosensors; glucose sensors; urea sensors; water quality monitoring; pH; dissolved oxygen; metal ions; heavy metals, organic pesticides, oil contamination monitoring; ammonia monitoring; integral water quality; biological oxygen demand; food quality monitoring; pH in foodstuff and beverages; vegetable extract monitoring; vegetables ready to use; phenol and dioxin in food; milk quality; milk fermentation; meat freshness; fish freshness; antibiotic residues in food; vitamin levels of oil and fat; electronic noses in food identification and classification; wine and beer qualification; biosensors in fermentation processes; ethylene monitoring in warehouses; smoke detection in cacao; volatile organic gas compound monitoring; ammonia detectors; environment gas detection; air pollutant detection; air quality, ventilation control, rain pH monitoring; airborne microbes; humidity sensing; relative humidity and dew-point sensor types; hygrometers; indoor climate control; condensation detection; alcohol tester; and residential sensor applications.

2. BIOMEDICAL APPLICATIONS[1]

The largest group of chemical sensors for measuring concentrations of various analytes in biomedicine are applied for *in vivo* or *ex vivo* monitoring of blood components, such as dissolved oxygen and CO_2 partial pressure, pH, and concentrations of ionic compounds. The oxygen saturation of blood hemoglobin is monitored by physical sensors (photosensors) using the methods of oximetry, which is different from dissolved blood oxygen measurements. Although the majority of sensors are applied for blood analysis, their use in monitoring secretions (gastric acid, sweat, etc.) is also becoming more and more important. Beyond the following description, some further details about chemical sensors applied in biomedicine can be found in the literature (Göpel et al. 1991; Fraser 1997; Spichiger-Keller 1998; Diamond 1998; Harsanyi 2000).

2.1. BLOOD CONTROL

Arterial blood gas analysis is one of the most frequently requested tests performed on critically ill patients both in the operating room and in the intensive care unit. The partial pressures of oxygen and carbon dioxide (pO_2 and pCO_2) in arterial blood, as well as pH, are determined to provide physicians with information about respiratory and metabolic imbalances as reflected by the adequacy of blood oxygenation and CO_2 elimination. With frequent measurement of blood gas parameters, the physician can correct these imbalances by adjusting mechanical ventilation, the ventilation gas composition, or administering pharmacological agents. Blood gas measurements are particularly important in prematurely newborn (preterm) babies who are vulnerable to respiratory illnesses because of immaturity of the lung or an inadequately developed mechanism for the control of breathing. If the arterial pO_2 falls to a low

[1]This section is reproduced in part from G. Harsányi, *Sensors in Biomedical Applications*, pp. 169–221, Technomic Publishing, Lancaster, PA, 2000, © 2000, Technomic Publishing Co., Inc., with permission of CRC Press, A Division of Taylor & Francis Books, Inc.

level for even a short period of time, brain damage or even death may follow. If, on the other hand, the arterial pO_2 goes too high as a result of the administration of oxygen, damage to the eyes, which can result in blindness, may occur. Moreover, rupture of blood vessels in the brain can happen as a consequence of fluctuations in cerebral blood flow caused partly by changes in pCO_2. The third parameter, the pH value, provides information about the general state of metabolism.

Individual blood samples give information at only one point in time—a snapshot of what is happening. It is known that large variations in blood gas levels can occur over the course of a few minutes. Traditional and new-type small bed-side analyzers suffer from the same shortcomings: loss of patient blood each time a test is requested and errors as well as loss of time during transfer of the blood sample from the patient to the analyzer. The ideal blood analyzer should allow continuous and accurate measurement that can only be realized by means of sensors. Continuous *in vivo* blood monitoring means collecting data directly from the body, and it can be performed either invasively or noninvasively. *Ex vivo* sensors are used to monitor blood samples which are continuously taken from the body by means of a catheter or cannula system.

Before starting with a detailed description, the background of blood gas transport and some related definitions have to be clarified. Oxygen is carried in blood in two forms: oxygen chemically combined with hemoglobin (Hb) as oxyhemoglobin (HbO_2) and oxygen physically dissolved in the plasma. The degree to which the oxygen is chemically combined with Hb is most commonly expressed as blood oxygen saturation and is defined according to the following expression:

$$SO_2 = (HbO_2)/(Hb + HbO_2) \tag{5.1}$$

It is generally given as a percentage. Normally, arterial blood saturation (S_aO_2) fluctuates around 98%, while venous blood saturation (S_vO_2) is approximately 75%. These definitions assume that no other hemoglobin species (e.g., combined with CO) is present in the blood.

Arterial pO_2 and S_aO_2 are both good indicators of blood oxygenation. Their relationship is given by dissociation curves (Parker 1987) that can be handled only with the exact measurement of blood pH. There are, however, a number of other factors, any one of which can lead to a displacement of the dissociation relationships. In the saturation ranges, the derivation of real S_aO_2 from the value calculated using pO_2 is likely to give large errors; thus, the direct measurement of S_aO_2 should be preferred. On the other hand, patients breathing 100% oxygen will have a pO_2 over 66.6 kPa (500 mm Hg), though pO_2 values cannot be practically derived from S_aO_2 measurements over 16 kPa (120 mm Hg). Therefore, the direct measurement of pO_2 values is desirable during anesthesia, for patients with artificial ventilation, and for babies in incubators. As a conclusion, ideally both parameters should be monitored directly. The needed measurement range of pO_2 is 4–80 kPa (30–600 mm Hg) with a resolution of 0.133 kPa (1 mm Hg) in the low concentration range and 1.33 kPa (10 mm Hg) in the high concentration range.

CO_2 is carried by the blood both in a dissolved state and in combination with Hb and plasma proteins as carbamino compounds. The normal pCO_2 of arterial blood is about 5.3 kPa (40 mm Hg), whereas that of venous blood is about 6 kPa (45 mm Hg). The needed measurement range is 2.7–13.3 kPa (20–100 mm Hg) with a resolution of 0.133 kPa (1 mm Hg).

Normal physiological pH fluctuates between 7.0 and 7.4. The generally accepted measurement range is rather narrow (6.8–7.8), but high precision (0.01) is needed. It is estimated that sensor-measured

pO_2 and pCO_2 values that are within 5–10% of blood gas analyzer values, and pH measurements that agree to within 0.02–0.04 pH units, will lead to equivalent clinical treatment decisions.

Blood gas and pH measurement is a serious challenge for sensorics. The most important technical requirements are an operating temperature range from 20 to 40°C, a response time less than 3 s, minimal cross-sensitivity among CO_2, O_2, and pH, and insensitivity to anesthetic gases (Soller 1994). A few further requirements are sterilizability, rapid and easy calibration, low cost and disposability (especially of intravascular elements), biocompatibility (thromboresistant and nontoxic), and the ability to fit inside 20-gauge (0.024-in.) catheters without compromising arterial pressure measurements.

2.1.1. Invasive and Transcutaneous Electrochemical Sensors

A general description of electrochemical cell-type sensors has been given previously. The special structures and operating principles of electrochemical pH, pO_2, and pCO_2 sensors adaptable for biomedical use are briefly summarized here.

- For pH, ion-sensitive potentiometric glass and polymer membrane electrodes are used. The basis of a pH-dependent Nernstian electrode potential measurable on these electrodes is the selective and reversible H^+ ion adsorption onto special-composition glass surfaces, as well as a complex formation with ionophores within polymer membranes, respectively. Metal/metal oxide electrodes can also be used for pH measurements, if the electrochemical oxidation of the metal is reversible.
- Dissolved oxygen concentration (partial pressure) can be measured by means of a Clark-type amperometric sensor (Clark 1958). The general structure contains a reference Ag/AgCl electrode. The cathode is generally made of platinum or gold (sometimes silver or graphite), and the electrolyte solution usually contains potassium chloride with a buffering agent. The whole electrochemical cell is separated from the sample liquid by a diffusion membrane that is permeable to oxygen but impermeable to water, ions, proteins, and blood cells. The most commonly used membrane materials are polytetrafluoroethylene (PTFE), polyethylene (PE), and polypropylene (PPy). The cathode reaction is the basis of operation. The cathode is usually polarized at -700 mV with respect to the reference electrode so that the sensor is operating in the diffusion-controlled limiting current region; thus, the cell current is theoretically a linear function of the external oxygen concentration.
- For the measurement of pCO_2, the Stow-Severinghaus type of potentiometric electrode can be used (Severinghaus and Bradely 1958). This is a glass pH electrode covered by a membrane that is freely permeable to carbon dioxide molecules but impermeable to water, hydrogen ions (H^+), and bicarbonate ions (HCO_3^-). A sodium bicarbonate ($NaHCO_3$) layer is placed between the pH electrode and the membrane. CO_2 molecules diffuse through the membrane to equilibrate with the electrolyte through the following reaction:

$$CO_2 + H_2O \Longleftrightarrow H_2CO_3 \Longleftrightarrow H^+ + HCO_3^- \tag{5.2}$$

In equilibrium, the pH value corresponds to the pCO_2:

$$pH = \text{constant} - \log pCO_2 \tag{5.3}$$

Thus, the glass electrode measures the external pCO_2 indirectly.

The invasive (intravascular) application of electrochemical sensors for blood gas and pH measurement seems easy to perform, since only electrodes (coated wires) have to be placed into the vessels. The practical problem arises from the fact that not only a single electrode but also electrode pairs or even whole electrochemical cells must be catheterized. Three main approaches may be distinguished in connection with the invasive application of sensors:

- Miniature membrane-covered cylindrical-shape electrochemical cells on catheter tips for *in vivo* monitoring
- Electrochemical cells mounted into the sample-taking cannula/catheter/dome system for *ex vivo* applications
- Miniaturized silicon sensors, made by the processing methods of microelectronics and micromachining, which are mounted either onto catheter tips or into cannula/catheter/dome systems.

The primary device for pH measurements is the glass electrode; catheter-size versions have been produced using pH-sensitive Pt wires covered with soda-lime glass. The main drawback of these electrodes, the possibility of the glass breaking, is the reason for searching for other solutions. Indium and palladium oxide electrodes can be used without this problem; however, their multiple valencies result in complicated pH-dependent characteristics. Today, silicon-based pH-sensitive ion-selective field-effect transistors (ISFETs) can be fabricated employing either SiO_2/Si_3N_4 or ion-selective poly(vinyl chloride) (PVC) membrane gate-insulating layers.

Sensors for measuring blood pO_2 are based almost exclusively on the Clark principle. Figure 5.1 shows a typical catheter-tip sensor structure fabricated for practical applications (Rolfe 1994). It contains the basis salt of the electrolyte in dry solid powder form between the surface of the electrodes and the plasticized PVC membrane. The membrane must allow water molecules to diffuse through it from blood into the electrolyte region and thereby dissolve the powder, forming a liquid electrolyte layer. The sensor uses a thimble cathode pushed over the tip of the catheter, with a hole at the anode Ag/AgCl reference electrode location. The type illustrated needs a water vapor–permeable membrane; this is the reason for using plasticized PVC. The silver cathode is used primarily because of its low cost.

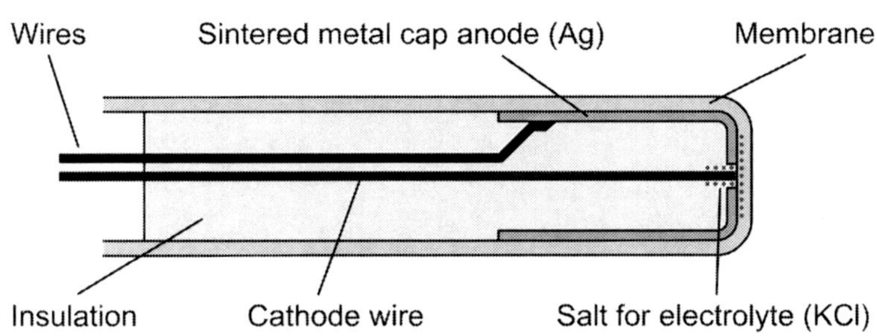

Figure 5.1. Conventional Clark-type blood oxygen sensor for catheter tip applications.

Very similar problems have to be solved when fabricating pCO_2 sensors for blood gas measurements. The most important differences are as follows:

- The cathode is a pH electrode.
- The electrolyte is generally $NaHCO_3$.
- Potentiometric or conductimetric measurements have to be used.
- The applicable membrane material is PTFE or silicone rubber.

Although intravascular monitoring remains an important part of intensive care, catheterization of arteries has some risks, especially in newborn babies. This is one of the reasons which have led to the advent and rapid growth of noninvasive blood gas monitoring. Of these, the transcutaneous technique is of much interest because it enables blood gas measurements through the skin. Human skin is practically impermeable to gas molecules at normal conditions. Baumberger and Goodfriend revealed in 1951, however, that if skin temperature is increased, it can behave as a semipermeable membrane. Near 45°C, pO_2 measured at the surface can approach the arterial value quite closely. The estimation of arterial pO_2 and pCO_2 based on this noninvasive method is called transcutaneous technique. The sensor elements are generally Clark or Stow-Severinghaus electrochemical cells equipped with heating elements and temperature sensors that enable operation within a well-controlled narrow temperature range.

Good estimation of arterial blood gas parameters is based on the correct choice of skin temperature (Huch et al. 1972). Heating the skin causes several effects to be considered for pO_2 measurements. According to empirical results, phenomena that cause a virtual change in pO_2 at the sensor site may balance each others' effects by an appropriate choice of temperature. A sensor temperature of 43.5 ± 0.5°C has been found to be optimal for reliable monitoring of arterial pO_2, ensuring good correlation between values measured by transcutaneous and invasive methods (Parker 1987).

A new type of transcutaneous pO_2 sensor structure is shown in Figure 5.2 (Harsanyi 2000), which is based on the multilayer ceramic technology that is used in the fabrication of high-density electronic interconnection systems. The electrodes are made by thick-film technology using small amounts of precious metals. The heating element is an integrated thick-film resistor, and a *p-n* junction is used for temperature sensing. The disc-shaped ceramic sensor body has a groove at the edge for packaging purposes and a deep cavity in its center for the temperature sensor chip. The sensor chip should be as close to the skin surface as possible for the necessary accuracy of temperature control. A diode chip is used for temperature sensing. The horseshoe-shaped heating resistor is screen-printed and fired on the top of the ceramic body using common thick-film processing. It consists of several segments to give more freedom for heating. The sensing electrode is fabricated of thick-film gold paste. The Ag film for the Ag/AgCl reference electrode is also formed by thick-film technique before being electrochemically chloridized in a 0.1 M HCl solution. The ceramic body is molded into a plastic package. The PTFE membrane, with a thickness of 10 μm, is fastened with a mounting ring to the surface of the sensor, as shown in the cross section in Figure 5.2. A conventional KCl solution with a buffering agent is used as electrolyte. A thin disc-shaped cellophane foil is placed between the electrodes and the membrane as a spacer. The application of multilayer wiring helps to make an appropriate electrode separation necessary to reduce the pickup of the hydrogen peroxide intermediate, which causes electrochemical cross-talk and electrode consumption. A very similar sensor structure based on thick-film technology was reported recently by Lam and Atkinson (2007).

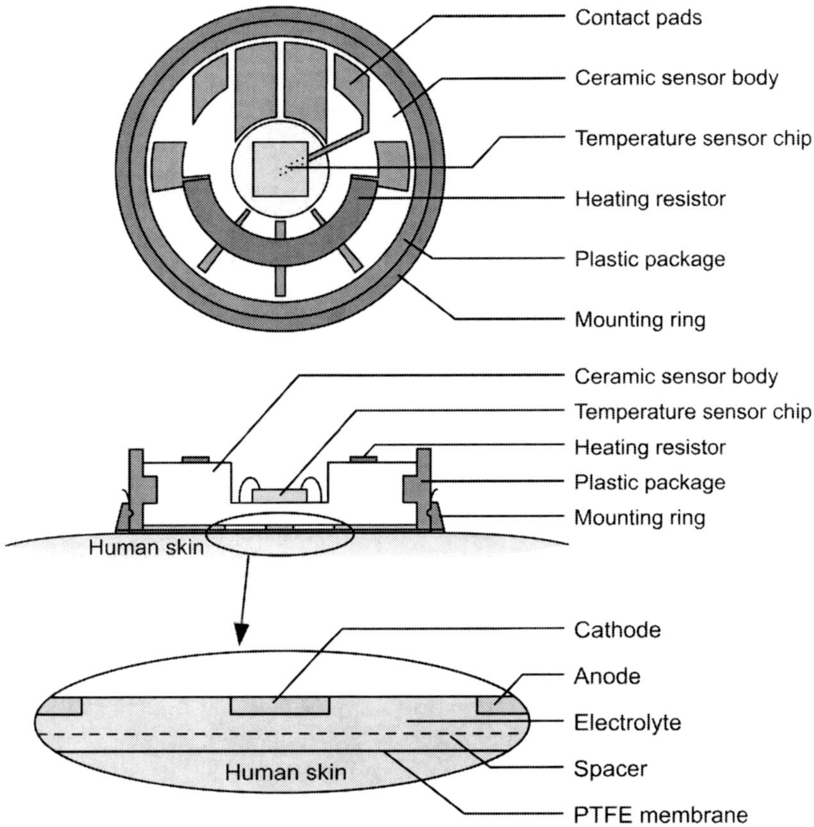

Figure 5.2. Structure of the ceramic transcutaneous blood O_2 sensor: (a) top view; (b) cross section. (Reprinted with permission from Harsányi 1995. Copyright 1995 CRC Press—A Division of Taylor and Francis Books Inc.)

Transcutaneous pCO_2 measurement relies on a similar technique. It is performed by means of heated Stowe-Severinghaus electrochemical sensors. A newer approach is the application of pH ISFETs instead of electrodes. A novel miniature transcutaneous pCO_2 sensor was developed recently using a pH ISFET, a conventional Ag/AgCl electrode, a hydrogel gate membrane containing 0.1 M $NaHCO_3$ electrolyte, a heater coil, and a *p-n* junction temperature sensor (Jinghong et al. 1995). The whole structure is covered with a CO_2 gas–permeable membrane. The sensor's measurement range is 0.53–40 kPa (4–300 mm Hg) and has a response time of less than 2 min.

2.1.2. Optical Fiber Sensors

Optical fiber sensors for blood gas and pH monitoring enable invasive, generally intravascular measurements and have optrode-type structure. There are both single-fiber and multiple-fiber realizations. In the case of a single-fiber optrode, the incident and outcoming light have to be separated by means of a beam splitter. In indicator-mediated sensors, the absorbance is measured; therefore, the reflection from the fiber tip should be increased by using reflecting surfaces. In fluorescent optrodes, a secondary light emission is excited; thus, the reflection of the primary light has to be prevented as much as possible. Currently, low-cost appliances generally operate with light-emitting diode (LED) sources and photodiode detectors (Wolfbeis 1991). Figure 5.3 shows a typical structure of a double-fiber, reflection-mode optrode catheter.

Optrodes for measuring blood pO_2 are generally based on fluorescent quenching of oxygen-sensitive fluorescent dyes such perylene dibutyrate (Soller 1994). The excited light intensity can be described by the Stern-Volmer equation. Neglecting the nonlinearity coming from the nonunit extension, the measurable intensity is

$$I = I_0/(1 + k\,pO_2) \tag{5.4}$$

where I_0 is the fluorescent emission intensity in the absence of the quencher and k is an empirical constant. The best sensitivity is achieved in the region 0–20 kPa (0–150 mm Hg), but it drops off considerably at higher levels, making it difficult to resolve small changes when the oxygen partial pressure is higher than 26.6 kPa (200 mm Hg). The biggest problem is interference by anesthetic agents: Narcotic halothane acts in a similar manner as oxygen in fluorescent quenching (Wolfbeis 1991).

Optical fiber blood pH sensors are mostly reflection-type optrodes based on acid–base indicator dyes. The absorbance is measured generally at two wavelengths: one with the maximal pH dependence and another which is insensitive to pH and can be used as a reference. Combining the Lambert-Beer equation with chemical equilibrium equations, the absorbance can be expressed as

$$\ln\frac{I_0}{I} = hdc_{\mathrm{dye}}\frac{K}{K + c_{\mathrm{H^+}}} \qquad (\mathrm{pH} = -\log c_{\mathrm{H^+}}) \tag{5.5}$$

where I is the output, I_0 is the incoming light intensity, h is the extinction coefficient, d is the optical path length, c is the concentration of the dye material, K is the chemical equilibrium constant of the protonation-deprotonation process occurring with the dye, and c_{H^+} is the hydrogen-ion concentration in the solution.

Blood pH measurements have to be performed in a narrow range with high accuracy (see above). One of the most effective ways to achieve resolution goals is to optimize the pK of the dye material

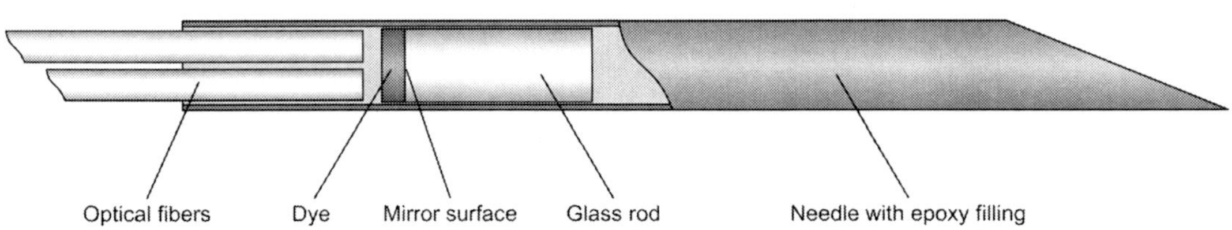

Figure 5.3. Typical structure of a double-fiber optrode for intravascular measurements.

(Soller 1994). When the pK of the dye is at the center of the range to be measured, 7.4 in the case of blood pH, maximum sensitivity will be achieved. The first blood pH optrodes employed phenol red indicator dye. A pH sensor with phenol red was realized for the physiological detection range with an accuracy of 0.01 and with a temperature coefficient of 0.017 pH units/°C (Peterson et al. 1980).

2.1.3. Oximetry

Optical oximeter devices measure oxygen saturation (SO_2) of blood or tissue by exploiting the differences between the absorption spectra of organic compounds transporting oxygen, such as hemoglobin (Hb). Most oximeter sensing appliances determine blood oxygen saturation [see Eq. (5.1)], based on the different absorption (extinction) spectra of Hb and HbO_2 (Parker 1987). This is actually the cause of the different colors of arterial and venous blood. The measurement employs a minimum of two different wavelengths: one in the red range (generally around 660 nm) and another in the infrared range (between 805 and 1000 nm). The extinction coefficient is the same for Hb and HbO_2 at 805 nm, so this (isobestic) wavelength can serve as a reference. In reflection-mode oximetry, back-scattered light from the specimen is sampled at two different wavelengths (λ_1 and λ_2) and the oxygen saturation is estimated with the reflectance [$R = \ln(I_0/I_r)$, where I_0 is the incident and I_r is the reflected intensity] ratio, according to the following expression:

$$SO_2 = A - B\,[R(\lambda_1)/R(\lambda_2)] \tag{5.6}$$

where A and B are empirical constants. In transmission mode, more complicated formulas are used (Takatani and Ling 1994; Mendelson 1991).

Originally, oximetry was used for the analysis of blood samples. Nowadays, sensor analyzer systems enabling continuous *in vivo* monitoring are the focus of interest. For intravascular oximetry, plain optical fibers are used to guide the light signal inside the vessel and the light reflected from red blood cells back to the light detector. Because of the well-known problems and risks of catheterization, methods of noninvasive oximetry have also been developed extensively.

Pulse oximetry is a noninvasive determination of blood S_aO_2. The basis of the technique is to measure the change in light transmitted through the skin that occurs as a result of arterial pulsation. The signal varies with plethysmographic pulsating changes in tissue blood volume (Mendelson 1991). It is assumed that the change in light transmitted through tissue during the inflow phase of the cardiac cycle (i.e., systole) is caused solely by the arterial blood and there is no pulse from the surrounding tissue and the pulse of venous blood is normally insignificant. Consequently, the pulsating component of the optical signal has to be measured. The optical signal is usually sampled at two wavelengths, one in the red (e.g., at 660 nm) and another in the infrared region (e.g., at the isobestic 805 nm, or at 940 nm). A conventional linear regression according to Eq. (5.6) is applied to obtain arterial saturation. S_aO_2 measured by pulse oximetry is often marked as S_pO_2. Most pulse oximeters are transmission type, in which forward-scattered light through the fingertip or the ear lobe is analyzed. Reflection-type pulse oximeters have also been developed, with more general applicability to any portion of the body, such as the forehead, cheek, calf, and thigh. However, in comparison to transmission, the reflection pulse oximeters have poorer signal-to-noise ratio. New S_pO_2 measurement appliances employ only two LEDs and a single photodiode on ceramic substrates in a variety of sensor heads for several applications. These include the adult finger glove, pediatric finger glove, and neonatal foot strap and earlobe clip, combined with a single measurement appliance (Kästle et al. 1997).

Since pulse oximeters rely on adequate arterial pulsation, a significant reduction in peripheral vascular pulsation, such as in hypotension, vasoconstriction, or hypothermia, can produce a signal too small to be processed reliably by the oximeter. Despite this limitation, pulse oximetry is becoming the most popular and useful technique for blood oxygen measurement and monitoring. Recently, it has been recommended as a standard of care for basic intraoperative as well as neonatal monitoring. This is due to its major advantages in comparison with all other methods used for measuring blood oxygenation:

- It is a noninvasive technique.
- It enables the direct determination of S_aO_2 with good accuracy.
- It is not necessary to heat up the skin.
- There is no need for complicated computing and calibration processes.
- The sensors can be easily and quickly attached, and the measurement also takes only a short time.
- The sensor elements do not contain any part that needs to be refreshed (such as electrolyte and/or membrane); thus, stable long term operation is assured.

2.2. ION-SELECTIVE SENSOR APPLICATIONS

Considering first the monitoring of blood compounds, a number of ion concentrations are of great importance in medical diagnostics, i.e., Na^+, K^+, Ca^{2+}, Mg^{2+}, NH_4^+, and Cl^-.

Hypernatremia ($Na^+ > 145$ mM) is usually a result of diseases that cause excessive urination. In these cases, water is lost, but sodium is still retained in the body. Symptoms include confusion and possibly convulsions and coma. Hyponatremia ($Na^+ < 130$ mM) provokes clinical problems only if serum sodium falls rapidly below 125 mM. The symptoms can be as mild as tiredness but may lead to convulsions and coma. Hyperkalemia ($K^+ > 5.0$ mM) is relatively uncommon, but sometimes occurs in patients with kidney failure who take potassium supplements. Hyperkalemia can result in abnormal beating of the heart (cardiac arrhythmias). Hypokalemia ($K^+ < 3.0$ mM), if it is mild, causes muscle weakness, while severe hypokalemia can cause paralysis, the inability to breathe, and cardiac arrhythmias. Hypercalcemia ($Ca^+ > 5.2$ mg/dL) results in the deposit of calcium phosphate crystals in the kidneys and eyes. Hypocalcemia ($Ca^+ < 4.4$ mg/dL) can cause depression and muscle spasms (tetany). Hypermagnesemia ($Mg^+ > 25$ mM) is rare but can occur with the excessive consumption of magnesium salts. Hypomagnesemia ($Mg^+ < 0.8–0.5$ mM) causes serum calcium levels to decline, which causes muscle twitching and convulsions. It can also result in hypokalemia and thereby cause cardiac arrhythmias.

The most important parameter to monitor in acid–base regulation is pH, which relates to the concentration of hydrogen ions. The value of pH is strongly connected to the levels of bicarbonate, ammonium, chloride, and phosphate in the body. Acidosis (pH < 7.2) and alkalosis (pH > 7.6) mean disorder of acid–base regulation because of failure of kidney, lung, gastrointestinal, hormonal, or central neuronal function.

The toxicity of chloride (Cl^-) results not from the chloride itself, but from the fact that the chloride occurs as an acid (hydrogen chloride). Hyperchloremia and acidosis (pH < 7.2) produce weakness, headache, sickness, and cardiac arrest. Hypochloremia and alkalosis (pH > 7.6) result in paralysis.

Hyperphosphatemia (PO_4^{3-} > 5 mg/dL) can mean failure of kidney or parathyreiod function. It can cause muscle spasms (tetany), convulsions, and cardiac arrest, because the phosphate makes complexes with free serum calcium, resulting in hypocalcemia. Hypophosphatemia (PO_4^{3-} < 2.0 mg/dL) can be caused by a shift in phosphate to intracellular tissues, drawing calcium away from the bloodstream via the formation of insoluble calcium phosphate crystals within cells, with consequent hypocalcemia.

Bicarbonate (HCO_3^-) is of concern because carbon dioxide (CO_2) is an end product of several catabolic processes; CO_2 converts spontaneously to carbonic acid (H_2CO_3), and some of the carbonic acid converts spontaneously to bicarbonate (HCO_3^-) plus hydrogen ion (H^+). The carbonic acid and bicarbonate in the bloodstream minimize (or buffer) any trend toward acidosis or alkalosis. A high level of bicarbonate results in acidosis (pH < 7.2) and means that carbon dioxide cannot escape from the lungs, because the rate of breathing is too low (hypoventillation). A low level of bicarbonate results in alkalosis (pH > 7.6) and means that the carbon dioxide is drawn off from the bloodstream and expelled in the breath at an increased rate (hyperventillation).

An ammonium test measures the amount of ammonium ions in the blood. Most ammonia in the body forms when protein is broken down by bacteria in the intestines. The liver normally converts ammonium ions into urea, which is then eliminated in urine. Ammonium levels in the blood rise when the liver is not able to convert it into urea. This may be caused by cirrhosis or severe hepatitis.

All these ionic compounds can be monitored by means of ion-selective sensors. Ion-selective electrodes for catheter-tip applications can be built using precious-metal wires covered by plasticized PVC membranes with ion-selective neutral carriers. With a reference electrode, potentiometric measurement can be realized, similarly to pH electrodes. A recent approach is the application of thin- and thick-film multilayer electrode systems and ISFETs integrated together for multisensing, with signal processing circuitry on the same chip or substrate.

An integrated micro-multi-ion sensor using platinum gate ISFETs with several polymeric membranes was presented by Tsukada et al. (1991). It consists of two kinds of ion sensors (K^+ and Na^+ ISFETs) and two CMOS unity-gain buffers. The ISFETs are buffered by high-impedance amplifiers. The configuration gives a linear dependence between the ion-selective membrane potential and the output voltage. In the chemically active area, the ion-selective membranes are formed on Pt/Ti/Al multilayer electrodes that are connected to NMOS gates. The platinum film is used as a protection layer against ion migration and hydration, and titanium is used as an adhesion layer. These are patterned with a lift-off process. SiO_2 and phosphosilicate glass are used as passivation films. Finally, a polyimide layer is deposited and patterned photolithographically to form a well. The ion-selective membranes consist of PVC, ionophore, plasticizer, and additives. Almost Nernstian sensitivities (49 mV/decade at 25°C have been detected with good selectivity factors.

Ion-selective sensors may have application not only in monitoring blood compounds, but in analyzing other secretions as well. As an example, an integrated probe for sweat analysis is mentioned here (Bezegh et al. 1987, 1988). Sweat, as a body fluid, can yield useful clinical information. One of the few viable sweat tests is the determination of chloride for diagnosis of cystic fibrosis. Although the concentration in sweat may vary within a wide range depending on many circumstances, the local concentration measured *in situ* may provide useful information. A solid-state integrated differential-type probe based on Na^+/Cl^- ISFETs was used for determination of sodium chloride in sweat. The two silicon chips were attached to a flexible printed wiring substrate. Then a blank membrane containing only the PVC polymer and plasticizer was applied as a continuous coating to the device surface. The electrochemical selectivity was introduced by doping this membrane with electroactive ingredients (ionophores). In practice, this was accomplished by dropping ionophores selectively onto the surface of the membrane. The device was attached to the surface of the skin, like a transcutaneous probe. The difference between the two transistor currents was measured as a function of the NaCl concentration.

The ambulatory monitoring of pH in the upper gastrointestinal tract is also being used increasingly for diagnostic and research purposes. This area offers a further application of inorganic component sensing beyond blood monitoring. Gastric acid measurements are currently performed using integrated glass and antimony electrodes. A severe drawback of these may be encountered if several sensors must be installed in the gastroesophageal tract. A new construction using ISFETs permits the realization of multiple mountings on a single catheter (Thybaud et al. 1990). It consists of several types of polymers; however, the ISFET itself is not polymer-based. The gastric probe is composed of a supple PVC catheter in which the ISFETs are mounted on a floppy substrate of polyimide. The passivating film is epoxy. The reference electrode is also installed at the catheter tip. The very small dimensions of the ISFET chips allow several ion-sensitive sensors to be mounted on a single catheter. Such multi-ISFET probes are expected to be very useful in the study of the gastroesophageal reflux that affects many people. The whole measuring system enables the 24-h ambulatory monitoring of pH and other ion concentrations in the gastrointestinal tract.

Another obvious approach for preparing gastric catheters is the application of optrode probes. The major inconvenience of optical pH sensors arises from their narrow pH range. A possibility for mitigating this disadvantage is the co-immobilization of several dyes (Boisde et al. 1991).

2.3. MEASUREMENT OF INHALED AND EXHALED GASES

Fast, real-time measurements of O_2 and CO_2 in human breath are fundamental tasks in therapeutic and diagnostic medicine. For example, the shape of curves obtained by recording O_2 and CO_2 in expired gas is modified by airway obstruction, and the analysis of the deformation provides an indirect measurement of lung disorders. For several diagnostic methods, the determination of carbon dioxide alone is not sufficient. Distribution disorders of gaseous diffusion to or out of the blood system cannot be analyzed by CO_2 measurement alone because of the high diffusivity of carbon dioxide (about 20 times higher than for O_2). Thus, the oxygen concentration in breath is an important additional parameter. Oxygen monitoring in pulmonary respiratory measurements involves a variety of clinical applications: adult and neonatal intensive care, anesthesia, pulmonary function diagnostics (breath-by-breath oximetry), and stress/exercise testing (sports medicine). In clinical monitoring applications (anaesthesia, intensive care), the measuring range covers 15–100 vol% of oxygen, but the oxygen concentration range in diagnostic medicine (15–21 vol%) in essentially limited to breathing of ambient air.

Generally, mainstream measurements (the gas mixture is analyzed in the primary flow directly at the patient's mouth) and side-stream measurements (a small stream of gas is extracted from the primary flow) can be distinguished. Oxygen monitoring of gas mixtures in clinical applications is presently performed by instruments based on a variety of methods. Conventional technical solutions may be classified as follows:

- Electrochemical sensors (polarographic or galvanic fuel cells) are commonly employed in slow-response applications and thus are restricted to determination of the average oxygen concentration (Teledyne 1994).
- Paramagnetic oxygen sensors are fast, but this technology is utilized successfully in side-stream configurations only.
- Zirconium oxide oxygen sensors show fast response times but cannot be used in the presence of anesthetic agents because of the high temperature (>650°C) inside the measurement cell. Thus, this principle in not frequently applied for medical purposes.
- Mass and Raman spectrometers are applied for side-stream measurements, with less suitability for widespread clinical applications because of their high price.
- Optical techniques for measuring oxygen in breath gas have also been suggested. For example, oxygen was measured based on its adsorption band at 145 nm. However, various substances such as water vapor or carbon dioxide were described as interfering. Another instrument has been presented which is based on the measurement of oxygen concentration at the very narrow absorption line at 760 nm (Moya et al. 2004).
- An application of an optochemical oxygen sensor to respiratory gas monitoring was presented by Kolle et al. (1997). The sensor is based on measurement of the quenching of the intensity of photoluminescence of the dye platinum(II)-octaethylprophyrin ketone by molecular oxygen. The dye was immobilized in various polymers with and without plasticizer. The suitability of optochemical oxygen sensors for clinical use in respiratory gas-exchange measurements has been demonstrated in combination with a prototype instrument. The sensors require no moving parts. Consequently, they are more stable and cheaper to construct than paramagnetic instruments. It is possible to apply the sensors in vibrating and harsh environments.

Side-stream CO_2 modules work with the mainstream CO_2 module and are designed for use with intubated adult and pediatric patients. It provides waveform and numeric values. The module samples an intubated patient's airway gases through a remote sample line. This places less weight on the patient's airway adapter and reduces risk of extubation. The remote sample cell is located on the front of the side-stream CO_2 module. From there, data are passed to the mainstream CO_2 module for analysis and display The side-stream CO_2 module provides the following information: CO_2 waveform, end-tidal ($etCO_2$) numerics, and inspired minimum ($imCO_2$) numerics.

Carbon dioxide gas sensors for biomedical uses have been fabricated by self-assembling single-wall nanotube films on a surface acoustic wave delay line operating at 286 MHz. Polymer functionalization was used to enhance the sensitivity of the carbon nanotubes to carbon dioxide. A pulse radar type of interrogation system was used to monitor the conductivity of the nanotube film by measuring the attenuation of the surface acoustic wave due to film conductivity changes. The conductivity of the film was found to decrease with carbon dioxide concentration (Sivaramakrishnan et al. 2008).

Anesthetic gas monitoring (AGM—e.g., Philips, Dräger) uses infrared absorption technology to measure the five most commonly used anesthetic gases (halothane, isoflurane, enflurane, sevoflurane, desflurane), plus N_2O and CO_2. Oxygen is generally measured using the paramagnetic or electrochemical technique. The most important functions include:

- Automatic alert when anesthetic agent mixture is detected
- Inspiration and expiration values for all gases
- Individual alarm limits for each parameter
- Respiratory rate derived from CO_2 waveform
- Flow rate measurements
- Airway respiratory rate (awRR) monitoring
- Hospital-based operating rooms
- Ambulatory surgical centers
- Low-acuity procedural sedation care settings

Bedside monitoring appliances have the same role; only the application site is different. A complete metabolic monitor appliance is a special bedside unit (e.g., Philips). It is ideal for use with a ventilator or a canopy and is equally suited to both spontaneously breathing and mechanically ventilated patients of all ages. The individual nutritional needs of critically ill patients vary extremely; both overfeeding and negative energy balance are harmful. The only reliable method of designing and maintaining optimal nutrition in these patients is to measure their true energy expenditure. This is exactly what the metabolic monitor's indirect calorimeter is designed to do. It measures caloric needs and estimates substrate oxidation of patients by continuously monitoring O_2 consumption and CO_2 production. The measurement of gas exchange makes it possible to study the complex interrelations among energy metabolism, nutrition, lung functions, and ventilator dependency.

Exhalation gas compound analysis is another potential application field of gas sensors. Certain gaseous constituents of exhaled human breath are associated with specific types of diseases: NO and CO with inflammation and oxidative stress (asthma, cystic fibrosis, chronic obstructive pulmonary disease); isoprene for blood cholesterol, NO_2 for bronchial epithelial infection resulting from tobacco smoking; acetone for diabetes; pentane for breast cancer; ammonia for uremia, kidney disorders, and certain stomach diseases. (Blood alcohol level is in close correlation with the exhaled breath concentration, but this is a topic for another section.)

The fact that an asthmatic person manifests a drastic increase in the concentration of nitric oxide (NO) in exhaled air is used for detection of the disease and for analysis of its severity. Having determined the NO concentration, one can take timely measures aimed not only at prevention of the beginning of the disease but also at the termination of the further process of the respiratory tract inflammation.

As mentioned, air exhaled by people may contain, together with nitric oxide, other gases, such as acetone and carbon monoxide, in microconcentrations; the level of oxygen in exhaled air may go down to 15 vol%. For this reason the analysis of exhaled air should be comprehensive. A multisensor approach was presented recently (Ivanova et al. 2004). The level of NO in exhaled air was measured by a semiconductor sensor based on WO_3; the presence of acetone in exhaled air can be determined using semiconductor sensors based on SnO_2. The level of oxygen was determined by an electrochemical sensor of the amperometric type operating on the basis of a two-electrode measuring system. The concentration of CO in air was measured with the help of an electrochemical cell made on the basis of a three-electrode measuring circuit. Because the characteristics of the sensors employed depend on such factors as temperature, humidity, and flow rate, it is necessary to check these parameters so as to take them into account during measurements. For this reason, the multisensor complex was fitted with sensors for temperature (a resistive-type sensor), humidity (a planar-type sorption-capacitance sensor), and flow rate (a thermoanemometer). The new multisensor complex is a multifunctional microprocessor-equipped measuring system intended for continuous measurements of the concentrations of nitric oxide, oxygen, acetone, and carbon monoxide in exhaled air, measurements of temperature and humidity of analyzed air in real time with further analysis, and data saving in a computer.

Fleischer et al. (2002) suggested the application of a complementary sensor system for breath analysis: work function measurements of phthalocyanine layers that allow the detection of NO at the parts-per-billion level with low cross-sensitivities to water and an array of quartz microbalance sensors that permits preferential detection of volatile organic compounds such as acetone, which is a biomarker for diabetes. (Use of an array enables screening of interferences.)

Measuring breath ammonia levels can be a fast diagnostic method for patients with disturbed urea balance, e.g., due to kidney disorders or ulcers caused by *Helicobacter pylori* bacterial stomach infection. A diagnostic breath analysis system for medical ammonia requires a rather low detection limit of 50 ppb. The sensor system should be very selective to ammonia. Furthermore, the system should respond to a change in ammonia concentration within a few minutes. The only ammonia sensors performing to these criteria are optical systems. These systems, however, are very large and expensive, making them less suited for routine clinical use. Miniaturization could solve this problem (Timmer et al. 2005).

Ehrmann et al. (2000) suggested the application of an electronic nose for fast detection of malodor components such as NH_3, H_2S, CH_3SH, indole, and putrescine—characteristic compounds in halitosis in elderly people, with little cross-sensitivity to ethanol and humidity. Halitosis is a general term applied to breath malodor and is a common condition in elderly people that may have oral or nonoral causes. It results from a variety of different etiologies and can have both social and medical consequences. In some instances, bad breath can reflect serious local or systemic conditions including gingivitis, periodontal disease, diabetes, hepatic failure, or respiratory infection. Odors are often distinct in quality, thus their characterization and determination of their changes with time can be useful. To allow self-diagnosis of human breath quality, a sensor system is required which allows highly sensitive gas detection combined with good selectivity at extremely low cost. Measurement of the electrical conductivity of semiconducting metal oxides represents one way of realizing cheap and sensitive detection of gaseous components. In the electronic nose of Ehrmann et al. (2000), breath odor quality was characterized with a metal oxide (SnO_2 and WO_3) sensor array consisting of 40 pieces whose different sensitivities were achieved by gradient technique: gradients of operating temperature and of thickness of a SiO_2 coating on the metal oxide. It was shown that using a gas sensor microarray based on a segmented metal oxide layer, with a temperature and a membrane gradient, allowed highly sensitive detection of gaseous components of bad breath. With WO_3 as the detection material, the cross-sensitivity to ethanol and changes in relative humidity could be suppressed.

Lin et al. (2001) also investigated an electronic nose in breath analysis for diagnostic purposes. Their system used an array of six 12-MHz AT-quartz crystals covered with probe peptides. In the clinical test, the sensor module discriminated breath from normal subjects, patients with uremia, patients with chronic renal insufficiency, and patients with chronic renal failure well, with a correct classification of 86%. In general, the normal subjects could be distinguished completely (with practically 100% likelihood) from the patients.

2.4. SENSORS FOR METABOLITES IN BLOOD, TISSUE, AND SECRETIONS

One of the most important applications of enzymatic biosensors is the monitoring of blood glucose levels in diabetics by means of a glucose sensor. Normally, fasting blood glucose levels are in the range of 3.6–6.4 mM/L, and the peak response to an oral glucose tolerance test will not exceed 11 mM/L, falling below 7.8 mM/L 2 h later. A diabetic will have a peak response exceeding 11 mM/L, and his or her blood glucose will remain above this level after 2 h. A person is diagnosed as diabetic when his or her fasting blood glucose level is ≥7.8 mM/L. The blood glucose level can reach even 30 mM/L in extreme conditions. One of the major medical advances of the last century was the development of insulin to regulate glucose metabolism. The insulin dose for an individual must be adjusted to minimize hyperglycemia while avoiding the serious conditions of hypoglycemia. This requires frequent and accurate testing to monitor the diabetic's blood glucose level.

It is now possible for diabetics to determine their blood glucose level at home using electronic home monitor appliances. Usually, the measurement should be done four times daily, before each meal and before bedtime. A drop of blood is obtained by pricking the finger. The blood sample is placed on a reagent strip, which is then analyzed by the instrument using reflectance photometry. This optical method is inherently nonlinear and requires somewhat more complicated signal analysis than other methods. Other disadvantages are the periodic recalibration necessity, and the interferences of various blood compounds such as erythrocytes and plasma proteins. In spite of these problems, the reagent strip method has been successfully implemented, and a number of instrument types have been commercially available for many years.

The goal of glucose biosensor research and development has been both to replace the reagent strip method by sensors and to enable the fabrication of portable instruments for *in vivo* or *ex vivo* blood glucose monitoring. Battery-operated portable appliances employed electrochemical cells equipped with an enzymatic membrane, based first on O_2, later on H_2O_2 transducer surfaces (Buerk 1993). The instrument is simple to use, operated by only one button, with instructions shown on a liquid-crystal display (LCD). The instrument prompts the user to supply either a calibration solution or a drop of blood into a small plastic well. The sample can be as small as 7 μL. The plastic well holds the sample over an electrochemical sensor, which is covered with the enzyme-activated membrane. The biosensor is linear with glucose concentration in the range from 0 to 30 mM/L. Since the current required for the electrochemical biosensor is much less than that needed to operate light-emitting diodes (LEDs), which are used in reagent-strip photometric-type instruments, the appliance can be battery-operated. Simpler electronics are required, allowing the instrument to be miniaturized to the size of a hand-held calculator.

It detects when the sample has been applied, and automatically times the analysis, which is completed in 30 s. After making the measurement, the LCD prompts the user to wipe out the sample and place a drop of cleaning solution into the well. A tight plastic cover is then closed to prevent evaporation or spillage, keeping the membrane hydrated until the instrument is used again. Measurement tests have shown good correlation with the results of conventional analytical methods. Furthermore, interferences of blood compounds are much less than with photometric instruments. The membrane has a lifetime of 500–1000 measurements, or 6 weeks before degradation in performance. The sensitivity of the instrument can be restored simply by replacing the enzyme-activated membrane. The only limitation of the glucose biosensor was found with measurements taken when the blood pO_2 was low, as anticipated from the rate-limited reaction. At the highest blood glucose levels, the instrument may underestimate the true level by approximately 10% when $pO_2 < 5$ kPa. When blood $pO_2 > 10$ kPa, there is practically no detectable error due to O_2 limitation. Nowadays, a number of pocket-portable instruments employ this measurement method.

Recent research and development has been directed toward devising implantable systems that are usable for either *in vivo* or *ex vivo* continuous monitoring of blood glucose. Ideally, a closed-loop system for regulating the delivery of insulin would be developed, using an implantable biosensor and an insulin pump actuator. More tightly regulated control of insulin during the early stage of the disease may spare the diabetic from more severe microcirculatory complications later in life. For widespread practical applications, a number of problems of glucose sensors have to be solved, such as:

- Miniaturization of the sensor elements and their integration with other sensors and circuitry components
- Improved biosensor lifetime and stability with sophisticated enzyme immobilization techniques
- Toxicology and stability problems due to mediator dissolution
- Biocompatible packaging and implantation methods
- Problems of enzyme refreshment, system recalibration, and sensor interchange ability

Glucose biosensors fabricated so far have been almost entirely based on the catalytic effect of the enzyme glucose oxidase (GOD). A discussion of biosensor operation is far beyond the scope of this section. The most important application fields for biosensors in biomedical areas include:

- Continuous monitoring in clinical practice
- Pharmacology: adjusting personal dosage and researching new medicines
- Portable and home monitors for patient self-control
- Implantable sensors for continuous monitoring in everyday life
- Implantable sensor-actuator systems that may replace the hormone-regulating function of one of the endocrine glands

The first two fields are still dominated by conventional analytical methods. The practical application of sensors is spreading in portable and home monitors, and implantable monitoring sensors are in the first stage of practical application, while the long-operating implantable regulating systems are in the research and development stage. It is important to mention here that miniature, electronically controllable medicine pumps have already been realized with silicon micromachining for continuous medicine delivery; however, there is a lack of long-term-operating, reliable biosensors. Thus, the realization of implantable regulating systems is hindered mainly by the instability and short lifetime of the biosensor elements. An important recent research area is concentrating on the body–sensor interface realization problems. Intravascular application, placing sensor elements into the bloodstream, is not an appropriate approach for long-term monitoring. Taking blood sample for analysis does not enable continuous operation of the system. Subcutaneous implantation seems to be the most promising approach, by which the sample-taking dialysis cell or the sensor element itself is placed into subcutaneous tissue, mainly in the abdominal fat of human bodies. The analyte is actually the tissue liquid and not the blood in this case.

There are three main versions of the subcutaneous technique: microdialysis, ultrafiltration, and direct implantation. Microdialysis, based on the principles of diffusion, is a method for removing chemical substances from extracellular fluid of the body without removing any liquid (Harsányi 2000). A double-lumen microdialysis probe using hollow fibers (e.g., polysulfonic, or cellulosic, generally with a molecular weight cutoff size of 15,000–20,000) functions as an artificial blood vessel; a physiological buffer solution is pumped at a constant flow rate (1–100 μL/min) into the fiber placed subcutaneously and retrieves compounds of low molecular weight. The dialysate of the tissue is subjected to analytical methods. If, for example, GOD is contained in the dialysate, the resulting pO_2 in it may be taken as a measure of glucose concentration using sensors for determining it (Mascini et al. 1992). In other approaches, the enzyme is immobilized onto the transducer surface, forming a real biosensor (Volpe et al. 1995; Steinkuhl et al. 1996). In both cases, however, the measurement is dependent on the flow rate, pressure, active membrane surface, etc. In recent practical solutions, the probe is connected to a specially designed micromachined silicon/glass chip flow-through cell containing the integrated biosensor together with a system of capillaries and flow channels through which the dialysate is pumped (Steinkuhl et al. 1996). The electrodes of the electrochemical-type biosensor are located in two pyramidlike microcontainments etched through the silicon by anisotropic etching. The amperometric current may be measured with miniaturized amplifier electronics with LCD display that has been developed to result in a hand-held monitoring system with the overall size of a cigarette packet. If an appropriately low pressure (vacuum) is applied to the subcutaneously inserted tubing of a sufficiently high-molecular-weight cutoff ultrafiltration membrane (Harsányi 2000), the interstitial fluid may be sampled and analyzed directly. According to the third version, the sensor element is implanted directly (Pickup 1993). Until now, nearly all devices have been laboratory-designed and -manufactured. Under consideration of time constants, the microdialysis probes may provide an excellent description of glucose kinetics in subcutaneous tissue over relatively short intervals. But any figure on the absolute glucose level can be obtained retrospectively only. In contrast, ultrafiltration technique may reveal the real glucose concentration in the subcutaneous interstitial fluid, but at relatively long intervals. All present sample-taking methods suffer from a slow alteration of the real subcutaneous conditions. On the other hand, for the practical application of direct implantation of the sensor itself, a number of biocompatibility problems, such as fibrosis, blood clotting, and immunogenic effects, as well as toxicity of the enzyme or mediator leakage have to be overcome.

A fiber optic sensor structure based on a percutaneous optical fiber that permits reliable spectroscopic measurement of chemical reactions in a nano-engineered polymeric matrix attached to the implanted end of the fiber was developed by Liao et al. in 2008. In this application, the goal was to measure interstitial glucose based on changes in fluorescence resonance energy transfer between fluorophores bound to betacyclodextrin and concanavalin A in a polyethylene glycol (PEG) matrix. *In vitro* experiments demonstrated a rapid and precise relationship between the ratio of the two fluorescent emissions and concentration of glucose in saline for the physiological range of concentrations (0–500 mg/dL) over several weeks. Chronic animal implantation studies have also demonstrated good biocompatibility and durability for clinical applications.

The detection of large-molecular-weight bioactive compounds is possible not only with biosensors, but with measurement techniques based on conventional analytical chemistry. The great advantage of these methods for biomedical applications is that they do not employ compounds of living things, which are the most critical lifetime-limiting factors. Their common disadvantage is poor selectivity. Although they are actually outside the scope of this chapter, a short summary is given here (more details can be found in the cited literature).

- Amperometric and, later, voltammetric methods have been used in the direct, three-electrode electrochemical technique. With amperometry, the oxidation current of glucose was measured at a constant potential; the greatest disadvantage of this method was nonselectivity. Special techniques of cyclic voltammetry (differential pulse or low-potential cyclic voltammetry) can assure better selectivity (Yao 1991).
- The measurement of glucose optical rotation activity on plane-polarized light has been used for the *in vitro* determination of D-glucose concentration in physiologic fluids. Amino acids and other blood or tissue fluid compounds having optical rotation effect may cause interferences, thus the efficiency of a perm-selective membrane has a significant effect on the operation (Burk et al. 1987).
- Noninvasive *in vivo* glucose monitoring is clearly the most attractive approach for patients with diabetes, allowing more frequent, at best even continuous, measurements without any pain or sensation. There are several categories of noninvasive glucose monitoring technologies, involving electromagnetic waves in near-infrared and and far-infrared spectroscopy. Since the interference of blood-constituting molecules is the major problem, a multiple-wavelength approach is necessary (Mendelson et al. 1990). The great advantage of multiple infrared attenuated total reflectance (ATR) spectroscopy is the possibility of *noninvasive* blood glucose determination through the skin using the skin's "transmission window," which exists in the near-infrared region from 0.7 to 2.5 μm (Arnold 1996). For screening interferences, the only solution may be multivariate calibration analysis using principal component regression with partial least-squares method or neural network analysis (Liem et al. 1992; Christison 1993; Pruna 1995; Arnold et al. 1998; Burmeister et al. 1998; Hazen et al. 1998; Small and Arnold 1998).
- Dielectric spectroscopy (DS) has also been tested for noninvasive glucose monitoring. Because of the specific reactions of blood and tissue cells to varying glucose concentrations, the electrolyte balance across the membranes of blood and underlying tissue is changed. This is the reason why both AC and DC conductivity are sensitive to these subtle changes in electrolyte balance, which is related to the blood's glucose levels. The typical impedance behavior of the capacitive sensor attached to the skin of a patient at different times, when significant changes of blood glucose have been induced, was found to be characteristic to the blood level (Caduff et al. 2003). Based

on these results, a commercial glucose-level monitoring device much like a watch is available (Pendragon Medical Ltd.).

2.5. HUMIDITY SENSORS IN BIOMEDICAL APPLICATIONS

Human skin plays an important role in protecting the body from environmental influences and is responsible for many physiological functions. One of them is to function as a water barrier. Healthy skin optimally keeps water inside the body and has a so-called transepidermal water loss (TEWL) factor in a range specified for age, sex, and actual activity. Instruments for measuring it use either surface skin impedance measurements or apply open or closed chambers equipped with sorption humidity sensors. The first solution is strongly dependent on personal skin differences and the sensor surface cleanliness. The second is not fast due to the relatively long sensor time constant.

A new idea for measuring the moisture content in human skin using a fast dew-point hygrometer sensor was presented recently (Jachowicz et al. 2004). The hygrometer construction is based on an integrated semiconductor sensor structure containing an interdigital impedance detecting the condensation at the dew point, a thermoresistor, and a heater. To speed up measurements, a fast silicon dew-point hygrometer with closed chamber can be applied. The semiconductor integrated sensor structure, located on a cooling Peltier couple module, was placed at an open side of the "closed chamber" measurement head. A flip-chip technology was used for electrical structure input/output bonding into flexible ribbon (PCB).

A special operation algorithm was used for very fast control of the detector surface temperature. After touching skin with the open side of the measurement head, the water molecules evaporated from the skin quickly increase hygrometer readings. During the first 15–20 s, fast growth of humidity in air over the skin surface (in the closed chamber) can be observed. Full humidity stabilization is attained after 100–150 s. The measurement value of saturated humidity as well as analysis of humidity changes with time (e.g., the humidity derivative on time) over the patient's skin can give much valuable information about the patient's skin conditions. It was demonstrated that the hygrometer follows the humidity changes with relatively low delay time (4 s for increase and 10 s for humidity decrease).

The water content of the stratum corneum, the outer part of the epidermis, could be one index with which to evaluate skin health. Among the various measurement techniques, the most widely used ones are based on measurement of skin impedance. A thin-film microsensor based on interdigital electrodes deposited on a glass substrate was developed by Sekiguchi et al. (2001). If the film side of the sensor chip is pressed onto human skin, the change of capacitance can be used as a measure for the water content of the skin; atopic, normal, and hydrated skin types can be differentiated sensitively, although the effect is frequency-dependent.

3. WATER QUALITY MONITORING

The most important areas of water quality monitoring today are the following.

- Oceanography: includes depth profiling of oxygen and benthic and coastal observations

- River monitoring: to monitor river water quality prior to extraction for treatment and distribution by water supply networks
- Wastewater: online monitoring of nutrient levels in wastewater treatment plants
- Wetlands: remote area monitoring of the impact of human and agricultural nutrient levels on wetlands
- Harbor monitoring: to monitor physical and nutrient levels and alert to the possible pollution of the waters
- Aquaculture: to assist farm operators to maintain the best possible levels of nutrients to maximize returns on their investments

The information to be collected is rather complex and can only be followed by a great number of components, including dissolved oxygen, pH, metal ions (Na^+, K^+, Ca^+, Mg^+) and other cations (NH_4^+), anions (chloride, phosphate, nitrate, sulfate), heavy-metal pollutants (Cd, Pb, Cu, Hg), organic pollutants (phenolic, chloroform), oil contamination, and organophosphate pesticides. Integral water quality (WQ) parameters have also been defined and followed by analytical methods, such as electrical conductivity (EC— characteristic of the total concentration of ionic compounds), oxidation–reduction potential (ORP—measures the ability of a fluid to be oxidized or reduced and is also dependent on the types and concentrations of ions present), turbidity (a measure of the amount of suspended solids in a solution), and biological oxygen demand (BOD—the amount of oxygen needed by microorganisms for the oxidative decomposition of all organic materials in the wastewater).

3.1. DISSOLVED OXYGEN AND pH

Dissolved oxygen (DO) refers to the amount of oxygen dissolved in the water. The DO can be significantly influenced by factors such as temperature, the presence of dissolved salts, and the biological or chemical demand of the water. For marine organisms the amount of oxygen available is at least as important as the amount of food. Thus, the monitoring of oxygen concentration is an important part of oceanographic studies.

The most widely used oxygen sensors in oceanography are Clark electrodes. These amperometric sensors are based on the reduction of oxygen at a working electrode. This electrode is immersed in an electrolyte and separated from interfering species by a membrane that is permeable to oxygen. Oxygen flow through this membrane leads to consumption of O_2 at the working electrode. The limiting current (when diffusion of oxygen is the determining reaction step) between the electrodes is proportional to the oxygen flow. In seawater, the diffusion properties of the membrane are subject to reserve modifications, mainly due to biofouling and temperature changes. This can result in signal drifts when the sensor is used *in situ* for a long period of time (one or two months) without recalibration (Hitchman 1978). Periodic membrane and electrolyte change with calibration is the general disadvantage of applying electrochemical cells as sensors.

Clark-type oxygen probes are commercially available (e.g., Imagiprobe from Imagiworks, Inc., or SBE sensor from Sea-Bird Electronics, Inc.), and they have experienced intensive development in recent years. These probes have a measurement range of 0–14 mg/L (ppm) of dissolved oxygen and provide temperature compensation; they have a fast response time, reaching 95% of full reading in 30 s. Calibration drift is caused primarily by chemical processes inside the sensor and by membrane fouling from ocean contaminants. If the membrane is kept clean, the steps taken to improve the new sensor's chemical stability yield demonstrated calibration drift rates of less than 2% over 1000 h. Not only do the new designs measure temperature in the right place, the temperature equilibration time is reduced to a few seconds. Hysteresis in oxygen measurements is caused by delays in a sensor's response to changing temperature, pressure, and oxygen. Slow temperature response and time mismatch of temperature corrections are responsible for most of the hysteresis in the upper 1000 m. These faults have been largely overcome in the new designs.

Some electrochemical water quality analyzers do not use the Clark-type amperometric method, but instead use Hersch-type galvanic cell oxygen sensors (e.g., the Minianalyser from Greenspan Technology, Inc.). The sensor in this case is a lead–oxygen battery, which has a lead anode and gold cathode. Oxygen molecules enter the cell through a semipermeable membrane. They then diffuse in the acid electrolyte and are reduced at the gold electrode, while the counter lead electrode becomes oxidized. The steady-state current is proportional to the oxygen concentration (Dilhan et al. 1995). A temperature-compensated voltage output proportional to the partial pressure of oxygen concentration is generated in most cases with appropriate circuitry.

The problems of electrochemical sensors can be theoretically overcome if optical sensing is used. Optical oxygen sensors are based mostly on the dynamic quenching of a fluorescent indicator by oxygen. Assuming dynamic quenching only, and under ideal conditions, the fluorescence intensity and lifetime are related directly to the oxygen concentration according to the Stern-Volmer relation:

$$\frac{I_0}{I} = \frac{\tau}{\tau_0} = 1 + K \cdot pO_2 \tag{5.7}$$

where I and I_0 are the intensities, τ and τ_0 are the lifetimes in the presence and absence of oxygen, respectively, and K is the Stern-Volmer constant. In comparison with electrodes, optrodes are more suitable for oceanography. First, the sensing does not consume oxygen. Second, the partial pressure of oxygen in the probe, where the fluorescent dye is located, is in equilibrium with the partial pressure outside the probe. Because the lifetime is related directly to oxygen concentration, and not to the flow of oxygen, a modification of the diffusion properties of the probe should only alter the response time. Furthermore, optrodes offer many advantages due to the utilization of optical fibers, such as remote sensing, small size, and sensor networking.

McFarlane and Hamilton in 1987 constructed an intensity-based optrode which has been used *in situ* for depth profiling. Commercially available fiber optic oxygen sensors (e.g., Foxy sensors from Ocean Optics, Inc.) are spectrometer-coupled chemical sensors for full spectral analysis of dissolved or gaseous oxygen pressure. The fluorescence method is used to measure the partial pressure of dissolved or gaseous oxygen. Optical fiber carries excitation light produced by the blue LED to the thin-film coating at the probe tip. Fluorescence generated at the tip is collected by the probe and carried by the optical fiber to the high-sensitivity spectrometer. When oxygen in a gas or liquid sample diffuses into the thin-film coating, it quenches the fluorescence. The degree of quenching correlates to the level of oxygen pressure. This solution is an alternative to the oxygen electrode, so changing membranes and filling solutions can be avoided. And because there is no membrane, the probe works in organic solvents and does not experience drift. Unlike electrodes, frequent calibrations are unnecessary, so the probe can be deployed for longer intervals. Furthermore, fiber optic sensors are unaffected by electrical interference or interference caused by pH change, ionic strength, salinity, or biofouling. Also, the operating temperature range is much wider; however, compensation is also necessary by the application of conventional temperature sensors. Because temperature affects both fluorescence intensity and the collision frequency of the oxygen molecules with the fluorophore, as well as the solubility of oxygen in analytes, accessories that monitor the temperature of the sensing environment and compensate for temperature fluctuation in the probe itself are parts of the measuring systems. The optrode is a fiber optic fluorescence probe with thin-film–coated tip. Several options are available, ranging from slender, needle-tipped probes for penetrating vial septa to 1/4–in.-OD stainless probes for process applications. The probe couples via a splice bushing to a bifurcated optical fiber assembly (fiber diameter varies according to the probe specified). Probe lifetimes are given as one year.

To overcome difficulties arising from the immobilization of the dye in a solid matrix, a special setup with an ideal dye solution environment was designed by Goouin et al. in 1997. Also, to address the problem of long-term stability of fluorescence intensity–based sensors in polluted environments, fluorescence lifetime measurements were made with two time windows set at two distinct positions along the fluorescence decay curve. The response time of the sensor is longer this way, but still compatible with special applications (e.g., in coastal water monitoring). The effect on the sensor of several parameters specific to the marine environment, such as salinity and hydrostatic pressure, was investigated. The salinity appeared to have no influence.

A new sensor based on a porous plastic probe has been developed by Guo et al. (2008) for the detection of dissolved oxygen in seawater. This probe was prepared by copolymerization of monomer, cross-linking reagent, porogent, and luminescent ruthenium(II) diimine complex. The porous plastic probe exhibits good response to dissolved oxygen and resistance to indicator leaching out due to its high hydrophobicity. This porous plastic probe serves as the analyte-sensitive function as well as the optical waveguide, which make it easy to assemble the sensor. The sensor has been applied to the determination of dissolved oxygen in seawater with satisfactory results compared with the standard methods.

As is well known, the pH of a solution is a measure of hydrogen-ion activity in a solution. It defines how acidic or alkaline the solution is, which may be very important in water quality monitoring. Water quality analyzers use glass electrode pH sensors with appropriate reference electrode constructions, mainly of the Ag/AgCl type, since water generally has chloride ion content. The glass has a special composition so that a potential is formed across it in the presence of hydrogen ions. Inside the glass is an electrolyte of constant pH and concentration and a reference electrode which gives a stable voltage as a comparison. The main task differs not significantly from industrial pH measurement (see Chapter 6), but some types are specially recommended for water quality monitoring purposes (e.g., those from Aanderaa Instruments). Between measurements the pH electrode is stored in moist air to prolong electrode life. Monitoring the pH value with a permanent installation over a relatively long period is feasible only if certain precautions are taken. This is due mainly to the fact that the pH probe itself has a limited lifetime. The lifetime is also related to how it is used and stored.

Planar thick-film electrode constructions became popular recently for their low cost. The ability to screen-print reproducible devices in high volumes makes them distinctly cheaper than their equivalents fabricated using traditional methods. Other advantages are their smaller size and the fact that, generally, many of the substrate materials used in thick-film production are more rugged. A typical example was demonstrated by Atkinson in 2004. The device occupies a single substrate measuring approximately 50 mm by 25 mm and is able to support sensors for the measurement of pH, temperature, conductivity, redox potential, and dissolved oxygen. These miniature and rugged pH sensors can give extremely good responses at significantly lower cost than conventional sensors. Typically, sub-Nernstain sensitivities of approximately 40–50 mV/pH unit have been obtained with lifetimes of over 20 days. The use of a solid-state reference electrode, however, does limit the useful lifetime, due to ionic salt loss from the electrode, and also imposes a limit on the hydration time of the sensor. There is a conflict between the need to maintain a stable salt matrix on the reference electrode surface and yet achieve rapid hydration of the electrode when it is immersed in the test medium. One way of avoiding this conflict is to supply the electrode prehydrated in a solution of the correct ionic strength for the given application.

3.2. ELECTRICAL CONDUCTIVITY, OXIDATION–REDUCTION POTENTIAL, TURBIDITY, AND SALINITY

Electrical conductivity (EC), oxidation–reduction potential (ORP), turbidity, and salinity have several features in common:

- They can be regarded as physical rather than chemical parameters.
- They can be measured with physical-type sensors.
- They are measurands that give some kind of integral quality of water.
- They are, however, strongly related to the chemical composition of the water.

Electrical conductivity is a measure of how well a solution conducts electrical current. It can be used to detect changes in the water system, i.e., new pollutants or rising salinity. To measure EC, analyzers generally use a flow cell with two electrodes and make a resistance/conductivity measurement using alternating current to prevent polarization occurring. The spacing and size of the electrodes determines the cell constant.

Oxidation–reduction or redox potential measures the ability of a fluid to be oxidized or reduced (give or take up electrons) and is dependent on the types and concentrations of ions present. Oxidation and reduction happen simultaneously; while one substance is being oxidized, the other is being reduced. Oxidation and reduction reactions occur until equilibrium is reached. Unreacted agents give the water the ability or "potential" to oxidize or reduce. This "potential" is what is measured. Water quality analyzers generally measure ORP using a noble-metal electrode. When a metal electrode is exposed to varying concentrations of chemicals, an electric potential is developed. Generally a platinum electrode is applied for this purpose. The potential of the platinum electrode is measured against that of a reference electrode made from Ag/AgCl. ORP is an aggregate measure incorporating all ions that are oxidizing or reducing agents. The result is given in millivolts.

Turbidity is a measure of the amount of suspended solids in a solution. In natural water, turbidity can be caused by clay, silt, small organic matter, and microscopic organisms. Turbidity is measured by passing light through a transparent cell containing the sample and measuring the amount of scattered light with a detector which is orientated at 90° to the emitter. This result is expressed in nephelometric turbidity units (NTUs) calibrated against standards.

Seawater salinity measurement is a very important parameter for oceanics, marine environment monitoring, seasonal climate prediction, and fishing. And there are widespread application backgrounds and great practical significance in a lot of industrial fields, including offshore oil exploration. A new optical fiber sensor was developed for simultaneous seawater temperature and salinity measurement by Zhao and Liao (2002). Temperature-dependent semiconductor absorption at the band edge is used as the principle of the temperature measurement, and the sensor exploits beam deviation caused by refraction due to the salinity of seawater at the receiving end face of the optical fiber array. The measured optical signals are reflected and transmitted through a sequentially linear arranged fiber array, and then the light intensity peak value and its deviant are detected by a charge-coupled device (CCD).

3.3. IONIC COMPOUNDS

The measurement of metal ions (Na^+, K^+, Ca^{2+}, Mg^{2+}) and other cations (NH_4^+), anions (chloride, phosphate, nitrate, sulfate), and heavy-metal pollutants (Cd, Pb, Cu, Hg) is essential in river water quality monitoring. Nowadays, chemical analysis is carried out mainly by gas and liquid chromatographic methods. However, these techniques have many problems, such as the complex procedures required for measuring the above components, and large and expensive instruments. In particular, monitoring by ion-selective sensors is more and more necessary for health-risk assessment of water quality in aquatic environments.

Ion-selective sensors applying PVC membranes with various ionophores have been developed and brought into practical application (e.g., valinomycin for K^+, ETH 2120 for Na^+, ETH 1001 for Ca^{2+}, nonactin for NH_4^+). Miniature versions of the ion-selective sensors have also been developed, such as ISFETs or thick-film electrodes using similar membranes. The best solution, however, would be the application of multichannel sensor arrays capable of sensing several ionic compounds at the same time (Harsanyi 1995).

Stripping analysis is a powerful tool for measuring heavy metals in environmental, clinical, and industrial samples. For these applications, mercury-based electrodes have been employed in order to obtain high reproducibility and sensitivity. However, the growing interest in decentralized analysis with sensor-based monitoring, coupled with environmental issues, requires new approaches involving mercury-free systems. Gold electrodes of different forms have been used successfully to detect heavy metals such as copper, mercury, lead, and cadmium by electrochemical measurements (Bonfil et al. 1999, 2002). Gold-sputtered and screen-printed electrodes were also recently used in lead detection by stripping voltammetry in a flow injection system. Laschi et al. reported in 2004 the application of gold-based screen-printed electrodes to heavy-metal detection using square-wave anodic stripping voltammetry (SWASV) as electrochemical technique. As well as the method to produce carbon-based strips, the working surface of these screen-printed electrodes is obtained by printing a thermoplastic ink which contains gold particles. Hence, no plating or sputtering steps are required, but only a short electrochemical pretreatment of the working surface. To perform the analysis, heavy metals are accumulated on the activated gold surface by applying a suitable reduction potential in an acidic medium (0.1 M HCl). Thereafter, SWASV measurement is performed using optimized conditions. The system has been applied to the detection of Cd, Pb, Cu, and Hg in standard solutions; detection limits of micrograms per liter were obtained for all the analytes, coupled with a short analysis time (about 3 min for each analysis). A differential pulse-mode anodic stripping voltammetric (DPASV) determination of lead ions in solutions was described by Santos et al. in 1998 utilizing a poly(vinylsulfonate)-doped polyaniline-modified glassy carbon electrode. The experimental signals observed were similar to those of typical stripping methods and showed excellent linearity within the micromolar concentration range studied. Yoo et al. (2003) applied the same method for trace mercury [Hg(II)] determination using polythiophene-quinoline/glassy carbon modified electrodes. Traces in concentration from 2 to 100 ppb could be detected. Chalcogenide glasses were used in potentiometric ion-selective electrodes for measurements in natural and wastewaters (Legin et al. 1995). Determination of copper(II), iron(III), chromium(VI), lead, cadmium, and mercury starting at the microgram concentration level has been carried out. The results of analysis with chalcogenide glass chemical sensors show good agreement with those of atomic absorption spectroscopy and other common methods.

Almost all phosphorous found in natural waters is in the form of phosphates. Phosphates are compounds which include PO_4^{3+}. They are normally produced from phosphoric acid (H_3PO_4) by removing a number of hydrogen atoms and replacing them with a metal. Phosphates are used commercially in water softeners and fertilizers. A colorimetric method is generally applied to measure phosphate. Reagents added to the sample react with the dissolved phosphate present to produce a blue color, the strength of which is relative to the amount of phosphate in the sample. A spectrophotometer is used to measure this color development. The turbidity of the sample is compensated for by measuring a reference sample. As there is no filtering of the sample, the phosphorus measured by this process is termed total reactive phosphorus. This method is suitable for classical analytical purposes. Optrodes that apply immobilized, reversibly operating indicator dyes for phosphates could be the solution for the future.

Biosensors are becoming more and more important in water quality monitoring. Pesticides are highly toxic compounds, and some of them (organophosphates, heavy-metal ions) are powerful inhibitors of enzymes involved in nerve functions. It is true that they show low environmental persistence, but they have acute toxicity and therefore there is a demand for fast screening methods for low concentrations of these pollutants.

Inhibitors have a great influence on enzymatic processes. They are substances that combine with the enzyme molecule to form an inactive complex, thus reducing the rate of enzyme-catalyzed reaction. Some of them inactivate the enzyme irreversibly, but others act reversibly; that is, their effect can be abolished by removing the inhibitor, which enables a sensor-type operation for the inhibitor. The inhibition can be expressed in terms of percentage according to the relation:

$$\text{Inhibition (\%)} = 100\,(X_0 - X_i)/X_0 \tag{5.8}$$

where X_i and X_0 are sensor responses with and without the presence of inhibitor with the same substrate concentration. At saturated substrate concentration, the sensor output is a function of the inhibitor concentration; thus, the biosensor can be used for measuring it. The application of biosensors for determining toxic compounds is becoming a dynamic trend in water quality monitoring. Strict regulations are being enforced in Europe up to a maximum of allowed concentration of 0.1 pg/ml for individual pesticide residues in drinking water. Contaminants such as heavy-metal ions (Zhylyak 1995), organophosphorous pesticides, and carbamates (Yuan et al. 1991; Cagnini et al. 1995), which have been widely used for pest control in agriculture for many years, can be detected by enzymatic biosensors based on their inhibition effect on enzymatic reactions. For operation, these sensor types need measurement media saturated with the substrate of the applied enzyme. Such enzyme types are useful for these sensor purposes that show reversible reaction with the inhibitor to be detected, or the enzyme can be refreshed easily for a subsequent measurement cycle. Figure 5.4 shows the typical characteristic of a Zn/Co/Cu-ion sensor based on enzyme inhibition (Satoh and Iijima 1995).

An enzyme-based sensor for the determination of organophosphate pesticides was described by Ristori et al. in 1996. It is based on the potentiometric determination of the inhibiting properties of the pesticides on acetylcholinesterase activity. The enzyme was deposited onto commercially available membranes placed on the surface of an ion-sensitive field-effect transistor. Use of this kind of membrane improves the behavior of the system with respect to the application of polymeric matrices chemically anchored to the surface of the sensor. The results presented here show good sensitivity for the pesticide used (0.1 mg/L) and the possibility of using the sensor for a preliminary screening of different pollutants on water samples simply by changing the sensitive membrane.

Immunosensors have also been applied in pesticide detection. Antibodies capable of specific recognition of certain classes of pesticides have been developed, and their use in competitive evanescent-wave sensors has been demonstrated. Schipper et al. (1995) investigated the possibility of developing an evanescent-wave sensor based on direct detection of these small molecules. The operating principle of a Mach-Zehnder interferometer sensor was applied. The light was split and coupled in reference and sensor channels with equal intensities. If in the sensor channel the analyte molecules could bind to the surface, the effective refractive-index contrast in this channel would change, thereby changing the intensity of the interference signal after outcoupling and combining the intensity of the static interference pattern at the position. Planar waveguides were fabricated using standard techniques ($Si/SiO_2/Si_3N_4/SiO_2$). The high refractive index of the waveguide layer (LPCVD Si_3N_4, n = 2.00), in contrast to those of the aqueous

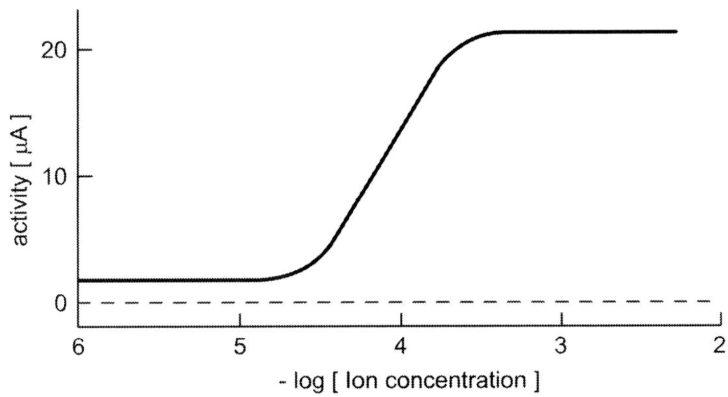

Figure 5.4. Typical characteristics of the Cu-ion sensor based on enzyme inhibition. (Reprinted with permission from Satoh and Iijima 1990. Copyright 1990 Elsevier.)

cladding (n = 1.33) and substrate layer (thermally oxidized SiO_2, n = 1.46), gave, at a waveguide-layer thickness of 170 nm, a TM_0 spatial electric field distribution with good confinement near the waveguide-layer surface. The Mach-Zehnder immunosensor has a proven detection limit of 1×10^{-3} nm thickness.

Taste-originating chemicals can be measured with a multichannel lipid membrane sensor, which responds to different taste qualities with unique patterns of sensor output signals (Hayashi et al. 1990). The lipid membranes are useful materials for transforming quality information about various solution samples into electric signals. The multichannel sensor utilizes lipid membranes as a transducer of taste substances and a computer as a data analyzer. The transducer plays the role of transforming taste information generated by chemical substances into electric potential changes. The method of taste sensing is discussed in detail in Chapter 2. The characteristics and applications of a multichannel electrode-type sensor were discussed by Sakai et al. in 2000. It was based on eight membranes using various lipids as a simple potentiometric sensor system for detecting river water quality by evaluating the results with principal component analysis. These results showed that the pattern of the output voltage signals from the sensor system can distinguish water from various rivers. Therefore, a map of the water quality in rivers based on the response of the sensor system and principal component analysis are useful for an estimation of water quality and water pollution.

Ammonium ions (NH_4^+) are formed during the breakdown of protein molecules from organic matter by bacteria and occur naturally in water systems. An important motivation for improving ammonia sensing is the fact that the presence of ammonia in drinking water or food is often an indication of organic decomposition processes potentially harmful to human health. Ammonium is measured generally with an ion-specific electrode (ISE) and a separate half-cell reference. The electrode has an ammonium ion-selective membrane. Test results are given in parts per million.

An attractive electrochemical protocol for measuring ammonium-ion concentration based on carbon nanotube (CNT)/copper composite paste electrodes was described by Valentini et al. (2007). The novelty of this work concerns the observed catalytic action of the CNT/Cu composite material, demonstrating the possibility to detect ammonia at a low potential (–0.120 V versus Ag/AgCl) and at higher current values. These optimized composite electrodes showed significantly better performance than those obtained from simple CNT or Cu paste electrodes (used here for comparison) in terms of linear range of concentration, lower detection limit, higher sensitivity, reproducibility, and response time. The sensors were applied to the determination of ammonia in real samples of tap water, spiking different ammonia concentrations directly into the water samples.

Ammonia (NH_3) is a pungent, colorless gas that is highly soluble in water. When ammonia is dissolved in water it reacts to form ammonium. Both species are normally present in a solution, but the relative amounts of each are influenced by the pH and, to a lesser extent, the ionic strength of the solution. A saturated water solution of ammonia contains about 30% ammonia at room temperature. Its main uses are as a fertilizer, a refrigerant (or freezing agent), and for manufacturing nitric acid. Analyzers generally use an ammonia electrode that has a membrane separating the sample from an internal solution. The membrane is permeable to gas and impermeable to water. Dissolved ammonia in the sample solution diffuses through the membrane until the partial pressure of ammonia is the same on both sides of the membrane. The diffused ammonia gas dissolves, and its concentration in the form of ammonia is related to the internal concentration of hydroxide ions. The measurement is then similar to that of a simple pH probe. An internal electrolyte with pH 4 buffer is used to "reset" the response of the electrode.

Nitrate (NO_3^-) is a very soluble ion and an oxidizing agent. It is often found in salts such as KNO_3 or $NaNO_3$. It is a common ingredient in agricultural fertilizers. Nitrate ion-selective electrodes are generally applied for measuring its concentration in solutions. (Problems of soil nitrate measurements are discussed in Chapter 6.)

Nitrate has a strong specific absorbance at 205 nm, so absorbance measurements can be used. Absorbance using photodiode arrays for spectral information in the ultraviolet and visible light region gives the possibility to construct low-cost multiparameter sensors. Unfortunately, some peaks in wastewater often remain hidden in the strong absorbance peaks from organic substances. However, nitrate analyzers based on absorbance at 205 nm are available on the market, all of which try to compensate for the organic substances by calculations using absorbance values from other wavelengths in the spectrum as well (Lynggaard-Jensen 1999). The principle of multivariate calibration of the spectra was demonstrated for nitrate by Karlsson et al. (1995), who also were able to measure total P, total N, ammonium, and iron simultaneously with the nitrate.

Nitrate biosensors based on immobilized bacteria are also starting to be commercially available. A microscale nitrate sensor used in sediments was able to work for 2–4 days (Larsen et al. 1997). This sensor has now been improved to work in activated sludge by the attachment of an ion-permeable membrane to the tip of the sensor, as shown in Figure 5.5.

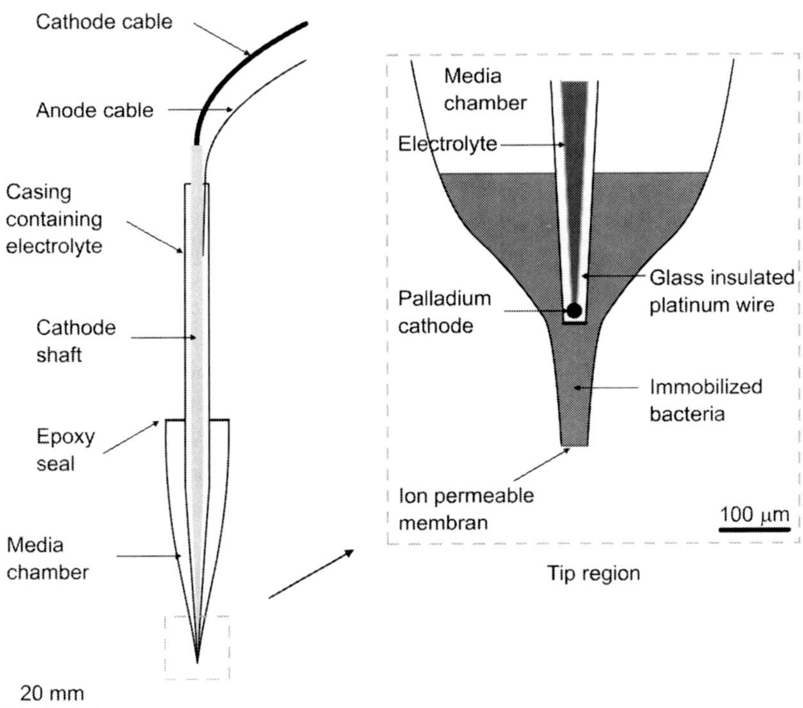

Figure 5.5. Nitrate biosensor based on immobilized bacteria for wastewater monitoring application. (Reprinted with permission from Lynggaard-Jensen 1999. Copyright 1999 Elsevier.)

The distance from the tip to the traditional N_2O microelectrode constitutes the reaction chamber, containing a strain of denitrifying bacteria lacking the enzyme N_2O reductase. The bacteria are then only able to reduce the NO_3^- coming through the membrane to N_2O, which in turn is measured electrochemically by the N_2O microelectrode (Lynggaard-Jensen 1999).

NADH is an important proton transporter for processes inside bacteria. NADH fluorescence is measured at 460 nm (excitation at 340 nm). Because the corresponding oxidized state is nonfluorescent, it is possible to measure biological activity expressed by the concentration of NADH, a fact that has been well known in the fermentation industry for several years. NADH fluorescence can be used to detect the end of the denitrification in an alternating activated sludge plant, as demonstrated by the fluorimeters of Norgaard et al. (1996) and Isaacs et al. (1994).

3.4. OIL AND OTHER ORGANIC POLLUTANTS

Near-infrared (NIR) photoacoustic spectroscopy is a sensitive spectroscopic technique for the detection of analytes in a wide range of applications. The basic principles of pulsed photoacoustic (PA) spectroscopy are as follow. In weakly absorbing liquids, the conversion of the pulse of optical energy into acoustic energy is based on the radiationless relaxation of the absorbed energy into thermal processes. The consequent thermal expansion generates an acoustic wave. Constituents of crude oil, such as pentane, methanol, and benzene, have recognizable photoacoustic spectral features. Hodgson et al. (1995) described an experimental study applying the PA technique in detecting crude oil contamination in water with a view to devising an online instrument for contamination monitoring in water discharge pipelines. At the lower end of the measurable range (0–60 mg/L) the accuracy is 10 mg/L with minimal data processing. It was also demonstrated that dispersed and dissolved hydrocarbon components give an additive photoacoustic response, whereas with most optical instrumentation the components must be measured separately. The first application of this sensor technology was in the continuous monitoring of hydrocarbons in return process water from oil production installations. The prototype measuring head utilized fiber optic delivery of diode laser sources and optimized acoustic detection to allow the legislative limit for the hydrocarbon concentration in process water of 40 ppm to be monitored.

Phenol and its derivatives are an important class of environmental pollutants. Phenolic compounds are induced to the environment as a result of their industrial applications. They appear as intermediates in many industrial processes, including those in the petroleum, pulp, and paper industries. Production of plastics, dyes, pharmaceuticals, and pesticides is related to phenolic products. They are widely used as wood preservative agents, pesticides, antiseptics, and disinfectants. The phenolic pollutants are toxic to fish and other aquatic life and also to many other mammals including humans. They affect the taste and odor of water and fish. Phenolic compounds are listed by both the U.S. Environmental Protection Agency (EPA) and the European Union as dangerous substances discharged into the aquatic environment. A poly(methyl methacrylate) (PMMA)-modified quartz crystal microbalance (QCM) has been tested for determination of phenol in solution by Mirmohseni and Oladegaragoze (2004). A solution casting method was used for modification of the quartz crystal with a thin layer of PMMA. Determination was based on frequency shifts of the PMMA-modified quartz crystal electrode due to the adsorption of phenol at the surface of the modified electrode. Measurements were carried out by exposure of the PMMA-modified electrode both directly to the phenol solutions and indirectly to the phenol vapors that exist in the headspace of phenol solutions (headspace analysis). The frequency shifts were found to be linear against the concentration of phenol in the range between 0 and 10 ppm for direct exposure and between 1 and 100 ppm for headspace analysis. Sensitivity factors of 37.87 and 1.47 Hz/ppm were found for direct exposure and headspace analysis, respectively.

Because volatile contaminants in water distribute between the liquid and gas phases, they can be detected not only in the liquid phase but also in the gas phase by continuously working gas sensor systems. In general, electronic noses possess various features which make them well suited for on-site water analysis. Goschnick et al. (2004) investigated the applicability of an electronic nose for monitoring volatile pollutants in waters with respect to sensitivity and gas discrimination power. Two model compounds of environmental importance but different chemical nature were chosen: ammonia, which is an end product in biodegradation and is often released from agricultural practices; and chloroform, as a representative of industrial pollutants. The electronic nose was based on a special gas sensor array of metal oxide gas-sensing elements. The headspace of water samples polluted with chloroform (as a representative for chloro-organic solvents) or ammonia (representing fecal contamination) was investigated to simulate stagnant and flowing waters. Discrimination and recognition of tap water, ammonia-contaminated water, and chloroform-contaminated water were successful. In both cases, response times of 1–3 s and recovery times less than 1 min were found. Detection limits were about 1 mppm (ppm by mass of the contaminant in water).

Sandia National Laboratories presented the development of a commercially available, real-time, microsensor-based monitoring system that can be used to detect and characterize volatile organic compounds in soil and groundwater (Ho and Hughes 2002). The system employs an array of polymer-based microsensors (chemiresistors) packaged in a waterproof housing that is designed to protect the sensor from harsh subsurface environments, including completely water-saturated conditions. The chemiresistor consists of a chemically sensitive polymer that is dissolved in a solvent and mixed with conductive carbon particles. The resulting ink is then deposited and dried onto thin-film platinum traces sputtered onto a solid substrate. When chemical vapors come into contact with the polymers, the chemicals absorb into the polymers, causing them to swell. The swelling changes the resistance of the electrode, which can be measured and recorded. The amount of swelling corresponds to the concentration of the chemical vapor in contact with the chemiresistor. A second unique feature is that a robust package has been designed and fabricated to house the chemiresistor array. This bullet-shaped package is small (~3 cm in diameter) and is constructed of rugged, chemically resistant material. Package designs have been fabricated from stainless steel. The package is completely waterproof, but gas is allowed to diffuse through a membrane that covers a small window to the sensor. The membrane prevents liquid water from passing through it, but the membrane "breathes," allowing vapors to diffuse through. Even in water, dissolved volatile organic compounds (VOCs) can partition across the membrane into the gas-phase headspace next to the chemiresistors to allow detection of aqueous-phase contaminants. The array of differing sensors can be used to identify different VOCs by comparing the resulting chemical signatures with calibration (or training) sets and pattern-recognition methods. The chemiresistor array has been shown to detect a variety of VOCs, including aromatic hydrocarbons (e.g., benzene), chlorinated solvents (e.g., trichloroethylene, carbon tetrachloride), aliphatic hydrocarbons (e.g., hexane, iso-octane), alcohols, and ketones (e.g., acetone). The on-board temperature sensor comprised of a thin-film platinum trace can be used not only to monitor the *in situ* temperature, it can also provide a means for temperature control. A feedback control system between the temperature sensor and on-board heating elements can allow the chemiresistors to be maintained at a fairly constant temperature, which can aid in the processing of data when comparing the responses to calibrated training sets. In addition, the chemiresistors can be maintained at a temperature above the ambient to prevent condensation of water, which may be detrimental to the wires and surfaces of the chemiresistor. The sensors and packaging have been tested in field and laboratory environments, and characterization methods are being developed that utilize contaminant transport models and time-dependent, *in situ* sensor data to identify the location of the contaminant source.

Atmospheric dispersion of malodors produced from wastewater treatment processes can cause a nuisance to adjacent populations. The nature and concentration of VOCs in wastewater varies from one situation to another, depending on several factors such as water quality and weather conditions. To monitor pollutants rapidly, a real-time monitoring system needs to be used. The ability of portable electronic noses for *in situ* monitoring of odors from a wastewater treatment plant under uncontrolled meteorological conditions was studied by Nake et al. (2004). They used e-noses based on principal component analysis [Cyranose 320 from Cyrano Sciences, applying 32 different conducting polymer (CP) chemiresistors and Pen-2 from WMA Airsense, consisting of 10 metal-oxide semiconductor (MOS) sensors] for determining VOCs in wastewaters. The experimental study showed that electronic nose technology based on the MOS sensors tested appears to be suitable for outdoor monitoring. On the other hand, CP sensors appear to be unacceptable for this application.

The proliferation in the manufacture and use of synthetic organic chemicals has resulted in their detection in most areas of the environment, especially in wastewaters, in increasing amounts. Thus discharges of wastewater, including those from sewage treatment works, represent a very important vector for contamination of the wider environment by these chemicals with increasing attention to recycling and reuse of wastewater. Chemicals of concern in relation to wastewater quality have been divided into a number of categories:

- Endocrine-disrupting chemicals (EDCs)
- Pharmaceuticals
- Personal care products (PCPs)
- Disinfection by-products (DBPs), e.g., *N*-nitrosodimethylamine (NDMA)

The assessment of the estrogenic activity of environmental samples including effluents from sewage treatment works can be determined by a number of assays, including both *in vitro* and *in vivo* assays. These have been extensively applied to wastewater discharges, including those from sewage treatment works and receiving waters, and biosensors based on these assays are now being developed (Rodriguez-Mozaz et al. 2004, 2005a). However, they give no indication of the chemicals responsible. Nevertheless, they are a good first approach for indicating whether chemical analysis for EDCs is required. A detailed discussion of the basis for these assays is outside the scope of this book.

Bisphenol A (BPA) is a weakly estrogenic phenol used in the production of plastics. Leaching from linings of food cans can result in one route of exposure. Because of its widespread use, it is a contaminant of the sewage system. Recent methods specific for bisphenol A include the application of optical immunosensors (Rodriguez-Mozaz et al. 2005b).

An annular column sensor was reported by Foster and Synovec (1996). The sensor is made by applying a thin polymeric film cladding of trifluoropropylpolysiloxane, nominally 0.1 μm thick, to a 200-μm-diameter bare glass fiber core and inserting the fiber into a short section of 250-μm-ID fused silica capillary tubing, producing an annular column. A He/Ne laser is directed into the end of the fiber optic. When a small sample volume is injected into a mobile phase flowing along the sensor, selective partitioning of an analyte into the thin polymeric film, i.e., the stationary phase, causes a local change in the cladding refractive index, and thus the aperture of the fiber, and a consequent change in the amount of light emerging from the side of the optical fiber is sensitively detected. Temporal selectivity is obtained simultaneously, as there is some chromatographic separation of analytes flowing through the annular column sensor. When the sensor signal is corrected for refractive index dependence, the revised signal is proportional to the partition coefficient into the polymeric cladding and is found to correlate well with the octanol/water partition coefficient for the analyte. The sensor can be used for the simultaneous separation and detection of organics in water and yields a detection limit of 500 ppb for cumene at 1 μL/min flow rate. The sensor also works well for chlorinated hydrocarbons and provides unique retention times and sensitivities, such as a 1.0-ppm detection limit for trichloroethylene in water.

3.5. BIOLOGICAL OXYGEN DEMAND AND INTEGRAL WATER QUALITY

Biochemical oxygen demand (BOD) (the amount of oxygen needed by microorganisms for the oxidative decomposition of all organic materials in wastewater) is one of the most widely used tests for the measurement of organic pollution. The main drawback of the conventional BOD test, however, is that it requires a 5-day incubation period. A more rapid and reproducible method is therefore required. Yeast (*Trichosporon cutaneum*) utilized for wastewater treatment was employed for a BOD sensor (Karube et al. 1995). The microorganisms immobilized on a porous acetylcellulose membrane were sandwiched between an oxygen-permeated Teflon membrane and a porous membrane. Then the membrane was fixed directly onto the surface of an oxygen probe's platinum cathode. When a sample solution containing glucose and acid was injected into the system, organic compounds were assimilated by the immobilized microorganisms. Consumption of oxygen by the immobilized microorganisms caused a decrease in dissolved oxygen around the membranes. As a result, the current of the sensor decreased with time until a steady state was reached. The steady-state current depended on the BOD of the sample solution. The time required for the current to reach a steady state depended on the kind of sample solution. Various microbial BOD sensors have since been developed and are now commercialized. They use generally the same idea with different bacteria. A BOD sensor employing a luminous bacterium, *Protobacterium phosphoreum*, was also developed. The apparatus consisted of a photodiode and a photosensor amplifier. The increase in luminescence intensity resulting from the cellular assimilation of organic compounds in wastewater was detected by a photodiode in a light-tight steel chamber. The photoelectric signal produced was amplified and then charted by a pen recorder.

The influence of pollutants on photosynthesizing objects was studied in order to reveal the capabilities and features of their application as controlled sensors in integral water quality monitoring (Yatsenko et al. 1995). The influence of herbicides, heavy metals, and some other pollutants (toxicants) on the operation of the photosynthetic apparatus of green plants was analyzed experimentally. An analysis performed pursuant to the results concerned the influence on the kinetic-fluorescent characteristics of plants, and it was shown that the green plant fluorescence induction curve adequately reflects the reaction of plants in response to unfavorable conditions present in the environment. This effect can be used for special sensory purposes.

3.6. DISSOLVED GASES IN WATER (CHLORINE, OZONE WATER TREATMENT MONITORING)

Total dissolved gas (TDG) pressure is an important parameter for monitoring water quality effectively. When the total pressure of all the gases in water exceeds the ambient atmospheric pressure at the water surface, supersaturation exists. The effect of excessive supersaturation on fish has been well documented, and if supersaturation exceeds the established safe levels, massive fish kills can occur. Dissolved oxygen measurements are not enough—oxygen levels may be acceptable while nitrogen levels may be lethal. A TDG sensor generally uses a pressure transducer mounted behind a rigid gas-permeable silicone membrane to measure the amount of total gaseous compounds dissolved in a liquid. Since this is a special application of pressure sensors, a more detailed discussion is outside the scope of this chapter.

Disinfection means a chemical or physical process that kills or inactivates microorganisms such as bacteria, viruses, and protozoa. Common disinfectants in water treatment include chlorine, chloramine, ozone, and ultraviolet light.

3.6.1. Chlorine Monitoring

Chlorine (Cl_2) is widely used both industrially and domestically as a disinfectant and bleach. In particular, it is widely used in the disinfection of swimming pools and is the most commonly used disinfectant and oxidant in drinking-water treatment. In water, chlorine reacts to form hypochlorous acid and hypochlorites. The use of chlorine in the treatment of drinking water has virtually eliminated waterborne diseases, because chlorine can kill or inactivate most microorganisms commonly found in water. Disinfection is essential to safeguard drinking water; the health risks from disinfection by-products are much less than the risks from consuming water that has not been disinfected. Free chlorine in drinking-water distribution systems ranges from 0.04 to 0.8 mg/L. Some definitions for relevant terms follow.

- *Chlorine residual:* the concentration of chlorine species present in water after the oxidant demand has been satisfied
- *Free chlorine:* the amount of chlorine present in water as dissolved gas (Cl_2), hypochlorous acid (HOCl), and/or hypochlorite ion (OCl^-) that is not combined with ammonia or other compounds in water
- *Combined chlorine:* the sum of the species resulting from the reaction of free chlorine with ammonia (NH_3), including monochloramine (NH_2Cl), dichloramine ($NHCl_2$), and trichloramine (nitrogen trichloride, NCl_3)

- *Total chlorine:* all chemical species containing chlorine in an oxidized state; usually the sum of free and combined chlorine concentrations present in water
- *Primary disinfection:* the application of a disinfectant in the drinking-water treatment plant, with a primary objective of achieving the necessary microbial inactivation
- *Secondary disinfection:* the subsequent application of a disinfectant, either at the exit of the treatment plant or in the distribution system, with a primary objective of ensuring that a disinfectant residual is present throughout the distribution system

Chlorinators incorporate state-of-the-art technology and include gas, liquid, and solid tablet chlorinators. For disinfection using chlorine, it is recommended that a continuous analyzer be used at the point of entry and throughout the distribution system or that grab samples be tested frequently. This ensures that the target chlorine level is being applied at all times and provides a comparison against residual levels observed throughout the distribution system. In addition, sample locations should be chosen to represent all areas of the distribution system. In general, water leaving a treatment plant should be tested daily for both chlorine residual and turbidity. In the distribution system, the presence of adequate chlorine residuals should be confirmed. All field measurements of chlorine in the distribution system should be carried out on site, since chlorine concentrations may decrease during transportation to the laboratory and storage. Eight standard analytical methods are given for determination of residual chlorine: two iodometric methods; amperometric titration and low-level amperometric titration methods; DPD (N,N-diethyl-*p*-phenylenediamine) or ferrous titrimetric method; DPD colorimetric method; syringaldazine method; and iodometric electrode technique. The discussion of these methods is outside the scope of this section.

Several portable field instruments are available for this purpose, which are based mainly on colorimetric, membrane-based polarographic, amperometric, or cyclic voltammetric methods and can be operated in sensory mode. On the other hand, efforts are being concentrated to develop solid-state sensors for continuous monitoring. Some examples are discussed in the following.

ATI's instrument is designed for on-line monitoring and control of chlorination systems. It is available in two versions: a free chlorine monitor for drinking and cooling water systems, and a combined chlorine monitor ideally suited for "chloraminated" drinking water. The combined chlorine system can also be used in wastewater effluents where sufficient ammonia is present prior to chlorination to result in a predominantly monochloramine residual. Unlike on-line colorimetric and most amperometric monitors, these systems utilize a membrane-covered polarographic sensor that does not require the addition of chemical reagents. This means no more costly reagents to purchase or flow cells that require frequent cleaning. All that is needed is a constant flow of water across the face of the sensor, provided by a simple constant-head flow cell. The residual chlorine sensors are generally available in two configurations: flow cell and submersion. The free chlorine sensor should always be used in the flow cell configuration. The constant-head flow cell is used to maintain a constant flow rate and pressure at the face of the sensor. The only requirement is that the sample inlet flow rate exceeds a given threshold. The submersion sensors require given flow velocities for proper operation. When the pH of the water stream varies regularly, a significant error in free chlorine concentration is introduced. With the simple addition of an optional pH sensor, the compensated systems will automatically correct for these changes up to a pH of 9.

Free chlorine levels are often measured by DPD colorimetric analysis (see above). However, the current DPD-based method has a narrow working range and suffers from interferences by organic material present in chiller water and chloramine formation. In addition, several significant problems are associated with the DPD-based method: The DPD reagent is not very stable; the oxidized colored products show fading within a few minutes; the color developed with DPD is reported to be temperature-sensitive; there is interference from combined chlorine, and it is typically a manual, off-line method. Researchers with Georgia Tech's Food Processing Technology Division are developing an electronic sensor that they believe will more accurately track chlorine compounds, including free chlorine, and their concentrations in process waters, particularly red-water chillers. The sensor uses an interferometric measurement principle. The technique can be used to detect both chemical and biological species, and is fast, has high sensitivity, and provides a direct measurement with no additional steps or consumable reagents. The sensor platform is designed for mobile, on-site field analysis with real-time results. For the current project, researchers have incorporated a novel reactive sensing film into the sensor to create an assay for rapid, on-line quantification of free chlorine (hypochlorous acid). The team has also designed and fabricated a waveguide chip with eight interferometers. With this new design, simultaneous multiple detection becomes achievable. In addition, researchers plan to integrate sensing chemistries for free chlorine, combined chlorine, and pH on a single-chip to provide real-time measurement for free chlorine, combined chlorine, and pH (http://atrp.gatech.edu/pt19v3f07/19-3_p3.html).

An optoelectrochemical approach to the detection of chlorine in water is also available (Baldini et al. 2004). In particular, Kasik et al. (2007) have shown that optical changes of a sensing layer caused by its interaction with chlorine are electrically reversed by using a thin conductive and transparent layer of indium tin oxide (ITO) deposited in between an optical substrate and the sensing layer acting as an electrode. This means that commercially available transducers applicable on ITO-coated planar substrates and/or silica optical fibers which are suitable for the detection of chlorine in water can be designed. Sensing layers of several absorption and luminescent transducers were applied onto ITO layers deposited on planar glass or fiber optic substrates by electropolymerization or by the sol-gel method. Kasik et al. (2007) found that the best response was observed from *o*-phenylenediamine and polyluminol layers prepared by electropolymerization. The sensitivity and limit of detection of polyluminol layers are comparable to hygienic limits for the chlorination of water. The parameters of *o*-phenylenediamine layers obtained up to now are close to these values. A damaging effect of chlorine in higher concentrations on the sensing layers was observed.

The feasibility of detecting chlorine in water using a fiber optic technique was investigated by Safaai-Jazi and Petersen (1991). The technique is based on evanescent field absorption through the color development of reacting diethyl phenylene diamine with chlorine, which forms an azo compound. Transmission attenuation results through color absorption of the fiber optic waveguide from the azo species surrounding the fiber. Sensor systems with single- and multifiber sensing elements have been designed, constructed, and evaluated for linearity of response, sensitivity, and reproducibility of results. The experimental results establish the feasibility of detecting low-concentration chlorine in water using the proposed fiber optic technique.

The integrated chlorine sensor of SILESENS is based on an electrochemical method with the possibility to easily replace the sensing head at a competitive price (www.silsens.com). The principle of the measurement is based on a Clark-type cell with working, counter, and reference electrodes. The key element is the photo-polymerized membrane deposited on the working electrode. This design has the advantage of increasing the lifetime of the sensor easily and cost-effectively. This is required in harsh environments or for low-cost chlorine-sensing applications.

A chlorine gas sensor was fabricated with the combination of divalent Mg^{2+} ion conducting $Mg_{1-2x}(Zr_{1-x}Nb_x)_4P_6O_{24}$ (x = 0.15) solid solution and divalent oxide anion conducting stabilized zirconia with calcium-doped lanthanum oxychloride as an auxiliary electrode (Imanaka et al. 2001). The response time necessary for the attainment of 90% response was within several minutes, and a rapid and reproducible response was continuously achieved. The relation between the sensor output and the logarithm of chlorine gas concentration was exactly 1 to 1 and obeyed the Nernst theoretical relation.

A residual chlorine sensor with a gold electrode was developed by Iketake and Yamada (2000) using voltammetry. The sensor consists of a working electrode (gold wire), a counter electrode (platinum wire), and a reference electrode [saturated calomel electrode (SCE)]. The electrochemical behavior of residual chlorine was measured by cyclic voltammetry. The reduction wave of residual chlorine was observed from −0.6 V versus SCE at pH 7. The measurements were carried out using the potential sweep method and were monitored at +0.1 V versus SCE. A linear calibration curve was obtained in the concentration range 0.1–200 mg/L residual chlorine. The relative standard deviation was 2.7% for 3 mg/L residual chlorine. The reduction current at the measurement potential was not dependent on the pH, and there was positive interference due to the existence of 0.1 mg/L copper ion. The working electrode was able to be used for over 300 continuous measurements. The sensor was applied to the determination of residual chlorine in tap water. The results agreed extremely well with those obtained by the photometric method.

3.6.2. Ozone Monitoring

Ozone is rapidly becoming the water treatment method of choice for bottled water, swimming pools, aquariums, and many other water purification applications. In comparison to chlorine, it creates less, if any, toxic by-products, it is generated on-site, and the finished water is generally more attractive. Ozone is generally dissolved in water to disinfect for bacteria, viruses and other microorganisms, to oxidize and thus create removable solids of dissolved minerals, and to neutralize undesirable chemicals. It is also used for other purposes, such as to allow lower-temperature wash water and to minimize residuals in wash processes ranging from laundry to semiconductors. The use of ozone for ultrapure water systems, cooling towers, aquatic farms, pharmaceutical-grade water, and potable water treatment has created the need to measure levels of dissolved ozone. For any application, ozone should be efficiently dissolved into the water, and the dissolved ozone should be measured at various points in the process—at the injection point, at the contact tank, and at the exit of the system. Dissolved ozone has generally been measured by the following methods (www.ecosensors.com/DO-100.pdf).

- Oxidation-reduction potential (ORP): A two-electrode ORP meter measures the oxidizing capacity in the water (Emerson 2008). It is not selective between different oxidizing agents such as chlorine and ozone.
- Electrochemical (amperometric): The ozone migrates through a permeable membrane and then is converted to oxygen. The relative quantity of oxygen is measured by an electrochemical cell. Electrochemical instruments are usually configured for online use. The probes are expensive, easily damaged, and require regular maintenance, but these instruments are much more ozone-selective than ORP meters, and are widely used in larger systems.

- Wet chemistries and test kits: Wet chemistries in use mainly are titrations of iodine solutions and indigo trisulfonate (ITS) test kits.
- Ultraviolet (UV) absorption: The ozone is removed or "stripped" from the water and the ozone is measured in air by a UV absorption analyzer. This method is very specific to ozone, is capable of high accuracy, and many concentration ranges can be accommodated. The equipment cost tends to be out of range for small systems, however.

Eco Sensors dissolved ozone instruments are commercially available and are designed to have many of the best features of the instrument types reviewed above. A key concept is to release all of the ozone from the sample water before measuring it by rapid agitation of the sample. This method is the basis of an instrument that is effective, inexpensive, and easy to use. Features include low cost, ozone specificity, and insensitivity to common concentrations of chlorine and salinity.

ATI has also developed a similar, membrane-based dissolved-ozone monitor that makes it possible to continuously measure dissolved ozone in the water stream without complicated maintenance or measurement interference from other sample components such as residual chlorine. The dissolved-ozone monitor is able to measure ozone over ranges from 0 to 2.00 and from 0 to 200 ppm. The sensing element exhibits excellent zero stability and is easy to maintain—it requires only a quick and easy membrane replacement two to four times a year.

A new type of sensor element to detect ozone in water was fabricated by Sano and Ohtsuki (2007) using coating film of single-wall carbon nanohorns (CNHs). This sensor is based on the phenomenon that the electric resistance of the CNH film decreased with the adsorption of ozone molecules thereon due to charge transfer from the surface of the CNHs to O_3 molecules. The shift of the electric resistance of the CNH film was monitored with varied concentrations of aqueous ozone at several temperatures. The shift of the electric resistance of the CNH film was correlated with the ozone concentration and temperature based on a monolayer adsorption model with consideration of activation energies of the relevant adsorption, desorption, and sensitivity of charge transfer.

4. FOOD QUALITY MONITORING

4.1. pH IN FOOD AND BEVERAGES

pH is one of the key parameters in food production; an optimally adjusted pH value determines the degree of enzyme activity in fermentation processes. A few examples include the following.

- Endopeptidase is responsible for decomposing protein in mash when brewing beer and requires a pH environment of 5.
- Rennin, which thickens milk in cheese production, requires a value of pH 6.2.
- In the production of acidified milk products, the pH must be maintained precisely to ensure the quality of the final product.
- In the production of yoghurt, the pH must not drop below 4–4.2. Therefore, the product is rapidly cooled down at a pH of 4.4–4.6 during yoghurt maturation to prevent overacidification.

- Before bottling, liquid mustard must not exceed a pH range of 3.5–3.7, otherwise quality is drastically impaired.
- At the milk acceptance point, milk collected from various farms is transferred from the tanker to the dairy. The key criterion is that the pH value of the incoming milk should not be less than 6.65 or more than 6.85. When milk is stored, it has a tendency to become more acidic. As soon as the pH falls below a specific value, the milk will no longer be marketed for direct consumption but rather will be processed into any of a large number of milk products.
- Because jam, jelly, marmalade, and fruit sauces require high and consistent quality, it is important to control the pH value during processing. The quality criteria greatly influence the flavor, color, and consistency. The pH value affects gelling behavior, flavor, and shelf life of the product. Generally, edible acids (e.g., citric acid) are added to lower the pH. The gel formation by pectin takes place at a certain pH range depending on soluble solids content. Therefore, the feed of edible acids is controlled by online pH measurement during the production of jam, jelly, marmalade, and fruit sauces.

Although the pH value in food processing is an important control parameter, it is usually measured in the laboratory and not inline in the process. However, laboratory measurements are, first, expensive, and second, inefficient compared with an inline pH control system, which can display the actual pH values and start or stop process steps directly. The problem is that continuous pH measurement in food processing has special requirements for the sensors that cannot be fulfilled with conventional versions such as those using glass electrodes:

- Contamination of the diaphragm with native or precipitated proteins is likely. This causes unpredictable changes in the calibration values (the zero point and slope) of the pH electrode. Therefore, cleaning ability requirements of the sensors and assemblies integrated in the process are extremely high.
- Chemical influences as well as cold acids and alkalis used during cleaning can dehydrate the pH-sensitive coating on the glass membrane.
- High temperatures during steam sterilization age the glass prematurely and lead to a loss of slope characteristics.
- On the reference side, the porous diaphragms (made mainly of ceramic) can become blocked by fat and proteins, which can lead to false measured variables.
- Connectors and cables must be spray-water-proof.

Commercially available food pH process electrodes (e.g., Endress + Hauser, Metler Toledo, ABB types), such as gel-filled or liquid KCl-filled types, apply a special pH-sensitive process glass membrane composition which supplies excellent measurement results in cold, weak acid, and acid pH ranges of the product (see also Chapter 6). Attack by acids and alkalis on the glass membrane during cleaning processes is negligible, i.e., the acid or alkaline error after cleaning is 0.1 pH maximum. This special glass recipe ensures long life, even when the electrode is exposed to extremely differing process conditions. Of course, these electrodes can be sterilized in steam and in autoclaves. Every electrode is provided with a diagram which shows the pressure settings to obtain the optimal electrolyte flow through the diaphragm at various process pressures and temperatures. In order to determine the pH value in the process with absolute accuracy, it is vital to measure the temperature and perform temperature compensation. Commercial electrodes have an integrated Pt100 temperature sensor which is located directly at the tip of the electrode behind the pH-sensitive glass. All instruments are available with conventional 4- to 20-mA output signal for the pH and temperature measured variables.

The search for new materials for pH sensor ion-selective membranes is still of great interest in the food industry because of the described weaknesses of glass membranes. Various oxides have been studied for metal oxide pH sensors, including molybdenum oxide, ruthenium, and iridium oxide. They are all based on redox reactions that are dependent on the redox potential of the solutions. Insulators such as Ta_2O_5 and Si_3N_4 have also been used in IFSET devices, but they remain at the laboratory level, without widespread commercial breakthrough. Lithium lanthanum titanate (LLTO) ceramics as sensitive material for pH sensors have been tested recently (Bohnke et al. 2003). The high lithium conductivity of this oxide at room temperature indicates a possible use of this material as a lithium ion–selective electrode. However, a strong interference in aqueous medium leads to a linear response of the electrode potential as a function of the pH of the solution and not as a function of the Li concentration. Two electrode configurations have been used: a membrane one with an aqueous internal reference and an "all solid state" one with a metallic titanium internal reference. Linear relationships were found, although sub-Nernstian responses were observed in both cases. The sensor is not sensitive to variations of the redox potential of the solutions. This sensor can be used in industrial processes such as milk fermentation, yoghurt making, and wastewater treatment, for example. Its good thermal (up to 1000°C) and the chemical stability of LLTO associated with the good mechanical properties linked to the ceramics suggest its use for *in situ* pH control during cleaning of fermentors, which uses hot acidic and alkaline media ("cleaning in-place" processes).

The nature of the food industry means that products to be analyzed are often not simple aqueous solutions. Meats, cheeses, fruits, conserves, creams, solids, protein-rich products—all pose a measuring challenge for pH. Measurements in such products demand special properties from the electrode:

- A robust pH electrode must penetrate easily into the sample.
- The product being measured must not be contaminated with electrolyte solution; the outflow of electrolyte solution through the diaphragm should be close to zero. A solid electrolyte solution should perfectly fulfil this purpose.
- Other challenges include protein-containing samples blocking the electrolyte diaphragm of conventional electrodes and the challenge of getting a suitable electrode into small containers or sachets.

A range of electrodes is commercially available to combat these various measurement challenges (e.g., Metrohm types). The type for pH measurement in semisolid samples has a robust electrode tip for penetrating semisolid samples such as meats, cheeses, fruits, etc. It contains maintenance-free solid polymer electrolyte to ensure long electrode life. The pinhole diaphragm is easy to clean. The very low electrolyte flow rate virtually eliminates product contamination. Special types are also available for measurements in protein-containing samples, with a capillary diaphragm that eliminates protein blockage.

The non-glass pH electrode is an unbreakable breakthrough, especially for the food industry, where the possibility and consequences of fractured glass present a highly dangerous threat. For example, sensors designed by Vci-Analytical Company (www.vci-analytical.com) are equipped with a solid-state pH-sensitive part (ISFET) and with a high-temperature reference system with gel electrolyte, making the electrode practically maintenance-free. The ISFET chip is integrated into a sensor body made of special polymer (polyether ether ketone), which is mechanically and chemically stable and is approved for applications in the food and pharmaceuticals industries. The glass-free system rules out any risk of glass breaking during operation—a decisive advantage in the mentioned sectors. It is pressure-resistant up to 6 bar at 130°C and is fully sterilizable, either *in situ* or in an autoclave. Another advantage is the fast response at low process temperatures, where classical glass electrodes respond fairly slowly. In contrast to the glass electrode, there is no ion flow between the fluid and the sensor. Thus, "chemistry" and electrical measurement are kept totally separate. As a result, the sensor material does not change and the need for recalibration is by no means as frequent as with glass electrodes. Since there is no source layer, ISFETs are also suitable for pH measurement in media with a low proportion of water. Modern ISFET materials are highly selective and follow the Nernst law to closer tolerance limits than glass. The sensor's extreme robustness comes from embedding the chip in a stable and unbreakable plastic body.

4.2. QUALITY CONTROL TESTS (MEAT FRESHNESS, MILK QUALITY, OIL, AND FAT)

At present, the quality control of meat products relies mostly on manual analysis by experts, and a simple and easy method to replace this manual analysis is eagerly desired. To attain optimal eating quality, meat needs to be subjected to an aging process before being distributed to consumers. The quality control of meat is therefore totally different from the case of fish, for which being fresh, with as short as possible a time lapse after catching or maintaining the prerigor state, is important for high quality. On the other hand, meat quality is judged by its eating quality, which is optimal just before bacterial spoilage begins during the aging process. Therefore the term "meat freshness control" should be regarded as a complex concept involving the aging and bacterial putrefaction of meat. Conventionally, in the method of counting colonies, the phase-difference microscope method is used to evaluate the degree of aging and bacterial putrefaction of meat. These methods are tedious and time-consuming for everyday quality control of meat. Enzyme-based sensors for putrecine and hypoxanthine were developed recently for monitoring the aging process of meat. However, even this system is based on a destructive analysis of the chemical components produced in meat. It has been shown recently that the gaseous components evaporating from food can be a good measure for judging food quality. The use of a gas sensor can provide a nondestructive, noncontact method of meat analysis. Ethyl acetate has been found to be a typical major product in the initial stage of bacterial putrefaction. Funazaki et al. (1995) dealt with the development of a semiconductor gas sensor for ethyl acetate vapor as well as its applicability for evaluating meat freshness in the initial stage of bacterial putrefaction. A semiconductor oxide–based, Rh-doped La_2O_3-In_2O_3 gas sensor resistor operated at 300°C was used for detecting ethyl acetate, a typical flavor volatile component produced in the initial bacterial putrefaction of meat. The sensor response signal increases with increasing meat storage time and reflects rather well the increase of bacteria determined by a conventional colony-counting method. The multiple correlation factor between the two methods was 0.8.

Milk is composed of a complex mixture of lipids, proteins, carbohydrates, vitamins, and minerals. A variation in the fat content, e.g., as a result of processing to achieve semiskimmed and skimmed milk, may result in a change in other nutrients, such as calcium or vitamin A, and these results may not be related solely to the variation of the fat content. Several standard methods can be used to detect the concentration of water or fat content in milk. These techniques are based on changes in the freezing point of the milk (cryoscopic method) or on changes in refraction of light of the whey component of milk after removal of the fat components. Recently, adulteration has led to considerable problems in the milk industry, and it has become necessary to develop new and improved instrumentation for the rapid and reliable detection of fraud.

Valuable information about the electrical conduction behavior of solid or liquid materials may be obtained by performing electrical impedance measurements over a wide range of frequencies. Impedance spectroscopy can also be used for quality control of a variety of foodstuffs. The electrical conductivity of dairy products has been studied for more than 40 years to provide accurate values of the fat and protein content. The method of electrical admittance spectroscopy has been used to study the water and fat content of milk by Mabrook and Petty (2002). Over the frequency range 5 Hz to 1 MHz, the electrical circuit was dominated by a single time constant. To eliminate the effect of electrode polarization, the conductance of the milk was measured at high frequencies, where it showed a saturation value. The characteristics at 100 kHz and 8°C for all milk samples revealed a linear decrease in conductance with increasing water content, as shown in Figure 5.6. Admittance data for full-fat, semiskimmed, and

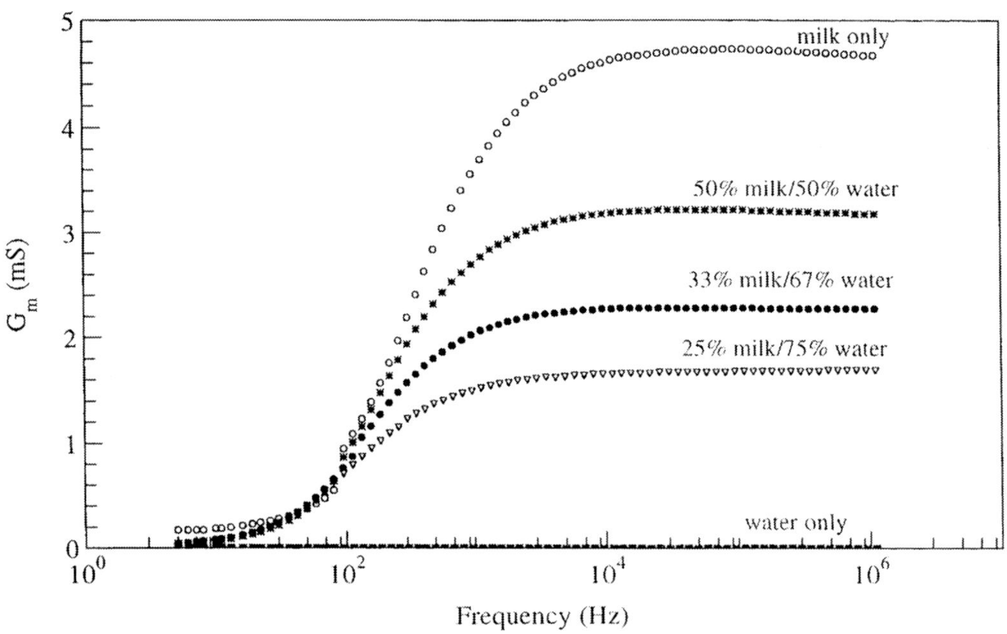

Figure 5.6. Frequency dependence of the measured conductance as a function of water content in a mixture of full fat milk and water at 8°C. (Reprinted with permission from Mabrook and Petty 2002. Copyright 2002 Elsevier.)

skimmed milk showed an increase in milk conductance with decreasing fat content. The sensing devices consisted of two gold electrodes (L-shaped), 15 mm × 6 mm, with a separation of 1 mm. The electrodes were deposited by vacuum evaporation onto clean glass microscope slides.

Quality control of oils and fats is also important in food industries. The level of antioxidant materials in vegetable oil products is of great importance in determining the stability of these products. Vitamin E represents one of the largest groups of natural antioxidants, which include several tocopherol and tocotrienol compounds. These phenol compounds are among the large number of naturally occurring substrates known to be electrochemically oxidizable. Electrochemical methods using linear sweep, differential pulse voltammetry with classical carbon were described for such purposes. The major inconvenience is the toxicity of the reagent, such as benzene or toluene, used as solvent. An electrochemical method using differential pulse voltammetry at an ultramicroelectrode in oil or fat/solvent solution was developed and applied to the quantification of vitamin E in vegetable oils and fats by Coatanea et al. 2001. *N*-methyl-pyrrolidone was chosen as a solvent of low toxicity. The electrolyte solution needed quaternary ammonium salts in concentrations less than 0.01 M in order to avoid biphasic emulsions. The use of a Pt disc ultramicroelectrode (50 μm in diameter) permitted the recording of voltammograms in solutions with very low proportions of solvent. The method was applied to the quantification of vitamin E in several vegetable oils and pork lards. For the pork lards, the results indicated that low levels of vitamin E can be measured (less than 5×10^{-5} M), and this allows the distinction of lards from pigs which have been fed with a supplemented diet.

4.3. SENSORS FOR FOOD PROCESS CONTROL

Food processing is an area where it is anticipated that there will be increasing commercial applications for sensors to monitor the quality of various production methods (Buerk 1993). Thus a growing interest in biosensor techniques, as routine methods, in foods and drinks analysis has been observed in recent years. Demands for sensitivity, speed, and accuracy of analytical measurements have stimulated considerable interest in developing sensor-based methodologies for nutrient, additive, and functional compounds determination in foods and drinks.

Fermentation processes are essential in a number of food production methods, which may result in the production of glucose and a number of similar metabolites (sucrose, fructose, lactate, maltose, ethanol, etc.). The continuous monitoring of these compounds may have great importance for testing the reproducibility and for the quality control of food processes. Another purpose of such measurements is to be able to learn at an early stage if the process has been infected by detrimental microorganisms. A fast and reliable method using sensor-based control minimizes the use of antibiotics. The limited lifetime of enzymatic biosensors as a limiting factor can be overcome if the application is designed appropriately. Also, the requirements are much looser than in biomedical applications. Most important possible combinations for obtaining such biosensors for various metabolites as analytes are summarized in Table 5.1.

A few examples of food processing applications of the various sensor types are listed below.

- Glucose: soft drinks, beer, wine, fruit and vegetable juices, fermentation broths
- Sucrose: soft drinks, cocoa, fermentation broths

Table 5.1. Summary of enzymatic biosensor types

Substrate	Enzyme(s)	Co-factor(s)	Transducer type	Detection limit in commercial types (mg/L)
Glucose	Glucose oxidase (GOD)	O_2	H_2O_2 or O_2	0.3
Glucose	Glucose dehydrogenase	NAD	pH, NADH Fluorescent or amperometric	
α-D-Glucose	Mutarotase + GOD	O_2	H_2O_2 or O_2	
Uric acid	Uricase	O_2	H_2O_2 or CO_2	
Xanthine	Xanthine oxidase	O_2	H_2O_2 or O_2	
L-Lactate	Lactate oxidase	O_2	H_2O_2 or O_2	0.4
L-Lactate	Lactate dehydrogenase	NAD	pH, NADH Fluorescent or amperometric	
Galactose	Galactosidase	O_2	H_2O_2 or O_2	2
Alcohol	Alcohol oxidase	O_2	H_2O_2 or O_2	Methanol 0.3 Ethanol 20
Alcohol	Alcohol dehydrogenase	NAD	pH, NADH Fluorescent or amperometric	
Sucrose	Invertase, mutarotase, GOD	O_2	H_2O_2 or O_2	2
Fructose	Hexokinase + glucose 6P dehydrogenase	O_2, NADP	O_2	
Maltose	GOD + glucoamylase	O_2	H_2O_2	4

NAD, nicotinamide adenine dinucleotide.

- L-Lactate: mashed tomatoes, milk, baby food, fermentation broths
- Ethanol: beer, wine, cider

Some families of useful sensors are commercially available (e.g., SIRE biosensors, Chemel AB). They generally use the same signal transduction: for example, the electrochemical determination of H_2O_2 evaluated in enzymatic processes. The instabilities caused by enzyme immobilization are avoided by the application of miniature flow-through injection (FIA) reaction chambers. During each measurement, a small amount of enzyme is transported to the end of the sensor probe by an internal buffer flow. Once the enzyme has reached the reaction chamber, the flow is stopped for a short period of time. The enzyme reacts with the substance to be detected, which causes formation of hydrogen peroxide, the concentration of which gives an electrical signal which is proportional to the analyte concentration. The effect of interfering substances is effectively minimized by a differential measuring procedure. The instrument is designed so that each sample is measured with and without enzyme present during every measurement.

The possibilities are almost endless, making it difficult to summarize them. One often-mentioned example is a fish freshness sensor developed in Japan (Watanabe and Tanaka 1991). The adenosine triphosphate (ATP) in the fish breaks down in a seven-step reaction to result in the formation of uric acid as tissue undergoes enzymatic autolysis and eventual spoilage. The depletion of high-energy phosphates (ATP, ADP, etc.) and accumulation of inosine and hypoxanthine alters the taste and quality of the meat. The chemical species in the last stages of the process were monitored by the sensor using four enzyme-modified O_2 electrodes, the signals of which were measured and digitized by computer. The fish freshness factor was defined as the ratio of inosine and hypoxanthine concentration over the overall concentration of all seven components taking part in the breakdown process.

Antimicrobial drugs (antibiotics) are widely employed in therapy of food-producing animals and, sometimes, as growth promoters. This massive use inevitably causes residues in food. The presence of antibiotic residues causes concerns for consumers' health, so the European Union has introduced in its legislation the concept of maximum tolerable concentrations of residues in foods, called maximum residue levels (MRLs). The main concerns are

- Possible allergic reactions
- Destruction of human gut microflora
- Reduced production of blood cells
- Possible carcinogenic effects after long-term exposure
- Selection of antibiotic-resistant bacteria

The application of low-cost, simple, and reliable analytical methods for screening analysis of antimicrobial drug residues in foods allows obtaining information and knowledge to solve specific problems of food safety. There are commercially available sensor kits for detecting antibiotics, chloramphenicol, clenbuterol, ractopamine, streptomycin, sulfamethazine, and sulfadiazine. Biacore AB has introduced a sensor system which is based on their patented direct (label-free) detection method of immunosensors applying the surface plasmon resonance (SPR) sensing effect (see more in Harsanyi 2000). This sensory kit gives real-time insights into biomolecular interactions and enables the detection of antibiotics residues in meat, shellfish, milk, and honey. The system also enables the detection of vitamins: vitamin B_2, biotin, folic acid, pantothenic acid, and vitamin B_{12}.

A microbial sensor solution based on specific carbon dioxide electrochemical sensor was presented (Pellegrini et al. 2004) for the detection and presumptive identification of quinolone and tetracycline residues in milk. The measurements were based on carbon dioxide production rate in relation to inhibition on sensitive bacteria growth, *Escherichia coli* in this case. The experimental approach and analytical method developed appear adequate for the purpose and, compared to older screening methods such as, for example, microbial inhibition assays and immunoassays, offers the advantage of relatively short analysis time (about 120 min); the sensitivity is ≤ 25 μg/L.

Polyphenol and phenol compounds have attracted much attention recently because of their purported health benefits. Polyphenols are responsible for the majority of the antioxidant capacity in most plant-derived products; with few exceptions, the antioxidants in foods are polyphenols. They influence the quality of vegetables in terms of color, aroma, and texture, and sensory attributes of beverages such as wine, juices, coffee and tea, and olive oil. Low-molecular-weight phenolic acids in foods such as chlorogenic, caffeic, and gallic acid do not have antioxidizing, antiviral, and stimulating functions, but do affect the activities of some enzymes. Therefore, the determination of polyphenols in tea and coffee is very important for the study of plant growth, investigation of natural sources of chemotherapeutic agents, and improvement of the quality of coffee and tea. Thus, their simple, low-cost, and rapid determination assumes particular importance. For this purpose, mainly two enzymatic biosensors groups have been developed within the last decade (Mello et al. 2003). One group is based on the biocatalytic effect of tyrosinase in oxidizing phenolic compounds. Either oxygen consumption in a dual Clark cell or electrochemical reduction of the quinone products can be followed amperometrically. The latter case needs an electrode covered by a membrane with the immobilized enzyme tyrosinase (Bulukin et al. 2004). The other main approach is the application of an amperometric biosensor utilizing horseradish peroxidase (HRP). Peroxidase-catalyzed oxidation of organic substrates can be reduced back electrochemically through direct or mediated electron transfer. By mediated processes, organic compounds such as phenols act as electron donors. The oxidized substrates are electrochemically regenerated and the reduction current can be related to the concentration of the donor.

Demand for fresh, minimally processed vegetables has led to an increase in the amount and variety of products "ready to use" (RTU) available to the consumer. Minimally processed vegetables may consist of trimmed, peeled, sliced/shredded, and washed and/or disinfected vegetables. To estimate food sanitary quality, the classic approach is based on the search for not only pathogenic microorganisms, but also for indicator microorganisms such as fecal coli forms, whose presence indicates possible pathogens and fecal food contamination of human and/or animal origin. The conventional detection methods for microorganisms are based on counting of colony-forming units (CFUs), which requires selective culture, or biochemical and serological characterizations. The growing market for fresh RTU vegetables, however, requires rapid, sensitive, and reliable monitoring of bacterial contamination. Ecole et al. in 2003 described the application of an antibody-based biosensor for the determination of *Escherichia coli* cells in vegetable food. Commercial samples were washed with peptone water to recover bacterial cells in the liquid medium. The presence of *E. coli*—as a bioindicator of bacterial contamination, fecal contamination in particular—was detected in this phase using a potentiometric alternating biosensing system, based on a transducing LAPS (light-addressable potentiometric sensor) element detecting pH variations due to NH_3 production by a urease–*E. coli* antibody conjugate. (For more about the LAPS operation principle, see Harsanyi 2000). Bacterial cells were found at concentrations as low as 10 cells/ml with detection times from 10 to 20 times shorter than using the conventional CFU procedure.

Ethylene gas causes fruits to ripen and decay. Fruits are stored in warehouses under controlled atmosphere (high humidity, low oxygen). Monitoring ethylene inside conservation chambers can extend the life cycle of fruits and increase quality. Ethylene gas is also used in ripening rooms to color up fruit, which is then moved to a regular cold storage room with other products. Some fruits treated with ethylene are citrus, bananas, tomatoes, and avocados. The monitoring of ethylene gas is very important for the fruit to reach the market in optimal condition. Nowadays, the problem of continuously measuring ethylene levels during storage of climacteric fruit remains unsolved, because ethylene detectors are bulky and very expensive. While ethylene gas is used under controlled conditions as a ripening agent, even small amounts of ethylene gas (20–100 ppm) during shipping and storage can cause fruit decay. Ivanov et al. (2004) reported a new type of sensor, able to selectively detect ethylene in the presence of ethanol (indicator of fruit stress) and ammonia (leakage from the cooling units). Their approach consists of designing an integrated sensor microarray with four metal oxide sensors. It is well known that metal oxides are poorly selective, however, overlapping sensitivities within a sensor array can be tuned by selecting appropriate materials loaded with noble metals. Using a screen-printing technique, sensitive layers of nanopowder tin and tungsten oxide were deposited on silicon micromachined micro-hotplate substrates with various doping metals; very promising results were obtained for selective ethylene detection.

4.4. DETERMINATION OF THE VARIETY OF FOODSTUFFS AND MANUFACTURED GOODS

The characteristic flavor and taste of foods enables their identification in an objective way. Biomimetic sensors, i.e., electronic noses (e-noses) and electronic tongues (e-tongues) have been introduced recently as technological attempts to mimic the functions of the human senses. In this configuration, sensors should not necessarily be specific for a target analyte; rather, they should be broadly selective or even not selective at all, with the only necessary requisite being to behave differently from one another, i.e., to provide partially correlated responses. The basic operating principles of these types of sensor arrays are described elsewhere. Here we concentrate on the special application possibilities.

The analysis of foods and beverages involves methods developed in many different basic sciences. Most of these measurements are concerned with safety (e.g., the search for contaminants), composition (to individuate the basic constituents), effects of food treatment and processing, and so on. The simultaneous utilization of electronic noses and electronic tongues has also been suggested on the basis of the idea that the cooperation of both systems is expected to increase the amount of information extracted from a sample. It is, indeed, a common experience that olfaction and taste cooperate in the formation of a judgment about, for instance, a certain food. The two sensor systems, when applied to the same liquid sample, do not sense the same features. Electronic tongue electrodes are immersed in the sample, while the electronic nose sensors are in contact with its headspace. For this reason, an electronic nose and electronic tongue are expected to provide more information than only one of the two systems (Natale et al. 2004). Table 5.2 gives a broad summary of suggested recent application examples of e-nose and e-tongue sensors in determination of a variety of foodstuffs and manufactured goods.

The application of electronic noses often needs a specially designed extraction process in order to extract odor materials from the solid or liquid sample. One example, developed by Barié et al. (2006), is illustrated in Figure 5.7. The samples were transferred to headspace vials with Teflon-coated septa and magnetic caps. They were preheated for a suited equilibration time, defined as the time necessary for the volatile substances of the sample to reach equilibrium with the headspace. Then the extraction fiber was inserted into the headspace above the sample in the vial. After exposure, the fiber was withdrawn from the sample and inserted directly, passing the septumless sampling head. Inside, for thermal desorption, the fiber was exposed to the heating zone and kept there until desorption was finished. The system was successfully applied for apple identification and fruit ripeness tests.

The same methods can be used in identification of different brands of manufactured goods with special flavors. The Chinese cigarette industry is losing millions of dollars per year due to counterfeit

Table 5.2. Recent application examples of electronic noses and tongues in the selection of foodstuff and manufacturing goods

Application field (indicator, tracer components)	Type[a]	Sensor array elements	Source
Red wine analysis (alcohol, SO_2, polyphenol, antocyanins, VOCs)	EN &	8 thickness-shear-mode quartz resonators coated with various films of metalloporphyrins & 5 glassy carbon electrodes chemically modified by electropolymerized metalloporphyrins	Di Natale et al. 2004
(K, Ca, Mg, pH, lactic acid, malic acid, etc.)	ET		
Milk classification (odor-active compounds, including dimethylsulfone, ethyl butanoate, ethyl hexanoate, heptanol, indole, nonanal, and 1-octen-3-ol)	EN	7 tin oxide–based gas sensors from Figaro Engineering, Inc.	Brudzewski et al. 2004
Codfish fillet freshness (fresh fish odor: long-chain alcohols and carbonyls, bromophenols, and N-cyclic compounds); microbial spoilage odor: short-chain alcohols and carbonyls, amines (trimethylamine), sulfur compounds, aromatics, N-cyclic compounds, and some acids	2 ENs	8 thickness-shear-mode quartz resonators coated with various films of metalloporphyrins & 5 electrochemical sensors for CO, NO, H_2S, SO_2, NH_3	Di Natale et al. 2001
Poultry, egg, milk dioxin toxicity (highest toxicity 2,3,7,8-tetrachlorinated dibenzo-*p*-dioxin and a group of 17 dioxins)	EN	Sensor array selective for the dioxins: piezoelectric quartz sensors coupled with synthetic oligopeptides	Mascini et al. 2004
Cacao—smoke of drying (undesired bitter smoky flavor tracers are monocyclic phenols: guaiacol, syringol, and creosol)	ET	A 3-electrode system used in cyclic voltammetry combined with multivariate signal evaluation (sample solved in distilled water)	Endres et al. 2004
Honey classification (rapid VOC aroma profiling) (zNose, a product of Electronics Sensor Technologies)	EN	Fast gas chromatography combined with PCA of retention times	Lammertyna et al. 2004
Mineral water, soft drink, and beer identification (alkali and acidic compounds, pH, metal ions)	ET	12 different highly cross-sensitive solvent polymeric membranes on screen-printed carbon paste electrodes on polyester substrate	Lvova et al. 2002
Wine discrimination (ethanol, aroma VOCs)	ET based on PCA and PNNs	Polymer-coated SAW device array	Santos et al. 2005

Fruit selection and ripeness (aroma VOCs)	EN	8 polymer-coated SAW devices plus one reference	Barié et al. 2006
Bread-baking aroma detection (as the baking starts, acetaldehyde and diacetyl; later, pyridines, pyrazines, and pyrrolines are produced)	EN, PCA	TiO_2, In_2O_3, WO_3, $MoWO_3$, chemiresistors	Ponzoni et al. 2008
White bread roasting (see above)	EN	ZnO/Pt and ZnO/Pd sensor resistors	Kohl 2003
Coffee ripening (aromatic VOCs)	EN	6 resistor films based on WO_3, SnO_2, SnO_2-In_2O_3	Falasconi et al. 2005
Verification of *Eurycoma longifolia* in herbal medicines in liquid form	ET based on ANN	Disposable array of 8 screen-printed lipid membrane–covered carbon electrodes with a Ag/AgCl reference	Rahman et al. 2004
Taste identification (bitter, salt, sweet, sour, water)	ET based on PCA	Dual shear horizontal surface acoustic wave (SH-SAW) device (measuring changes in the velocity, frequency, amplitude, or phase of the acoustic wave)	Sehra et al. 2004
Cigarette brand identification	EN (based on ANN)	32 carbon black/conducting polymer composite gas-sensitive resistors based on swelling of polymers due to gas absorption	Luo et al. 2004

[a]EN, electronic nose; ET, electronic tongue.
ANN, artificial neural network; PCA, principal component analysis, PNN, probabilistic neural network.

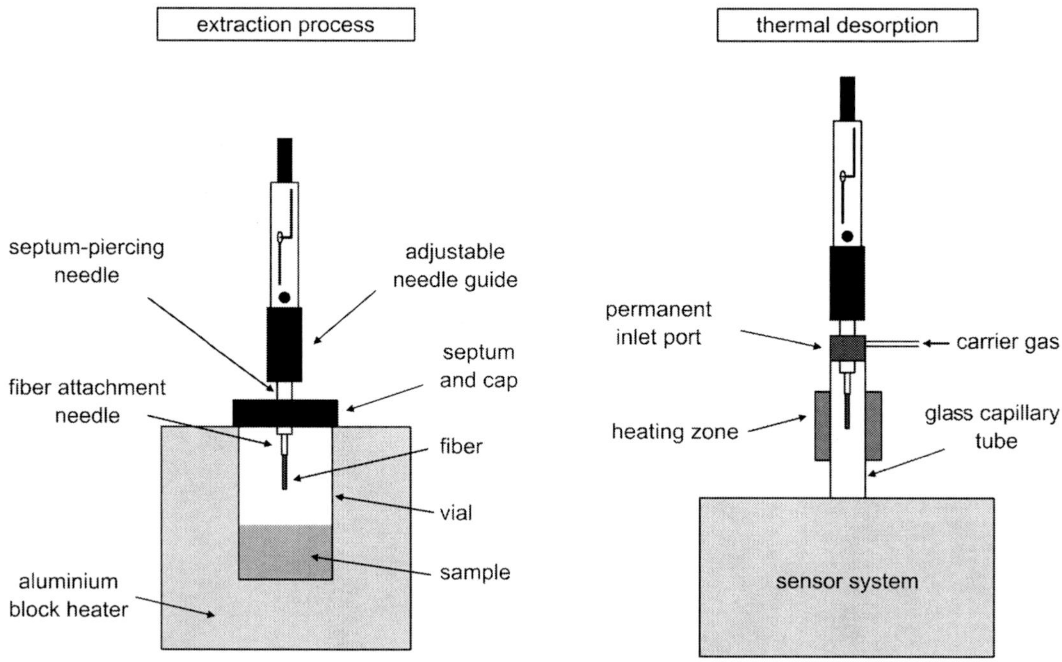

Figure 5.7. Extraction process (left) and process of thermal desorption (right) for headspace of an electronic nose applied in food quality testing. (Reprinted with permission from Barié et al. 2006. Copyright 2006 Elsevier.)

cigarettes. Detecting illegal cigarettes in the field is difficult, but may be possible using portable electronic noses. However, the differences between odors from counterfeit and genuine cigarettes are small, and detection may prove difficult. Luo et al. (2004) proposed a practical approach to increase the performance of e-noses in cigarette brand identification. A portable e-nose was employed to collect and classify aroma signals from different brands of cigarettes. Artificial neural networks (ANNs) were employed and trained with raw data and extracted features from the data collected by the e-nose to identify the cigarettes. This preliminary investigation succeeded in identifying four different types of cigarettes.

5. ENVIRONMENTAL CONTROL

5.1. ATMOSPHERE MONITORING (AIR POLLUTION AND HUMIDITY CONTROL)

Environmental monitoring is mainly concentrated on the continuous measurement of pollutant concentrations in water, soil, and, of course, air. Air quality monitoring has become a problem of great interest during recent decades. Air pollutants from industrial and transportation emissions are responsible not only for some environmental but also for several human health problems. Air quality monitoring in the environment means measuring various gaseous compounds, including oxygen, ozone, and toxic or explosive compounds. These compounds include the following groups that require somewhat different approaches for detection: oxygen, ozone, CO_2, CO, compounds of S (SO_2, H_2S), N (NO, NO_2, NO_x, NH_3, N_2O), hydrocarbons (HCs), volatile organic compounds (VOCs), and others.

The principles of various gas sensor types, gas compound monitoring strategy for industrial applications, including workplace safety (toxic and explosive gas monitoring and gas leakage detection), industrial process control, combustion control, automotive exhaust control, as well as room air quality monitoring (gas leakage, combustion) in household and garage safety (residential, ventilation control) applications are discussed in other chapters.

Urban atmospheric pollutants are considered responsible for the increased incidence of respiratory illness in citizens, and some of these pollutants (e.g., benzene) are known to induce cancers in case of prolonged exposure. Precise estimation of pollutants distribution is hence relevant for traffic management in municipalities and more generally for the definition of integrated mobility plans designed to address these problems. Nowadays, urban air pollution monitoring is primarily carried out by means of networks of spatially distributed fixed stations. This equipment, mostly based on industrial spectrometers, can selectively and precisely estimate the concentrations of many atmospheric pollutants, but their costs and sizes seriously hamper the deployment of adequately dense measurement networks. Unfortunately, pollutants diffusion is heavily affected by atmosphere dynamics, and the availability of a limited number of measurement nodes may lead to misevaluation of the real distribution of gases and particles concentrations in a complex and turbulent environment such as a city.

Primary emissions to the environment are CO, NO, and unburned HCs. However, upon interaction with atmospheric oxygen and sunlight, secondary pollutants such as NO_2 and O_3 are also generated. These secondary pollutants can be dominant far away from the primary sources; this is particularly true for tropospheric ozone.

As a response of the Fifth Action Programme targeting the protection of public health and the quality of ambient air, the European Commission published a new Directive, the 96/62/CE. The principal objectives of this Directive are (1) to establish and define ambient air quality objectives, (2) to monitor the ambient air quality in each state member, and (3) to obtain the best available set of data on atmospheric pollution. The pollutants to be monitored are sulfur dioxide, nitrogen dioxide, PM10 particles (particles measuring 10 μm or less), lead, ozone, carbon monoxide, and benzene. More details are available in the literature (Santos-Alves and Patier 1999).

Ozone is essential for life on the earth in its function as a protective layer against UV radiation and is present in concentrations of up to 8 ppm at a height of approximately 30 km. On the other hand, ozone is one part of summer smog and resulting toxicity. In the office environment, photocopiers and laser printers produce ozone. Most countries have established threshold values for long- and short-time exposure periods. Regarding ozone, the environmental air quality is classified as follows: very good (0–50 ppb), good (50–100 ppb), poor (100–200 ppb), and very poor (>200 ppb). The natural background level is 30 ppb. Therefore, it is very important to monitor the concentration of ozone in the atmosphere. The direct determination of ozone in the gas phase in the higher parts-per-billion and parts-per-million range is performed with electrochemical or chemiluminescent sensors or resonant fluorescent spectroscopic analyzers, but it is also possible with sensors based on semiconducting oxide films, as shown in Table 5.3. Atmospheric ozone profile is measured with radiosonde from ground level up to 20–30 km altitude.

Table 5.3. Recent application examples of sensors in atmospheric air pollution monitoring

Sensor type	Monitored compounds	Source
Sol-gel pure and Pd-doped SnO_2 thin-film resistors	CO (13 ppm) and NO_2 (100 ppb)	Rella et al. 1999
Thin- and thick-film SnO_2 resistor films in a microreactor system	O_3, CO, NO, NO_2, CH_4	Becker et al. 2000
Thick, later nanostructured SnO_2, TiO_2, In_2O_3, and $LaFeO_3$ resistor films	O_3, CO, NO, NO_2	Carotta et al. 2000, 2001
ZnO, TiO_2, and In_2O_3 semiconductor-based resistors	O_3 and ozonosphere compounds' monitoring	Gutman et al. 1995
A commercially available metal oxide gas sensor (TGS2106, Figaro Engineering)	NO_2	Tsujita et al. 2005
An array of three screen-printed SnO_2 thick films, neural calibration	NO_2 and O_3	Kamionka et al. 2006
Monolithic *n*-type epitaxial InP resistor films	O_3 and/or NO_2	Talazac et al. 2001
Screen-printed thick-film SnO_2 and TiO_2 resistor films	CO and NO_x with H_2O cross-sensitivity compensation	Carotta et al. 2007
Metal oxide chemoresistors (SnO_2, TiO_2, In_2O_3, WO_3) in an e-nose based on an ANN	Benzene estimation in an urban pollution monitoring system	De Vito et al. 2008
$SmFeO_3$-based *p*-type and *n*-type SnO_2 (TGS2600, Figaro Engineering) gas sensors	VOCs and NO_2 contamination level	Tomoda et al. 2004
$SmFeO_3$-based *p*-type resistors	O_3	Hosoya et al. 2005
Spray pyrolysis–deposited SnO_2, In_2O_3	O_3	Korotcenkov et al. 2004, 2007
SnO_2, TiO_2, In_2O_3, $LaFeO_3$, and $SmFeO_3$ thick films	CO, CO_2, NO, NO_2, O_3	Traversa et al. 2000
Amperometric sensor array of Nafion/Pt (or Au)/ceramic plate microelectrodes	NO_x, CO, O_2, and SO_2	Do and Chen 2007
Amperometric sensor: precious metals (Pt, Pd and Au) combined with solid polymer electrolytes	SO_2	Chiou and Chou 2002
Amperometric: based on the self-assembly of (3-mercaptopropyl)-trimethoxysilane over a glassy carbon electrode	SO_2	Shankaran et al. 2002
Potentiometric sensor combining NASICON with V_2O_5-doped TiO_2 electrode	SO_2	Liang et al. 2008

The ozone module is attached to the sonde, which travels through the troposphere up to the upper stratosphere. The ground equipment receives data through a radio-frequency link. The sonde may travel through a harsh environment—clouds, rain, snow, fog—and it needs a shield. The shield may convert ozone molecules into oxygen; thus an appropriate sample-taking pumping mechanism should be applied. Vaisala has investigated the application of metal oxide semiconductor gas sensors (In_2O_3, In_xO_y nanowire and Au-doped Sn_xO_y) in comparison with electrochemical cells and reached the conclusion that they would be a very cost-effective solution but they still suffer from stability and cross-sensitivity problems; atmospheric moisture has the largest impact on sensitivity. The application possibility of spray pyrolysis In_2O_3 and SnO_2 thin films was investigated by Korotcenkov et al. (2004, 2007).

Environmental air monitoring is performed by stationary and mobile stations. Conventional measurements of pollutants and undesired compounds include analytical methods such as IR, UV, x-ray fluorescence, atomic absorption spectrometry, gas chromatography, flame photometry or ionization, gravimetric monitoring, chemiluminescent reactions, etc. Monitoring of pollutants is more and more becoming a part of normal meteorological monitoring; however, the high cost of the instruments previously is a major hindrance. The widespread application of low-cost sensors for continuous monitoring is highly desirable, but some reliability and selectivity problems have still to be overcome in the near future.

A miniaturized air analyzer for low-cost environment monitoring applications with a modular setup was described by Meckes et al. (1996). It is built with components all fabricated by silicon micromachining. The active parts are a thin-film gas sensor, a chemical filter element to calibrate the sensor, and two microactuators, i.e., a three-way valve and a pump. High integration of these components can be done by mounting them all to a common channel substrate, also fabricated by silicon micromachining. The design of the active parts is determined by the requirements of this kind of integration.

A multisensor microsystem for organic gas sensing in air, containing 39 metal oxide detectors, 2 temperature sensors, and 4 heaters on an area of 8 × 9 mm, was described by Althainz et al. (1996). The row of four heaters on the back of the chip can be used for inhomogeneous heating, providing different operating temperatures for each sensor element. A resulting lateral temperature gradient can be controlled by temperature measurements made with the two platinum strips at opposite ends of the array. Concentrations between 10 and 20,000 ppm were measured. The chip is equipped with 150-nm tin dioxide gas detector films. The sensor signal s is taken to be the conductivity change caused by the test gas with respect to the conductivity in pure air ($s = \Delta\sigma/\sigma_0$). To obtain specific signal patterns from the 39 sensor elements free of quantitative information, normalization was done by dividing all of the signals by the length of the signal vector:

$$S = \frac{s}{\sqrt{s_1^{\,2} + s_2^{\,2} + \cdots + s_n^{\,2}}} \tag{5.9}$$

These signal vectors were composed into a polar coordinate system where the temperature range of the sensor elements is designated as the polar angle. Such patterns are characteristic for the different chemical species and independent of concentration, as shown in Figure 5.8. Recognition of different air contaminants is achieved by pattern recognition of the signals obtained, using temperature gradients and inhomogeneous ceramic membranes to tune the selectivity of the sensor elements. The overall analysis time, including measurement of all signals and pattern evaluation, is about 100 s.

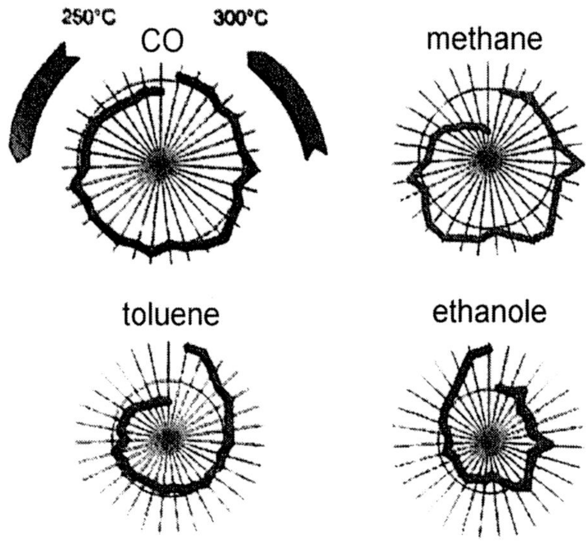

Figure 5.8. Signal pattern for various gases obtained with a temperature gradient between 250 and 300°C over a gas sensor array (Reprinted with permission from Althainz et al. 1996. Copyright 1996 Elsevier.)

The investigation of a variety of carbon-containing gases shows the chemical state of carbon to be relevant for the specificity of the signal pattern. Not only is the identification of a single air contaminant feasible using signal-pattern evaluation techniques, but also the analysis of a mixture of contaminants. A simple deconvolution procedure was developed to extract the mixing ratio from superimposed patterns of two components.

Recent examples for air quality monitoring sensor systems are summarized in Table 5.3. The majority of them use the e-nose approach because of the necessity to detect several compounds in a mixture with nonselective sensor elements. In this case, sensor fusion algorithms, when used in regression schemes, need to be properly tuned via supervised training, i.e., using reference data. Unfortunately, a neural calibration obtained via training with synthetic mixtures was revealed to be unsuccessful, strongly suggesting the need to use on-field recorded data (De Vito et al. 2008). Therefore, several attempts to verify the possibility of obtaining a calibration with on-field campaigns have been made. Environment monitoring stations are used generally as references in the on-field learning processes. Kamionka et al. (2006) used an array of three solid-state sensors (screen-printed tin dioxide thick layers) and obtained a neural calibration with on-field data for estimation of NO_2 and O_3. The calibration obtained showed good performances at parts-per-billion levels for 1-week intervals, after which sensor drift necessitated a new calibration. Carotta et al. (2000) achieved an on-field univariate calibration with least-square methodology for real-time NO_x and CO concentration estimation, presenting results for a 3-day-long test. Tsujita et al. (2005) performed a calibration for their multisensor device capable of estimating NO_2 with a 10-ppb mean absolute error, increasing up to 40 ppb within 2 months. They showed how a simple automatic recalibration procedure based on the base resistance, if performed every 2–4 weeks, could fix the error under 10-ppb levels. De Vito et al. (2008) presented the results of the evaluation of an on-field, multivariate calibration of a multisensor device based on seven metal oxide semiconductor sensors. The overall work was targeted at the estimation of CO, NO_x, NO_2, and benzene concentrations.

The conductance changes of a $SmFeO_3$-based *p*-type gas sensor and *n*-type SnO_2 gas sensor (TGS2600, Figaro), with volatile organic compounds and NO_2 in air were examined by Tomoda et al. (2004). Prediction of the contaminated levels with VOCs generally becames difficult when NO_2 levels are changed and also vice versa. The use of one sensor was poor to detect a level of contaminated air's quality. By using at least two $SmFeO_3$ sensors operating at different temperatures, prediction of each contaminated level with VOCs and NO_2 became possible. Furthermore, by the combination with both sensors, sub-parts-per-million levels of contamination with VOCs and NO_2 was interpretable due to the higher sensitivity of TGS2600 for NO_2 and of $SmFeO_3$ for VOCs.

SnO_2, TiO_2, In_2O_3, $LaFeO_3$, and $SmFeO_3$ thick films have been prepared by Traversa et al. (2000) using screen-printing technology on alumina substrates with comb-type Au electrodes, starting from either commercial or laboratory-synthesized powders. An array of thick-film prototype sensors was placed beside a conventional station for environmental monitoring. Field tests were performed by measuring the change in conductivity of the thick films exposed to real atmosphere. Their electrical response was compared with the results of the analytical instruments for environmental monitoring, which were approved by international standards. The same trend was observed for both systems, with very promising results, when nanostructured oxides were used for preparation of the thick films. This allows us to consider as feasible the use of these sensors for cheap, innovative stations for air quality control.

Exposure to high ammonia concentrations is a serious health threat. Concentration levels near intensive farming can be higher than the allowed exposure limit, as discussed in another chapter. Another interesting point in environmental control is the formation of ammonium salt aerosols. Sulfuric acid and nitric acid react in the atmosphere with ammonia to form ammonium sulfate and ammonium nitrate. These salts are condensation nuclei, forming several nanometer-sized airborne particles. These ammonia aerosols have a sun-blocking function, as can often be seen above large cities or industrial areas. Ammonia levels in the natural atmosphere can be very low, down to sub-parts-per-billion concentration levels above the oceans. Very accurate ammonia detectors with a detection limit of 1 ppb or lower are required for measuring such concentrations. (Near intensive farming areas, ammonia concentrations are much higher, up to more than 10 ppm, which can be handled much more easily). Some optical gas-sensing methods (colorimetric, absorption spectroscopy) are suitable with a sufficiently low detection limit (Timmer et al. 2005). However, the optical gas sensors are large and expensive, making them less suited. A smaller system would be beneficial, but low-cost metal oxide or polymer chemiresistors, as well as electrochemical cells, cannot detect reliably such low concentration levels.

Sulfur dioxide (SO_2) is produced mainly by burning of substances containing sulfur, such as coal and fuel oil, and also comes from natural reactions such as volcanic eruptions and forest fires. It is the most voluminous chemically active gas emitted by volcanoes and is readily oxidized to sulfuric acid, normally within weeks. However, even trace amounts of SO_2 exert significant influence on climate. Global climate change, prior to the twentieth century, appears to have been initiated primarily by major changes in volcanic activity. Large volumes of SO_2 erupted frequently appear to overdrive the oxidizing capacity of the atmosphere, resulting in very rapid warming (Ward 2009). Sulfuric acid produces acidity in rainwater and fogs, and is a major source of corrosion of buildings. The major health concerns associated with exposure to high concentrations of SO_2 are effects on breathing, respiratory illness,

Table 5.4. Relations between psychrometric properties

Dew point (°C)	Relative humidity at 20°C (%)	Water content (ppm)
-75	0.002	~1
-45	0.3	70
-20	4.5	10^3
0	25	6.1×10^3
10	50	1.25×10^4
20	100	2.34×10^4

alterations in the lungs' defenses, and aggravation of existing cardiovascular disease. Hence, the development of SO_2 sensors is very important for monitoring the concentration of sulfur dioxide, which can help reduce the damage from acid rain (Chiou and Chou 2002). Conventional methods employed for the determination of SO_2 have limitations due to the expensive and bulky instrumentation techniques and time-consuming procedures. To date, SO_2 sensors with semiconductors, liquid electrolytes, and solid electrolytes have been proposed (see also Table 5.3). Among these, the sensors fabricated from solid electrolytes have strong advantages such as high selectivity, rapid and reproducible response, and low-concentration detection (Shankaran et al. 2002).

As humidity is a very common and continuously changing component of our environment, measurement and/or control of humidity are important not only for human comfort but also for a broad spectrum of application fields including environment monitoring and meteorology. The measurement of relative humidity has always been a difficult problem because of the interaction of other variables such as temperature and gas components in the air and because of the lack of stable and reliable low-cost sensors.

Humidity can be described by humidity content, relative humidity, and dew point. Relative humidity (RH) is defined as the ratio of actual partial pressure of water vapor to the saturated vapor pressure of the water at the given temperature. The dew point can also be used as a characteristic parameter for the water vapor content in the air. It is defined as the temperature at which the gas becomes saturated during cooling. The connections between the mentioned parameters are given by the laws of psychrometry (Norton 1982). A few examples are given in Table 5.4.

There are three major groups of humidity-measuring methods: those which measure mechanical property changes; psychrometric measurements which compare the latent heat of evaporation of a saturated environment to the environment in question; and those which respond to an electrical or optical property change such as resistance, capacitance, color, etc., and can be related to sensors. Due to research efforts of the last few years, a lot of different methods and types have been developed, which can be categorized as follows:

- Temperature methods
  - Dew-point hygrometers
  - Psychrometers
- Sorption methods
  - Impedance types (based on capacitance or resistance changes or both, including dielectrics, electroconducting materials, and electrolytes)

- o Transistor structures based on material interactions
- o Mass-sensitive devices (bulk or surface acoustic wave resonators)
- o Electrochemical cells
- o Fiber optic types
- Propagation methods
  - o Infrared spectroscopy
  - o Microwave propagation
  - o Acoustic wave propagation

Dew-point sensors are essentially physical sensors that detect water condensation, either optically or electronically (by means of interdigital impedance) on a cooled surface and recording the dew-point temperature in this way.

A lot of materials are known to adsorb and absorb moisture. If there are pores, capillary condensation of water takes place in addition to absorption. The capacitance of absorbing layers changes sensitively with the absorption of water (Boltshauser et al. 1991). The first commercially available polymer humidity sensors applied cellulose acetates as sensitive layers. In this case the permittivity value is roughly proportional to the ambient humidity; it shows a small nonlinearity in the range of >80% RH. Recent improvement in polyimide processing for integrated circuit and interlay dielectric applications enables humidity sensors to be fabricated on integrated circuit structures. Using this technique, small, low-cost humidity sensors are available, and, most important, it becomes possible to integrate the humidity sensor with other sensors and signal-processing circuitry on the same substrate. The application of photosensitive polyimides reduces the number of patterning process steps. Various types of polymer electrolytes have also been used for humidity sensors. If the polymer electrolyte material sorbs water vapor from the atmosphere, its electrical resistivity decreases. As a result, one can tell the humidity of the air surrounding the polymer by measuring the resistance impedance. The humidity sensors utilizing the variation of resistivity of the materials with humidity have been referred to as resistive-type humidity sensors. The ionic conductivity in polymer electrolyte increases with an increase in water adsorption due to increases in the ionic mobility and/or charge carrier concentrations.

Resonator-type (BAW or SAW) devices are attractive for chemical microsensor applications because of their small size, low cost, high sensitivity, and good reliability. The sensors are based on the piezoresonance effect: The resonance frequency changes depending on the amount of gases absorbed by a polymer coating.

Optical fiber systems are also of interest as a possible technique to detect gas species in the atmosphere. Several approaches have been reported for fiber optic sensors to detect the humidity of the environment. Fiber optic sensors operate based on colorimetric, fluorescence, and/or luminescence effects of indicators that may be embedded into a polymer material. Generally, the latter is just a host material.

5.2. RAIN COMPOSITION MONITORING

Rain is the predominant route for the atmospheric deposition of environmentally relevant compounds. Deposition of acidic pollutant compounds is one of the present global environmental problems.

Therefore, detailed investigations of rainwater composition and atmospheric washout effects are of increasing importance. The analysis of the chemical components of rain, including pH, is usually carried out only on bulk rain samples, which have been collected over a time period of days, weeks, or even months. The measurement is performed with conventional glass electrode sensors on bulk rain samples A minimum sample volume of 20 ml is usually required for the analysis of many chemical components in each sample. However, such long-time integrated samples cannot reflect the large variations during and between the deposition events. Hence, the obtained results are mainly useful for the calculation of deposition budgets, but difficult to utilize for detailed studies of rain events. Since the pH can change significantly during the rain event, a quantitative analysis on a well-defined time scale as well as determination in small sample volumes is necessary. Microanalytical measuring systems that consist of chemical microsensors can offer such an analytical tool. Their main advantage is the possibility to record the pH value both at short time intervals and in small volumes (raindrop size is often <5 μl) in order to reduce dilution effects. In this way one can get more information about the pH changes during a rain event. For this purpose, the ion-sensitive field-effect transistor (ISFET) delivers a detection device that is an encouraging alternative to conventional ion-selective electrodes with additional advantages, such as reduced size, faster response time, and low output impedance. Poghossian et al. demonstrated for the first time in 2001 that ISFETs can be used not only for pH measurements in bulk rain samples, but also, by choosing appropriate experimental conditions, for quantitative pH determination in single rain droplets. The results obtained were in good agreement with those from investigations using highly sophisticated electroanalytical methods. Thus, pH ISFETs can serve as tools for *in situ* raindrop monitoring, which is in the forefront in the near future.

5.3. DETECTION OF AIRBORNE MICROBES

The real-time detection of microorganisms such as virus, bacteria, and fungi is an emerging and rapidly evolving field of research. The spread of airborne pathogens such as measles, anthrax, *Legionella*, influenza, smallpox, and rhinovirus is often regarded as a major threat to public health since they cause severe airborne infectious diseases with high mortality rates.

Conventionally, the detection of airborne bacteria has been achieved by collecting and culturing, a method which is very effective but requires a long incubation time (at least 24 h). However, to prevent airborne infectious disease, real-time detection and monitoring systems of airborne pathogens are needed.

Since adenosine triphosphate (ATP) is a major biological energy source, the detection of ATP in aerosols reflects the existence of living microbes. A new biosensor to detect ATP from aerosols in real time using an aerosol condensation system, a microfluidic channel, and an ATP-bioluminescence transducer was developed by Lee et al. (2008). The condensation system, which is illustrated in Figure 5.9, enabled aerosol microbes to be hydrosolized (0.2 ml) in 2 min. The bacterial intracellular ATP was then extracted in the passage through the microfluidic channel. The concentration of ATP could be determined by a bioluminescence sensor integrated into the channel.

There are many other sources of contamination of the environment by biologically active substances. Different types of biological particles exhausted into the air from brewery, dairy, biotechnological plants,

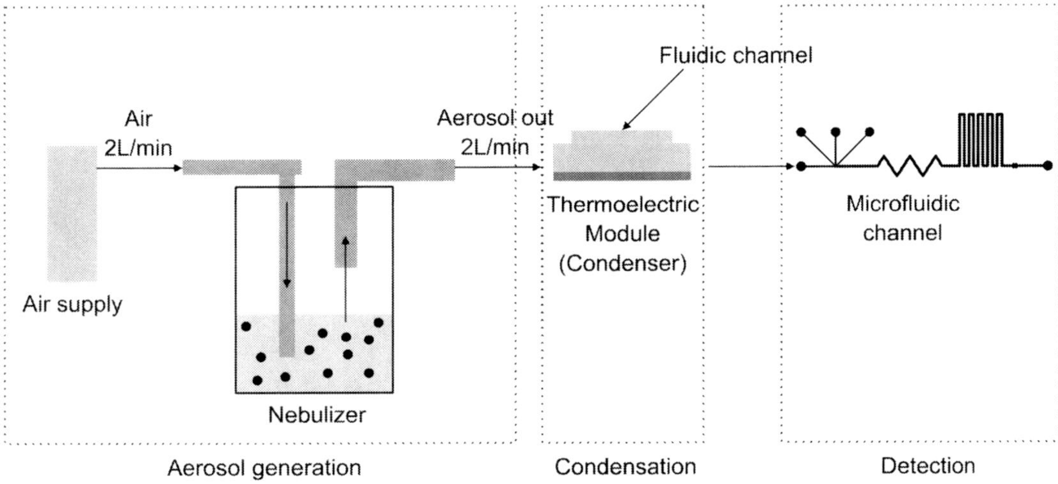

Figure 5.9. Schematic diagram of the airborne microbe detecting system consisting of a pneumatic aerosol generation module, a high-efficiency condenser, and a microfluidic chip for ATP extraction and detection of bioluminescence. (Reprinted with permission from Lee et al. 2008. Copyright 2008 Elsevier.)

etc., cause numerous diseases, including different types of allergies, among people working at those units and living in the surrounding areas. Optical immune sensors for monitoring biological pollutants discharged into the air from a biotechnological plant producing lysine were tested by Starodub et al. (2000). One type of sensor was based on the photoluminescence of porous silicon, and a second type involved the use of fiber optics in combination with chemical luminescence. The proposed sensors were found to be comparable in sensitivity to the ELISA method.

6. HOUSEHOLD APPLICATIONS

The household appliance industry has grown into one of the largest markets for electronic products. This industry can be classified into heating and climate control, security devices, ventilation control, large household appliances, and small household appliances. Over the last decades, the technical requirements for appliances have been increased, and customers expect improved functionality, as well as reduced and carefully controlled consumption of water, energy, and detergents. Here the use of sensors and microsystem products started slowly. The use of sensors was dominated in the past by those for temperature, level, and flow. Now, increasingly more and new microsystems have been introduced in household appliances, and even more are expected in the near future, although the application of physical sensors still dominates the area. However, new and improved types of gas sensors, as well as multigas sensors and electronic noses, are expected to find their way into future kitchens and future climate control systems. Several sensors are expected to be introduced soon for water quality, like those for turbidity, and some more will follow, including, e.g., sensors for color, surface tension, detergent concentrations, pH, etc. Air quality control is also available for comfort application in air conditioning and ventilation systems.

Table 5.5. Examples of sensors and microsystems in household and home safety/comfort appliances, with their technologies

Appliance/function	Sensor type	Technology used
Cooker/automatic baking	Multigas sensors Electronic noses	Thin-film or metal-oxide resistors
Washing machine Detergent control Water break	Conductivity, pH Turbidity Humidity	Electrodes, ISFETs Optical Impedance
Gas leakage, explosion safety	Hydrocarbons (propane, butane, methane)	Metal oxide resistors, pellistors
Toxic-gas safety (combustion, heating, garage)	CO	Metal oxide resistors, electrochemical cells
Combustion control	O_2	Electrochemical cells
Air quality, comfort Ventilation, air conditioning	CO_2 Humidity	IR absorption Impedance
Fire	Smoke	Ionization, infrared absorption, scattering
Breath alcohol tester	VOC	Metal oxide resistors

Handheld appliances are also available for alcohol self-tests in order to avoid unsafe driving. The function of an appliance determines the choice of sensors. An overview of chemical sensors and microsystems in home safety/comfort and household appliances is summarized in Table 5.5.

6.1. CLIMATE CONTROL

The main target of household heating and climate control is to keep the temperature of the indoor air, and the heating/cooling medium (air, water, cooling gas, etc.) in a given range, while also preventing overheating/overcooling effects. This task needs the application of temperature-sensing elements. For better comfort, the air flow rate should also be limited, which requires flow sensors. Chemical sensors are needed to control humidity. Humidity is an important component of comfort and other physiological processes, since the evaporation efficiency of self-cooling systems of living things are strongly influenced by the humidity of the air. We feel a dry hot environment as being much "cooler" than a humid one at the same temperature.

Over the past few decades, an increasingly large number of people have been employed either in offices or in light industry where activities are mostly sedentary. During this period people have also become more aware of, and concerned with, creating the best possible working conditions. Improving the working environment requires greater control of the environment and a deeper understanding of its impact on people. Most people spend more than 90% of their lives in artificial thermal environments, and of course they like to be comfortable. Consequently, anyone staying for long in a particular building or vehicle—whether for work, residence, or recreation—wants the thermal environment to suit them. The thermal comfort of the environment can determine, for example, why passengers will choose one transport operator instead of another, or whether a particular swimming pool is popular with users. Lack of comfort can even be the reason why people working in an office don't work as effectively as they might. Thermal comfort is the state of mind in which satisfaction is expressed with the thermal environment. A thermally neutral person does not know whether he would like to be warmer or cooler. The subjective and physiological reaction of a person to the thermal environment is determined by the rates of bodily heat generation and emission, and these in turn are functions of six parameters: air velocity, mean radiant temperature, air temperature, water vapor pressure, the person's metabolic rate, and the thermal insulation of his or her clothing.

Localized thermal differences affect many people, causing discomfort. The most frequent reason for complaint is draft, which is defined as an undesirable, local convective cooling of the body. Thus, draft is caused by air velocities that are too high relative to the air temperature. Other reasons for local thermal discomfort are radiant temperature asymmetry, air temperature difference between head and ankle level, and too low or too high floor temperature. All these factors influence the heat loss from one or more parts of the body.

Absolute humidity has a major influence on the evaporative exchange that takes place between a human being and its environment. It is often expressed in the form of the partial pressure of water vapor and is thus related to the actual quantity of water vapor in the air. Sensors are needed that measure the air humidity very accurately, by measuring the dew-point temperature or the relative humidity and temperature. The water vapor pressure can be calculated from these measurements, and if the air temperature is also known, their relationship is determined as shown in Table 5.4. This is done automatically by indoor climate analyzer appliances available commercially nowadays. The operator can, therefore, choose to have the air humidity expressed as dew-point temperature, water vapor pressure, or relative humidity in percent. Appropriate controlling units modify the climate conditioning system of the room or car accordingly.

6.2. HOME SAFETY (GAS LEAKAGE, FIRE ALARM)

The most developed area of residential applications is home safety. Various hazardous materials are used (propane, butane, methane) in households, and CO can be formed during incomplete burning of cooking or heating systems. These problems need low-cost gas sensors that ensure the safe alarm, but false alarm does not cause big problems in households, since it has no high-cost consequences. Household fires are one of the major sources of residential catastrophes, thus, reliable fire (smoke) detection has the same demand. A few important issues in connection with the gas compound sensor types must be emphasized here.

- A CO detector is a kind of wall-hung household alarm designed to detect the concentration of carbon monoxide. It provides visual and audible signals when the CO concentration reaches a preset level, reminding you take to prompt action: You might start up other, linked equipment, ventilate gas, or shut off the gas supply to avoid suffocation and toxic accident. CO detectors are installed not only in homes, but also in many other potentially hazardous areas. They should be installed in rooms where burning processes occur, around 0.5 m below the ceiling.

- Gas leakage detectors detect the leakage of explosive hydrocarbons and are installed near potential sources. They are recommended where fires are present which are fueled by HC gases (propane, butane, methane).
- One group of fire detectors detects the presence of smoke (small particles in the air) and sometimes also heat. These do not detect CO and combustible gases. They should be installed in each room on the ceiling.
- Gas sensors as fire detectors give better results in comparison with standard smoke particle detectors because they can detect toxic gases—the main cause of loss of life during fires.
- CO_2 detectors are used to measure air quality to assure comfort. They can be installed close to the floor of the room. Safety applications include wine cellars, etc.

Examples are shown in Table 5.5.

6.3. VENTILATION (INDOOR AIR QUALITY) CONTROL

The increasing interest in indoor air quality (IAQ) monitoring and ventilation control is mainly due to the growing incidence of a new class of diseases, identified as building-related illnesses and "sick building" syndromes, arising from long-term occupancy of confined living spaces, such as office buildings or homes/apartments. The occurrence of these diseases is partially related to the presence of chemical contaminants inside the building. The effects on human health due to indoor pollution also include pathologies for which the symptoms arise from the concomitant effect of several pollutants which are all present in the ambient, e.g., from low "air quality" (Zampolli et al. 2004). In buildings with inadequate ventilation, these can accumulate with time. Typically, we can find several hundreds of indoor chemical contaminants, including by-products of combustion (NO_2, SO_2, CO, etc.), cigarette smoke, and a number of volatile organic compounds. In spite of the very low concentrations, some of these compounds, such as NO_2 and CO, are extremely toxic; some others, such as benzene and formaldehyde, have been proved to be carcinogenic. Therefore, monitoring of air quality is of paramount importance to maintain safe and healthy conditions. Several attempts to implement demand-controlled ventilation have been reported in literature. Many of them rely on the monitoring of CO_2, used as a tracer of human occupancy in confined living spaces or as a surrogate for inhabitant-generated pollution. However, this approach is inadequate to monitor air quality, since several toxic compounds released by building materials or generated by human activities may also be present in the indoor environment. Some recent sensor types applied in indoor air quality and ventilation control are listed in Table 5.6.

There are several indoor air quality monitoring and ventilation control system types for residential or workplace safety applications commercially available or close to being commercialized. Two novel silicon-based sensor arrays for online air quality monitoring are advertised by Extanza, Inc., the first based on films deposited onto bulk silicon wafers and the second a novel micromachined device with

Table 5.6. Recent application examples of sensors in indoor air quality monitoring

Sensor type	Monitored compounds	Source
Potentiometric electrochemical cell based on $Na_2Zr_3Si_2PO_{12}$ (NASICON) as a solid electrolyte and Li_2CO_3 as a carbonate phase	CO_2	Kaneyasu et al. 2000
Array of eight polymer-coated SAW devices, PARC	VOCs	Bender et al. 2003
Thin Au-doped SnO_2 and WO_3 films on MEMS hot plates (constant or fast pulsed temperature mode, PARC evaluation)	CO (5 ppm) and NO_2 (20 ppb) (VOC and H_2O compensation)	Zampolli et al. 2004
Carbon nanotube–coated SAW device	CO_2	Sivaramakrishnan et al. 2008

low power consumption. The micromachined version reduces the power consumption from 1.6 W to 0.26 W at 350°C working temperature. The sensing elements ($Cr_{1.8}Ti_{0.2}O_{3+z}$, ZnO, SnO_2, WO_3, V_2O_5) examined in the arrays include four different metal oxides deposited by thin-film technologies. The measured responses in lab and application tests provide patterns which correlate with air quality and are thus appropriate for online air quality monitoring.

Fraunhofer IPM has developed a semiconductor thin-film metal oxide gas sensor array and integrated it into a microsystem. The sensor array can be used for indoor air quality monitoring. The Webnose measures CO, NO_x, and C_xH_x. These compounds are typical for cigarette smoke. The sensor array consists of four metal oxide resistors realized onto a 3 × 3 mm chip, which enables the screening of cross-sensitivities and compensation for humidity dependency. The custom design enables detection of the following gases: NO_2, CO_x, H_2, CH_4, C_2H_5OH, H_2O, and NH_3. The microsystem technology enables integration of CMOS electronics for signal processing and temperature control. The detection limits are in the range of 0.1 × MAK/TWA (see next chapter) values (MAK 2000), the operating temperature range is 150–450°C, and the power consumption is from 50 mW to 1 W.

An electronic nose (e-nose) based on a solid-state gas sensor array for the identification and quantification of two typical indoor air quality tracers was presented by Zampolli et al. (2004). The simplicity of the proposed approach, which uses stand-alone gas sensors, simple driving electronics, and fuzzy-logic pattern-recognition algorithms, aims to realize a low-cost tool suitable for monitoring some of the compounds of interest for indoor air quality monitoring. CO and NO_2 were taken into consideration as target compounds, while some VOCs and the humidity were considered as interfering species. The optimization of the sensing-layer materials and of the operating conditions allowed reaching the sensitivity necessary for the reliable detection of CO down to concentrations of 5 ppm and of NO_2 down to 20 ppb. To detect CO, a SnO_2 +SiO_2 sensor with Au catalyst operated at 400°C was used, while the sensor targeted for NO_2 was a SnO_2 sensor with Au catalyst operated with pulse heating at a temperature of 375°C. Operating the sensors under optimized conditions for the detection of a given compound does not make them fully selective, but they do provide unique response patterns, which can be used to classify and quantify the single pollutants inside complex gas mixtures, as found in real sensors for CO detection and WO_3-based sensors for NO_2 detection.

6.4. MULTISENSOR SYSTEMS FOR SMART HOMES

In the near future, all electronic appliances in a "smart" home will be networked: PCs, telephones, stereos, refrigerators, and even washing machines. In former days, only toxic gases could be checked for human safety inside houses; nowadays, however, room atmosphere aims at comfortable life. Heating and air conditioning, previously controlled by a single, fixed, manual thermostat, can now be managed by an intelligent controller with remote-access capabilities. Recently, with increasing living standards and expectations of comfort, the use of residential air conditioning is becoming widespread. The control and monitoring of indoor atmosphere conditions represents an important task with the aim of ensuring suitable working and living spaces for people. However, comprehensive air quality monitoring, including monitoring of humidity, temperature, CO_2, flying dust particle density, etc., is not easy to accomplish and controlled. An important further goal of smart home automation is to improve the user's comfort and security along while reducing overall energy consumption. A core element of this strategy is an automated, decentralized indoor climate control system. So, a wireless multisensor module for monitoring indoor air is very important technology for near-future home networking technology (Chung and Oh 2006). Widespread application of the smart home concept needs low-cost and reliable sensors for the particular application, including some chemical sensor types.

A smart home concept with a multigas sensor module as its main part was described by Ivanov et al. in 2002. The following parameters of indoor air are measured: concentration of pollutants CO_2, CO, O_3, H_2, H_2S, fire gas detection (methane concentration at LEL, the lower explosion limit), dust particle concentration, air temperature, and relative humidity. The air composition in the room may vary considerably; therefore the use of a broad-band electrochemical gas sensor was preferred over selective gas sensors (e.g., based on IR absorption). Electrochemical gas sensors are available for a large variety of gases. The sensitivity and selectivity of the gas sensors are controlled by the value of the external voltage. Their advantages are mainly small dimensions, no heat radiation, low energy consumption, and low price. Several conventional Pt1000 platinum resistance thermometers are used as temperature sensors. The nonlinear temperature characteristics are modeled by a third-order polynomial. The capacitance humidity sensors are built in a double oscillator configuration with a reference branch in order to reduce temperature drift. The humidity value follows from the difference between the two oscillator frequencies. The dust sensor is based on IR light absorption, which depends on the particle concentration. By using the same sensor principle for all gas sensors, sensor output evaluation can be largely unified. The necessary signal processing is carried out by an inexpensive microcontroller.

The term "smart home" is also used with a very special meaning nowadays that may better be called the concept of a "health-smart home," one designed to monitor frail or ill people at home in order to avoid hospitalization, thus limiting hospital stays to short acute care or fast, specific diagnostic investigations (Demongeot et al. 2002). In the era of information technology, the elderly and disabled can be monitored with numerous intelligent devices. Sensors can be implanted into their homes for continuous mobility assistance and nonobtrusive disease prevention. Modern sensor-embedded houses—"health-smart houses"—not only can assist people with reduced physical functions but can also help resolve the social isolation they often face. These homes are capable of providing assistance without limiting or disturbing the resident's daily routine, giving him or her greater comfort, pleasure, and well-being. This technology offers the prospect of significant improvements in the quality of life and level of independence for elderly or disabled people. The desired effect is therefore to equip with the home environment with intelligent and autonomous technology which subsequently extends the period of time people can remain in their own homes. Although this area represents a new application field of smart sensors including chemical and biosensors, it is obviously outside the scope of common household applications (the scope of this section); thus, we refer to the literature for further information (Chan et al. 2008; Hong et al. 2008).

6.5. CONTROL OF FOOD PREPARATION

Control of food preparation processes in household application is a dream of housewives. Although a great number of sensors are used in various appliances nowadays, such as temperature measurement in baking or cooking media—water/oil, contact or contactless measurement of food temperature, weight loss of the food, etc.—they are dominated by physical sensors. There are many areas for applying chemical sensors, but their widespread exploitation is still to come. A few examples as potential applications are given here.

- Electronic noses for the food quality (flavor) detection during food preparation (which is already used in the food industry; see next chapter)
- Detection of poisoning agents that are injected by the process tools during the preparation process

A novel type of metal oxide gas sensor array was used by Ehrmann et al. (2000) in order to develop an "intelligent cooking pot." The sensor array consists of 40 monolithic sensor elements whose different sensitivities were achieved by a gradient technique. Metal oxide conductivity sensors were prepared based on the segmentation of a monolithic metal oxide layer by a set of parallel electrodes. A large number of sensor elements is formed in one simple production step, each of them consisting of two adjacent electrode strips and the metal oxide segment in between. To obtain an array of sensors with different sensing properties, gradient techniques are used. One or more sensor-relevant parameters are continuously varied across the array of metal oxide segments. A temperature gradient and a thickness gradient of a SiO_2 coating on the metal oxide are used. These two features have a substantial impact on the selectivity of the sensor elements, both exhibiting a different response to changes of atmospheric composition. Due to the simplicity of the design, manufacturing costs can be kept very low. The sensor array was installed in the cover of a pot. With this setup the feasibility of monitoring cooking processes by gas analysis was shown using the example of frying beef. The distinction of five different stages of the frying process was achieved by principal component analysis of the signal patterns.

Copper is an essential element but toxic to humans at high levels. Aluminum is not an essential element to humans, and is considered to be a toxic metal ion. Since Cu and Al pans are the most commonly used cooking utensils in rural areas, an acoustic wave sensor was developed by Verissimo et al (2005, 2006) to quantify the amount of these metals leaching when cooking some recipes. The sensitive layer of the sensor consists of a membrane with a specific ionophore for the aluminum or copper, in a PVC/plasticizer matrix. Results showed that neutral liquids, such as water, do not experience any noticeable contamination of copper, but that acidic elements, such as tomato, extract as much as 17.5 μg of copper per gram from a pan in which they were cooked for 1 h. Red cabbage samples cooked with different acidic additives (lemon juice, wine vinegar, and apple cider vinegar) showed that low pH values increased leaching of aluminum. Results obtained with the new piezoelectric copper sensor are not statistically significantly different, both in terms of accuracy and precision, from results obtained by analytical methods.

7. NOMENCLATURE

AC	alternating current
ADP	adenosine diphosphate
ANN	artificial neural network
ATP	adenosine triphosphate
ATR	attenuated total reflectance
awRR	airway respiratory rate
BAW	bulk acoustic wave
BOD	biological oxygen demand
CA	cluster analysis
CFU	colony-forming unit
CMOS	complementary metal oxide semiconductor
CNH	carbon nanohorn
CNT	carbon nanotube
CP	composite polymer
DC	direct current
DO	dissolved oxygen
DPASV	differential pulse mode anodic stripping voltammetry
DPD	N,N-diethyl-*p*-phenylenediamine
DS	dielectric spectroscopy
E	activation energy
EC	electrical conductivity, electrochemical
ELISA	enzyme-linked immunosorbent assay
F	Faraday's constant
FET	field-effect transistor
FIA	flow-through injection
GOD	glucose oxidase
Hb	hemoglobin
HbO_2	oxyhemoglobin
HC	hydrocarbon
HRP	horseradish peroxidase
I	light intensity
IAQ	indoor air quality
IR	infrared
ISE	ion-selective electrode
ISFET	ion-selective field-effect transistor
ITO	indium tin oxide
k	Boltzmann's constant
K	Stern-Volmer constant
LAPS	light-addressable potentiometric sensor
LCD	liquid-crystal display
LED	light-emitting diode
LEL	lower explosive limit
LLTO	lithium lanthanum titanate
LPCVD	low-pressure chemical vapor deposition
M	molecular weight
MAK	Maximale Arbeitsplatz-Konzentration
MDA	multiple discriminant analysis
MOSFET	metal oxide semiconductor field-effect transistor
MRL	maximum residue level
MVA	multivariate analysis
NIR	near-infrared
ORP	oxidation–reduction potential
PA	photoacoustic
PARC	pattern recognition
PCA	principal component analysis
PCB	printed circuit board
pCO_2	partial pressure of carbon dioxide
pO_2	partial pressure of oxygen

PLS	partial least square	SWASV	square-wave anodic stripping voltammetry
PMB	partial model building	τ	lifetime
PMMA	polymethyl methacrylate	T	absolute temperature
pO_2	partial pressure of oxygen	TEWL	transepidermal water loss
PTFE	polytetrafluoroethylene	THC	total hydrocarbons
PVC	poly(vinyl chloride)	TDG	total dissolved gas
QCM	quartz crystal microbalance	TWA	time-weighted average
R	universal gas constant	UEL	upper explosive limit
RH	relative humidity	UV	ultraviolet
RTU	ready-to-use	V_{om}	maximum fractional volume of absorption
S_aO_2	arterial blood saturation	VNA	vector network analyzer
SAW	surface acoustic wave	VOC	volatile organic compound
SCE	saturated calomel electrode	W	free energy of absorption
SO_2	blood saturation	WQ	water quality
SPR	surface plasmon resonance		
S_vO_2	venous blood saturation		

REFERENCES

Althainz P., Goschnick J., Ehrmann S., and Ache H.J. (1996) Multisensor microsystem for contaminants in air. *Sens. Actuators B* **33**, 72–76.

Arnold M.A. (1996) New developments and clinical impact of noninvasive monitoring. In: Kost G.J. and Welsh J. (eds.), *Handbook of Clinical Automation, Robotics, and Optimization,* John Wiley, New York, 631–647.

Arnold M.A., Burmeister J.J., and Small G.W. (1998) Phantom glucose calibration models from simulated noninvasive human near-infrared spectra. *Anal. Chem.* **70**(9), 1773–1781.

Atkinson J. (2004) Packaging and deployment of commercial thick film screen printed chemical, biological and environmental sensors. In: *Proceedings of the XXVIII International Conference of IMAPS Poland,* Wroclaw, Poland, 3–9.

Baldini F., Homola J., Lieberman R.A., and Miler M. (eds.) (2007) *Optical Sensing Technology and Applications. Proc. SPIE* **6585**.

Barié N., Bücking M., and Rapp M. (2006) A novel electronic nose based on miniaturized SAW sensor arrays coupled with SPME enhanced headscape-analysis and its use for rapid determination of volatile organic compounds in food quality monitoring. *Sens. Actuators B* **114**, 482–488.

Becker T., Mülberger S., Braunmühl C., Bosch V., Müller G., Ziemann T., and Hechtenberg K.V. (2000) Air pollution monitoring using tin-dioxide-based microreactor systems. *Sens. Actuators B* **69**, 108–119.

Bender F., Barié N., Romoudis G., Voigt A., and Rapp M. (2003) Development of a preconcentration unit for SAW sensor micro array and its use for indoor air quality monitoring. *Sens. Actuators B* **93**, 135–141.

Bezegh K., Bezegh A., Black P.G., and Janata J. (1988) Integrated probe for sweat analysis. *J. Clin. Lab. Anal.* **2**, 16–18.

Bezegh K., Bezegh A., Janata J., Oesch V., Xu A., and Simon W. (1987) Multisensing ion-selective field-effect transistors prepared by ionophore doping technique. *Anal. Chem.* **59**, 2846–2848.

Bohnke C., Duroy H., and Fourquet J.L. (2003) PH sensors with lithium lanthanum titanate sensitive material: applications in food industry. *Sens. Actuators B* **89**, 240–247.

Boisde G., Blanc F., and Machuron-Mandard X. (1991) PH measurements with dyes co-immobilization on optrodes: Principles and associated instrumentation. *Int. J. Optoelectron.* **6**(5), 407–423.

Boltshauser T., Chandran L., Baltes H., Base F., and Steiner D. (1991) Humidity sensing properties and electrical permittivity of new photosensitive polyimides. *Sens. Actuators B* **5**, 161–164.

Bonfil Y., Brand M., and Kirowa-Eisner E. (1999) Determination of sub-$\mu g l^{-1}$ concentrations of copper by anodic stripping voltammetry at the gold electrode. *Anal. Chim. Acta* **387**, 85–95.

Bonfil Y., Brand M., and Kirowa-Eisner E. (2002) Characteristics of subtractive anodic stripping voltammetry of Pb and Cd at silver and gold electrodes. *Anal. Chim. Acta* **464**, 99–114.

Brudzewski K., Osowski S., and Markiewicz T. (2004) Classification of milk by means of an electronic nose and SVM neural network. *Sens. Actuators B* **98**, 291–298.

Buerk D.G. (1993) *Biosensors, Theory and Applications.* Technomic Publishing, Lancaster, PA.

Bulukin E., Bettazzi F., Palchetti I., and Mascini M. (2004) Sensitive and rapid bioelectrochemical monitoring of phenols in food and environmental matrices using enzyme based biosensors. In: *Digest of Technical Papers of the 18th European Conference on Solid State Transducers, EUROSENSORS XVIII,* Rome, Italy, P1.14.

Burk E.D., Arrieta I.C., and Batich D.C. (1987) The feasibility of an implantable opto-electronic glucose sensor. In: *Proceedings of the 9th Annual Conference of the IEEE Engineering in Medicine and Biology Society,* 13–16 December, Boston, MA, 788.

Burmeister J.J., Arnold M.A., and Small G.W. (1998) Spectroscopic considerations for noninvasive blood glucose measurements with near infrared spectroscopy. *Leos Newsletter*, April/1998, 6–9.

Caduff A., Hirt E., Feldman Y., Ali Z., and Heinemann L. (2003) First human experiments with a novel non-invasive, non-optical continuous glucose monitoring system. *Biosens. Bioelectron.* **19**, 209–217.

Cagnini A., Palchetti I., Lionti I., Mascini M., and Turner A.P.F. (1995) Disposable ruthenized screen-printed biosensors for pesticides monitoring. *Sens. Actuators B* **24–25**, 85–89.

Carotta M.C., Martinelli G., Crema L., Gallana M., Merli M., Ghiotti G., and Traversa E. (2000) Array of thick film sensors for atmospheric pollutant monitoring. *Sens. Actuators B* **68**, 1–8.

Carotta M.C., Benetti M., Ferrari E., Giberti A., Malagú C., Nagliati M., Vendemiati B., and Martinelli G. (2007) Basic interpretation of thick film gas sensors for atmospheric application. *Sens. Actuators B* **126**, 672–677.

Chan M., Esteve D., Escriba C., and Campo E. (2008) A review of smart homes—Present state and future challenges. *Comput. Meth. Prog. Biomed.* **91**, 55–81.

Chiou C.Y. and Chou T.C. (2002) Amperometric SO_2 gas sensors based on solid polymer electrolytes. *Sens. Actuators B* **87**, 1–7.

Christison G.B. (1993) Summary of near infrared glucose sensing research towards a noninvasive measurement technique. *Chemical Sensors for In Vivo Monitoring* (a newsletter published by a European Union Concerted Action) **11**, March 1993, 8–10.

Chung W.Y. and Oh S.J. (2006) Remote monitoring system with wireless sensors module for room environment. *Sens. Actuators B* **113**, 64–70.

Clark L.C. (1958) Monitoring and control blood and tissue oxygen. *Trans. Am. Soc. Artif. Intern. Organs* **2**, 41–48.

Coatanea M., Darchen A., and Hauchard D. (2001) Electroanalysis at ultramicroelectrodes of oils and fats Application to the determination of vitamin E. *Sens. Actuators B* **76**, 539–544.

Demongeot J., Virone G., Duchene F., Benchetrit G., Herve T., Noury N., and Rialle V. (2002) Multi-sensors acquisition, data fusion, knowledge mining and alarm triggering in health smart homes for elderly people. *C. R. Biol.* **325**, 673–682.

De Vito S., Massera E., Piga M., Martinotto L., and Di Francia G. (2008) On field calibration of an electronic nose for benzene estimation in an urban pollution monitoring scenario. *Sens. Actuators B* **129**, 750–757.

Di Natale C., Paolesse R., Burgio M., Martinelli E., Pennazza G., and D'Amico A. (2004) Application of metallo-porphyrins-based gas and liquid sensor arrays to the analysis of red wine. *Anal. Chim. Acta* **513**, 49–56.

Diamond D. (1998) *Principles of Chemical and Biological Sensors*. John Wiley, New York.

Dilhan M., Esteve D., Gue A.M, Mauvais O., and Mercier L. (1995) Electrochemical oxygen microsensors. *Sens. Actuators B* **26–27**, 401–403.

Do J.S. and Chen P.J. (2007) Amperometric sensor array for NO_x, CO, O_2 and SO_2 detection. *Sens. Actuators B* **122**, 165–173.

Ercole C., Del Gallo M., Mosiello L., Baccella S., and Lepidi A. (2003) *Escherichia coli* detection in vegetable food by a potentiometric biosensor. *Sens. Actuators B* **91**, 163–168.

Ehrmann S., Jungst J., Goschnick J., and Everhard D. (2000) Application of a gas sensor microarray to human breath analysis. *Sens. Actuators B* **65**, 247–249.

Emerson Process Management (2008) Fundamentals of ORP measurement. *Application Data Sheet, ADS 43-014/ rev.B* (www2.emersonprocess.com/siteadmincenter/PM%20Rosemount%20Analytical%20Documents/Liq_ADS_43-014.pdf).

Endres H.E., Wetzel S., Sandmaier D., Rieblinger J., and Ziegleder G. (2004) Detection of smoke in cacao with an electrochemical sensor. In: *Digest of Technical Papers of the 18th European Conference on Solid State Transducers, EUROSENSORS XVIII*, Rome, Italy, P3.36.

Falasconi M., Pardo M., Sberveglieri G., Ricc'o I., and Bresciani A. (2005) The novel EOS835 electronic nose and data analysis for evaluating coffee ripening. *Sens. Actuators B* **110**, 73–80.

Fleischer M., Simon E., Rumpel E., Ulmer H., Harbeck M., Wandel M., Fietzek C., Weimar U., and Meixner H. (2002) Detection of volatile compounds correlated to human diseases through breath analysis with chemical sensors. *Sens. Actuators B* **83**, 245–249.

Foster M.D. and Synovec E. (1996) Liquid chromatographic sensing in water on a thin-clad optical fiber by mode-filtered light detection. *Anal. Chem.* **68**(8), 1456–1463.

Fraser D.M. (ed.) (1997) *Biosensors in the Body: Continuous in vivo Monitoring*. John Wiley, New York.

Funazaki N., Hemmi A., Ito S., Asano Y., Yano Y., Miura N., and Yamazoe N. (1995) Application of semiconductor gas sensor to quality control of meat freshness in food industry. *Sens. Actuators B* **24–25**, 797–800.

Göpel W., Jones T.A., Kleitz M., Lundström I., and Seiyama T. (eds.) (1991) *Sensors: A Comprehensive Survey*, Vol. **2–3**, *Chemical and Biochemical Sensors*. John Wiley, New York.

Göpel W. and Schierbaum K.D. (1991) Chemical and biochemical sensors. In: Göpel W., Hesse J. and Zemel J.N. (eds.), *Sensors, A Comprehensive Survey*. VCH, Weinheim, Germany, Part 1. 119–158.

Goschnick J., Koronczi I., Frietsch M., and Kiselev I. (2004) Water pollution recognition with the electronic nose KAMINA. *Sens. Actuators B* **106**, 182–186.

Gouin J.F., Baros F., Birot D., and André J.C. (1997) A fibre-optic oxygen sensor for oceanography. *Sens. Actuators B* **38–39**, 401–406.

Guo L., Ni Q., Li J., Zhang L., Lin X., Xie Z., and Chen G. (2008) A novel sensor based on the porous plastic probe for determination of dissolved oxygen in seawater. *Talanta* **74**, 1032–1037.

Gutman E.E., Perov S.P, and Timashev S.F. (1995) Biosphere dynamics and chemical sensors. *Sens. Actuators B* **24–25**, 162–165.

Harsanyi G. (1995) *Polymer Films in Sensor Applications*. Technomic Publishing, Lancaster, PA.

Harsanyi G. (2000) *Sensors in Biomedical Applications*, Technomic Publishing, Lancaster, PA.

Hayashi K., Yamanaka M., Toko K., and Yamafuji K. (1990) Multichannel taste sensor using lipid membranes. *Sens. Actuators B* **2**, 205–213.

Hazen K.H., Arnold M.A., and Small G.W. (1998a) Measurement of glucose and other analytes in undiluted human serum with near-infrared transmission spectroscopy. *Anal. Chim. Acta* **371**, 255–267.

Hazen K.H., Arnold M.A., and Small G.W. (1998b) Measurement of glucose in water with first-overtone near-infrared spectra. *Appl. Spectrosc.* **52**(12), 1597–1605.

Hitchman M.L. (1978) *The Measurement of Dissolved Oxygen*. John Wiley, New York.

Ho C.K. and Hughes R.C. (2002) *In-situ* chemiresistor sensor package for real-time detection of volatile organic compounds in soil and groundwater. *Sensors* **2**, 23–34.

Hodgson P., Quan K.M., MacKenzie H.A., Freeborn S.S., Hannigan J., Johnston, E.M., Greig F., and Binnie T.D. (1995) Application of pulsed laser photoacoustic sensors in monitoring oil contamination in water. *Sens. Actuators B* **29**, 339–344.

Hong X., Nugent C., Mulvenna M., McClean S., Scotney B., and Devlin S. (2009) Evidential fusion of sensor data for activity recognition in smart homes. *Pervasive Mobile Comput.* **5**(3), 236–252.

Hosoya Y., Itagaki Y., Aono H., and Sadaoka Y. (2005) Ozone detection in air using $SmFeO_3$ gas sensor. *Sens. Actuators B* **108**, 198–201.

Huch A., Huch R., and Lübbers D.W. (1972) Quantitative continuous measurement of partial pressure (PO_2 measurement) on the skin of adults and newborn babies. *Pflügers Arch. Ges. Physiol.* **337**, 185–198.

Iketake H. and Yamada A. (2000) Fabrication of a residual chlorine sensor using a gold electrode. *Bunseki Kagaku (Abstr.)* **4–9**/1–2.

Imanaka N., Okamoto K., and Adachi G. (2001) A chlorine gas sensor based on the combination of Mg^{2+} cation conducting and O^{2-} anion conducting solid electrolytes with lanthanum oxychloride as an auxiliary electrode. *Electrochem. Commun.* **3**(2), 49–51.

Isaacs S. and Henze M. (1994) Fluorescence monitoring of an alternating activated sludge process. *Water Sci. Technol.* **30**(4), 229–238.

Ivanov B., Zhelondz O., Borodulkin L., and Ruser H. (2002) Distributed smart sensor system for indoor climate monitoring. In: *Proceedings of KONNEX Scientific Conference*, München, 10–11 Oct. 2002.

Ivanov P., Llobet E., Stankova M., Vergara A., Vilanova X., Hubalek J., Gracia I., Cané C., and Correig X. (2004) Towards a microsystem for monitoring ethylene in warehouses. In: *Digest of Technical Papers of the 18th European Conference on Solid State Transducers, EUROSENSORS XVIII*, Rome, Italy, P3.05.

Ivanova O., Chuprin M., Krutovertsev S., Pislyakov A., Redina O., Vdovichev S., Kalinovsky V., and Konovalov V. (2004) A multisensor system for detection the composition of exhaled air. In: *Digest of Technical Papers of the 18th European Conference on Solid State Transducers, EUROSENSORS XVIII*, Rome, Italy, P1.57.

Jachowicz R., Weremczuk J., and Tarapata G. (2004) Fast dew point hygrometer for measurements of human skin evaporation factor. In: *Digest of Technical Papers of the 18th European Conference on Solid State Transducers, EUROSENSORS XVIII*, Rome, Italy, A2.4.

Jinghong H., Dafu C., Yating L., Jine C., Zheng D., Hong Z., and Chenglin S. (1995) A new type of transcutaneous pCO_2 sensor. *Sens. Actuators B* **24–25**, 156–158.

Kamionka M., Breuil P., and Pijolat C. (2006) Calibration of a multivariate gas sensing device for atmospheric pollution measurement. *Sens. Actuators B* **18**, 323–327.

Kaneyasu K., Otsuka K., Setoguchi Y., Sonoda S., Nakahara T., Aso I., and Nakagaichi N. (2000) A carbon dioxide gas sensor based on solid electrolyte for air quality control. *Sens. Actuators B* **66**, 56–58.

Karlsson K., Karlberg B., and Olsson R.J.O. (1995) Determination of nitrate in municipal waste water by UV spectroscopy. *Anal. Chim. Acta* **312**, 107–113.

Karube I., Nomura Y., and Arikawa Y. (1995) Biosensors for environmental control, *Trends Anal. Chem.* **14**, 295–299.

Kasik I., Mrazek J., Podrazky O., Seidl M., Tobiska P., Matejec V., Kovacs B., Markovics A., Szili M., and Vickova K. (2007) Preparation and characterization of transducer layers for opto-electrochemical detection of chlorine in water. In: Baldini F., Homola J., Lieberman R.A., and Miler M. (eds.) (2007) *Optical Sensing Technology and Applications. Proc. SPIE* **6585** (doi:10.1117/12.723642).

Kästle S., Noller F., Falk S., Bukta A., Mayer E., and Miller D. (1997) A new family of sensors for pulse oximetry. *Hewlett-Packard J.* February 1997, 39–53.

Kohl C.D. (2003) Electronic noses. In: Waser R. (ed.), *Nanoelectronics and Information Technology*. Wiley-VCH, Weinheim, Germany, Part VII, 853–863.

Kolle C., Gruber W., Trettnak W., Biebernik K., Dolezal C., Reininger F., and O'Leary P. (1997) Fast optochemical sensor for continuous monitoring of oxygen in breath-gas analysis. *Sens. Actuators B* **38–39**, 141–149

Korotcenkov G., Cerneavschi A., Brinzari V., Vasiliev A., Ivanov M., Cornet A., Morante J., Cabot A., and Arbiol J. (2004) In_2O_3 films deposited by spray pyrolysis as a material for ozone gas sensors. *Sens. Actuators B* **99**, 297–303.

Korotcenkov G., Blinov I., Ivanov M., and Stetter J.R. (2007) Ozone sensors on the base of SnO_2 films deposited by spray pyrolysis. *Sens. Actuators B* **120**, 679–686.

Lam Y.Z. and Atkinson J.K. (2007) Biomedical sensor using thick film technology for transcutaneous oxygen measurement. *Med. Eng. Phys.* **29**, 291–297.

Lammertyna J., Veraverbeke E.A., and Irudayaraj J. (2004) zNose^{TM} technology for the classification of honey based on rapid aroma profiling. *Sens. Actuators B* **98**, 54–62.

Larsen L.H., Kajaer T., and Revsbech N.P. (1997) A microscale NO_3^- biosensor for environmental applications, *Anal. Chem.* **69**, 3527–3531.

Laschi S., Palchetti I., and Mascini M. (2004) Screen-printed gold electrodes for heavy metal detection. In: *Digest of Technical Papers of the 18th European Conference on Solid State Transducers, EUROSENSORS XVIII*, Rome, Italy, A3.1.

Lee S.J., Park J.S., Im H.T., and Jung H.I. (2008) A microfluidic ATP-bioluminescence sensor for the detection of airborne microbes. *Sens. Actuators B* **132**, 443–448.

Legin A.V., Bychkov E.A., and Vlasov Y.G. (1995) Analytical applications of chalcogenide glass chemical sensors in environmental monitoring and process control. *Sens. Actuators B* **24–25**, 309–311.

Liang X., Zhong T., Quan B., Wang B., and Guan H. (2008) Solid-state potentiometric SO_2 sensor combining NASICON with V_2O_5-doped TiO_2 electrode. *Sens. Actuators B* **134**, 25–30.

Liem K.D., Hopman J.C.W., and Oeseburg B. (1992) Method for the fixation of optrodes in near infrared spectrophotometry. *Med. Biol. Eng. Comput.* **30**, 120–121.

Liao K.C., Hogen-Esch T., Richmond F.J., Marcu L., Clifton W., and Loeb G.E. (2008) Percutaneous fiber-optic sensor for chronic glucose monitoring *in vivo. Biosens. Bioelectron.* **23**, 1458–1465.

Lin Y.J., Guo H.R., Chang Y.H., Kao M.T., Wang H.H., and Hong R.I. (2001) Application of the electronic nose for uremia diagnosis. *Sens. Actuators B* **76**, 177–180.

Luo D., Hosseini H.G., and Stewart J.R. (2004) Application of ANN with extracted parameters from an electronic nose in cigarette brand identification. *Sens. Actuators B* **99**, 253–257.

Lvova L., Kima S.S., Legin A., Vlasov Y., Yang J.S., Chaa G.S., and Nama H. (2002) All-solid-state electronic tongue and its application for beverage analysis. *Anal. Chim. Acta* **468**, 303–314.

Lynggaard-Jensen A. (1999) Trends in monitoring of waste water systems *Talanta* **50**, 707–716.

Mabrook M.F. and Petty M.C. (2002) Application of electrical admittance measurements to the quality control of milk. *Sens. Actuators B* **84**, 136–141.

MAK (2000) Werte Kommission (Die Senatskommission zur Prüfung gesundheitsschädlicher Arbeitsstoffe der Deutschen Forschungsgemeinschaft) (2000) MAK und BAT-Werte-Liste (TRGS 900).

Mascini M., Moscone D., and Bernardi L. (1992) In vivo continuous monitoring of glucose by microdialysis and a glucose biosensor. *Sens. Actuators B* **6**, 143–145.

Mascini M., Macagnano A., Scortichini G., Del Carlo M., Diletti G., D'Amico A., Di Natale C., and Compagnone D. (2004) Biomimetic sensors for dioxins detection in food samples. In: *Digest of Technical Papers of the 18th European Conference on Solid State Transducers, EUROSENSORS XVIII*, Rome, Italy, P1.23.

McFarlane R. and Hamilton M.C. (1987) A fluorescence based dissolved oxygen sensor. In: *Proc. SPIE* **798**, 324–330.

Meckes A., Behrens J., Hausner M., and Benecke W. (1996) Miniaturised air analyser. In: *Proceedings of the Conference on MicroSystem Technologies 1996*, Potsdam, Germany, 181-186.

Mello L.D., Del Pilar M., Sotomayor T., and Kubota L.T. (2003) HRP-based amperometric biosensor for the polyphenols determination in vegetables extract. *Sens. Actuators B* **96**, 636–645.

Mendelson Y. (1991) Invasive and noninvasive blood gas monitoring. In: Wise D.L. (ed.), *Bioinstrumentation and Biosensors.* Marcel Dekker, New York, 249–279.

Mendelson Y., Clermont A.C., Peura R.A., and Lin B.C. (1990) Blood glucose measurement by multiple attenuated total reflection and infrared absorption spectroscopy. *IEEE Trans. Biomed. Eng.* **37**, 458–465.

Mirmohseni A. and Oladegaragoze A. (2004) Application of the quartz crystal microbalance for determination of phenol in solution. *Sens. Actuators B* **98**, 28–36.

Nake A., Dubreuil B., Raynaud C., and Talou T. (2004) Outdoor in situ monitoring of volatile emissions from wastewater treatment plants with two portable technologies of electronic noses. *Sens. Actuators B* **106**, 36–39.

Norgaard P., Helmo K., and Sorensen E.W., (1996) Treatment process for nitrogen removal controlled by NADH. *Vand Jord.* **3**(3), 125–129 (in Danish).

Norton H.N. (1982) *Sensor and Analyzer Handbook.* Prentice Hall, Englewood Cliffs, NJ.

Parker D. (1987) Sensors for monitoring blood gases in intensive care. *J. Phys. E: Sci. Instrum.* **20**(9), 1103–1112.

Pellegrini G.E., Carpico G., and Coni E. (2004) Electrochemical sensor for the detection and presumptive identification of quinolone and tetracycline residues in milk. *Anal. Chim. Acta* **520**, 13–18.

Peterson J.I., Goldstein S.R., Fitzgerald R.V., and Buckhold D.K. (1980) Fibre optic pH probe for physiological use. *Anal. Chem.* **52**, 864–869.

Pickup J.C. (1993) Developing glucose sensors for in vivo use. *Trends Biotechnol.* **11**, 285–291.

Poghossian A., Baade A., Emons H., and Schöning M.L. (2001) Application of ISFETs for pH measurement in rain droplets. *Sens. Actuators B* **76**, 634–638.

Ponzoni A., Depari A., Falasconi M., Comini E., Flammini A., Marioli D., Taroni A., and Sberveglieri G. (2008) Bread baking aromas detection by low-cost electronic nose. *Sens. Actuators B* **130**, 100–104.

Pruna S. (1995) Development of noninvasive method for *in vivo* sensing of glycaemia. *Chemical Sensors for In Vivo Monitoring* (a newsletter published by a European Union Concerted Action), **19**, December 1995, 30–33.

Rahman A.S.A., Yap M.M.S., Shakaff A.Y., Ahmadc M.N., Dahari Z., Ismail Z., and Hitam M.S. (2004) A microcontroller-based taste sensing system for the verification of *Eurycoma longifolia. Sens. Actuators B* **101**, 191–198.

Rella L., Siciliano P., Capone S., Epifani M., Vasanelli L., and Licciulli A. (1999) Air quality monitoring by means of sol-gel integrated tin oxide film. *Sens. Actuators B* **58**, 283–288.

Ristori C., Del Carlo C., Martini M., Barbaro A., and Ancarani A. (1996) Potentiometric detection of pesticides in water samples. *Anal. Chim. Acta* **325**, 151–160.

Rodriguez-Mozaz S., Marco M.P., de Alda M.J.L., and Barcelo D. (2004) Biosensors for environmental monitoring of endocrine disruptors: a review article. *Anal. Bioanal. Chem.* **378**, 588–598.

Rodriguez-Mozaz S., de Alda M.J.L., Marco M.P., and Barcelo D. (2005a). Biosensors for environmental monitoring—A global perspective. *Talanta* **65**, 291–297.

Rodriguez-Mozaz S., de Alda M.J.L., and Barcelo D. (2005b). Analysis of bisphenol A in natural waters by means of an optical immunosensor. *Water Res.* **39**, 5071–5079.

Rolfe P. (1994) Intra-vascular oxygen sensors for neonatal monitoring. *IEEE Eng. Med. Biol.* June/July 1994, 336–346.

Safaai-Jazi A. and Petersen J.V. (1991) Fiber optic chlorine sensor based on evanescent field absorption. *IEEE Proc. Digital Object Identifier* **2**(10), 1139–1142.

Sakai H., Iiyama S., and Toko K. (2000) Evaluation of water quality and pollution using multichannel sensors. *Sens. Actuators B* **66**, 251–255.

Sano N. and Ohtsuki F. (2007) Carbon nanohorn sensor to detect ozone in water. *J. Electrostat.* **65**, 263–268.

Santos J.H., Smyth M.R., and Blanc R. (1998) Mercury-free anodic stripping voltammetry of lead ions using a PVS-doped polyaniline modified glassy carbon electrode. *Anal. Commun.* **35**, 345–348.

Santos J.P., Fernández M.J., Fontecha J.L., Lozano J., Aleixandre M., García M., Guitérrez J., and Horrillo M.C. (2005) SAW sensor array for wine discrimination. *Sens. Actuators B* **107**, 291–295.

Santos-Alves J.S.G.D. and Patier R.F. (1999) The environmental control of atmospheric pollution. The framework directive and its development. The new European approach. *Sens. Actuators B* **59**, 69–74.

Satoh I. and Iijima Y. (1995) Multi-ion biosensor with use of a hybrid-enzyme membrane. *Sens. Actuators B* **24–25**, 103–106.

Schipper E.F., Kooyman R.P.H., Heideman R.G., and Greve J. (1995) Feasibility of optical waveguide immunosensors for pesticide detection: physical aspects. *Sens. Actuators B* **24–25**, 90–93.

Sehra G., Cole M., and Gardner J.W. (2004) Miniature taste sensing system based on dual SH-SAW sensor device: An electronic tongue. *Sens. Actuators B* **103**, 233–239.

Sekiguchi N., Komeda T., Funakubo H., Chabicovsky R., Nicolics J., and Stangl G. (2001) Microsensor for the measurement of water content in the human skin. *Sens. Actuators B* **78**, 326–330.

Severinghaus I.W. and Bradely A.F. (1958) Electrodes for blood pO_2 and pCO_2 determination. *J. Appl. Physiol.* **13**, 515–520.

Shankaran D.R., Uehera N., and Kato T. (2002) Determination of sulfur dioxide based on a silver dispersed functional self-assembled electrochemical sensor. *Sens. Actuators B* **87**, 442–447.

Sivaramakrishnan S., Rajamani R., Smith C.S., McGee K.A., Mann K.R., and Yamashita N. (2008) Carbon nanotube-coated surface acoustic wave sensor for carbon dioxide sensing. *Sens. Actuators B* **132**, 296–304.

Small G.W. and Arnold M.A. (1998) Data handling issues for near-Infrared glucose measurements. *Leos Newsletter* April/1998, 16–17.

Soller B.R. (1994) Design of intravascular fibre optic blood gas sensors. *IEEE Eng. Med. Biol.* June/July 1994, 327–335.

Starodub V.M., Fedorenko L.L, and Starodub N.F. (2000) Optical immune sensors for the monitoring protein substances in the air. *Sens. Actuators B* **68**, 40–47.

Steinkuhl R., Sundermeier C., Hinkers H., Dumschat C., Cammann K., and Knoll M. (1996) Microdialysis system for continuous glucose monitoring. *Sens. Actuators B* **33**, 19–24.

Takatani S. and Ling J. (1994) Optical oximetry sensors for whole blood and tissue. *IEEE Eng. Med. Biol.* June/July/1994, 347–357.

Talazac L., Brunet J., Battut V., Blanc J.P., Pauly A., Germain J.P., Pellier S., and Soulier C. (2001) Air quality evaluation by monolithic InP-based resistive sensors. *Sens. Actuators B* **76**, 258–264.

Teledyne Brown Engineering/Sensor Technologies (1994) *Oxygen Monitors and Sensors for the Health Care Industry,* Product Catalog.

Thybaud L., Depeursinge C., Rouiller D., Mondin G., and Grisel A. (1990) Use of ISFETs for 24h pH monitoring in the gastrooesophageal tract. *Sens. Actuators B* **1**, 482–485.

Timmer B., Olthuis W., and van den Berg A. (2005) Ammonia sensors and their applications—A review. *Sens. Actuators B* **107**, 666–677.

Tomoda M., Okano S., Itagaki Y., Aono H., and Sadaoka Y. (2004) Air quality prediction by using semiconducting gas sensor with newly fabricated $SmFeO_3$ film. *Sens. Actuators B* **97**, 190–197.

Traversa E., Sadaoka Y., Carotta M.C., and Martinelli G. (2000) Environmental monitoring field tests using screen-printed thick-film sensors based on semiconducting oxides. *Sens. Actuators B* **65**, 181–185.

Tsujita W., Yoshino A., Ishida H., and Moriizumi T. (2005) Gas sensor network for air-pollution monitoring. *Sens. Actuators B* **110**, 304–311.

Tsukada K., Miyahara Y., Shibata Y., and Miyagi H. (1991) An integrated micro multi-ion sensor using platinum-gate field-effect transistors. In: *Proceedings of the International Conference on Solid State Sensors and Actuators, Transducers'91*, San Francisco, CA, 218–221.

Valentini F., Biagiotti V., Lete C., Palleschi G., and Wang J. (2007) The electrochemical detection of ammonia in drinking water based on multi-walled carbon nanotube/copper nanoparticle composite paste electrodes. *Sens. Actuators B* **128**, 326–333.

Verissimo M.I.S., Oliveira J.A.B.P., and Gomes T.M.S.R. (2005) The evaluation of copper contamination of food cooked in copper pans using a piezoelectric quartz crystal resonator. *Sens. Actuators B* **111–112**, 587–591.

Verissimo M.I.S., Oliveira J.A.B.P., and Gomes T.M.S.R. (2006) Leaching of aluminium from cooking pans and food containers. *Sens. Actuators B* **118**, 192–197.

Volpe G., Moscone D., Compagnone D., and Palleschi G. (1995) In vivo continuous monitoring of L-lactate coupling subcutaneous microdialysis and an electrochemical biocell. *Sens. Actuators B* **24–25**, 138–141.

Ward P.L. (2009) Sulfur dioxide initiates global climate change in four ways. *Thin Solid Films* **517**, 3188–3203.

Watanabe E. and Tanaka M. (1991) Determination of fish freshness with a biosensor system. In: Wise D.L. (ed.), *Bioinstrumentation and Biosensors,* Marcel Dekker, New York, 39–73.

Wolfbeis O.S. (1991) Biomedical application of fibre optic chemical sensors. *Int. J. Optoelectron.* **6**(5), 425–441.

Yatsenko V.A., Goushcha A.O., Dobrovolsky A.A., Privalko A.V., and Sushko V.I. (1995) Intelligent sensor for integral estimation of water quality. *Sens. Actuators B* **29**, 332–338.

Yoo K.S., Woo S.B., and Jyoung J.Y. (2003) Trace mercury determination by differential pulse anodic stripping voltammetry using polythiophene-quinoline/glassy carbon modified electrode. *Bull. Korean Chem. Soc.* **24**(1), 27–30.

Yuan Z.Y., Hong M., Wlodarski W.B., and Ji X.S. (1991) A novel organophosphorous pesticides sensitive EN-FET. In: *Proceedings of the International Conference on Solid State Sensors and Actuators, Transducers'91*, San Francisco, CA, 703–705.

Zampolli S., Elmi I., Ahmed F., Passini M., Cardinali G.C., Nicoletti S., and Dori L. (2004) An electronic nose based on solid state sensor arrays for low-cost indoor air quality monitoring applications. *Sens. Actuators B* **101**, 39–46.

Zhao Y. and Liao Y. (2002) Novel optical fiber sensor for simultaneous measurement of temperature and salinity. *Sens. Actuators B* **86**, 63–67.

Zhylyak G.A., Dzyadevich S.V., Korpan Y.I., Soldatkin A.P., and El'skaya P. (1995) Application of urease conductometric biosensor for heavy-metal ion determination. *Sens. Actuators B* **24–25**, 145–148.

CHAPTER 6

Chemical Sensors in Industry, Agriculture, and Transportation

G. Harsanyi

1. INTRODUCTION

This chapter provides a broad survey of the practical applications of chemical sensors in industry, agriculture, and transportation. In considering the application-oriented chapters we concentrated not on the measured parameter, but on the application fields. This chapter is devoted to those areas where mainly professionals may meet sensors: automotive, industrial, agricultural, fire alarm, low enforcement, airspace, and military applications.

The main competitor of chemical sensors in practice is the application of conventional analytical chemical instrumentation with its always emphasized disadvantages: large instruments, high-cost measurements. Sensors offer the possibility of continuous monitoring using low-cost devices, but the always returning problem is the accuracy and reliability of the measurements. This is an important argument for increasing the number of sensors and continuing their development. New sensing principles or existing ones applying new materials or sensor structures may result in better properties in terms of sensitivity, selectivity, stability, and reliability. Although the development of sensor materials and devices needs multidisciplinary knowledge, this is even more true for the applications: Each application field means a "vocation" with very special challenges; both from the application environment and from the requirements on sensor parameters. The general strategy is to select or develop the appropriate sensor type for the given application at the lowest possible cost.

The following main areas are discussed in this chapter: automotive applications of sensors; sensors for exhaust control; potentiometric, amperometric, and semiconductor-based lambda sensors; sensors for passenger-compartment air quality; car odor sensors, oil and brake fluid quality sensors; chemical sensors in industrial settings; pH monitoring in chemical industry; sensors in power plant monitoring; pH monitoring in galvanic processes; combustible gas detection for safety; volatile organic gas compound monitoring; ammonia detectors; gas leakage detectors; workplace safety control; humidity sensing in industry; wood and paper dryness control; drying process control; clean-room technology control; combustion control of gas turbines and boilers; gas sensing in fuel cells; smoke detectors; particle sensors; fire sensing, gas sensors in fire sensing; agricultural sensor applications; soil moisture measurements; heavy-metal and nitrate content control; ammonia control in animal facilities; nutrition quality monitoring; chemical sensors in law enforcement; alcohol detection in breath; drug detection; cocaine analysis; low enforcement applications; breath analysis; alcohol testing; sensors against terrorism; airspace applications; detecting hydrogen leakage is spacecraft; emission control; spacecraft air quality control; military applications; detection of landmines and explosive residues; warfare agent detection.

2. AUTOMOTIVE APPLICATIONS

2.1. SENSORS FOR EXHAUST CONTROL

Sensors are used to monitor the exhaust gas mixture in automobiles and to provide information on the status of the exhaust gas content to the engine control unit (ECU). The control strategy allows the ECU to vary the air–fuel mixture around stoichiometry (14.7 air to fuel ratio, A/F) to increase catalytic converter efficiency, prolong the life of the converter, and provide increased fuel economy. To ensure the fastest response time, sensors are placed in the exhaust manifold at points where they can respond quickly to exhaust changes with minimal lag time (Figure 6.1). The operating temperature inside the

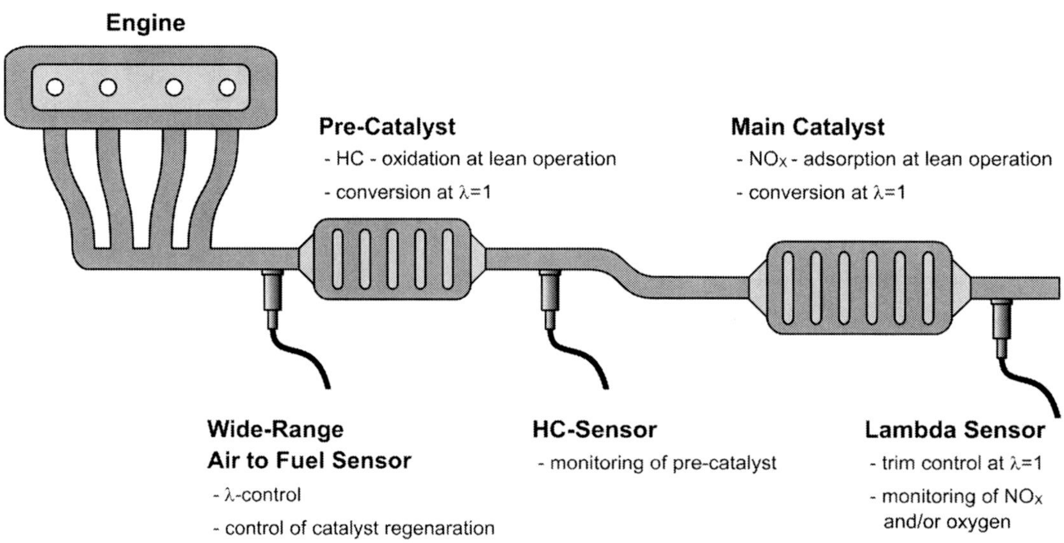

Figure 6.1. Application of various sensors in the exhaust system of a car.

manifold can easily exceed 900°C, and in fact, a minimum operating temperature (about 300°C) is necessary to ensure proper performance of existing sensors. External electrical power is frequently applied to an integral heating element in the sensor so that closed-loop operation can be initiated within 30 s of starting and during idle (Linsmeier 1998; Jurgen 1999; Zabler et al. 2001).

In the subsequent text we use A/F for the air-to-fuel ratio, which is 14.7 in the ideal case. For the ideal stoichiometric chemical oxidation reaction, λ (lambda) is defined as a normalized A/F ratio $\lambda = 1$ for the above condition; $\lambda > 1$ nears lean condition, while $\lambda < 1$ yields rich engine operation conditions.

2.1.1. Zirconia Lambda Cells

In principle, the lambda sensor operates as a solid-electrolyte galvanic oxygen-concentration cell (Figure 6.2a). A ceramic element consisting of zirconium dioxide and yttrium oxide is employed as a gas-impermeable solid electrolyte. This mixed oxide is an almost perfect conductor of oxygen ions over a wide temperature range. The solid electrolyte is designed to separate the exhaust gas from the reference atmosphere. Both sides feature catalytically active platinum electrodes. Basic principles of zirconia solid-electrolyte sensors are given elsewhere. The active sensor ceramic (ZrO_2) is heated from inside by means of a ceramic W heater so that the temperature of the sensor ceramic remains above the 350°C function limit regardless of the exhaust-gas temperature. In equilibrium (potentiometric) mode, the

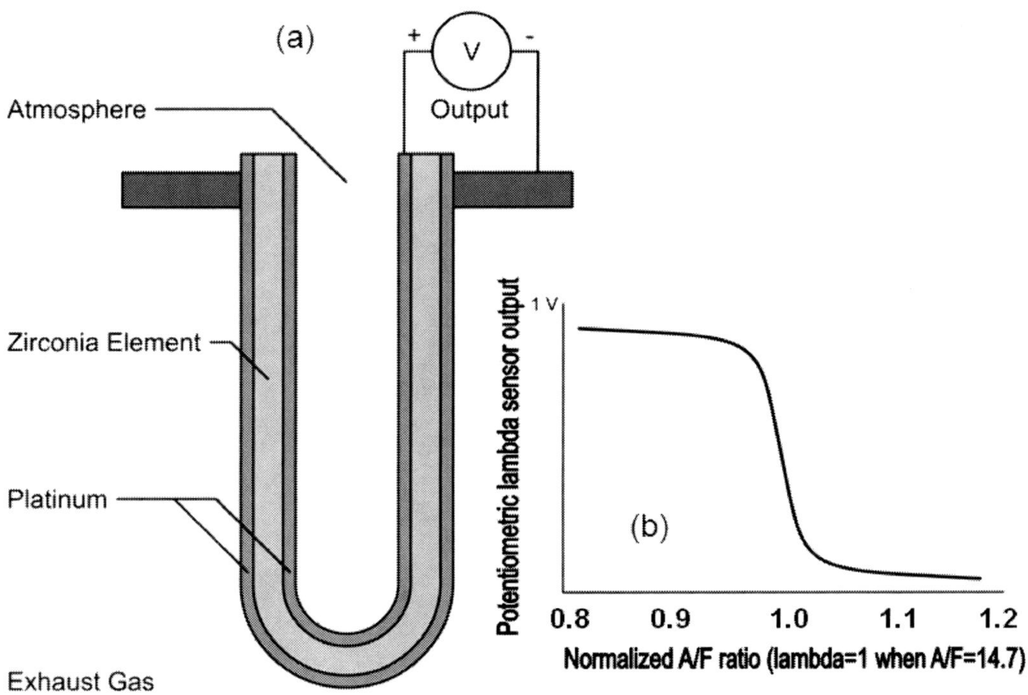

Figure 6.2. (a) Structure and(b) characteristics of the potentiometric (Nernst-type) lambda sensor.

sensor output voltage (Figure 6.2b) is a logarithmic function of the partial pressures of oxygen on either side of the cell wall (Docquirer and Candel 2002; Mukundan and Garzon 2004). One side of the cell is exposed to exhaust gas and the other to atmospheric air. When the engine is operating in a lean mode, the exhaust gas oxygen increases and the partial pressures of oxygen on both sides of the ceramic are almost equal, causing the sensor's output voltage to be near zero volts. As the exhaust gas is enriched, it passes stoichiometry, which is the ideal burn point. Just beyond stoichiometry, oxides (e.g., CO and hydrogen) increase dramatically, which increases the pressure differential across the electrolyte, causing a voltage to be generated.

Planar-structure Nernst-type lambda sensors have also been designed and realized for commercial use. These use multilayer ceramic technology in their production (Zabler et al. 2001). The structure contains either a chamber in contact with the ambient air or a hermetically closed reference chamber for O_2.

In amperometric (limiting-current, also called pumping-type) lambda sensors an external electrical voltage is applied to the two electrodes on a heated ZrO_2 electrolyte to pump O_2 ions from the cathode to the anode (electrochemical O_2 pumping cell; see Figure 6.3). When a diffusion barrier impedes the flow of O_2 molecules from the exhaust gas to the cathode, the result is a current saturation beyond a certain pumping voltage threshold (limiting current condition; Figure 6.4). Thus, the resulting current is a function of the exhaust gas's oxygen concentration (Mukundan and Garzon 2004).

In lean exhaust gas, the signal drops in response to a reduction in lambda, but below $\lambda = 1$ there is a radical surge. As a result, this sensor is only suitable for use in lean gases.

An appropriate combination of a limiting-current sensor with a Nernst concentration cell on a single substrate will produce a dual-cell wide-range lambda sensor (Zabler et al. 2001). The pumping and concentration cells are made of ZrO_2. Each cell is coated with two porous platinum electrodes, and they are arranged with a measurement gap of approximately 10–50 μm in height between them. A gas-intake opening in the solid electrolyte connects this measurement gap to the exhaust gas, while the gap serves as a diffusion barrier to control limiting current. An electronic circuit regulates the current applied

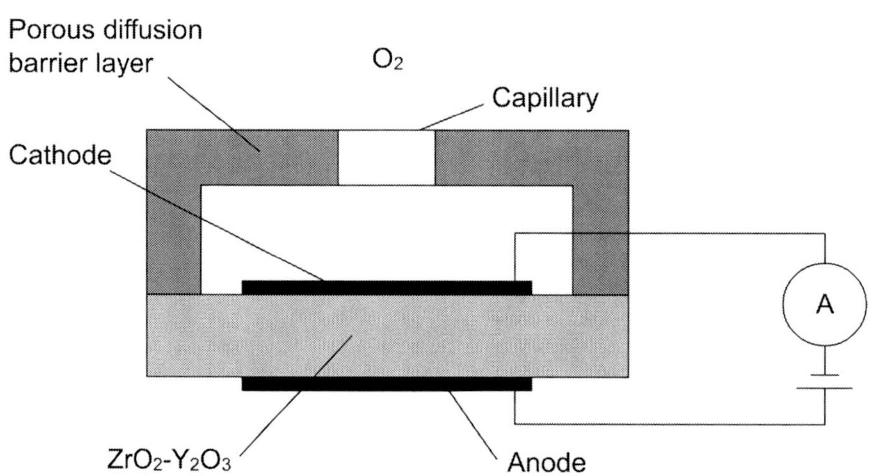

Figure 6.3. Structure of the amperometric (pumping-type) lambda sensor.

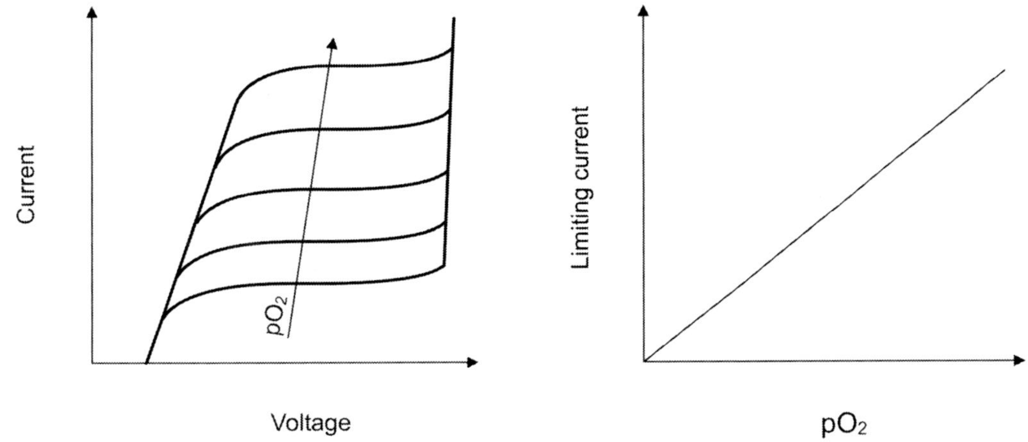

Figure 6.4. Typical current versus voltage and limiting current versus O_2 concentration characteristics of the amperometric (pumping-type) lambda sensor.

to the pumping cell to maintain a constant gas composition (for instance, $\lambda = 1$) in the measurement gap. This corresponds to a Nernst cell voltage of $U_s \approx 450$ mV. If the exhaust gas is lean, the pumping cell drives the oxygen outward from the measurement gap. If the exhaust gas is rich, the flow direction is reversed and oxygen from the surrounding exhaust gas (e.g., from decomposition of CO_2 and H_2O) is pumped into the measurement gap. The pumping current is a function of the oxygen concentration (or oxygen requirement). With the attendant electronic control circuitry, an unambiguous signal increase over a wide lambda range of $0.7 < \lambda <$ infinite (air) can be obtained if the current is calibrated with respect to the production scattering.

2.1.2. Semiconductor Lambda Sensors

Recent developments include efforts to reduce the response time (to $\leq$10 ms) so that individual cylinders and not just the average composition can be detected. One approach uses a thin (1-µm) film semiconducting metal oxide which is deposited on an aluminum oxide substrate. Oxide semiconductors such as TiO_2 and $SrTiO_3$ rapidly achieve equilibrium with the oxygen partial pressure in the surrounding gas phase at relatively low temperatures. These oxygen-sensitive materials show both p-type and n-type conductivity, which depend on the value of the partial pressure of oxygen. This will lead to an ambiguous response of the sensors in specific applications, but this ambiguity can be avoided by the selective incorporation of dopants. A doped sensor with low acceptor concentration results in a reduction of the response time and improvement of the selectivity as well as an increase in the oxygen sensitivity and temperature stability. Changes in the partial pressure of the adjacent oxygen produce variations in the oxygen vacancy concentration of the material (TiO_{2-x} respectively $SrTiO_{3-x}$), thereby modifying its volume conductivity. This effect, which is exploited to determine lambda, is also superposed by the conductivity's temperature dependence. Electrical resistance and sensor response times are both inversely proportional to temperature, as equilibrium is achieved more rapidly through diffusion in the oxygen vacancies.

The ability to dispense with an O_2 reference allows an extremely simple design featuring an integrated heater. The porous semiconductor thick-film layer is generally positioned and sintered onto a planar Al_2O_3 substrate between two electrodes. Thin-film layers are also being developed as an alternative. At $\lambda = 1$, the sensor layer displays an extreme change in conductivity due to the large pO_2 variation. New TiO_2 sensors provide essentially the same response in resistance value as ZrO_2-based $\lambda = 1$ probes in voltage output characteristics (Zhu et al. 1996). Variations in the rates of increase for lean and rich resistance and for response time occur over the lifetime, with the emission-control system undergoing a significant shift toward lean.

The diffusion of the oxygen vacancies occurs quickly enough, but the surface reaction of the $SrTiO_3$ thin film with the other components of the exhaust also proceeds so that a thermodynamic equilibrium is established fast enough for a reliable measurement of the λ value, thus allowing a response time of less than 10 ms to be reached in the engine (Meixner et al. 1995).

A new type of semiconductor TiO_2-based lambda sensor was demonstrated recently (Francioso et al. 2003). The semiconductor thin films were deposited by the sol-gel method on 3 mm × 3 mm alumina substrate chips between gold interdigitated electrodes on the front side. A Pt thick-film heater was applied on the back side. As shown in Figure 6.5, a complete set of measurements performed at 16 different λ values demonstrated a resistance jump around $\lambda = 1$, which enables the application of the sensors in automotive exhaust systems. These types of sensors are expected to have much better dynamic behavior than conventional zirconia-based lambda sensors that contain bulk electrochemical processes.

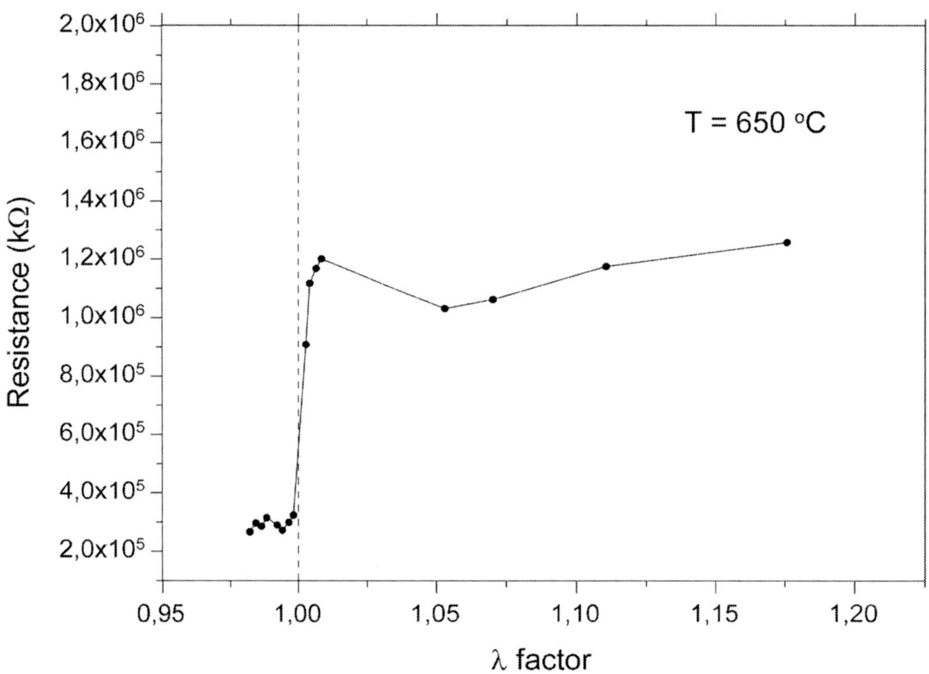

Figure 6.5. Resistance versus lambda characteristics of a TiO_2 semiconductor fast lambda sensor. (Reprinted with permission from Francioso et al. 2003. Copyright 2003 Elsevier.)

2.1.3. Other Sensors in Exhaust Control

All previous sensor types discussed are essentially from the solid-state lambda sensor family and measure the oxygen content of the exhaust gas, which becomes very low in the case of stoichiometric burning conditions.

Ion current probes or ionization sensors provide alternative possibilities in combustion control. These elements are used to characterize ion currents associated with the flame. This technique is well suited to spark ignition engines because one may use the spark plug as the ion current detector. As a result, time-resolved combustion-related information may be obtained at a low cost on a cylinder-to-cylinder basis. The main source of ions in flames is provided by chemiionization. This process occurs during elementary reactions when the energy of the reactant together with the reaction energy are large enough to ionize one of the products of a chemical reaction involving neutral species (Docquier and Candel 2002).

Another important application is to follow the combustible and toxic gas component concentrations within the exhaust gas. In monitoring the catalytic converter, the most important compounds are CO, NO_x, and hydrocarbons, as the total HC content (THC). The operating temperature is significantly lower, and short-term temperature peaks must not destroy the sensor. For this purpose, it must react to the relevant components in the exhaust as sensitively and selectively as possible. Semiconductor metal oxide–based resistors can be used for such purposes, but their nonselective behavior limits their application in the separate detection of components considerably. There are, however, some successful compositions (Meixner et al. 1995).

The Al_2O_3/V_2O_5 system has proved to be a gas-sensitive sensor material for measuring the nitrogen oxides. It became apparent that a compound of composition $AlVO_4$ has the best gas-sensing properties. This material can be manufactured as a thin film by reactive sputtering. Measurements show very small cross-sensitivity with respect to the other components of the exhaust. Thin films of Ga_2O_3 have proved to be most suitable for measuring the hydrocarbons. Ga_2O_3 is an *n*-conducting metal oxide with high chemical and thermal stability. It can be manufactured easily as a thin film by reactive sputtering on a ceramic target and can be stabilized in a subsequent annealing process. This material has very low catalytic activity to CO; thus, it has rather good selectivity properties for detecting hydrocarbons.

A portable electronic nose system was fabricated and characterized using an oxide semiconductor gas sensor array and artificial neural network by Hong et al. in 2000. The sensor array consisted of such thick-film oxide semiconductor sensing materials as Pd-doped WO_3, Pt-doped SnO_2, TiO_2–Sb_2O_5–Pd-doped SnO_2, TiO_2–Sb_2O_5–Pd-doped SnO_2-Pd-coated layer, Al_2O_3-doped ZnO, and $PdCl_2$-doped SnO_2. As an application, the system was used to identify 26 CO/HC car exhaust gases in the concentration range of CO 0%/HC 0 ppm to CO 7.6%/HC 400 ppm, and the identification was successfully demonstrated.

A novel concept using the calorimetric principle for a combined multipurpose sensor was demonstrated recently (Wu and Micheli 2004). The deterioration of converter catalysts could be determined using a calorimetric sensor in combination with various diagnostic strategies. The principle of the "ancient" calorimetric sensor relies on the fact that heat generated by catalytic exothermic reactions, within certain limits, is proportional to the concentration of hydrocarbons in the exhaust. A thermocouple can measure the temperature rise caused by exothermic reactions. Compared to other hydrocarbon-sensing techniques, this type of sensor offers a number of advantages for exhaust applications. The device measures essentially the heating values of these hydrocarbon molecules, giving rise to a signal that is proportional to the total hydrocarbon (THC) content in the automobile exhaust. The calorimetric hydrocarbon sensor also is ideally suitable for diesel exhaust applications, where the engine is operated in the far lean region (10–14% oxygen) and the exhaust flow rate is fairly constant. The high oxygen concentration in the diesel exhaust provides an oxidative environment for catalytic combustion, which is needed for the calorimetric sensor. The multipurpose calorimetric hydrocarbon/CO sensor utilizes thermocouples and highly selective oxidation catalysts. A thermocouple consists of two thick-film junctions made of Nb_2O_5-doped TiO_2 and a lithiated Ni, which are applied on a planar alumina substrate (Zhu et al. 1996). The highly selective catalyst is coated on one of the junctions to form a calorimetric sensor. A Pb-modified Pt catalyst exhibited superior selectivity in CO oxidation in preference to hydrocarbon and hydrogen oxidation reactions. Thus, a calorimetric CO sensor can be built with one junction thermocouple coated with a Pb-modified Pt catalyst operated at 350°C. The calorimetric hydrocarbon sensor uses a two-junction design. Pt catalyzes one junction, while the other is deposited with a Pb-modified Pt catalyst. The Pt-covered junction catalyzes both the CO and hydrocarbon oxidation reactions, while the catalyzed junction coated with Pb-modified Pt promotes only the CO oxidation reaction. The two-junction thermocouple detects the difference in temperature due to the different catalytic reactions over the two junctions. Although an effort was made to generate equal amounts of heats from the CO oxidation (thus equal temperature rises) at the two junctions by adjusting the amount of the catalysts, the complete suppression of the CO signal was not achieved for many practical reasons. Therefore, a three-junction design was finally applied, in which Pt catalyzes one junction, the second junction is covered with the Pb-modified Pt catalyst, and the third junction remains noncatalyzed. An external circuit is used to subtract the contribution from the exothermic CO oxidation reaction from two sensors. The resulting sensor produces information both approximate CO and total HC concentration.

Within a few years, so-called urea selective catalytic reduction (SCR) systems will help to minimize nitrogen oxide (NO_x) emissions from commercial vehicles in order to meet future standards. Such systems operate with water-diluted urea that is injected into the exhaust pipe. A simplified implementation of such a closed loop is illustrated in Figure 6.6 (Moos et al. 2002). In front of the SCR converter, the urea decomposes into ammonia, which serves as a reducing agent for NO_x. In such systems, not only

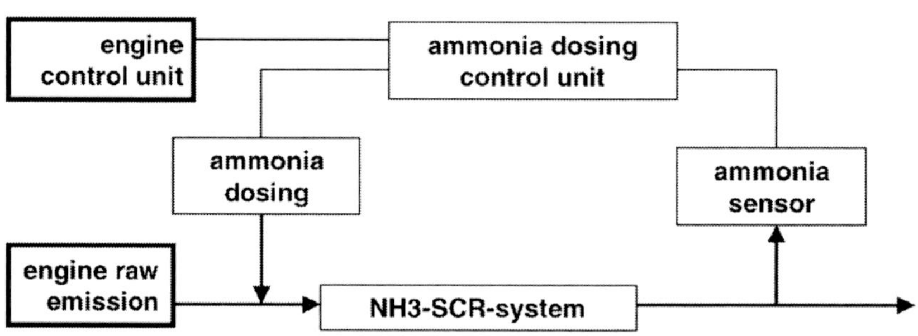

Figure 6.6. Principle of selective catalytic reduction using an ammonia sensor. (Reprinted with permission from Moos et al. 2002. Copyright 2002 Elsevier.)

is the selective detection of NO_x required, but in order to optimize the injected amount of urea and to assure that no ammonia emissions occur, a closed control loop control with an ammonia exhaust gas sensor should be applied (Broecker et al. 1992; Daudel et al. 1994).

The engine control unit (ECU) receives the ammonia signal of the ammonia sensor. Based on this and using other engine operation status parameters, the ECU can modify the urea injection in order to eliminate the presence of ammonia within the exhaust gas. Such a closed control loop system will control the ammonia dosing system.

Several types of ammonia sensors have been developed, but the special conditions of exhaust gas control need special considerations. Differently doped semiconducting metal oxides such as tungsten oxide, tin oxide, or indium oxide heated resistors can be used for ammonia detection purposes and are inexpensive enough for automotive applications, but their lack of selectivity, or rather their sensitivity to other exhaust gas compounds, humidity, and oxygen content, hinders their application here. Sensors based on functional films of organic materials or solid electrolytes do not meet the requirements for the harsh conditions.

Another very interesting attempt is the use of zeolite thick film as a functional film, suggested by Moos et al. (2002). The zeolite material changes its complex electrical impedance very selectively and sensitively with the ammonia concentration. Because the applied type of zeolite is stable at high temperatures, 800°C and above, and because the thick-film technology, which is already in use for automotive exhaust gas sensors is very inexpensive, reliable, and cost-effective, exhaust gas ammonia sensors can be manufactured. The sensor structure follows the usual thick-film arrangement: The bottom of the alumina substrate comprises the platinum heater, while on the top of the substrate a thick-film interdigital capacitor with a zeolite thick-film dielectric film is screen-printed. The heater keeps the sensor at operating temperature. There, the zeolite film, which is the heart of the sensor, selectively changes its electrical properties as a function of the ammonia content of the ambient gas. The type of zeolite used in the ammonia sensor has been selected from different zeolite modifications, with respect to selectivity, temperature stability, and chemical stability. Although the optimal operating temperature with the best compromise between sensitivity and response time was determined to be 420°C, the zeolite is long-term stable up to more than 800°C due to its excellent thermal and chemical stability. With proper packaging, the sensor can be used with appropriate long-term stability and will stand against reducing interaction with soot and other critical components of the exhaust gas.

2.2. PASSENGER COMPARTMENT AIR QUALITY SENSORS

Regulations for automobile exhaust emission have been tightened year by year with the increase in awareness of environmental pollution. Still, every driver must have experienced irritating exhaust odors from vehicles ahead while driving on highways. The situation gets worse as traffic becomes more congested on urban roads. Automobile exhaust contains toxic components such as CO, NOx, series of HCs, and particulates, some of which are cancer-stimulating. The emission of harmful components is unavoidable insofar as an internal combustion engine is powered by fossil fuels. Exhaust gas is mixed with the clean air immediately after the release in free space and is diluted. The concentrations are estimated to be 30–100 ppm for CO, 3–10 ppm for HCs, and 2–10 ppm for NOx. The real constitution of NO_x within the tailpipe is believed to be NO, but it is oxidized to NO_2 upon exposure to air. Gas sensors to detect exhaust gas require sensitivities of this level or higher (Reavell et al. 1997).

On the other hand, the concentration may increase rapidly by several orders of magnitude when waiting behind a truck in a traffic jam. In most commercial solutions, only the external air composition is monitored with sensors; the lock inside the air ventilation system closes the external air inlet, and only internal circulation is allowed. Semiconductor oxide-type sensors are generally used for these purposes, either commercially available versions, or specially developed thick-film types in which the sensor and heater resistors are deposited onto alumina substrates. As is well known, sensors based on these materials are not very selective, so a careful selection for the proper type and/or material as well as operating temperature is required. The classical material, SnO_2, has been proven to be a good candidate for a combined CO/HC sensor, while another semiconductor metal oxide can serve as a NO_x/NO_2 sensor material. Commercially available versions are also equipped with humidity and temperature sensors in order to compensate for their cross-sensitivities. The CO sensor should operate in the range of 10–100 ppm, and the NO_x sensor in the range of 0.5–5 ppm (Oto et al. 2001).

In future ventilation systems, the quality of the internal air should also be monitored. Both oxygen reduction and CO increase are harmful and are said to result in a decrease in the ability to concentrate on driving. These changes may be detected by either O_2 or CO gas sensors. It is also likely that passengers may object to the smell of smoking materials or bad breath. An "odor" gas sensor can be applied to monitor the air quality inside for amenity reasons. The concept is demonstrated in Figure 6.7 (Nakagawa et al. 2000). A combined sensor consisting of a NO_x sensor and a CO/HC sensor monitors the quality of the outside air, and is placed at the entrance of the air intake. An odor sensor monitors the inside air and is placed under the dashboard. Signals from the three gas sensor elements are transferred

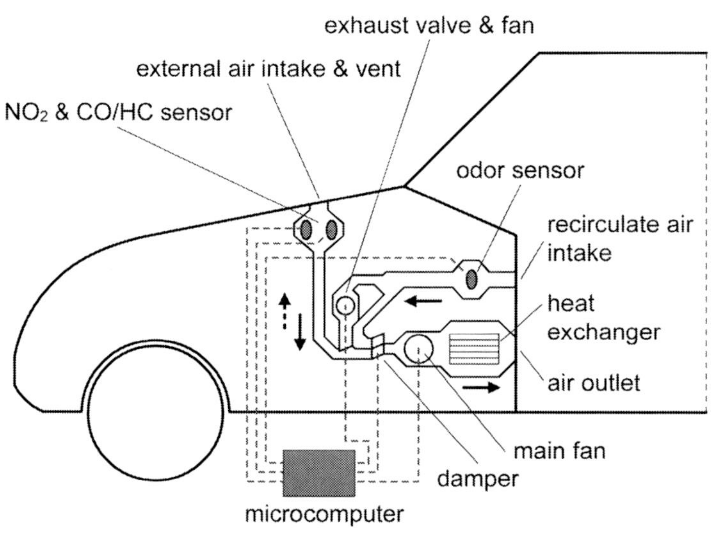

Figure 6.7. Schematic illustration of an automated car ventilation system. (Reprinted with permission from Nakagawa et al. 2000. Copyright 2000 Elsevier.)

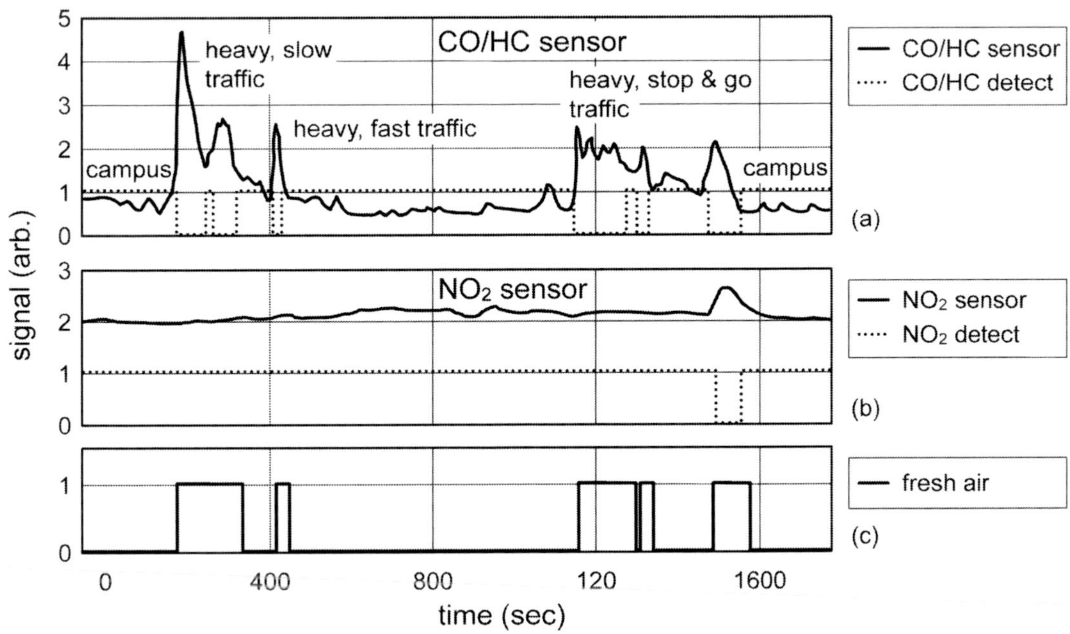

Figure 6.8. Operating data from the controlled system on a highway. (Reprinted with permission from Nakagawa et al. 2000. Copyright 2000 Elsevier.)

into a microcomputer that controls the entire ventilation system. An exhaust valve and exhaust fan are controlled to remove the polluted air introduced into the system. If outside-air is judged to be cleaner than inside-air, outside air is introduced. In the opposite situation, the vane of the damper is lowered to set to the internal mode, and inside air is recirculated. Polluted air may be partly introduced into the ventilation system immediately before the pollution is detected, because of the relatively slow response of the gas sensors, but the introduction of polluted air into the cabin may be avoided if the main fan speed is low. When the outside air becomes clean, the system activates the exhaust valve and the exhaust fan. The actual running data on a highway in a congested traffic is presented in Figure 6.8, with comments on traffic conditions. The logic signals "CO/HC alarm," "NO_2 alarm," and "Fresh air" are presented with a negative logic. In other words, the damper is opened and outside air is introduced when the "Fresh air" logic signal is low.

CuO nanowires (NWs) were grown by the thermal oxidation of Cu foil at 400°C, and gas sensors were fabricated by the deposition of a solution containing the CuO NWs, as reported by Kim et al. (2008). At 300–370°C, the sensor resistance increased upon exposure to 30–100 ppm CO. This was explained by the gas-sensing characteristics of the *p*-type oxide semiconductor. In contrast, the sensor showed two opposite NO_2 sensing behaviors: The resistance decreased upon exposure to 30–100 ppm NO_2, but increased upon contact with ≤5 ppm NO_2. The increase in resistance upon contact with both reducing CO and a low concentration of oxidizing NO_2 gases provided a simple and novel sensing algorithm for air quality control in automotive cabins.

2.3. ODOR SENSORS FOR QUALITY CONTROL IN THE AUTOMOTIVE INDUSTRY

For today's cars, comfort is an important part of customers' preferences. In addition to the traditional visual and tactile requests, a new dimension has lately appeared: smell. The "new car smell" may be found to be unpleasant or overpowering. Car odor is a result of a mixture of volatile compounds coming from different interior trim parts and materials. In order to reduce and improve this characteristic, off-odor compounds must be identified. Traditionally, this identification has been carried out using gas chromatography (GC), perhaps combined with mass spectrometry (MS) analysis. However, these methods are time-consuming, costly, and appear to be poorly adapted for quality control requirements. A novel rapid and objective method for odor evaluation appeared during the last decade, the so-called electronic nose (e-nose), which is discussed in more detail later. This technology is based on nonselective gas sensors associated with pattern-recognition techniques in order to recognize gas compound mixtures. The objective of new studies is to investigate the ability of e-nose technology to recognize car odors for quality control monitoring. Gas sensor technologies based on electrical resistance variation, metal oxide semiconductor and conducting polymer resistors, as well as MOSFET arrays were tested in order to select the most appropriate one for automotive interior-trim polymer applications. The effectiveness of gas sensors in discriminating between foams can be observed using principal component analysis (PCA) and multiple discriminant analysis (MDA). Results showed that composite polymer sensors are not well suited for discriminating foams, while MOS sensors and MOSFET in association with MOS sensors present good efficiency in separating samples according to their suppliers (Morvan et al. 2000). There is, however, still debate, since recent, modified versions of the GC/MS method seem to be more effective.

2.4. OIL QUALITY SENSORS

Changing the oil in an engine regularly protects the system and pays off in the long run. But it also involves throwing out a lot of serviceable oil from vehicles that see less strenuous use, and a schedule cannot predict exactly when oil will break down. That is why great efforts are aimed at developing on-board systems to measure oil quality in real time or, at least, to base predictions on specific, observable conditions. A reliable means of monitoring the condition of engine oil will permit a new kind of just-in-time maintenance, and that, according to some forecasters, could save millions of quarts of oil a year.

Engine oil life sensors are intended to alert drivers to the need to change engine oil and avoid costly engine failure, while saving oil form sometimes unnecessary changes. The most important advantages are

- Prolonged oil-change intervals—leading to an environmentally friendly system due to longer oil use and reduced filter waste
- Protection of the engine by early detection of possible engine failure
- Harmonized maintenance intervals
- No oil dipstick checking periods necessary

Several car manufacturers use capacitive sensors that measure the shift in the dielectric constant of the oil as an indicator of viscosity and therefore oil condition. The sensor is designed to measure the aging of oil by monitoring the dielectric constant, which correlates to the acidity, an indicator of oil degradation. The dielectric constant of oil depends on the length of the service period and the composition of the molecules. As friction and temperature cause the molecules to break down and as additives are depleted, the dielectric constant increases, measured by increased voltage. The sensor structure is illustrated in Figure 6.9. It consists of two cylindrical capacitors, one reference, which shifts only with permittivity changes, and the other also changing with the oil level; thus, both the oil level and the quality are monitored with the same device. Since the parameters are temperature-dependent, a thermoelement is also used for temperature-compensation purposes. One difficulty in measuring the dielectric constant is its sensitivity to temperature. As the temperature rises, the dielectric constant also rises. To cancel out this temperature dependency, the sensor signal conditioner is designed with compensation circuitry. The circuit is mounted adjacent to the sensing element that is in contact with the oil so temperature can be accounted for. The sensor is generally mounted on the bottom of the oil container. The measurement of the impedance over a given range of frequency, the so-called impedance spectroscopy, allows for a significant improvement in the evaluation of the oil condition since the results correlate to the viscosity of the oil samples. In this way, the influence of oil dilution with gasoline can also be detected. However, the dielectric constant of different oils varies considerably. As a result, the oil type should be controlled at the car using this approach (Lincmeier 1998).

Impedance, which is the resistance and capacitance of the oil, gives an indication of conductive particles, such as salts or metal particles. The open-circuit voltage gives information about antioxidants that might be present. The pH relates to the total acid number of the oil. A combination of measurements will present a truer picture of oil degradation than just one measurement. The combined sensor uses two electrodes—one of iridium with iridium oxide, and the other of silver with silver chloride—to measure pH. The silver electrode works with a third, made of platinum, to gauge open-circuit potential. The iridium and platinum electrodes work in tandem to measure the impedance of the oil.

Another approach measures oil viscosity by determining the time it takes for a sample to flow through a small chamber, which is a more direct indication of the actual state of the engine's oil (ZPlus

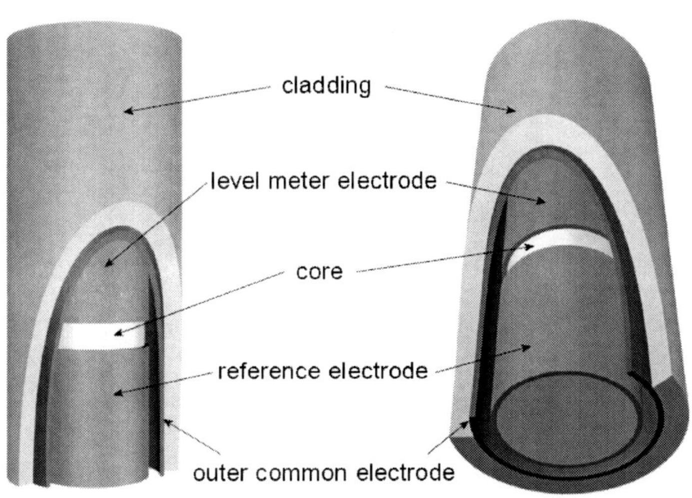

Figure 6.9. Structure of the capacitive oil quality/oil-level sensor.

2008). It has proven to be too costly for automobiles but it may be applicable for trucks. Another possibility is to measure the acoustic impedance of the fluid, which depends on its mechanical properties including the density and the sound propagation speed. A piezoelectric transmitter capacitor can generate ultrasound waves. The reflection coefficient at material interfaces depends on the acoustic impedances of both sides, similarly to the telegraph equations. Changes in the mechanical properties of the oil can be followed by the ultrasound coupling efficiency changes (Ju et al. 2010). The same piezoelectric elements can be used as ultrasound sensors in the receiver mode to detect the reflected wave intensity.

A large number of other sensor principles are also under investigation; in some cases, prototypes are already available. These methods include the following.

- *ATR-spectroscopy* (attenuated total reflection) is already established on a laboratory scale (Schulz et al. 2002). The basic principle is as follows. The sensor element is an optical waveguide crystal in which infrared light propagates through total reflections. At the reflection spot, a so-called evanescent field is generated on the other medium, which in this case is the oil. An attenuated radiation will be reflected this way, and the attenuation depends on the optical properties of the oil. For an online sensor, the signal stability and sensitivity as well as the sensor durability were investigated for a zinc selenide crystal (Wang et al. 2011). The ATR-sensor showed good sensitivity to soot, water, glycol, and sulfation. No chemical degradation of the crystal surface was observed after 12 months of measurement, although a signal decrease was measured.
- *Semiconductor metal oxides as sensitive transducer layers:* The gas sensitivity of these materials has been discussed many times. Thin-film resistors formed on alumina substrates can also show sensitivity for changes in liquid media. The resistance of tin dioxide and wolfram oxide layers was measured in fresh and used oil with high acid number. There was a slight change in resistance, but the sensitivity and stability has to be improved, however, prototypes are already available at Fraunhofer Institut für Physikalische Messtechnik.
- *Oil resistance measurement using the transmission line model (TLM)* is a method to obtain some information about the sensor surface itself by the adaptation of a method for measuring the contact resistance (Booth et al. 2010). The basic idea is to form several electrodes with the same area but at various distances in order to form several resistors inside the oil material with various length values. This enables measuring oil resistance versus a length function: The intersection of the linear curve on the R axis gives the doubled value of the metal–oil contact resistance, and the steepness of the curve is a function of the oil resistivity. Both parameters can be used for oil qualification (Patil et al. 2010).
- *Water cross-sensitivity* is a major interference in determining the condition of the oil by means of capacitive or conductive measurements. Zeolites were tested for their suitability for the detection of water in oil to compensate for the interference (Baker and Senaratne 1992). It was shown that zeolites are suitable for water detection in lube oil at temperatures above 50°C.

Functional fluids in automotive applications other than engine oil may also need a quality monitoring sensor. In order to ensure the safe operation of a vehicle, the quality, level, and temperature of the brake fluid should be continuously monitored. As of now, only a dip below a minimum level can be detected, so the fluid is changed arbitrarily every two years. The goal is therefore to monitor the brake system completely, thus making longer service intervals possible. A simple gold thin-film interdigital capacitor structure was suggested by Binder et al. (1996) in order to measure the break fluid complex impedance as a function of the water content. The Cole-Cole plot showed considerable changes with various water contents. The capacitance showed 1.5 pF/vol%, while the conductance showed 0.05 mS/vol% sensitivity. The accuracy enabled detection of the critical 4–6% water content with proper reliability.

3. CHEMICAL SENSORS IN INDUSTRIAL PROCESSES

Industrial sensor solutions must cover a wide range of process conditions found in chemical processes. The products should work in even the most demanding chemical environments—high temperature and pressure—and the design has to ensure high safety and reliability in order to comply with plant safety requirements. The application solutions have to be able to be integrated into plant-wide process control and asset management systems.

3.1. INDUSTRIAL pH AND OXIDATION–REDUCTION POTENTIAL SENSORS

Industrial pH and oxidation–reduction potential (ORP) measuring sensor solutions try to optimize quality and throughput by including several types of combination electrodes, transmitters, housings, and cleaning systems. Conductivity measurement allows optimal process control, resulting in a reduction of costs and downtimes. Crystallization, phase separation, and emulsion processes are examples where continuous turbidity measurements are carried out. Conductivity and turbidity are important parameters and are handled together with pH and ORP, as in the case of water quality monitoring (see Chapter 5 of this volume), but pH and ORP are monitored with physical sensors, so in this section we concentrate on those measurements that are generally performed with potentiometric devices: ion-selective electrodes (ISEs), ion-selective field-effect transistors (ISFETs), and other redox electrodes.

Monitoring of pH and ORP is very important in many different chemical industry processes to reduce cost and optimize yield. These processes include a diverse range of applications, from checking and maintaining product quality in the chemical, pharmaceutical, food, and beverage industries through to helping water industry users ensure they meet regulatory limits for acidity and alkalinity when discharging water (see Chapter 5 of this volume). The requirements of many industry processes are different. In chemical processes the material has to withstand strong acids or caustic solutions under, for example, high temperatures. The electrodes must also be certified for use in hazardous areas. Biochemical and pharmaceutical industries frequently produce products in batch processes that have a very high added value. Inaccurate pH measurement can lead to the product curing during the process. The entire batch of production must then be thrown away; it is not possible to rework it. It is very expensive to have to discard a complete batch. Safety must also be taken into consideration and reflected by installing accurate and reliable sensor systems. In pharmaceutical and food processes, measuring systems have to meet requirements for hygienic design, must withstand many sterilization cycles, and must provide traceability for validation purposes. Examples of such industrial applications include the following.

- Chemical industries: chemical production monitoring, chrome detoxification colorants, metal finishing and mining, petrochemicals, refineries, power plants, boiler feedwater, paper, resin production, starch, gelatine
- Pharmaceutical industries: biotechnology, blood plasma fractioning, fermenters, pharmaceutical chemical production
- Food and beverages: beverages, brewing water, cheese production, dairy, milk and yoghurt processing, drinking water, fish farming, sauces/ketchup, yeast propagation, marmalade, puree, drinks bottling, sugar
- Water treatment: gas scrubbers, boiler feedwater, cooling water, ultrapure water (semiconductor plants), neutralization, drinking water, municipal wastewater, industrial wastewater, wastewater treatment, tannery, detoxification, flue gas cleaning, galvanic wastewater, heavy-metal precipitation, sludge draining, ground sanitation

The latest generations of automatic pH measuring devices have brought pH measurement into a completely new cost range. pH sensors had previously been considered costly to operate because the electrode had to be regularly serviced, cleaned, calibrated, and even replaced. That is now all a thing of the past. An automatic pH measuring point does not just fulfil its measuring tasks, it also takes care of all the servicing tasks, i.e., it checks, cleans, and calibrates the pH electrode.

A great variety of pH ion-selective electrodes are produced by numerous companies and are commercially available (e.g., Mettler Toledo, Endress+Hauser, ABB, etc.). The various types apply different glass membrane materials according to the application temperatures and chemical environments, conventional liquid or solid polymer or gel-type internal electrolytes as well as solid or flowing-type reference electrodes. Generally, pH sensor electrodes operate within the overall pH range (0–14) and the temperature range 0–130°C; some types work in more limited ranges, pH 2–12 and temperature range 0–100°C. Recently, glass-free sensor types with ISFETs have also emerged in industrial applications. High-temperature glass membranes have special application fields, because high-temperature applications can degrade general-purpose pH membranes.

PH/redox electrodes with liquid electrolyte allow refilling of the electrolyte reservoir, and are especially suitable for applications where precise measurement values are important. With correct maintenance, they are very well known for their long service life, even under harsh operating conditions. Temperature compensation of the measurements (RTD) can be handled as standard. PH/redox electrodes with a gel- or liquid-type electrolyte or solid polymer reference system, all with open junctions, obviate the need to pressurize the electrode through the housing. A solid polymer electrolyte (no refilling and pressurizing of electrolyte) is in direct contact with the process medium via an open junction. It allows reliable process control in highly demanding chemical processes, particularly suitable for protein-containing solutions, emulsions and suspensions, hot caustic solutions, sulfide-bearing media, etc. The gel electrode is prepressurized (e.g., up to 2.5 bar) and the polymer reference system is (process) pressure-resistant. The internal overpressure ensures gradual electrolyte outflow, efficiently cleaning the diaphragm and preventing influx of process medium. These types of electrode cannot be refilled and therefore require reduced maintenance.

As is well known, pH sensors always need a built-together counter reference electrode. While for many applications a simple gelled reference is adequate, solid reference electrodes provide additional protection. A solid gel or potassium chloride (KCl)–impregnated wood provides a barrier preventing contact between the sample liquid and the reference electrode, reducing the risk of contamination and greatly prolonging sensor life. Solid reference electrode sensors are ideal for most industrial and municipal wastewater applications, where high levels of sulfides are present that could contaminate the reference electrode of a standard pH sensor. They are also ideal for use in pressurized environments such as tanks and pipelines. Their one main drawback is their limited life in pure waters. For that application a flowing reference electrode is recommended. Flowing reference electrodes are the best choice wherever pH monitoring is required for high-purity water applications, such as steam raising for power plants or for use in semiconductor manufacture.

The nonglass pH electrode is an unbreakable breakthrough, especially for the food industry, where the possibility and consequences of fractured glass present a highly dangerous threat. The sensor is equipped with a solid-state pH-sensitive part (ISFET) and a high-temperature reference system with gel electrolyte, making the electrode practically maintenance-free. The ISFET chip is integrated into a sensor body made of special polymer (polyether ether ketone), which is mechanically and chemically stable and is approved for applications in the food and pharmaceuticals industries. The glass-free system rules out any risk of glass breaking during operation—a decisive advantage in the mentioned sectors. Another advantage is the fast response at low process temperatures, where classical glass electrodes respond fairly slowly. In contrast to glass electrodes, there is no ion flow between the fluid and the sensor. Thus, "chemistry" and electrical measurement are kept totally separate. As a result, the sensor material does not change and the need for recalibration is by no means as frequent as with glass electrodes. Since there is no source layer, ISFETs are also suitable for pH measurement in media with a low proportion of water. Modern ISFET materials are highly selective and follow the Nernst law with closer tolerance limits than glass. The sensor's extreme robustness comes from embedding the chip in a stable and unbreakable plastic body.

Thick-film electrodes have also become popular low-cost, glass-free potentiometric sensors (Fraunhofer Institute, IKTS). In these systems, both the reference and the pH sensitive electrodes are realized on the same alumina ceramic or enameled steel substrate; both liquid electrolytes in reservoirs and solid polymer electrolytes have been applied. Custom-designed structures and special applications are available without the need for large-batch production, as in the case of ISFETs.

3.2. HUMIDITY SENSING IN INDUSTRY

Conventional sensors for the relative humidity of air that are commercially available apply capacitive thin-film sensor elements (or sometimes resistive ones) extended with a temperature sensor both for recording temperature and compensating the temperature dependency of the humidity sensor element. The sensors are mounted either directly into a miniature display appliance that can be mounted onto a wall or into a handheld probe that contains the preamplifier units, with the sensors located in the gas transparent ending. Probe-type arrangements can be tested and calibrated with special tools more frequently. These types of sensors generally have a limited operating window: The real limits are near 100% and 100°C, and the edge of this corner cannot be reached by the sensors.

Many commercially available dew-point or capacitive-type sensors can be operated in a wide range of humidity/temperature conditions, such as from –30 to +100°C and 5–98% RH (with special probe

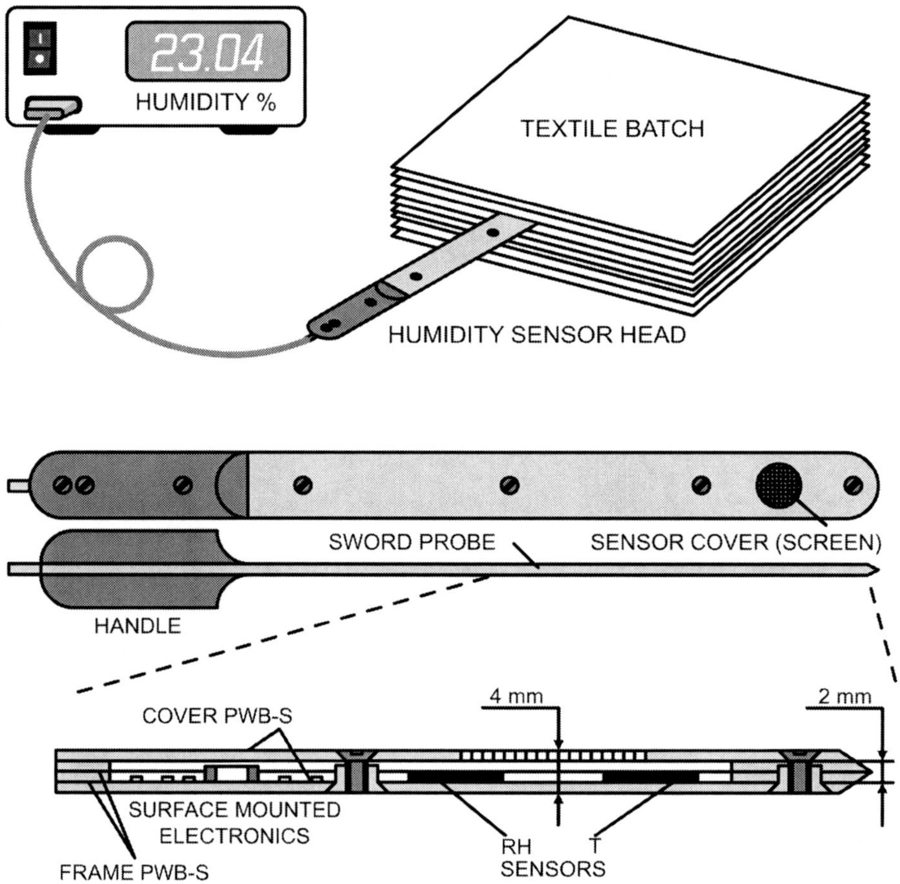

Figure 6.10. Sword probe for measuring textile or paper humidity in stacks.

and handling materials, even from −50 to +150°C and 0–98% RH). Insensitivity to harsh gas environments (50 ppm CO, 50 ppm NO_x, 70 ppm SO_2, even humidity in the newest sensor versions) enables applications in compressed air technology, drying processes, and clean-room technology.

Conventional humidity sensors, however, are sensitive to extreme process and environmental conditions. Directly acting acoustic gas sensors use as primary sensing effect the variability of airborne sound properties, e.g., the speed of sound. These sensors are in principle distinguished by high reliability, long-term stability, and insensitivity to extreme temperatures, contamination, or aggressive media, and by short response time. This opens a wide field of industrial application. However, directly acting acoustic sensors are nonselective and suitable only for analyzing mixtures of two gas components, e.g., humid air (water vapor and dry air). Zipser and Frane (2004) described the development of a robust acoustic humidity sensor which can be used even under extreme conditions, such as high temperature up to 400°C, aggressive, explosive, or toxic pollutions in the gas to be analyzed, dust or steam, electromagnetic disturbances, nuclear radiation, or very fast changes of the humidity concentration. An important application of the acoustic sensor will be the *in situ* measurement of humidity in the exhaust air of industrial driers. In general, two types of such gas sensors can be distinguished: electric-acoustic and fluidic-acoustic. The first type does not need a pressure supply and is insensitive over a wide range to pressure fluctuations in the unknown gas mixture. On the other hand, the electric-acoustic sensors require precise temperature control, and they are sensitive to high gas temperature. Fluidic-acoustic gas sensors are usable even under extreme measuring conditions (high temperature, contamination, aggressive media, etc.) and have short response times, but for operation a stabilized pressure difference between sensor inlet and outlet must be realized.

In several industrial applications, the humidity content of solid media should be determined. This can be performed in two ways: measuring either the impedance of the given medium (wood, soil, paper, etc.) or the relative humidity inside a small closed cavity which is in equilibrium with the given medium. An example is given in Figure 6.10: a special sword-shaped probe for measuring textile or paper water content. The probe can be inserted between sheets in paper or textile stacks.

3.3. GASES IN INDUSTRY (PROCESS CONTROL, LEAKAGE, AND WORKPLACE SAFETY)

A great variety of gas sensors can be applied in industrial settings. The types and application problems in air quality monitoring and ventilation as well as gas composition monitoring in workplace safety applications are discussed in other sections. Sensor applications in industry are of several kinds:

- Hazardous material emission monitoring in order to limit it (see Section 3.4)
- Gas composition control in chemical processes for optimal production
- Leakage detection in supplies and transporting pipe/tank systems (for both gases and volatile organic vapors)
- Workplace atmosphere monitoring to limit hazardous (explosive or toxic) compound concentration [both ventilation-controlling units and personal handheld (lower-cost) monitoring devices are commercially available]

A wide range of industries need to use and analyze process gases. The breadth of use may be illustrated by listing the key market sectors that produce, use, or analyze process gases:

- Chemical and petrochemical industries
- Environmental monitoring, including both ambient air and stack emission monitoring
- Scientific and engineering research organizations, including universities and national laboratories
- Medical institutions, including hospitals
- Food processing and drinks industries, where gases such as nitrogen are used to enhance the shelf life of products by reducing oxidation, and carbon dioxide is widely used in soft drinks and alcoholic beverages
- Microelectronics industries, including semiconductor manufacturing and telecommunications
- Fabrication industries, including motor, ship, and aircraft industries

- Power generation, particularly the nuclear industry, e.g., advanced gas reactors
- Instrument manufacturers

Here is a list from the point of view of gas types, where specific gas sensors are widely applied:

- *Ammonia (NH_3):* fertilizer plants, fish and meat packaging plants, industrial refrigeration and cold storage, the semiconductor industry, water and wastewater treatment plants, munitions plants and storage
- *Benzene (C_6H_6):* petrochemical industries, refineries, oil and gas distribution, feedstock and production, gas stations
- *Carbon monoxide (CO):* furnaces: combustion control, gasoline generators/engines, grain storage silos, lumber drying kilns, mining and metallurgy, parking garages
- *n-Hexane (C_6H_{14}):* electricity generation, gas stations, peanut oil extraction plants, solvent distribution centers
- *n-Butane (C_4H_{10}):* petrochemical industries, aerosol propellant filling docks, feedstock, storage tanks and filling docks
- *Carbon dioxide (CO_2):* breweries and wineries, carbonated beverage bottling plants, dry ice plants, food processing plants, fruit storage and ripening chambers, greenhouses, indoor air quality and ventilation control, mushroom farms stack gas, oil well injection
- *Chlorine (Cl_2):* mining and metals industries, nuclear reactors, pulp and paper mills, PVC plastics manufacturing, semiconductor etching facilities, swimming pool chlorinization plants, water treatment plants, chlorine gas manufacturing
- *Hydrogen (H_2):* battery charging stations, semiconductor plants for water furnaces, underground vaults containing transformers, vegetable oil hydrogenation plants
- *Hydrogen cyanide (HCN):* plating and mining, nylon manufacturing
- *Methane (CH_4):* petrochemical industries, oil and gas refining and distribution, mining
- *Nitrogen dioxide (NO_2):* boilers and furnaces (combustion control), semiconductor plants, mining
- *Oxygen (O_2):* combustion control, cargo holds and storage tanks, grain storage silos, inert atmospheres, liquid-nitrogen storage, liquid-nitrogen–cooled laser facilities and cooled telescopes, sewer maintenance, underground vaults (utilities)
- *Ozone (O_3):* oxidizing gas in various technical processes—food, pharmaceutical, textile, and chemical industries, water treatment, purification of gases.
- *Propane (C_3H_8):* petrochemical industries, aerosol propellant filling lines, feedstock, propane-powered forklifts, storage tanks and filling docks
- *Hydrogen sulfide (H_2S):* leather tanneries and paper mills, mining and metals industries, oil fields and refineries, sewage treatment plants, sewer maintenance
- *Nitric oxide (NO):* semiconductor plants, mining
- *Sulfur dioxide (SO_2):* combustion control, circuit-board etching, paper mills, sulfur processing plants, water treatment
- *Volatile organic compounds (VOCs):* petrochemical industries, airline maintenance facilities (jet fuel), leaking underground storage tanks, petrochemical manufacturing, aerospace manufacturing,

automotive manufacturing (paint booths), high-tech electronics manufacturing, pharmaceutical manufacturing, printed circuit-board fabrication, transportation (spills from tankers, cars), printing (lithography and flexography), fibreglass-reinforced plastics production, hazardous waste facilities, coating (cans, coils, and fabrics), component manufacturing (parts spray painting, resin components, adhesive components, miscellaneous metalworking), military (base closures), marine chemistry (ships and barges), fire departments—arson investigation, coke oven off gases, halogen refrigerants

Within the natural gas industry "fiscal monitoring," i.e., the analysis of oil and gas steams to determine calorific value, is a key part of the quality system requiring traceable gas calibration standards. These analyses have a significant and direct impact on oil and gas companies' profitability and the consumer. Furthermore, there is a widespread need for analysis to monitor occupational exposure levels to gases in the workplace. Toxic gases, such as silane, arsine, and phosphine, are widely used in the manufacture of semiconductor devices in gas cabinets and containment chambers. The whole operation is, however, conducted under clean-room conditions, with recirculated air, so gas monitoring is essential.

The various gas compounds have very different physicochemical properties and therefore require somewhat different approaches for detection. The main groups from this point of view are somewhat different from the groupings of application fields: oxygen, ozone; inert CO_2; reducing agents (H_2; CO); sulfuric compounds (SO_2, H_2S); nitrous gases (NO, NO_2, NO_x, NH_3, N_2O); hydrocarbons (HCs: methane, ethane, butane, propane, hexane, etc.); volatile organic compounds (VOCs: alcohols, acetone, benzene, toulene, etc.); chlorine (Cl_2); and others.

In workplace safety and leakage detection, the task is in most applications the same: to detect a certain level of gas compounds above which their presence is either harmful (toxic) or dangerous (explosive). The primary risk associated with combustible gases and vapors is the possibility of explosions. Explosion, like fire, requires three elements: fuel, oxygen, and an ignition source. Each combustible gas or vapor will ignite only within a specific range of fuel/oxygen mixtures. Too little or too much gas will not ignite—these conditions are defined as the lower explosive limit (LEL) and the upper explosive limit (UEL), respectively. Any amount of gas between the two limits is explosive. It is important to note that each gas has its own LEL and UEL, as shown in Table 6.1. The gas concentrations are shown by percent of total volume, with the balance as normal air. Between these two limits, explosions can occur under some conditions, with the maximum explosive energy available at approximately the midpoint. In environments with combustible gas hazards, it is important to know long before the gas concentration reaches the LEL. Typical safety standards require that a gas detection unit give warnings at 10–20% of the LEL.

A great number of compounds are toxic directly with inhalation, so their maximal concentration must be limited, especially in workplaces where they can appear. German standards (MAK-Werte Komission) therefore give the so-called MAK (Maximale Arbeitsplatz-Konzentration) values of gas and vapour compounds; examples are listed in Table 6.2. Other recommendations (NIOSH, OSHA) define the *threshold limit value* (TLV), a term used to express the maximum airborne concentration of a material to which most workers can be exposed during a normal daily and weekly work schedule without adverse effects. TLVs can be expressed in three ways:

Table 6.1. Explosive limit values of various combustible gas and vapor compounds

Compound	Composition	Lower explosive limit, LEL (%)	Upper explosive limit, UEL (%)
Hydrogen	H_2	4.0	75.0
Ethane	C_2H_6	3.0	12.5
Methane	CH_4	5.0	15.0
Propane	C_3H_8	2.1	9.5
n-Butane	C_4H_{10}	1.6	8.4
Pentane	C_5H_{12}	1.5	7.8
n-Hexane	C_6H_{14}	1.1	7.5
Octane	C_8H_{18}	1.0	6.5
Ethylene	C_2H_4	2.7	36.0
Propylene	C_3H_6	2.0	11.1
Acetylene	C_2H_2	2.5	100
Acetone	$(CH_3)_2CO$	2.5	12.8
Benzene	C_6H_6	1.2	7.8
Isopropyl alcohol	$(CH_3)_2CHOH$	2.0	12.7
Methanol	CH_3OH	6.0	36.0
Ethyl alcohol	CH_3CH_2OH	3.3	19.0
Propanol	$CH_3(CH_2)_2OH$	2.1	13.5
Toluene	C_7H_8	1.1	7.1

Source: Flammable and Combustible Liquids, OSHA 29 CFR 1910.106), www.osha.gov.

- C (or CEL): ceiling exposure limit. Not to be exceeded at any time for any purpose
- TWA: time-weighted average. Usually expressed in exposure limits for 8 h per day and a 40-h work week
- STEL: short-term exposure limit. Maximum concentration for a continuous exposure period of 15 min (with a maximum of four such periods per day, with at least 60 min between exposure periods, during which C and TWA values are not exceeded)

Extraordinary care must be taken when considering the above limits, since the values may be different in various countries and they are also revised from time to time. Even organizations in the same country may recommend different values. The most standardized lists are given by the U.S. Occupational Safety and Health Administration (OSHA). In the United States, The Occupational Safety and Health Act of 1970 created both the National Institute for Occupational Safety and Health (NIOSH) and the Occupational Safety and Health Administration (OSHA). OSHA is in the U.S. Department of Labor and is responsible for developing and enforcing workplace safety and health regulations. NIOSH is in the U.S. Department of Health and Human Services and is an agency established to help assure safe and healthful working conditions for men and women by providing research, information, education, and training in the field of occupational safety and health. OSHA has created a number of gas exposure limit values, but these were revised according to the recommendations of NIOSH and other similar

Table 6.2. Maximum workplace concentration (MAK, Maximale Arbeitsplatz-Konzentration) values of gas and vapor compounds

Compound	Maximum workplace concentration (ppm)
SO_2	5
CO	33
NO_2	9
N_2O	180
NH_3	35
Halothane	40
Acetone	500
Ethyl alcohol	1000
Methanol	200
Toluene	50
Xylene	100

institutions. Table 6.3 gives values of STEL and TWA for the most toxic gases commonly used in practice according to OSHA comments from the January 19, 1989, Final Rule on Air Contaminants Project extracted from 54FR2332 et. seq. This rule was remanded by the U.S. Circuit Court of Appeals and the limits are not currently in force. Former recommendations are indicated in parentheses. These values are sometimes very low concentrations, at the level of a few parts per million, which must be detected very reliably by sensors and analyzers operating in workplace safety applications.

In office environments, photocopiers and laser printers produce ozone. Most countries have established threshold values for long- and short-time exposure periods. Regarding ozone, the environmental

Table 6.3. OSHA recommendations for short-term exposure limit (STEL) and time-weighted average (TWA) maximum values for toxic gases (old values in parentheses; see text for explanation)

Gas	Formula	STEL (ppm)	TWA (ppm)
Carbon monoxide	CO	— (400), C = 200	35 (50)
Hydrogen sulfide	H_2S	15 (20)	10
Sulfur dioxide	SO_2	5	2 (5)
Nitric oxide	NO	—	25
Nitrogen dioxide	NO_2	1 (5)	— (3)
Chlorine	Cl_2	1	0.5 (1)
Hydrogen cyanide	HCN	4.7	— (10)
Ammonia	NH_3	35	25 (50)
Phosphine	PH_3	1	0.3
n-Butane	C_4H_{10}	—	800
n-Hexane	C_6H_{14}	—	50

air quality is classified as follows: very good (0–50 ppb), good (50–100 ppb), poor (100–200 ppb), and very poor (>200 ppb).

A great number of gas-monitoring appliances are based on conventional analytical methods, or versions of these that are very much specialized for the given task and contain a signal transducer that can be called already as a "sensor," although its operation cannot fulfil all requirements of a sensor's definition (e.g., gas chromatography–mass spectrometry, GC-MS). Other appliances are based on real sensors, but the recovery of the sensor elements often needs the application of flow-through cells, or sample-taking mechanisms. Thus, their practical operation is often very similar to that of the analyzers. For this reason, it is very difficult to draw a real interface. We discuss the various approaches here in the wide sensor application meaning. Table 6.4 gives a survey of examples from commercially available gas-monitoring appliances and modules, sensing principles and characteristics for various gases are demonstrated. We are far from the goal of being able to list all available gas sensor types; we only try to give a picture of the wide variety of sensing approaches and their application possibilities.

The various sensing principles are briefly explained as below.

- *Infrared spectroscopy:* The measuring principle is based on the molecule-specific absorption of bands of infrared radiation. The absorbed wavelengths are characteristic to the individual gases, but may partially overlap. This results in cross-sensitivities which can be reduced by the application of gas-specific filtering. Many types of multichannel gas analyzers operate according to the nondispersive IR (NDIR) two-beam alternating light principle and measure gases highly selectively whose absorption bands lie in the infrared wavelength range, such as CO, CO_2, NO, SO_2, NH_3, H_2O, CH_4, and other hydrocarbons.
- *Raman spectroscopy* has been used to monitor chlorosilane and phosphorus trichloride synthesis.
- The *Laser diode gas sensor* measuring principle is also based on the specific light absorption of different gas components. A diode laser emits a beam of near-infrared light, which is sent through the process gas and detected by a receiver unit. The wavelength of the laser diode output is tuned to a gas-specific absorption line.
- *UV absorption* can be used for the determination of ozone; it has an absorption maximum at the wavelength of 254 nm.
- *UV fluorescence* detectors can be used for sulfur dioxide (SO_2, ppb) determinations.
- *Chemiluminescent* detectors for oxides of nitrogen (NO_x, ppm, and ppb).
- *Flame ionization* detectors can measure the total hydrocarbons in a gas stream. After ionizing the hydrocarbons in a burning hydrogen flame, the carbon ions are attracted to a charged grid and the resulting pulses are counted.
- *Photo ionization detectors (PIDs)* measure VOCs and other toxic gases in low concentrations of 0.1–2000 ppm. A PID is a very sensitive broad-spectrum monitor, not unlike a low-level LEL monitor based generally on pellistors. A PID uses an ultraviolet (UV) light source to break down chemicals to positive and negative ions that can easily be measured with a detector.
- *Thermal conductivity detectors* can be used to measure concentrations of gas components if their thermal conductivities differ significantly from those of the residual gases. The measuring principle is based on the different thermal conductivities of the components. Some gas analyzers use this principle for quantitative determination of H_2 or He in binary or quasi-binary gas mixtures (Dräger).

- *Paramagnetic sensors:* In contrast to almost all other gases, oxygen is paramagnetic. This property is utilized as the measuring principle in oxygen gas analyzers. Oxygen molecules in an inhomogeneous magnetic field are drawn in the direction of increased field strength due to their paramagnetism.
- *Catalytic sensors* are based on the thermal heat-of-reaction principle. The flammable gas or vapor to be measured diffuses into the sensor onto a heated detector element (catalytic-based) and is catalytically burned. The required oxygen to support this burning is provided by the oxygen in the ambient air which diffuses into the sensor. As a result of this burning, the detector element is heated. Recently, an interesting micro-differential scanning calorimeter for detecting combustible gases has been fabricated using a CMOS-compatible process followed by post-CMOS etching (Cavicchi et al. 2004). The device uses a thermopile to measure the temperature difference between a sample and a reference zone, analogous to differential thermal analysis.
- *Photo-acoustic gas detection* is based on the measurement of the pressure variation generated in a cavity by selective absorption of IR radiation. In newer configurations, the photo-acoustic cell is prefilled with the actual target gas and sealed during manufacture. When modulated IR light is passed into the absorption cell, a photo-acoustic pressure signal is generated. If the target gas is introduced in the absorption path outside the cell, a reduction of the cell pressure signal is observed. This reduction is nearly proportional to the concentration of target gas in the absorption path.

The above methods are characteristic for relatively high-cost analyzer appliances that can be used in industrial, environmental, and analytical applications. The sensor elements are generally physical sensors with relatively long lifetimes. On the other hand, micro-system technologies today allow the design of electrochemical cells:

- *Electrochemical cells (based on liquid or solid-state electrolytes) and*
- *Semiconductor-based solid-state sensors (metal oxide semiconductors or silicon-based IC-compatible versions)*

The basic principles of the two latter categories have been discussed in previous chapters.

Explosion-proof detection of combustible gases is of crucial importance in environments where the risk of exceeding the lower explosive limit is possible. A wide range of applications in oil and chemical industries requires such arrays of gas sensors with minimized power consumption. Pellistors operate on the calorimetric transduction principle by detecting the heat generated during catalytic oxidation. They measure gas concentrations up to several volume percent without a significant effect of relative humidity on the response. Therefore, they seem to be ideally suited for the given purpose. A new type of integrated micro-pellistor array for hydrocarbon detection (methane, propane, butane, and hexane) was demonstrated by Kulinyi et al. (2004). In their device, manufactured by one-sided porous silicon micromachining, Pt microfilaments embedded in nonstoichiometric silicon nitride are suspended over a cavity. The thermal isolation gap is realized by selective dissolution of the electrochemically formed porous silicon beneath the hotplates. The size of the hotplates is determined by the targeted power consumption (<20 mW at 500°C for a single element) and the thermal properties of the construction materials. A porous matrix containing finely dispersed Pt catalyst is formed on the top of the micro-heater. The dropped material is dried and baked at 600°C by self heating of the micro-hotplate. A pair of

Table 6.4. Examples of commercially available gas monitoring appliances and modules: sensing principles and characteristics for various gases

Analyte	Sensor principle	Detection range	Operating temperature range (°C)	Cross-sensitivities	Sources (company)
O_2	Electrochemical	0–100%	0–50	—	Gellert
	Amperometric electrochemical	0–100%	-40 to +65	Reducing gases as disturbances	Dräger
	Paramagnetic alternating pressure	0–100%	0–45/50	NO, NO_2	Siemens Panametrics
	Fuel cell electrochemical	0–100%	+5 to +45	—	Siemens
O_3	Three-electrode electrochemical	0–1 ppm	-40 to +65	Minimal	Dräger
	UV absorption	0–10 ppm	5–40	—	MLU
	Semiconductor resistor ($InSnO_2$)	0–10 ppm	Indoor conditions	—	FIS
H_2 (or He in binary mixtures)	Thermal conductivity	0–100%	+5 to +45	He	Siemens
CO_2	Nondispersive infrared (NDIR)	0–10,000 ppm	0–50	—	Gellert
	IR absorption	0–5%	-25 to +55	—	Dräger
			5–40	—	Dynament
		0–100 ppm	5–40	—	MLU
		300–2000 ppm	5–40	—	Steinel
	Semiconductor resistor (La-doped SnO_2)	400–2000 ppm	Indoor conditions	Humidity-compensated	FIS
CO	Three-electrode electrochemical	0–300 ppm	-40 to +65	Minimal for H_2 and some HCs with filter	Dräger
	Electrochemical	0–1000 ppm	-10 to +60	H_2	Figaro
	Photo-acoustic	0–100 ppm	0–40	Selective	MSA
	Semiconductor resistor	0–200 ppm	0–40	VOCs	Hanwei
NH_3	NA	0–200 ppm	0–50	—	Gellert
	Three-electrode electrochemical	0–100 ppm	-40 to +65	Methyl- and ethylamines	Dräger
NO_2	Three-electrode electrochemical	0–10 ppm	-40 to +65	Minimal for some HCs, H_2, disilane	Dräger
NO	Three-electrode electrochemical	0–50 ppm	-40 to +65	CO, H_2S, ethanol	Dräger
$NO/NO_2/NO_x$	Chemiluminescent	0–20 ppm	5–40	—	MLU
NO_x	Semiconductor resistor	0–10 ppm	Indoor conditions	—	FIS
H_2S	Three-electrode electrochemical	0–50 ppm	-40 to +65	Mercaptanes	Dräger

SO_2	Three-electrode electrochemical	0–10 ppm	−40 to +65	Minimal for some HCs, H_2S, NO_2	Dräger
Cl_2	Three-electrode electrochemical	0–50 ppm	−40 to +65	F_2, Br_2, ClO_2	Dräger
CH_4	Catalytic pellistor	1–100% LEL	−20 to +55	H_2, CO, NH_3, HCs, VOCs	Dräger
HCs	IR absorption	1–100% LEL	−20 to +60	Minimal among HCs	Dräger
	ZrO_2/Y_2O_3 electrochemical	1–100% LEL	0–40	Minimal for H_2O	Zirox
	Photo-acoustic	1–10 ppm	0–40	Selective	MSA
CO_2/HC	NDIR absorption	HC 1–100% LEL CO_2 0–5%	0–40	—	City Techn. Dynament
Total HC	Flame ionization detection	Different	+5 to +45	Low for O_2	Siemens
Flammable gases and VOCs	Catalytic pellistor	0–100%	−40 to +55	Among them	Dräger
VOCs	Photo-ionzation	0–2000 ppm	0–40	Among VOCs	Dräger RAE
Benzene, toluene, xylol	Photo-ionization	0–150 ppm	5–40	—	MLU
	Photo-acoustic	1–10 ppm	0–40	Selective	MSA
PFC (polyfluorinated compounds)	Three-electrode electrochemical	0–5 ppm	5–40	Minimal	Dräger
Multigas from CO, CO_2, NO, SO_2, NH_3, H_2O, CH_4	Nondispersive infrared (NDIR)	CO 10 ppm/30% CO_2 5 ppm/30% NO 100 ppm/30% SO_2 5 ppm/30% NH_3 100 ppm/30% H_2O 1000 ppm/3% CH_4 50 ppm/30%	0–50	Very low	Siemens
Multigas from HF, HCl, NH_3, H_2O, O_2, CO, CO_2	Laser near-infrared absorption	HF 0–2 ppm HCl 0–10 ppm NH_3 0–10 ppm H_2O 0–50 ppm O_2 0–1% CO 0–1% CO_2 0–1%	+5 to +45	Very low	Siemens
Refrigerants, hydrochlorofluorocarbons (HCFCs), hydrofluorocarbons (HFCs), chlorofluorocarbons (CFCs), and ammonia	Photo-acoustic	0–10 ppm 0–1000 ppm	0–40	Selective	MSA

an active and a passive resistor element in a Wheatstone bridge arrangement is heated up to 300–500°C with a power dissipation of 20–40 mW in the environment to be monitored. Explosion-proof detection of hydrocarbons of the sensor was demonstrated by measuring concentrations between the lower and upper explosion limits (LEL, UEL) without protective encapsulation.

Volatile organic compounds (VOCs) are widely used in residential and industrial locations and can present serious medical, environmental, and explosion dangers. VOC sensors are therefore commonly used to monitor oil drilling platforms, fuel storage tanks, air pollution sources, and other locations where VOC leaks might occur. VOC sensors are also useful in civil defence and military applications, because many chemical warfare agents and explosives can be detected with VOC sensor technology. A low-cost, low-power VOC sensor has been constructed from an array of micro-machined parallel-plate capacitors, recently (Patel et al. 2003). A micro-machined parallel-plate chemicapacitor structure was constructed that, unlike previous parallel-plate chemicapacitors, has a fixed gap width and makes it possible to realize an array of different capacitors on a single chip. These micro-machined chemicapacitors are significantly smaller than previous chemicapacitors, so more of them can be integrated on a single sensor chip. In place of a standard dielectric, the individual capacitors are filled with selectively absorbing polymers. Absorption of a target vapor alters the permittivity of the polymers and thereby changes the capacitance of the elements in the array. A variety of polymers have been used, including polyethylene-co-vinylacetate, which is sensitive to nonpolar hydrocarbons, and siloxanefluoro alcohol, which is highly sensitive to polar VOCs and chemical warfare agent simulants. The response magnitude for each element depends on a combination of different phenomena such as the dielectric constant of the analyte and polymer swelling. The measured sensitivity of the sensor to most VOCs was found to be in the low parts-per-million range. The sensor typically responded within a second but frequently required 5–10 min to reach equilibrium. The sensor has demonstrated detection of many VOCs well below the lower explosive limits and could be used in industrial leak-monitoring applications.

3.4. COMBUSTION CONTROL

The level of oxygen/CO/CO_2/hydrocarbons in industrial exhaust gases is characteristic for the combustion process and is important in emission control. Applying combustion monitoring and direct control of boiler settings, the power plants of today can gain in power-to-fuel ratio and reduce the amount of pollution generated. This monitoring requires quite expensive instrumentation today (such as instruments with a paramagnetic cell for determining the oxygen concentration, a flame ionization detector to measure the total amount of hydrocarbons (THC), and an IR detector to register the carbon monoxide concentration) and sometimes 24-hour surveillance by operators, requirements which naturally cannot be met in the many small boilers installed in local residential heat distributing facilities, such as schools and homes. A possible solution that enables more efficient combustion in small boilers might be a low-cost sensor-based system with software that automatically controls the combustion. The exhausts produced in combustion are complex mixtures of compounds, with some key components such as CO, HCs, and NO_x. The actual composition depends on the fuel types and the phase of the burning process.

Common practice today in combustion control is the open loop without feedback of information to the injection system. It is well recognized that this does not allow optimization of the process. It is therefore believed that new combustion technologies will integrate feedback controllers. Automotive engines already use closed-loop concepts (see Section 2). The adjustment of operation with sensors to monitor the exhaust gases is considered in many current applications including biomass combustion (Beedie et al. 1996) and gas turbines (Docquier et al. 2001). While they share some of features and objectives, it is convenient to divide combustion control into two categories: operating-point control (OPC) and active combustion control (ACC) (Docquier and Candel 2002). In ACC, the controller output is used to modulate the flow properties (e.g., fuel flow rate modulation) to avoid or limit pressure oscillations, or to improve the combustion characteristics. In OPC, the injection of fuel is regulated in order to maintain certain flame parameters such as the equivalence ratio (lambda) in a prescribed range of values.

Active control may have potential applications in combustion enhancement. Instability control is one major area of concern. Active control is specifically considered for ramjets, lean premixed combustors, afterburners, and segmented solid-propellant rocket motors (Docquier and Candel 2002).

Control of the operating point is part of every combustion monitoring device designed to date. From the early problem of load control to the somewhat more recent concern with reducing pollutant emissions, any engine or combustion process relies on a more or less complex control system in order to operate in an acceptable way in terms of power, safety, or emissions. Requirements to limit pollutants emissions concern the gas turbine industry similarly to automotive applications as discussed earlier. Here we concentrate on other applications. Regulations are becoming so stringent that they have often resulted in a radical redesign of combustion systems employed on high-pressure-ratio aero-derived gas turbines. Reduction of emissions is also a permanent objective for jet engine manufacturers. Important progress has been obtained with premixed combustors. It is known that oxides of nitrogen may be notably reduced by the mixing of reactants before burning. This diminishes the flame temperature and the related production of NO_x, which depends exponentially on temperature. NO_x and CO emissions follow opposite trends: Whereas low flame temperature is favorable to NO_x reduction, it rapidly prevents the complete oxidation of CO and HCs. Control of the operating point of such a combustion process has to include a trade-off in reducing these pollutants. In an ideal case, one will try to drive the system toward the optimum point where both CO and NO_x levels will be minimized.

3.4.1. Optical Methods

Figure 6.11 shows a generic combustion process with available or promising sensing techniques. Control of a combustion process relies on several elements which are essentially similar for automotive and gas turbine applications. In both cases, combustion takes place between fresh reactants supplied (at least for the fuel) by flow-metering injectors. Exhaust gases are produced and pass downstream of the combustor through a manifold. The figure illustrates the application of many type of sensors (pressure, temperature, flow velocity, mass flow, etc.), but here we concentrate on chemical sensors or on sensors for monitoring chemical quantities.

The development of near-IR absorption techniques is closely related to the progress in room-temperature, tunable, and low-cost diode lasers in the telecommunications industry. Commercially available units allow exploitation of absorption bands of species such as H_2O, NO_x, CO, or CO_2. The light is provided by diode lasers, and the detection system includes optical isolators, fibers, and photodetectors which are most of the time photodiode elements. Diode laser–based absorption sensors are

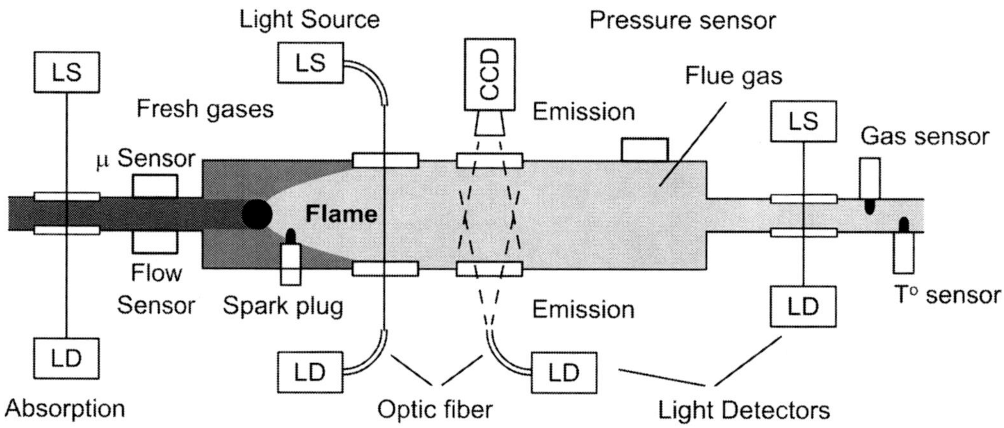

Figure 6.11. Typical application of sensors and sensing methods in a combustion control system (LS, light source; LD, light detector). (Reprinted with permission from Docquier and Candel 2002. Copyright 2002 Elsevier.)

quite promising. *In situ* measurements in the flame region or in the exhaust gases of major combustion species are well demonstrated. Although the detection of minor species such as NO_x and CO has been carried out in combustion environments, strong interferences with high-temperature water vapor limit the sensitivity (e.g., ~140 ppm for NO). Sampling the combustion gases into a low-pressure, low-temperature multipass absorption cell eliminates many of the interference problems associated with *in situ* measurements (Docquier and Candel 2002).

The light emitted by the flame may also be used to monitor and control combustion. Information may be gathered from UV to IR wavelengths. Several phenomena contribute to the light emission process. Discrete emission spectra correspond to the chemical reactions producing electronically excited radical in the flame front in the UV-visible range (chemiluminescence). Combustion products in thermal equilibrium such as H_2O, CO_2, as well as CO are also responsible for rotation-vibration emission bands in the near-IR and IR range (IR emission). Chemiluminescence sensors have already been designed for operating-point and archive control purposes. In active control applications, chemiluminescence has been used mainly to track heat-release fluctuations, but control schemes have also been developed to perform operating-point control of boilers.

3.4.2. Solid-State Sensors and Other Technologies

Combustion control applications based on solid-state gas sensors are used mainly for automotive engines, but oxygen probes are now used for continuous emission monitoring in other areas as well. While the technology developed for detection of minor species is not completely mature, it has considerable potential and in the future could find a wide range of applications in combustion optimization.

Ion current probes also have potential for combustion control applications. Although this technique seems particularly well adapted for spark-ignition engines (localized combustion process, spark plug sensing, fast cylinder-to-cylinder measurements), it could certainly be adapted to other situations provided studies are carried out to determine the best ion sensor location. Ion current signals are generally difficult to interpret, but they clearly correlate with parameters such as pressure and equivalence ratio.

A high-temperature electronic nose (HT e-nose) with 16 metal oxide sensors and 16 metal insulator silicon carbide FET (MISiCFET) devices was proposed for chimney surveillance by Petersson and Holmberg (2004). Metal oxide sensors and MISiCFET devices are known to function in harsh environments such as in combustion exhausts (Spetz et al. 2000). Furthermore, these devices are sensitive to a wide range of compounds, but by producing them with individual variations in oxides and gate metals, they can be tuned to be more or less sensitive to compounds such as CO, hydrocarbons, NO_x, H_2, and NH_3. By analyzing the signal patterns that are continuously produced by the HT e-nose array, an estimate of the actual exhaust mixture, and thereby also the actual status of the combustion, may be retrieved in order to alter boiler settings such as air flows so as to continuously optimize the fuel-to-power ratio and pollutant emissions. Principal component analysis (PCA) was used to summarize the characteristics of each data set and to visually inspect data set similarities (see Chapter 1 of this volume).

Getino et al. (1999) suggested the application of sensor arrays consisting of 16 sputtered thin-film metal oxide resistors for combustion gas monitoring. The films were deposited using SnO_2 and TiO_2 sputtering targets. Some oxide films were doped with metals such as Pd, Pt, and Al from metallic targets in an argon atmosphere. Different amounts of dopants were introduced, varying the deposition time during the sputtering process. Several pattern-recognition techniques were applied for detecting hydrocarbons, SO_2, and NO_x.

3.5. OTHER APPLICATIONS

The variety of industrial processes is very large, and the application possibilities of the various sensors are almost endless. Here some special additional examples illustrate this.

Sulfur has detrimental influences on the performance of steel when the sulfur content is beyond the specification. For ultraclean steel, the sulfur content is required to be rigidly controlled. Hence, a sensor used to monitor the sulfur content in molten steel *in situ* at high temperature is highly desired. However, no sulfur sensors are available which can be used in steelmaking, owing to the fact that the materials used at present to assemble sulfur sensors, such as $CaS(TiS_2)$, $MgS(ZrS_2)$, and ZrO_2-CaO-CaS as electrolytes, as well as Na_2S and CaS as auxiliary electrodes, are liable to be oxidized under the conditions of steelmaking. A novel electrochemical sulfur sensor using Laβ-Al_2O_3 as solid electrolyte and $La_2O_2S + La_2O_3$ as auxiliary electrode material has been developed, however (Hong et al. 2002). It can be applied not only in molten iron but in molten steel as well, where the oxygen concentration is in the range from 2.1 to 51.5 ppm, because a rather stable auxiliary material of La_2O_2S is used in the sensor. The electromotive force values of the electrochemical cell–type sensors when measuring sulfur in liquid metal containing oxygen over the range from 2.1 to 51.5 ppm have shown fairly good reproducibility and sensibility to the sulfur activities in the melt. The influence of the oxygen activity in molten metal on the lower limit of measuring sulfur activity with the sensor was limited. The lower limit for measuring sulfur activity is dependent on oxygen activity in the melt, according to $a_{[S]} = 3.84a_{[O]}$.

Cooling plants, refrigerator systems, and heat pumps have to be monitored for coolant leaks to enable alarms and precautions if certain concentration levels are exceeded. Nowadays ammonia is increasingly used as coolant as a substitute for substances with high potential for endangering the ozone layer. However, ammonia is a toxic and flammable gas, and therefore, according to many safety regulations, certain alarm levels for ammonia are specified. According to some standards in a few countries, if a concentration level of 200 ppm is exceeded, monitors have to switch on the ventilation and have to signal a pre-alarm. At concentrations above 1000 ppm, the main alarm has to be signaled and the cooling system has to be switched off. Available monitors mostly suffer from poor long-time stability or insufficient gas identification capability, and both are reasons for numerous false alarms. Ammonia sensors are also discussed in Section 5.1 in connection with agricultural applications, where the reliability is less critical because of the nature of the application—a late alarm will not cause any human health problems. Workplace safety applications need highly reliable detection, without much potential for false alarms.

A prototype of an intelligent field-suitable ammonia leakage monitor was developed which may overcome this drawback (Jerger et al. 2002). The microprocessor-supported hardware operates a sensor head with three tin oxide–based gas sensors simultaneously. Gas identification and concentration determination of such gas sensors could be considerably improved by periodic variation of the working temperature of the sensor elements. Simultaneous sampling of the sensitive-layer conductance and application of a rather simple new algorithm for analysis of the resulting conductance–time profiles was applied when heating and cooling the sensor elements between 100 and 400°C according to linear temperature–time profiles. These curves are very reproducible and represent the molecular specific temperature behavior of the conductance. A simple and quite economic numerical procedure was applied for the analysis that was found to be capable meeting all the requirements of ammonia identification. A remarkably good accuracy of ammonia monitoring at various concentrations (in the range from 100 ppb to 1000 ppm) and humidity values (from 500 to 2500 Pa partial pressure) and the excellent capability of ammonia identification without recalibration was demonstrated by long-term test experiments over several weeks.

Nitroethane is an organic toxicant which is present in gas outbursts of plants producing nitroethers, nitrolacquers, and perfumery preparations. A multisensor system was proposed by Koremann and Kalach (2003) that provided recognition of graphic images (imprints) of nitroethane vapor. This unit operates with a high speed in the range of concentrations of nitroethane vapor from 5×10^{-3} to 7.5×10^{-2} g/m^3; it is simple to operate and is easily calibrated. The possibility of an express qualitative and quantitative analysis of the samples is attained due to the use of an algorithm for pattern recognition of nitroethane. Time instability of the frequency of piezo-quartz resonators that are applied (~10 Hz/15 min) allows determining nitroethane at levels for ambient air standards in a residential zone.

In the coming conversion from fossil fuel energy sources, safety and a clean source of hydrogen will be paramount in the utilization of hydrogen as a source of clean, abundant energy. Effective hydrogen sensors and ancillaries, able to respond accurately and quickly to hydrogen gas leaks, will be critical in the safe deployment of mobile, hydrogen-based vehicles and stationary hydrogen-derived power sources. The economics of hydrogen fuel cells are based on three large markets: automotive fuel cells, stationary or residential fuel cells, and portable fuel cells. The hydrogen gas must be kept as pure as possible to avoid degradation of the catalysts and to avoid premature performance degradation of the fuel cell system itself. All fuel cells require two types of sensors: The first type monitors the quality of the hydrogen feed gas and the second, more important sensor system, is for leak detection. As hydrogen fuel cells are an emerging technology, the codes and standards for the fuel cells and the sensors supporting the fuel cells are still in development by a number of bodies, including European and American regulators. Sensor specifications for these applications are still being written, with no definite standard in place.

Hydrogen sensors vary from other gaseous sensors in that they must detect over the general level of ambient hydrogen levels, and in various environments. There will always be hydrogen in any local environment. Automobile lead–acid batteries evolve hydrogen routinely, and a sensor must be able to discriminate between low-level sources of hydrogen unrelated to fuel cell operation and that which is generated by a hydrogen leak. Commercial hydrogen leak detectors for the fuel cell market will depend on their reliability, reliable sensing, calibration and self-testing, minimal false alarms, integrated autonomous fuel cell cutoff and venting procedures, and back up and failsafe system.

Several types of chemical hydrogen sensors are available: catalytic bead, semiconductor, electrochemical, and resistive palladium alloy sensors, and hydrogen field-effect transistors. More information on each of these types is available in other chapters. A small, low-cost, durable thick-film palladium sensor developed by Oak Ridge National Laboratory and DCH Technology is under economic evaluation for the automotive market, use in personnel safety badges, and as a remote monitoring device when coupled with a radio tag (Lauf et al. 1994). Another small, low-cost, reliable, and reusable device, which needs minimal power to operate, was developed by the University of California at Irvine and Montpellier University in France (Favier et al. 2001). This device uses ultrathin palladium wires to detect the presence of hydrogen. Research with the use of fiber optic technology for sensing the presence of hydrogen is being conducted by the National Aeronautics and Space Administration, DCH Technology, and the National Renewable Energy Laboratory. A small detector specifically designed for automotive applications was engineered by RKI and UTC (Clark and Knight 2005). The device weighs less than 220 g and is intended to provide output which is then incorporated into the vehicle manufacturer's warning system. Its measurement range is 0–5000 ppm with an output signal from 0.5 to 4.5 V.

Palladium can absorb hydrogen to become PdH_x, and the storage of hydrogen increases the electrical resistance and volume of Pd materials. The use of this material is attracting growing interest as a reliable, cheap, ultracompact, and safe hydrogen sensor for use in fuel cells. Oharal et al. (2008) selected polyacrylic acid (PAA) as a polyelectrolyte and reduced the palladium ions on the PAA by using ascorbic acids in order to form a unique spherically shaped mosslike hybrid nanoparticle. They showed that the utilization of the Pd–PAA hybrid nanoparticles allowed them to build a highly sensitive hydrogen sensor that exhibited a switch response depending on volume expansion in a cyclic atmosphere exchange.

The performance of hydrogen fuel cells using polymer electrolyte membranes depends on the moisture level of the membrane. The membrane only conducts protons when it is hydrated or moist. The moisture is provided by water vapor in the incoming air or hydrogen stream, so real-time measurements of the amount of water vapor in these gas streams within a fuel cell is necessary. A VIASPACE instrument is commercially available for such purposes.

Hydrogen generation can come from a variety of methods, including the reforming alcohols (e.g., methanol). For example, Integrated Sensing Systems Inc. (ISSYS) has developed small methanol concentration sensors for the direct methanol fuel cell market (www.mems-issys.com/fuelcell.shtml). The sensors are also intended to be used in other applications where a low-cost, accurate, small, inline density or chemical concentration sensor is required. ISSYS general fuel cell sensing technology can be used with: direct methanol fuel cells, reformed methanol fuel cells, ethanol–water fuel cells, and ethylene glycol–water fuel cells. Direct methanol fuel cells benefit when they employ a methanol concentration sensor; efficiency is improved when a constant methanol concentration can be maintained over time and temperature.

4. FIRE ALARM SYSTEMS

4.1. SMOKE DETECTION

Although particle sensors are in practice physical sensors, their application as smoke sensors puts them with fire alarm systems including industrial and residential building safety uses and in other applications. For this reason, a short description is given here about this topic for comparison with chemical sensors.

Smoke is the totality of the airborne visible or invisible particles of combustion. Smoke detectors are designed to sense the presence of particles, but depending on the sensing technology and other design factors, different detectors respond to different types of particles. An area smoke detector is a device that will detect visible or invisible particles of combustion. It is also called a spot-type detector. A duct smoke detector is a device located within the duct, protruding into the duct, or located outside the duct that detects visible or invisible particles of combustion flowing within the duct. While the primary role of area smoke sensors is early-warning detection and is part of a building's regular fire detection system, the purpose of duct smoke detection, in particular, is to prevent injury, panic, and property damage by reducing the spread (recirculation) of smoke (NFPA Standards: 90A, Installation of Air Conditioning and Ventilating Systems; 92A, Recommended Practice for Smoke Control Systems; 72, National Fire Alarm Code). Duct smoke detection can also serve to protect the air conditioning system itself from fire and smoke damage, and can be used to assist in equipment-protection applications, for example, in the ventilation/exhaust duct work of mainframe computers and tape drives. Three main principles are used in smoke detectors that are widely available commercially.

- Ionization smoke detectors use the principle of ion flow within a chamber to detect visible and invisible particles of combustion within a size range normally encountered as a result of fire. Detectors based on ionization detection technology are most responsive to the smaller, invisible, submicrometer-sized particles.
- Light beam smoke detection devices operate on the smoke obscuration principle, thus, light extinction technology, and respond to both visible and invisible particles.
- Photoelectric smoke detectors use the principle of optical detection of visible particles of combustion and are based on light scattering.

Multisensor detectors incorporate two or more sensors of different types—ionization, optical, heat, CO—and process the signals to produce one output on which the decision fire/no fire is based. The majority of multisensor detectors have microprocessors in the detector head to carry out this processing. The driving force behind the development of multisensor detectors was the perceived shortcomings of ionization detectors combined with the growing influence of the "green" lobby in a number of European countries. Ionization detectors were by far the most widely used, but their sensitivity to small-particle smoke meant that they tended to produce more false alarms than optical detectors. Optical detectors, on the other hand, although recognized in standards and in practice as good general-purpose detectors, were not sensitive enough to small-particle smoke as to be adopted without reservation as substitutes for ionization detectors. It seemed that a gap had opened up—a gap that could not be ignored, since heat detectors did not offer the degree of fire protection afforded by smoke detectors. The issue of ionization detectors was also influenced by environmentalists, who saw the radioactive foil in these detectors as a hazard which should be eliminated as soon as possible. If elimination was not immediately possible, then the storage, transportation, and use of ionization detectors should be controlled by government agencies. Detectors using radioactive foils had been developed and modified over the years with a view to eliminating erstwhile high levels of radioactivity and had attained the OECD standards of less than 1 μCi (microcurie) by the early 1980s. This activity, which is contained inside the detector, is less than that of a color television set. Nonetheless, because of fears of radiation, regulations by various governments on the circulation and storage of ionization detectors became increasingly restrictive, making designers and installers of fire protection equipment less willing to use them, even though their effectiveness in detecting fires was undisputed. Multisensor detectors are invariably a combination of a heat and at least one type of smoke sensor. Detectors which combine more than one smoke sensor are treated by the regulating bodies as smoke detectors having more than one type of sensor. Multisensor detectors that incorporate CO sensing as well are planned.

4.2. GAS SENSORS AS FIRE DETECTORS

Current fire detector systems based on smoke sensors can cause a relatively large number of false alarms because many types of airborne particles may lead to similar effects as smoke particles. Oppelt (1999) identified some environments where classical smoke detectors run into difficulties: smokers' offices, discotheques, sites with dust sources or hovering lint rolls, solvent vapors (formation of small droplets), steam formation, soldering (fumes), car exhaust gases (particles), automatic burglar defense systems with fog generators, and also cooking in kitchens. Therefore a technological push is being provided by new gas sensor technologies for automatic fire detection by gas sensors or by combined smoke and gas sensing with appropriate signal processing concepts for discriminating reliably a threatening fire from a nonthreatening condition. Applications were successful mainly in fossil fuel power plants, aircraft, and building safety (Kohl et al. 1995). Other application fields are less well penetrated, but now extensive tests are under way. Research issues concentrate on new sensor elements, sensor arrays, sophisticated signal processing, systems integration, regulations, and testing. Important goals of research are a lower ratio of false-positive to actual fire indications, more foolproof installations, and the development of maintenance methods.

An important advantage of the application of chemical sensors in fire detector systems is the direct detection of toxic gases that are the main cause of loss of life during fire. The majority of toxic compounds are CO, CO_2, Cl_2, SO_2, various hydrocarbons or chlorinated hydrocarbons, chemically active radicals, and others, depending on the kind of burning materials. This fact excludes the possibility of using any selective gas sensors sensitive to one of the toxic products of a fire. For the reliable detection of any fire, independent of combustible material, the gas sensors for alarm system must be nonselective and sensitive to the majority of toxic gases liberated at the fire. As is well known, the semiconductor chemoresistor gas sensors are exactly such.

Korotchenkov et al. (1999) investigated the application possibility of SnO_2 thin-film resistors as smoke-detecting gas sensors. Spray-deposited films possess smoke sensitivity (R_{air}/R_{smoke}) on the level of 15–60 depending on the origin of the smoke. In comparison with CO sensors, the smoke sensors have a wider temperature range of sensitivity with some maxima, a greater hysteresis of and a longer time of restoration of initial condition. The value of sensitivity is higher at the operating temperature $T = 120°C$. Determined exploitation gas-sensing parameters of SnO_2 gas sensors indicate that they can be used successfully as sensitive elements for fire alarm systems.

On the other hand, conventional fire-detecting systems based on temperature, optical, and smoke sensors are able to detect real fires in the "firing stage." Gas sensors could detect the presence of compounds that are characteristic in the initial stage of fires and with appropriate detection and alarm systems the break out of the fire could be prevented. Hydrogen is a basic constituent of gases produced during smoldering of most building materials—wood, textiles, plastics—as a result of pyrolysis. Later, aromatic HCs and CO are produced. When flame occurs, CO_2 is produced in an increasing amount. Experimentally verified concentrations for a reliable fire alarm should be around 0.002% for both hydrogen and CO.

A gas sensor module containing several solid-state gas sensor elements was used in power plants for detecting smoldering lignite fire (Kohl 1996). A smoldering lignite fire emits partially oxidized gaseous products, carbon monoxide, unsaturated hydrocarbons, aldehydes, organic acids, and alcohols, because there is not enough oxygen available for complete combustion. At elevated temperatures, water molecules absorbed in the lignite dust and humidity of the ambient air feed the water gas shift reaction:

$$CO + H_2O \leftrightarrow CO_2 + H_2 \tag{6.1}$$

Because of the high dilution factor of the emitted gas molecules, concentration values as low as of 4–20 ppm CO and 1–3 ppm H_2 must be detected. SnO_2-based chemically sensitive resistor sensors were used, one sensor for CO and another for H_2. A third sensor was added to discern the presence of open fires (NOx) and solvents. The H_2 sensor was doped with Pd and covered by a protective layer on its surface, impeding the access of larger molecules. It was operated at 400–450°C. The cross-sensitivity was therefore low to CH_4 and H_2O; CO was found to be low. The CO sensor was also Pd-doped, but without the protective layer, operated at 120–150°C. At that temperature the cross-sensitivities to H_2 and H_2O were both low. However, cross-sensitivity to NO_x remained. If the CO sensor is operated continuously at such a low temperature, humidity is accumulated within its pores and its sensitivity decreases monotonically. To avoid this effect, the CO sensor had to be annealed every 8 h for 1 min at 280°C for regeneration. The NO_x + solvent sensor was undoped, operated at 180–250°C with some cross-sensitivity to H_2O. High detection reliability was achieved by applying a multicriteria alarm system.

There are several locations in aircraft fuselages in which smoke detectors and other fire protection measures are installed. The forward cargo compartment is ventilated and used for the transport of animals and other goods requiring fresh air. The other compartment lacks any climatization. Since ventilation occasionally whirls up some dust, 90% of the nonfire alarms arise in the forward compartment. Generally, fire detectors measuring ionization current with a radioactive source were installed in the current aircraft generation. Later, only photoelectric detectors were chosen. These rugged detectors use light scattering and a special algorithm able to discriminate particle sizes and to compensate for drift (as described in the previous section). Today the main problem with the installed photoelectrical detectors, based on light diffraction by particles, is the false alarm rate in the cargo compartment caused by dust and moisture, e.g., from perishables or livestock. The consequences of a fire alarm are the activation of fire-extinguishing halon systems, "land as soon as possible," eventual aircraft evacuation, and the "aircraft on ground" case because the halon bottles have to be recharged. An approach to new fire detectors within the pressurized fuselage area is to monitor fire parameters other than only smoke. Besides temperature, radiation, and visual inspection with cameras, gas sensors were considered and investigated (Freiling and Schmötzer 2000). One example is the specific detection by gas sensors of smoldering fires in equipment bays from overheated cables of electronic components. In early stages no particles are formed, but the insulation is already decomposing to detectable gaseous products.

A combined CO_2/CO sensor is of interest for aeronautic fire safety applications. The detection of CO_2 is, like the detection of oxygen, based on the use of a solid electrolyte. A significant improvement will be the use of Nasicon (sodium super ionic conductor) as the solid electrolyte. Nasicon is an ionic conductor which is sensitive for CO_2 detection. The preparation of this material is performed using a sol-gel technique (Hunter et al. 1998).

5. AGRICULTURAL APPLICATIONS

5.1. *AMMONIA CONTROL IN ANIMAL FACILITIES*

Ammonia is a common substance that plays an important role in the nitrogen cycle. Recently, agricultural ammonia emission has become one of the major worldwide air pollution problems and has attracted more and more attention. It is believed that excessive emissions of NH_3 from agriculture to the atmosphere have caused direct and indirect damage to the ecosystem in some regions with intensive animal production. Experts estimated that 80–95% of the total NH_3 emission in some areas originates from agricultural practices. Others estimates also indicated that ammonia emission from animal excreta contributed over 80% and emissions from use of fertilizers contributed less than 20% of the total NH_3 emission. High concentrations of NH_3 inside the animal houses also represent potential health hazards to humans and the animals themselves; chronic respiratory diseases of swine production facility workers have been attributed to dust and NH_3. Animal respiratory diseases, such as sneezing, coughing, or pneumonia, increased when NH_3 concentrations were 20–40 ppm as compared with 5–15 ppm. On the other hand, the TLV (threshold limit value) and TWA (time-weighted average maximum) are both 25 ppm, while the STEL (short-term exposure limit) value is 35 ppm for ammonia, both relatively small concentrations for detection.

Although studies of agricultural NH_3 have increased in recent years, reliable field measurements of NH_3 at animal facilities (animal houses, and manure storage and treatment) are still a major need. Understanding and control of NH_3 at animal facilities depend on sampling/measurement techniques, including devices, instruments, and procedures. The most desirable method should be based on the long-term operation of reliable sensor devices.

A large number of methods based on classical chemical analysis have been applied in animal houses (Ni and Heber 2001). The "wet methods" rely on collecting gaseous NH_3 into a suitable acid solution (acid trap, or scrubber) and then performing concentration determination in the laboratory. The volume of air passed through the scrubber is recorded and the NH_3 concentration in the air can be calculated. The most commonly used acid traps for measuring NH_3 at animal facilities include boric acid (H_3BO_3), sulfuric acid (H_2SO_4), and orthophosphoric acid. The evaluation methods include colorimetric techniques based on color changes of indicators; noncolorimetric wet-chemical techniques include acid conductimetry and titrimetry.

Spectroscopic methods are also used. In Fourier-transform infrared (FTIR) spectroscopy, the output signal is known as an interferogram, and is produced by an interferometer. As the movable mirror is gradually displaced, a cycle of maximum and minimum intensity recurs. It yields specific information about the chemical structure of organic and inorganic compounds based on the unique vibration modes of different chemical bonds (Johnston 1991).

In photo-acoustic spectroscopy (PAS), the gas to be measured is irradiated by intermittent light of preselected wavelengths. The gas molecules absorb some of the light energy and converts it into an acoustic signal, which is detected by a microphone. In the ultraviolet differential optical absorption spectroscopy (UV DOAS) method, an emitter–receiver set creates a light path in a detection zone. Light is generated by a xenon lamp in the emitter and projected to the receiver. Each gas in the detection path absorbs different parts of the light spectrum in a unique way. The absorption is recorded using a spectroscope (Ni and Heber 2001).

Electrochemical sensors consist of a minimum of two electrodes and detect NH_3 with the following electrochemical reactions (Hoy and Willig 1994):

$$2NH_3 \rightarrow N_2 + 6H^+ + 6e^- \quad \text{on the measuring electrode} \tag{6.2}$$

$$(3/2)O_2 + 6H^+ + 6e^- \rightarrow 3H_2O \quad \text{and on the counter electrode} \tag{6.3}$$

Electrochemical sensors provide direct readout and continuous measurements. Several types of sensors have been tested or used at animal facilities. Their main problem is the stability of the liquid electrolyte.

The application of real solid-state electronic NH_3 sensors is a relatively new measurement method in agriculture. There are several advantages to solid-state sensors, including simplicity, low price, quick response, and automatic measurement. Several types of NH_3 sensors have been tested in animal houses. The general requirements are a response time <10 s, a detection range of 4–500 ppm, and accurate detection between 20 and 30 ppm. Thick-film semiconducting metal oxide sensors consisting of a heater element, a dielectric layer, a contact layer, and a gas-sensitive semiconducting metal oxide layer were applied. The conductivity of semiconducting metal oxide films with special additives in a certain temperature range was influenced by the presence of NH_3 gas in the surrounding atmosphere. The sensors' optimum operating temperature was around 350–400°C. Their limitations included low accuracy, drift, and interference by humidity and other gases. This is a general problem of the semiconducting oxide–based sensing resistor, especially when operating in harsh environments, as animal houses are in comparison with room conditions (Göpel and Schierbaum 1991)

Lechuga et al. (1992) reported results with Schottky barrier devices, Pt/*n*-GaAs, with discontinuous platinum films that are sensitive detectors of ammonia over a wide temperature range. When a polymer layer of polyetherimide (PEthI), an amorphous polyimide, is combined with the device, the diode sensitivity increases and also its selectivity can be improved. The diodes were fabricated on Si-doped GaAs epitaxial layers grown by molecular-beam epitaxy. The devices have a dual metallic configuration: a platinum dot is first deposited for electrical contact. Then, a thin, porous platinum layer is evaporated by electron beam over and outside the contact dot. Back-ohmic contact was made previously by Au-Ge thermal evaporation. PEthI films are applied by casting a solution of PEthI in chloroform and subsequent annealing. The response of the devices was tested by the capacitance changes of the Schottky barrier. The effect of the PEthI coating resulted in a considerable improvement in the sensitivity for ammonia.

Zhou et al. (1989) used a porous optical fiber structure with bromocresol purple indicator for detection of ammonia vapor. Operation was based on the detection of ammonium ions originating from ammonia vapor. These ions are present in the water adsorbed in the porous polymer structure. Porous polymer optical fiber can be made by heterogeneous cross-linking polymerization of methyl methacrylate and triethylene glycol dimethacrylate. After polymerization, the inert solvent can be removed, leaving an interconnected porous structure. The structure exhibits very high gas permeability and liquid impermeability, so it can be used for the detection of gases. Vapors permeating into the porous zone can produce a spectral change in transmission. For better sensitivity and selectivity, colorimetric reagents can be trapped within the polymer matrix. The observed light intensity is a function of the concentration of the analyte to be measured. The ammonia sensor produced in this way had a detection limit 50 ppm.

A highly sensitive and selective ammonia sensor for agricultural monitoring and workplace safety applications was reported by Harsanyi et al. (1999). Electroconducting polymers (ECPs) have been found to sense gases and vapors by monitoring the change in conductance upon exposure of the polymer to the gas sample. This type of gas sensor should operate at room temperatures or close to room temperatures. Their selectivity behavior, sensitivity, and temperature dependency seem to be favorable in comparison with inorganic material–based gas-sensitive resistors. A cost-effective polyaniline (PANi)–based ammonia gas sensor was designed by Harsanyi et al. (1999). The sensing polymer film was electrochemically deposited on thin- or thick-film electrodes. The isolation gap between the electrodes was formed by laser engraving, which resulted in a gap size of 10–30 μm. The polymer film had to grow through this gap to form a resistor element between the separated electrodes. The electrochemical deposition and subsequent doping–dedoping of the resistor film was performed until a desired resistance of 1–10 kΩ was achieved. Deposition parameters were optimized to achieve the best selectivity and temperature dependence at the same time. The sensors were investigated in air containing various concentrations of ethanol, methanol, NO_x, CO, and ammonia gases. The sensitivity for ammonia was found to be extremely high compared to that for other gases. For the critical sensing concentration of 50 ppm, the sensor gave 2–5 R_{gas}/R_{air} for ammonia and practically no change for the other gases investigated. This enabled very sensitive and selective detection of ammonia even within as low a range as 20–50 ppm. This is a much higher selectivity than had previously been achieved with semiconductor-based sensors. The main advantages over inorganic semiconductor predecessors are as follows:

- High sensitivity for NH_3
- High selectivity
- No deterioration from a number of toxic gases
- Operation at room temperature (no need for heating)
- Easy measurement (resistance change)
- Small geometry, low cost

The most important parameters are summarized in Table 6.5.

5.2. SOIL QUALITY CONTROL

The quest for an ever-growing agricultural output has resulted in an uncontrolled use of fertilizers worldwide (Roblin and Barrow 2000). Additionally, intensive agriculture methods have also generated environmental problems including soil acidification, an increasing presence of toxic elements, the pollution of surface waters, and in some instances, a reduction in biodiversity. Nitrogen is an essential

Table 6.5. Most important parameters of the electroconducting polymer–based ammonia sensor

Gas to be detected	Ammonia, alarm possibility at levels from 20 to 50 ppm
Sensitivity, R_{gas}/R_0	
To 25-ppm NH_3—TLV[1]–TWA[2]	1.6–1.8
To 35-ppm NH_3—STEL[3]	1.7–2.2
To 50-ppm NH_3	2.1–2.6
No false alarms by	10,000-ppm methanol
	10,000-ppm ethanol
	100-ppm CO
	1,000-ppm hydrocarbons
No deterioration from	50-ppm NO_x
	50-ppm H_2
	50-ppm SO_2
Ambient temperature	−5 to +40°C
Operating temperature	At ambient temperature (no heating required)
TCR[4]	0 ppm/°C at 25°C
Wind velocity	0–1.5 m/s
Response time	10 s (at 50-ppm NH_3)
Dimensions	5 × 12 mm

[1]TLV, threshold limit value; [2]TWA, time-weighted average; [3]STEL, short-term exposure limit; [4]TCR, temperature dependence of resistance.

nutrient for the development of crops. It is present in soil in several forms, but mainly as ammonium and nitrate naturally bonded. To satisfy the requirements of current agricultural practices, farmers add fertilizers with high nitrate content. The term *precision farming* was developed in the United States at the start of the 1980s, and described a new philosophy that prioritizes the optimization of agricultural production processes by monitoring and controlling key agricultural parameters. These aims involve the physical and chemical characterization of soils and also the monitoring of nutrient delivery, measuring their influence on crop growth. This implies the development of new analytical procedures and new instrumentation capable of supplying the relevant information. Knowledge of the chemical and physical conditions of soil enables the optimized addition of fertilizers based on the results of the monitoring processes while minimizing their impact on the environment. The monitoring of nitrates in soil is a key parameter in the context of precision farming.

Standard methods to measure nutrient levels in soil are complex and time-consuming due to the extraction and pretreatment processes involved. In addition, the instrumentation used for these measurements is also expensive. The control of nutrients with sensors will permit an optimization of irrigation and fertilization management systems and thus will be useful for reducing the environmental impact caused by the runoff of nutrients into surface and ground waters. The use of chemical sensors based on ion-sensitive field-effect transistors (ISFETs) for soil analysis was proposed by Artigas et al. (2001). Fabrication of pH, Ca^{2+}, K^{+}, and NO_3^- ISFETs with photocurable polymeric membranes and their evaluation in aqueous solutions was reported. The purpose of their work was the development of an "in-soil" analysis system applying the ISFET-based sensors directly in soil. The response of the ISFETs had been evaluated previously in aqueous solutions to test their performance—sensitivity, selectivity, linear range, etc. Further studies by inserting ISFETs directly into the soil were performed in different types of soil for horticultural crops. Measurements of response to different ion concentrations and comparison with standard methods were also performed to establish the viability of ISFET application to soils.

In addition, a nitrate probe based on ion-selective thick-film electrode sensors was developed recently (Artigas et al. 2003). The probe incorporates three sensors, each at a different depth and featuring a copper plate as the reference electrode. The insertion of metals into the soil causes its superficial surface oxidation, resulting in metal/metal oxide electrode behavior. The probe also contains all necessary instrumentation required for processing the sensor signal and to transmit it via radio. The sensor elements consist of an ion-selective electrode structure and are produced by screen-printing of graphite–epoxy composites and PVC membranes. The ion-selective PVC membrane contains dibutylphthalate plasticizer and a special quaternary ammonium salt as nitrate ionophore. A PVC tube is used to build the probe. The tube is joined to a T on the top, and the lower part is closed using a plug. The probe contains three thick-film sensors to measure nitrates at depths of 15, 30, and 60 cm. All connections are inside the tube to protect them from wetting and are isolated with silicone. Results demonstrate the viability of thick-film sensors for agricultural applications and validate the instrumentation developed to measure the potential, store the data, and transmit them via radio. The designed calibration tubes permit the calibration of the sensors *in situ* and predict Nernstian behavior of these sensors.

Another important application field is keeping toxic elements in soil under control. Although arsenite and mercury have natural origins, humans and other organisms are normally not exposed to toxic levels of these heavy metals. However, several severe environmental accidents during recent years called attention to the possibility of a toxicity problem in soil.

Major sources of arsenic pollution are the burning of coal and industrial metal smelting. Mercury is still being released to the environment, mainly during mining and smelting of ores, burning of fossil fuels, as contaminants of fertilizers, and naturally, by weathering of rocks. Arsenic toxicity is dependent on the state of oxidation. Arsenite (3+) is 25–60 times more toxic than arsenate (5+). A change in the ratio of As^{5+} to As^{3+} can be brought about purely by an inorganic mechanism with pH changes, but the presence of microorganisms can also influence the reaction and equilibrium. All chemical compounds of mercury are toxic to humans. Methyl mercury is of particular significance because it is produced by microorganisms from Hg^{2+} in various natural environments.

Biosensors based on living cells were developed using genetically engineered microorganisms that recognize and report the presence of specific environmental pollutants. With the help of plasmids, genes that code an enzyme (e.g., luciferase) synthesis are placed into bacteria cells under the control of a promoter that recognizes the presence of the analyte. The presence of the analyte results in synthesis of the enzyme, which than catalyzes reactions resulting in the production of detectable products, such as hydrogen peroxide, or a chemiluminescent radiation in the case of luciferase.

Microbial biosensors have been used for several years to measure general toxicity, in which the receptor bacteria are nonspecifically sensing heavy metals. An example is the luminescence toxin-sensitive system, with the *lux*-gene construct residing in the soil bacterium *Pseudomonas fluorescens*. Specific biosensors, which are based on inducible promoters fused to reporter genes, are more sensitive than both chemical analysis methods and nonspecific toxicity biosensors. The biosensors are more sensitive than chemical methods, and they require less expensive equipment. However, only specific biosensors are likely to be useful for development of methods for *in situ* measurement of contaminant bioavailability. Two specific heavy-metal (arsenite and mercury)–detecting soil bacteria–based sensor strains (*P. fluorescens* OS8, pTPT11 and *P. fluorescens* OS8, pTPT31) were developed, and their applicability in biosensors for analysis of arsenite- or mercury-polluted soils was tested and compared to the traditional methods, such as cold vapor atomic absorption spectrometry (Petänen and Romantschuk 2002). The lowest mercury concentration measured using a biosensor (0.003 μg/kg) was considerably lower than by a chemical method (0.05 μg/kg). Although some of the arsenite samples showed higher variation between methods, it was concluded that bacteria can be used as an alternative to traditional methods for some types of samples. These results mean a possibility of using biosensors as low-cost detectors in place of chemical methods. Moreover, *Pseudomonas fluorescens* OS8 is, contrary to many other candidates, a soil bacterium, which will prove to be an advantage when developing direct *in situ* bioavailability measurements in natural conditions in the future.

Soil moisture is measured by sensing the dielectric constant of the soil. Soil moisture is the major contributor to dielectric constant, which can be monitored with simple capacitive sensors using, for example, interdigital structures and complex impedance spectrum analysis, like oil sensors in automotive applications (see Section 2). Using several sensor elements, the sensor probe averages moisture over its length when placed vertically or measures moisture at various specific depths when placed horizontally. The soil moisture sensor accurately monitors and controls soil moisture, which means that bioremediation systems can be maximized and the ground water can be protected. Optional salinity and temperature sensors can be added to track salt buildups or contamination inflows. Soil temperature can be monitored for plant growth studies and control of bioremediation. The so-called global soil moisture sensor has a high-level signal output for direct interface to global water control, data logging, and telemetry systems (Omni Controls Inc.).

6. LAW ENFORCEMENT APPLICATIONS

6.1. DRUG DETECTION

The market launch of various drug test and sensing systems is a response to calls from criminal prosecution authorities and other users, e.g., in the areas of rehabilitation and substitution, law enforcement, and workplace investigations, for systems to be made available for the quick, reliable, and straightforward detection of illegal drugs. The need to effectively screen passengers, baggage, and cargo at airports for a wide range of illegal substances including drugs increases daily. A variety of agencies are involved and concerned with their detection, including customs, police, transport security, immigration, airlines, airport authorities, manufacturers, and insurance companies. The legal and moral obligation to detect and prevent the movement of these substances through ports and airports also grows with each passing day. A new tool is now available for the detection of drugs (and explosives): the application of biosensors. These sensors are able to screen unambiguously for specific substances, chemicals, or materials. As such, they will greatly improve airport safety and security.

Traditional methods of screening people and baggage at airports have relied on a wide range of low- and high-tech solutions. The operational effectiveness of some of these methods and technologies is often random and can be less than reliable. These include imaging technologies (e.g., x-ray, gamma ray, pulsed fast neutron), nonimaging technologies (e.g., gamma backscatter, radioactivity detectors, metal detectors, human detectors, and dogs), and chemical identification (e.g., trace detectors, nuclear quadruple resonance, x-ray diffraction). Each of these has specific characteristics and limitations. Despite the best efforts of the agencies involved, traditional screening methods also inhibit the flow of passengers, baggage, and cargo. Thus, there is a pressing need to make the screening process more reliable, effective, efficient, and less intrusive.

The application of biosensors is a new weapon for the detection of illicit substances. It is an air sampling method, which is able to detect very low concentrations of a given substance. The concept employs the same principle as immunosensors (see Chapter 5 in this volume), in which specially tailored antibodies react with a given material or substance. The antibody is mounted on an extremely sensitive piezoelectric quartz microbalance system, for example. In the presence of the substance, the antibody attaches itself to the target molecule, leading to a weight addition on the crystal which can be measured extremely accurately by the microbalance system. The antibody is very specific—it reacts only with the substance in question and no other. The antibody is also very large in comparison with the airborne molecule, leading to a magnification effect. The application of this biosensor is essentially a three-step process: air sample collection, a concentration (or compression) phase, and the detection and analysis phase. Air sampling systems rely on the absorption of the airborne molecules. The collection system should be able to concentrate 100 L of air into 10 μl of solution of drug. The use of an adsorption filter combined with a cooling trap prevents dust and particles from entering the sensor unit. Performance can be measured in four ways: yield, sensitivity, and speed and memory effect. Existing air sampling systems typically achieve a yield of 5%. The Bio-Sensor (Advanced Aviation Technology Ltd.) has already achieved yields of over 15%, and production systems are expected to achieve 50% (www.aatl.net/publications/Bio-sensor.htm). This is a major advance over all other air sampling technologies. Up to 600 L/min can be collected, allowing detection over large volumes. Most substances can be detected if the appropriate antibody has been developed for it, including chemical and biological agents. Currently, a sensor for amphetamine has been fully developed, and sensors for cocaine, heroin, and cannabis could soon be available in commercial versions (AAT).

Immunoassay test systems are also available for the qualitative detection of amphetamines, designer amphetamines, cocaine metabolites, opiates, and cannabinoids in oral fluid samples (Dräger, Craig Medical, etc.). The technology used is based on combining selective immunochemical components with types of fluorescent particles to indicate the detection reaction. The use of these fluorescent particles and a new method of optoelectronic analysis achieve the required detection sensitivity in the range of just a few nanograms of drugs per milliliter of saliva. The system consists of two main components: the test kit for oral fluid samples and the analyzer (test evaluation reader and data management system). A core component of the immunological detection process is antibodies which are "primed" to their target substance (e.g., cocaine) and bind with the substance with a high degree of specificity. The degree of binding, i.e., the number of complexes formed by the antibodies and the target substance, depends on the concentration of the target substance. The antibodies are fixed to fluorescent particles to allow analysis of the immunological binding reaction. Drug molecules are affixed to the test strip membrane in a test zone, creating an area in which the detection reaction takes place. The cocaine-specific antibodies are coupled with fluorescent particles and deposited upstream to the test strip. As the sample is transported across the test strip, the antibodies cross the membrane and bind with the membrane-fixed drug molecules in the test zone if the sample does not contain any cocaine. If, on the other hand, the sample does contain cocaine, the antibodies react directly with the drug molecules in the sample and form an immune complex. This means that the binding pockets of the antibodies are already occupied, so no reaction can take place with the membrane-fixed drug molecules in the test zone. In other words, there is an inverse relationship between the cocaine concentration and the signal strength in the test zone. Test strips with distinct and physically separate test zones, in each of which different drug molecules are fixed, can also be realized. As a result of the immunological detection reaction, the antibodies coupled to the fluorescent-labeled particles will bind exclusively in the test zone corresponding to their drug specificity (cocaine, cannabis, amphetamine). To determine the drug concentration, the degree of complex formation needs to be quantified. To this end, the test strip in the test cassette is scanned optically with the aid of the analyzer (Polzius and Manns 2002).

The system analyzes samples of saliva, which is a faster, simpler, and more discreet method of detection than previously used techniques involving blood, urine, or hair samples. A result is available in just 15 min. The system is portable and can therefore be used directly on site. Sample taking is minimally invasive, nonembarrassing, allows precise control of the collection process, and avoids the possibility of sample adulteration. For the procedure, which lasts approximately 1 min, the subject's mouth is slightly open, and the sample is taken from under the tongue and the inside cheek surfaces. The sample preparation cartridge is placed in the test cassette and the oral fluid sample is transferred. After a brief incubation period and subsequent test development, it is placed in the analyzer. The analyzer is a handheld electronic and optical scanning instrument capable of reading, interpreting, and storing results of the saliva drug of abuse test within a 60-s operation cycle. Because urine collection is not necessary, there are certain advantages to using such a system for on-site drug screening. Additionally, because the test subject is present for the entire sample collection and testing process, the possibility of sample adulteration and/or substitution is eliminated. The detection limit for marijuana, cocaine, opiates, and methamphetamine is in the range of 10–50 ng/ml.

The development of new methods for the determination of cocaine has taken on increasing importance in recent years, due to the worldwide diffusion of this drug. Many analytical techniques may be used for detecting cocaine, such as nuclear magnetic resonance, Fourier-transform infrared spectroscopy, high-performance liquid or gas chromatography, mass spectrometry, and continuous flow immunoassay. However, many of these techniques present the disadvantages of being expensive, time-consuming, and requiring the sample to be manipulated in several steps. Recently, piezoelectric sensors (Guilbault and Schmid 1991) and classical PVC-based ion-selective membrane electrodes (Elnemma et al. 1992) have been developed and applied to cocaine determination. On the basis of the experience acquired in the development of ISFET sensors, a cocaine-sensitive ISFET device was developed by Campanella et al. (1995). The device is based on cocaine-reineckate or -tetraphenylborate complexes as exchanger materials, dispersed in a PVC matrix. The slope values of the characteristics are –53.9 and –51.3 mV/concentration decade, respectively. Selectivity coefficient values for several possible interferences have been determined, and the application of these sensors to the analysis of authentic samples containing cocaine and other substances commonly associated with it were also reported. This all shows that the proposed sensors can certainly be used to perform direct analyses on authentic samples containing cocaine hydrochloride, even when the latter has been cut with the more common local anesthetics and sugars. It is not necessary to perform any pretreatment of the sample except to dissolve it in an aqueous solution, and if necessary, to dilute the latter. Other advantages are the low cost of the sensors and of the measuring equipment. Furthermore, the small size of the latter means that it is extremely portable and thus allows tests to be carried out *in situ*. One final consideration is the rapidity and simplicity with which determinations can be made. Compared with the other ion-selective liquid or polymeric membrane electrodes, the ISFETs usually display lower values for the minimum detection limits and faster response times.

A fiber optic biosensor has been reported for rapid detection and quantification of coca alkaloids as cocaine equivalents in leaf extracts (Toppozada et al. 1997). Benzoylecgonine-fluorescein (BE-FL) was used as an optical signal generator. When cocaine is present, it competes for the monoclonal antibody (mAb) and interferes with the binding of BE-FL, hence reducing the fluorescence transmitted by the fiber. This sensor can ultimately be used for plant toxonomic identification, which can prove useful in estimates of illicit drug yields.

6.2. BREATH ANALYSIS (ALCOHOL TESTING)

Although the number of accidents caused by drunk driving is decreasing from year to year, drivers under the influence of alcohol remain one of the greatest risks in road traffic. Precise, quick, and conclusive proof of alcohol is often the only and last possibility of eliminating risks, preventing danger, and guaranteeing the greatest possible degree of safety of our roads. Breath analysis is used on all five continents, for control of alcohol at workplaces in industry, for monitoring or medical diagnosis of patients, and for traffic control by police authorities. Nowadays new approaches have emerged: the car immobilizer and personal alcohol test devices. A wide variety of appliances are commercially available (Dräger, Hanwei, Craig Medical, etc.), which can be grouped into three main categories:

- Alcohol testers for professional use, mainly for biomedical and law enforcement applications. Since severe conclusions may be drawn from the results, the appliances must operate with high reliability and accuracy within the measurement range 0.00–1.50 mg/L; 0.00–3.00‰ blood alcohol concentration (BAC), which is in close correlation with the breath alcohol concentration if exhalation is coming from the deep lung areas. Testers generally have a flow-through system in which the investigated person must apply a forced blow: The appropriate flow rate and flow time is indicated by LEDs on the appliances, and a gas sample is taken with a pump system for analytical purposes. The analysis is performed by ethanol sensors. In professional appliances, highly reliable and selective electrochemical and fuel cell types of sensor elements are used. Roadside breath alcohol measurements on individuals are almost always performed under stressful conditions. Law enforcement officers performing the test must be prepared for unpredictable, insecure, and sometimes aggressive subjects. At the same time, they are under time pressure, because of the traffic. Equipment which is easy to use, with a short reaction time, and that delivers 100% reliable data is a must. Professional appliances have been adapted to determine breath alcohol concentrations in medical applications as well. They may be used, for example, to diagnose vascular injuries (among others, in urology during transurethral resections of the prostate) after previous application of an ethanol-containing rinsing solution. They can also be used to test patients for previous alcohol consumption at the emergency reception area or before methadone substitution.
- A second type is a breath-alcohol measuring instrument with a vehicle "immobilizer." New automobiles and other vehicles are being equipped with such tools, which will soon be a worldwide practice . After taking a breath alcohol measurement, the appliance prohibits a driver who has consumed alcohol from starting the engine. Before starting a vehicle equipped with such a tool, the driver has to provide a breath sample. The tester measures the breath alcohol concentration and stores the measured value, the starting attempt, and any indication of tampering or violations of the system in the data memory. The vehicle can only be started when the measured breath alcohol concentration lies below a preset limit. Since the consequences are not so direct and severe, accuracy and measurement speed parameters are not as critical as in the previous case. On the other hand, a causeless start block may result in anger and lack of attention while driving. Thus, reliability remains an important issue.
- Personal alcohol detectors measure the blood alcohol level anytime people need to know if they are crossing the line between safe and not, and act as a reminder to avoid any possibly dangerous action. Compact, lightweight, and easy to operate, they can be used over and over again without a mouthpiece and associated hygiene problems. And they are quick: A digital reading shows within seconds after blowing into the mouth vent. Reliability and accuracy are not very critical, but low cost, low power consumption, and pocket size are. Low-cost hot-wire or semiconductor oxide alcohol sensor elements are used, which can detect breath alcohol concentration with limited accuracy and selectivity.

The development of a convenient, low-cost, and reliable method for detection of alcohol has always been in great demand in the biotechnology and chemical industries, as well as in law enforcement, as indicated above. Titanium dioxide (TiO_2) is a transition-metal oxide with a variety of applications, including great importance in the field of gas sensing, based on its surface interactions with reducing or oxidizing gases, which can be enhanced with decreased grain size of the material. Therefore, nanoparticle TiO_2 thin films appear to be a promising candidate for alcohol sensing. The properties of nanosized doped and undoped TiO_2 thin films prepared using a sol-gel method were reported recently (Sberveglieri et al. 2000). Experimental results also highlighted that the highest sensor response was obtained at 400°C for methanol and at 500°C for ethanol and propanol. The different doping enhanced the response toward different gases: Nb with Pt for the sensing of ethanol and propanol, and only Nb for methanol sensing. These films are able to monitor alcohols with sensitivities at parts-per-million levels with negligible cross-sensitivity to interfering gases such as benzene, CO, and NO.

6.3. SENSORS AGAINST TERRORISM

Surface plasmon resonance (SPR) sensor systems are valuable analytical tools with a broad range of applications, including studies of biomolecular interactions and discovery of special explosive materials. The real-time analytical capabilities of SPR may be useful for environmental monitoring, including detection and identification of agents of bioterrorism and biowarfare. Currently, most commercially available SPR instrument systems are very expensive and not portable. Compact, inexpensive sensor systems capable of rapid and sensitive detection of analytes are urgently needed. The following challenges need to be met to provide such sensor systems: The size and cost need to be reduced to allow portability; the optics need to be ruggedized for field use; and temperature regulation is required to detect target analytes at varying ambient temperatures. The first three challenges can be addressed by eliminating lasers and expensive and fragile optics in SPR sensor. A small, integrated SPR module is produced by Texas Instruments, Inc., for example. The fourth challenge can be addressed by implementing temperature regulation of the sensor modules and introducing a reference channel which compensates for nonspecific binding, bulk changes in refractive index, as well as small fluctuations in temperature (Naimushin et al. 2003). A temperature-controlled enclosure maintains the sensor at the optimal binding temperature, independent of variations in ambient temperature. The inexpensive, temperature-regulated, portable SPR sensor system described by Naimushin et al. (2003) provides an excellent system for field use in monitoring, detection, and identification of biological agents, drugs, as well as environmental pollutants such as explosive materials used by terrorist groups. It also serves as an affordable, rugged, and reliable general laboratory system for the characterization of intermolecular interactions. While the reference channel alone of the two-channel SPR sensor provides only marginal compensation for large temperature fluctuations outside a controlled laboratory environment, enclosing the sensor in the metal housing and controlling the temperature with a *thermoelectric (*TE) module overcame instabilities due to fluctuations of ambient temperature.

New types of biosensors are under development as part of Sandia's (Sandia National Laboratories 2002) emerging role in defending against biothreats at home and on the battlefield. Their biosensor is a part of the U.S. antiterrorist strategy. One aspect of this is prompt and accurate detection of a biological threat. Efforts are concentrated on developing an anthrax detector using microfabricated components that can recognize the deadly biotoxin in only a few minutes. Identification of anthrax in 5 min, rather than the hours currently needed, is a critical step in alerting building occupants to flee the deadly threat. The technique involves preconcentrating airborne particles on a tiny heating element that vaporizes fatty acids within the anthrax cell walls to derive fatty acid methyl esters that are unique "fingerprints" for bacteria and are distinctive to a detector.

Fatty acids are found in all living organisms with cell membranes. Gases derived from these acids have been used to identify a number of pathogens. A computer program compares the mass of each molecule type in the gas at a particular time to known data indicative on anthrax or other pathogens. Another approach to detection makes use of ultra-thin double layers of fat molecules, which resemble and act much like soap bubbles. This dime-sized sensor has the potential to identify a number of viruses and bacteria (www.sandia.gov). A sol-gel process is used to build up a durable glasslike film for chemo-resistor purposes.

In the past few years, concern about the possibility of terrorist actions against domestic and military water supplies has become very important. A thimble-sized portion of a nerve toxin such as sarin, soman, or tabun can kill a person in minutes; a few particles can produce death in 24 h. The technology currently being used for detecting these types of agents, such as gas chromatography–mass spectroscopy and high-performance liquid chromatography, is large in instrumentation (not portable), expensive, and often requires extensive analysis procedures. Antibodies, peptides, and enzymes have been used as molecular recognition elements in biochemical sensors; however, such devices often have storage and operational stability issues that limits their success. Molecularly imprinted polymers (MIPs) have been successfully demonstrated as an alternative in pesticide applications by Jenkins and Bae (2005). The MIP materials possess selective molecular recognition properties that are complementary to the analyte in the shape and positioning of functional groups. These polymers have high selectivity and affinity constants while maintaining excellent thermal and mechanical stability. The sensors are based on imprinted polymers that have been directly polymerized onto a fiber optic probe. A luminescent lanthanide (europium) is incorporated into the polymer to act as a signal transducer. The use of lanthanide ions as spectroscopic probes of structure and content is an established technique. These sensors have been shown to provide limits of detection in the parts-per-trillion range. Detection of the analyte is based on the changes that occur in the spectrum when the pesticide or insecticide is coordinated to Eu^{3+}. Direct imprinting of chemical agents such as sarin and soman was performed, and the limits of detection calculated. The ability of these sensors to work in various water matrices including tap, deionized, and reverse-osmosis water with residual chlorine was also evaluated. Typical detection limits for these MIP sensors are approximately 50 ppt with wide linear dynamic ranges (ppt–ppm). Sensor response time is approximately 15 min.

7. AIR AND SPACE APPLICATIONS

The needs of space and aeronautic applications require the development of sensors with capabilities beyond those of commercial sensors. These requirements include operation in harsh environments, high sensitivity, and minimal size, weight, and power consumption. Areas of interest include launch-vehicle safety monitoring, emission monitoring, and fire detection. Sensor technology is being developed to address these requirements using fiber optics, microfabrication/micromachining technology, as well as high-temperature semiconductor materials such as SiC.

7.1. DETECTING HYDROGEN LEAKAGE

In the field of aerospace applications, rocket engines are cryotechnic ones which use liquid hydrogen as combustible and liquid oxygen as combustive. Due to its extremely low temperature, liquid hydrogen induces a contraction of materials in contact with it, which, combined with its very low viscosity, leads to the potential for leaks. In launch-vehicle safety applications, detection of low concentrations of hydrogen at potentially low temperatures is important for applications involved with, for example, operation of the Space Shuttle. In 1990, leaks on the Space Shuttle while it was on the launch pad temporarily grounded the fleet until the leak source could be identified. The method of leak detection used was a mass spectrometer connected to an array of sampling tubes placed throughout the region of interest. Although it was able to detect hydrogen in a variety of ambient environments, the mass spectrometer had a delay time associated with its detection of a leak, and pinpointing the exact location of the leak was problematic (Hunter et al. 1998).

One component of hydrogen sensor technology development is silicon-based Schottky diodes using Pd alloys as the hydrogen-sensitive metal. The first generation of these sensors used palladium silver (PdAg). Palladium silver has advantages over Pd: It is more resistant to damage from exposure to high hydrogen concentrations, and it has faster response times than pure Pd. Although the PdAg sensor showed excellent properties for a number of applications, at higher temperatures and higher hydrogen concentrations (exposed to 100% hydrogen as in the Space Shuttle application), its calibration changed and the sensor occasionally failed (Hunter et al. 1994). This led to the development of the next generation of sensors, which use PdCr as the hydrogen-sensitive alloy The PdCr sensor is much more stable than the PdAg sensor after exposure to 100% hydrogen at 100°C.

An optical fiber sensor has been developed for the detection of hydrogen leaks on the cryotechnic engine of the European rocket Ariane V (Bévenot et al. 2000). The principle of this sensor is based on the variation of reflectivity of a palladium micromirror deposited at the output end face of a multimode optical fiber. An experimental setup (see in Figure 6.12) using a monitored high-power laser diode provides optical heating of the palladium layer, which allows it to perform in the wide range of temperatures between −196°C and +23°C. Detection responses between 1% and 17% with response times shorter than 5 s were demonstrated. Fast and accurate detection of hydrogen concentrations well below the explosive limit is possible.

7.2. EMISSIONS CONTROL

The development of a new class of sensors is also necessary for monitoring emissions from aircraft engines. The ability to monitor the type and quantity of emissions being generated by an engine is important not only for controlling those emissions (as discussed in Section 3.4), but also for determining the status of the engine. However, the harsh conditions and high temperatures inherent near the reaction chamber of the engine render most solid-state sensors inoperable. A notable exception to this limitation in sensor technology is the commercially available zirconia oxygen sensor presently in use in automobile engines (see Section 2).

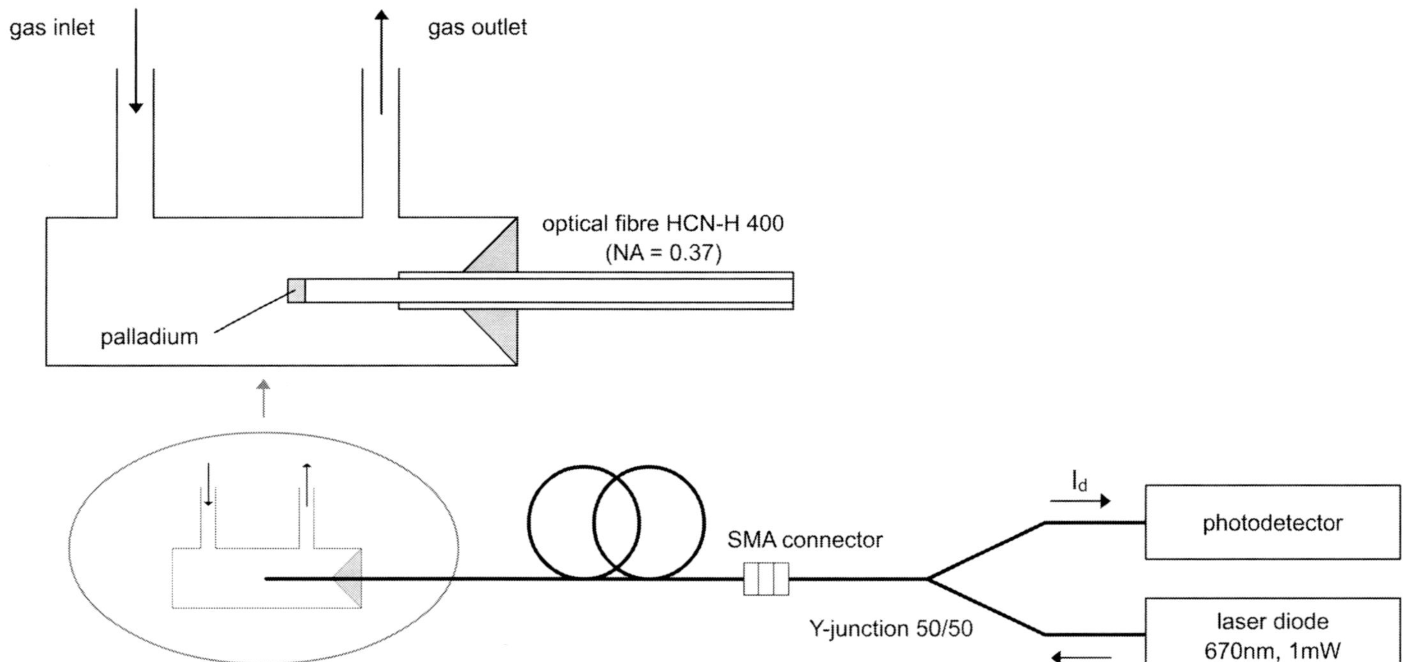

Figure 6.12. Hydrogen leak-detection system using palladium-coated optical fiber sensor. (Reprinted with permission from Bévenot et al. 2000. Copyright 2000 Elsevier.)

The development of high-temperature semiconductors, especially silicon carbide, to provide electrical components and sensors operable at high temperatures is of great interest. Using these technologies, point-contact sensors have been developed to measure hydrogen (H_2), and hydrocarbons (C_xH_y) at high temperatures (100–400°C). Pd/SiC and PdCr/SiC Schottky diodes are good candidates for hydrogen, while the Pd/SnO_2/SiC MIS (metal insulator semiconductor) diode shows good sensitivity properties for hydrocarbons (Hunter et al. 1998).

Thin heated and catalytically doped SnO_2 resistor films deposited on Si microfabricated heaters have also been tested for NO_x and CO aircraft engine emission-sensing purposes, but their application is limited by the well-known stability problems of metal oxide semiconductor resistors. The latter may be overcome with the application of nanostructured SnO_2 films in the near future.

7.3. SPACECRAFT AIR QUALITY

The ability to monitor air contaminants in a closed environment, such as the Space Shuttle, the International Space Station, and future human missions, is important to assure the health and safety of astronauts and equipment integrity. Postmission analyses of "grab" air samples from the shuttle have confirmed the occasional presence of on-board volatile organic contaminants. The spacecraft maximum allowable concentration (SMAC) was established by NASA as a guideline to maintain the air quality in spacecraft. To assure compliance to SMAC, air monitoring must be performed, and specific compound identification is required. A light, low-power, miniature instrument capable of reliably identifying abnormalities and contaminants at trace levels in specific operating environments would fulfil these needs. Continuous air monitoring could provide notification of adverse events such as spills or leaks. Continuous air monitoring could also provide notification of an impending fire. Fire in a closed environment such as a spacecraft can be disastrous, and electrical fire is a major concern.

A large number of commercially available "electronic noses," including KAMINA (Karlsruhe Research Center—38 MOS), i-Pen2 (Air Sense—7 MOS), SamDetect (Daimler-Benz Aerospace—5 MOS), Cyranose (Cyrano Sciences—32 CP), Jet Propulsion Laboratory (JPL—32 CP), Early Fire Detection System (Marconi Applied Technologies—4 SAW), VaporLab (SawTek—4 SAW), were tested by Young et al. (2003) for space application purposes. Although numerous applications in the space program exist that could benefit from the unique, powerful analytical capability of the e-nose, three specific examples have been identified: a general air monitor for organic vapors in closed cabins; hypergolic fuel monitoring, and a precombustion alarm. These represent specific needs within current NASA space programs, and each presents a unique challenge for the e-nose.

A smart gas sensor array (e-nose) based on conducting polymer gas sensors and quartz microbalance sensors was constructed and tested extensively as reported by Persaud et al. (1999). The system was launched as part of the EuroMir '95 mission, was successfully operated for 6 months, and was again reactivated as part of the DARA Mir mission in 1997. It was capable of monitoring changes in the air quality of the space station in real time as the cosmonauts carried out their daily duties. It successfully monitored deliberate atmospheric contamination, as well as leakage, spillage, and combustion events during the missions.

8. MILITARY APPLICATIONS

8.1. DETECTION OF LANDMINES AND EXPLOSIVE RESIDUES

The global landmine crisis is a topic of great interest among the international community. It is believed that at least 125 million unexploded landmines have been buried around the world as a result of war and terrorist activities, which render millions of acres of productive land dangerous and unusable in addition to directly threatening life. Currently, physical detection methods and dogs are used for landmine detection. Physical detection methods are less selective or less accurate. Dogs are highly effective in landmine detection but have several problems, such as extensive training costs, fatigue, behavioral variation, and changing moods, which had led to difficulty in practical monitoring of landmines. Thus, there is an urgent need for the development of highly sensitive, innovative, and user-friendly analytical methodologies capable of on-site detection of landmines. Recently, extensive research has been actively pursued in the development of chemical sensors for the detection of 2,4,6-trinitrotoluene (TNT) and the degraded products of landmines in environmental samples. Despite a number of sensors demonstrated for detection of nitroaromatic compounds based on different analytical techniques, immunosensors seem to provide a great potential for sensitive and selective detection for on-site application. With the aim of developing new immunosensors with highly improved analytical characteristics for landmine detection, Shankaran et al. (2004) carried out initial studies with 2,4,6-trinitrophenol (TNP), a nitroaromatic compound having structural resemblance to TNT. As mentioned above, the surface plasmon resonance (SPR)–based immunosensors have been receiving considerable attention in recent years due to easy and efficient recognition of antibodies (antigen–antibody interaction), which provides both sensitivity and specificity for identification and quantification of TNT and other nitroaromatic explosive compounds in a variety of environmental samples. SPR detects changes in the refractive index at the sensor surface due to binding of an analyte to an immobilized ligand (antigen–antibody interaction) and therefore is generally more sensitive to large molecules. Since most of the nitroaromatic explosive compounds are small molecules, the increase in the refractive index produced by the adsorption of these molecules may not be sufficient for detection by direct SPR sensors. Hence, it is preferred to use an indirect method (competitive assay) for detection of small molecules. Shankaran et al. (2004) demonstrated a SPR immunosensor for the detection of TNP based on indirect competitive immunoreactions using TNP–BSA (BSA = bovine serum albumin) conjugate and anti-TNP antibody. In the indirect competitive immunoassay, the analyte is incubated together with antibody and the mixture is passed over a conjugate immobilized surface for binding of free antibody. The proposed immunosensor showed very high sensitivity up to the parts-per-million level for the detection of TNP and has promising analytical characteristics, which can be extended for the development of immunosensors for on-site landmine detection.

A portable fiber optic biosensor has been reported and employed for the on-site analysis of explosive trinitrotoluene (TNT) in ground water (Bakaltcheva et al. 1998). Over the years, ground water at military bases has been contaminated with explosives, such as TNT. Proper documented remediation and clean-up of a site are necessary prior to base closure. In order to save cost and time for clean-up, on-site analysis is necessary. The fiber optic biosensor employs fluorescent evanescent wave sensing with antibodies used as recognition molecules. The sensor is capable of monitoring four probes simultaneously with a detection limit of TNT of 20 μg/L.

The TNT-based landmines also contain 2,4-DNT as a manufacturing impurity, which forms more vapor in the vicinity than TNT itself, and hence, due to its high vapor pressure, forms a distinctive "chemical signature" indicative of a landmine. The concentration of these compounds in the air over landmines is extremely low (sub-ppb level). A polymer (carbowax)–coated surface acoustic wave (SAW) sensor for the detection of buried/hidden explosives based on this compound was suggested and tested by Kannan et al. in 2004. Response was observed during this study in the range of 0.56–1.1 Hz/ppb of 2,4-DNT under laboratory conditions.

The presence of explosive residues in water, mostly caused by the use and disposal of explosive substances for the past 100 years both as expended ammunition and as waste from the manufacture of military ordinance, represents a significant environmental problem. Also, increasing antiterrorism surveillance activity has made the detection of TNT in marine environments a high priority. A fast, simple, sensitive, reliable, and cost-effective *in situ* remote sensing system for detecting TNT in the field (in both marine environments and in other natural waters) is in high demand. A mobile remote underwater electrochemical sensing system was described by Fu et al. (2005) for detecting TNT in marine environments. The detection technology uses a carbon fiber as the working electrode and applies a rapid and sensitive square-wave voltammogram for scanning. This submersible, pulsed-voltammetric, three-electrode electrochemical sensor is mounted to a remotely operated surface vehicle with visual detection capability. The use of vision enhances the sensor package by allowing detection of and placement below detection limits. A wireless communication unit provides the ability to monitor and detect in real time. The detection scheme is based on a stepwise reduction of the nitro aromatic groups on the explosive molecule first to hydroxylamines, and then conversion of the latter to amine groups. The reduction potentials provide the method selectivity for a specific explosive, and the current required per unit time determines the concentration of the targeted explosive in aqueous media. Successful field test data have been obtained which are comparable to those measured with laboratory instruments.

8.2. WARFARE AGENT DETECTION

Chemical warfare agents (CWAs) are very dangerous for health because of their colorlessness and toxicity. Therefore, fast and correct detection of CWAs is essential to protect human beings. CWAs are divided into four types: blood, nerve, vesicant, and choking agents. Since CWAs cannot be directly tested due to safety issues, simulant gases that generally have less toxicity are used for testing gas sensors because the structure of their molecules as well as their physical and chemical properties are similar to those of the CWAs. Test gases are dimethyl methyl phosphonate (DMMP), dipropylene glycol methyl ether (DPGME), acetonitrile, and dichloromethane, which are used as simulants of talbun, nitrogen mustard, HCN, and phosgene (Choi et al. 2005).

Implementation of any chemical weapons treaty depends on the existence of the necessary technology to monitor and verify compliance. A critical issue is the need for on-site inspection of factories and facilities suspected of making chemical warfare agents or precursors, and the associated requirement to detect the presence and amount of specified toxic chemicals. In contrast to many other applications, sensor units for chemical weapons treaty verification will likely be custom units produced in very small numbers, so an expensive unit acquisition cost would be acceptable. However, the issues of selectivity and sensitivity are of key importance. These compounds are complex organic molecules without a distinctive characteristic that would allow easy and rapid identification. In addition, they can be masked by, or mistaken for, other organic compounds that may be present in higher concentrations. Limits of detection on the order of 1 ppb are required (i.e., high sensitivity) together with high selectivity in order to avoid incorrect sensor response to common, legitimate chemicals, such as pesticides.

A number of candidate sensor types could potentially meet the requirements, including gas chromatograph–mass spectroscopy, ion mobility spectrometry, patch chemical reactions showing color changes, fiber optic and related methods, electrochemical sensors, acoustic-wave chemical sensors, and immunochemical assays.

The state of the art in development of chemical detectors for chemical warfare agents and for toxic industrial chemicals is toward multifunctional, low-power-consuming, and portable devices, which can be used for environmental monitoring in a fast and reliable way. Available portable-size detectors typically apply arrays of chemical sensors such as electrochemical sensors as well as surface acoustic wave sensors or resistor-type sensors that are typically suitable for some inorganic toxic gases as discussed elsewhere, but not for most of the chemical warfare agents, such as organophosphorous nerve agents or blister agents such as mustard gas. Therefore, a leading technology for detecting the widest variety of nerve agents and highly toxic industrial chemicals in a reliable way is still the sufficiently mature ion mobility spectrometry (IMS).

Conventional IMS detectors are based on a time-of-flight principle and a drift tube construction that would be difficult to miniaturize for portable devices. A small aspiration condenser type of ion mobility cell (IMCell) in combination with a metal oxide (WO_3) semiconductor cell (SCCell) sensor was applied by Utriainen et al. in 2003. The characteristic features of this ion mobility detector are that it has a direct interface with ambient air, it uses a flow-through construction, and, furthermore, different from the time-of-flight measurement, it detects the spatial distribution of ion flow. The chemical detector is a modular flow-through system containing an air inlet and dust filter, the IMCell, the SCCell, an air pump, a flow sensor and electronics for flow regulation, a humidity sensor, and an air outlet as shown in Figure 6.13.

The ionization is performed by an Am-241 5.9-MBq radiation source. A heater element is embedded to keep the cell at 30–45°C constant temperature. The flow of ions is interfered by the electric fields, which can be set individually among eight electrode strip pairs. Typical electric field strengths are 5–8 kV/m. The output of the IMCell is electric current (pA) collected on each electrode separately. The polarity of the electric field is altered sequentially over short time intervals. The SCCell is also the flow cell, where the sensor is a screen-printed ceramic WO_3 gas sensor. The two sensors are orthogonal in their response. This is supported by the fact that the IMCell provides sensitive and specific output to highly toxic chemical vapors such as organophosphonates, isocyanates, and many toxic organic acids, while the SCCell provides high sensitivity to the less toxic VOCs, such as alcohols and hydrocarbons. These features enable reliable and sensitive detection especially for highly toxic chemicals such as nerve agents and other classic chemical warfare agents. Furthermore, it was also demonstrated that the chemical recognition of toxic substances by the IMCell sensor is improved when the signal from the MOS sensor is available for the pattern recognition. Long-term stability studies were carried out, too, and the results indicate that a minimum 3000 h operational lifetime could be obtained for the whole detector system.

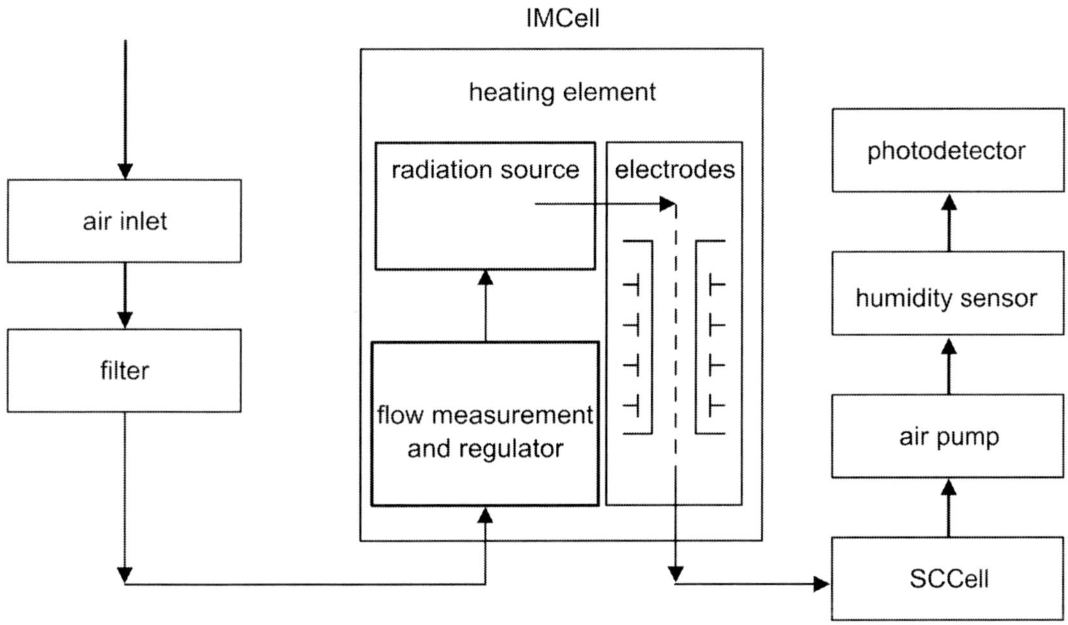

Figure 6.13. Construction of the chemical warfare detector utilizing the combination of an ion mobility cell (IMCell) and a semiconductor gas sensor cell (SCCell). (Reprinted with permission from Utriainen et al. 2003. Copyright 2003 Elsevier.)

In another approach, four different semiconductor thick-film sensors based on tin oxide were fabricated (after a selection process) and their responses to the four simulant gases (see above) were examined with different amounts of added materials, with different operating temperatures, and with different gas concentrations by Choi et al. (2005). These sensors showed sensitivities higher than 50% in 500 ppb of test gases and also good repetition behavior. PCA was adapted to the array consisting of four thick-film sensor elements to classify the object gas among the four simulant gases. Park et al. (2008) optimized the composition (transition metal–Pt, Pd, In–doped, nanosized SnO_2/TiO_2 powder mixture) and operating temperature (250°C) of the thick-film semiconductor resistor sensor for sensing dichloromethane.

9. NOMENCLATURE

AC	alternating current
ANN	artificial neural network
ATR	attenuated total reflectance
BAC	blood alcohol concentration
BE-FL	benzoylecgonine-fluorescein
BSA	bovine serum albumin
CEL	ceiling exposure limit
CMOS	complementary metal oxide semiconductor
CP	composite polymer
CWA	chemical warfare agent
DC	direct current
DOAS	differential optical absorption spectroscopy

E	activation energy
ECU	engine control unit
F	Faraday's constant
FET	field-effect transistor
FTIR	Fourier-transform infrared
GC	gas chromatography
HC	hydrocarbon
HCN	hydrogen cyanide
I	light intensity, or current
IMS	ion mobility spectrometry
IR	infrared
ISE	ion-specific electrode
ISFET	ion-selective field-effect transistor
k	Boltzmann's constant
K	Stern-Volmer constant
LCD	liquid-crystal display
LED	light-emitting diode
LEL	lower explosive limit
M	molecular weight
MAK	Maximale Arbeitsplatz-Konzentration
MDA	multiple discriminant analysis
MOSFET	metal oxide semiconductor field-effect transistor
MIP	molecularly imprinted polymers
MS	mass spectrometry
MVA	multivariate analysis
NDIR	nondispersive infrared
NIOSH	National Institute for Occupational Safety and Health
NIR	near-infrared
ORP	oxidation–reduction potential
OSHA	Occupational Safety and Health Administration
PANi	polyaniline
PARC	pattern recognition
PAS	photo-acoustic spectroscopy
PCA	principal component analysis
PCB	printed circuit board
pCO_2	partial pressure of carbon dioxide
PEthI	polyetherimide
PID	photo-ionization detector
pO_2	partial pressure of oxygen
PTC	positive temperature coefficient
PVC	poly(vinyl chloride)
R	universal gas constant
RH	relative humidity
RTD	resistance temperature detector
SAW	surface acoustic wave
SMAC	spacecraft maximum allowable concentration
SPR	surface plasmon resonance
STEL	short-term exposure limit
τ	lifetime
T	absolute temperature
THC	total amount of hydrocarbons
TLM	transmission-line model
TLV	threshold limit value
TNP	trinitrophenol
TNT	trinitrotoluene
TWA	time-weighted average
UEL	upper explosive limit
UV	ultraviolet
VOC	volatile organic compound

REFERENCES

Artigas J., Beltran A., Jimenez C., Baldi A., Mas R., Dominguez C., and Alonso J. (2001) Application of ion sensitive field effect transistor based sensors to soil analysis. *Comput. Electon. Agric.* **31**, 281–293.

Artigas J., Jimenez C., Lemos S.G., Nogueira A.R.A., Torre-Neto A., and Alonso J. (2003) Development of a screen-printed thick-film nitrate sensor based on a graphite-epoxy composite for agricultural applications. *Sens. Actuators B* **88**, 337–344.

Bakaltcheva I.B., Shriver-Lake L.C., and Ligler F.S. (1998) A fiber optic biosensor for multianalyte detection: Importance of preventing fluorophore aggregation. *Sens. Actuators B* **51**, 46–51.

Baker M.D. and Senaratne C. (1992) Trace analysis in solution using zeolite-modified electrodes. *Anal. Chem.* **64**(6), 697–700.

Beedie D., Bowen P.J., O'Doherty T., and Syred N. (1996) Cyclic variations and control of a batch-loaded biomass gasifier-combustor. *Combustion Sci. Technol.* **113–114**, 529–556.

Bévenot X., Trouillet A., Veillas C., Gagnaire H., and Clément M. (2000) Hydrogen leak detection using an optical fibre sensor for aerospace applications. *Sens. Actuators B* **67**, 57–67.

Binder J., Henning F., Hofmann T., Kalinowski T., Zacheja J., Bort M., and Lampe T. (1996) Automotive microsensor for control of the brake system and monitoring of the long-term degradation of different types and brand of brake fluids. In: Reichl H. and Heuberger A. (eds.), *Proceedings of the Microsystem Technologies 96.* VDE-Verlag, Berlin-Offenbach, 169–173.

Booth J.C., Orloff N.D., Mateu J., Janezic M., Rinehart M., and Beall J.A. (2010) Quantitative permittivity measurements of nanoliter liquid volumes in microfluidic channels to 40 GHz. *IEEE Trans. Instrum. Meas.* **59**(12), 3279–3288.

Broecker F., Ruppel W., Wuensch M., Agar D., Daudel H., Gaertner U., Koechel O., and Morsbach B. (1992) Process for Selective Catalytic Reduction of Exhaust Gases from Diesel Engines. European Patent EP 0515857.

Campanella L., Colapicchioni C., Tomassetti M., Bianco A., and Dezzi S. (1995) A new ISFET device for cocaine analysis. *Sens. Actuators B* **24–25**, 188–193.

Cavicchi R.E., Poirier G.E., Tea N.H., Afridi M., Berning D., Hefner A., Suehle J., Gaitan M., Semancik S., and Montgomery C. (2004) Micro-differential scanning calorimeter for combustible gas sensing. *Sens. Actuators B* **97**, 22–30.

Choi N.J., Kwak J.H., Lim Y.T., Bahn T.H., Yun K.Y., Kim J.C., Huh J.S., and Lee D.D. (2005) Classification of chemical warfare agents using thick film gas sensor array. *Sens. Actuators B* **108**, 298–304.

Clark T. and Knight B. (2005) Development of sensors for automotive fuel cell systems. *Final Report for DOE Sensors Contract DE-FC04-02AL67616.* United Technologies Corporation (www.lanl.gov/orgs/mpa/mpa11/FinalReportforDOESensorsContractUTRC.pdf).

Daudel H., Marquard K.H., and Gaertner U. (1994) Exhaust gas after treatment device for internal combustion engines. U.S. Patent US 5.369.956.

Docquier N. and Candel S. (2002) Combustion control and sensors: A review. *Prog. Energy Combustion Sci.* **28**, 107–150.

Docquier N., Lacas F., and Candel S. (2001) Operating point control of gas turbine combustor. In: *Proceedings of the AIAA 39th Aerospace Sciences Meeting,* Reno, NV, January, Paper 2001-0485.

Elnemma E.M., Hamada M.A., and Hassan S.S.M. (1992) Liquid and poly(vinylchloride) matrix membrane electrodes for the selective determination of cocaine in illicit powders. *Talanta* **39**, 1329–1335.

Favier F., Walter E.C., Zach M.P., Benter T., and Penner R.M. (2001) Hydrogen sensors and switches from electrodeposited palladium mesowire arrays. *Science* **293**, 2227–2231.

Francioso L., Presicce D.S., Taurino A.M., Rella R., Siciliano P., and Ficarella A. (2003) Automotive application of sol–gel TiO_2 thin film-based sensor for lambda measurement. *Sens. Actuators B* **95**, 66–72.

Freiling A. and Schmötzer K. (2000) Improving safety for aircraft: New fire detection technologies for aircraft. In: Olbert F. and Lüttgen H. (eds.), *Conference Proceedings of the Deutscher Luft- und Raumfahrtkongress 2000 der Deutschen Gesellschaft für Luft und Raumfahrt (DGLR),* Leipzig, Germany, September.

Fu X., Benson R.F., Wang J., and Fries D. (2005) Remote underwater electrochemical sensing system for detecting explosive residues in the field. *Sens. Actuators B* **106**, 296–301.

Getino J., Ares L., Robla J.I., Horrillo M.C., Sayago I., Fernandez M.J., Rodrigo J., and Gutierrez J. (1999) Environmental applications of gas sensor arrays: Combustion atmospheres and contaminated soils. *Sens. Actuators B* **59**, 249–254.

Göpel W. and Schierbaum K.D. (1991) Chemical and biochemical sensors. In: Göpel W., Jones T.A., Kleitz M.,

Lundström I., and Seiyama T. (eds.), *Sensors: A Comprehensive Survey.* VCH, Weinheim, Germany, Part 1, 119–158.

Guilbault G.G. and Schmid R.D. (1991) Biosensors for the determination of drug substances. *Biotechnol. Appl. Biochem.* **14**, 133–145.

Harsanyi G., Lepsényi I., Reichardt A., Réczey M., Dobay R., Schön A., Illyefalvi-Vitéz Z., Van den Steen J., Vervaet A., Reinert W., Urbancik J., Livovsky L., Petrikova H., Guljajev A., Visy C., Inzelt G., and Bársony I. (1999) New perspectives of selective gas sensing: Combining electroconducting polymers with thick and thin films. In: *Proceedings of the 1999 International Symposium on Microelectronics,* Chicago, IL, 207–213.

Hong H.K., Kwon C.H., Kim S.R., Yun D.H., Lee K., and Sung Y.K. (2000) Portable electronic nose system with gas sensor array and artificial neural network. *Sens. Actuators B* **66**, 49–52.

Hong Y.R., Jin C.J., Li L.S., and Sun J.L. (2002) An application of the electrochemical sulfur sensor in steelmaking. *Sens. Actuators B* **87**, 13–17.

Hoy S. and Willig R. (1994) Results of continuous measurement of ammonia in the air of pig houses, using electrochemical sensors. (Ergebnisse von Verlaufsmessungen der Ammoniak-Konzentration in Schweinestallen mit Hilfe der Sensortechnik). *Monatshefte für Veterinärmedizin* **49**, 37–41.

Hunter G.W., Chen L.Y., Neudeck P.G., Knight D., Liu C.C., Wu Q.H., Zhou H.J., Makel D., Liu M., and Rauch W.A. (1998) Chemical Gas Sensors for Aeronautic and Space Applications II. NASA/TM-1998-208504.

Hunter G.W., Liu C.C., Wu Q.H., and Neudeck P.G. (1994) Advances in hydrogen sensor technology for aerospace applications. In: *Advanced Earth-to-Orbit Propulsion Technology.* NASA, Huntsville, AL.

Jenkins A.L. and Bae S.Y. (2005) Molecularly imprinted polymers for chemical agent detection in multiple water matrices. *Anal. Chim. Acta* **542**, 32–37.

Jerger A., Kohler H., Becker F., Keller H.B., and Seifert R. (2002) New applications in oxide gas sensors II. Intelligent sensor system for reliable monitoring of ammonia leakages. *Sens. Actuators B* **81**, 301–307.

Johnston S. (1991) *Fourier Transform Infrared: A Constantly Evolving Technology.* Horwood, New York.

Ju H., Gottlieb E., Augenstein D., Brown G., and Tittmann B. (2010) An empirical method to estimate the viscosity of mineral oil by means of ultrasonic attenuation. *IEEE Trans. Ultrason. Ferroel. Freq. Control* **57**(7), 1612–1620.

Jurgen R.K. (ed.) (1999) *Automotive Electronics Handbook.* McGraw-Hill, New York.

Kannan G.K., Nimal A.T., Mittal U., Yadava R.D.S., and Kapour J.C. (2004) Adsorption studies of carbowax coated surface acoustic wave (SAW) sensor for 2,4-dinitro tolune (DNT) vapour detection. *Sens. Actuators B* **101**, 328–334.

Kim Y.S., Hwang I.S., Kim S.J., Lee C.Y., and Lee J.H. (2008) CuO nanowire gas sensors for air quality control in automotive cabin. *Sens. Actuators B* **135**, 298–303.

Kohl D. (1996) Semiconductor and calorimetric devices and arrays. In: Kress-Rogers E. (ed.), *Handbook of Biosensors and Electronic Noses: Medicine, Food and the Environment.* CRC Press, Boca Raton, FL, 533–562.

Kohl D., Kelleter J., and Petig H. (1995) Detection of smoldering fires by gas emission. In: *Proceedings of AUBE '95—10th International Conference on Automatic Fire Detection,* April 5, Duisburg, Germany, 223

Koremann Y.I. and Kalach V. (2003) Application of multi-sensor system for determination of nitroethane in the air. *Sens. Actuators B* **88**, 334–336.

Korotchenkov G., Brynzari V., and Dmitriev S. (1999) SnO_2 thin film gas sensors for fire-alarm systems. *Sens. Actuators B* **54**, 191–196.

Kulinyi S., Brandszajsz D., Amine H., Ádám M., Fürjes P., Bársony I., and Dücső C. (2004) Olfactory detection of methane, propane-butane and hexane using conventional transmitter norms. In: *Digest of Technical Papers of the 18th European Conference on Solid State Transducers, EUROSENSORS XVIII,* Rome, Italy, P1.54.

Lauf R.J., Hoffheins B.S., and Fleming P.H. (1994) Thin-film hydrogen sensor. U.S. Patent 5,367,283.

Lechuga L.M., Calle A., Golmayo D., Briones F., De'Abajo J., and De La Campa J.G. (1992) Ammonia sensitivity of Pt/GaAs Schottky barrier diodes. Improvement of the sensor with an organic layer. *Sens. Actuators B* **8**, 249–252.

Linsmeier, K.D. (1998) *Sensorsysteme für das Auto*. Verlag Moderne Industrie, Landsberg/Lech, Germany.

Meixner H., Gerblinger J., Lampe U., and Fleischer M. (1995) Thin-film gas sensors based on semiconducting metal oxides. *Sens. Actuators B* **23**, 119–125.

Moos R., Müller R., Plog C., Knezevic A., Leye H., Irion E., Braun T., Marquardt K.J., and Binder K. (2002) Selective ammonia exhaust gas sensor for automotive applications. *Sens. Actuators B* **83**, 181–189.

Morvan M., Talou T., Gaset A., and Beziau J.F. (2000) Electronic-nose systems for control quality applications in automotive industry. *Sens. Actuators B* **69**, 384–388.

Mukundan R. and Garzon F. (2004) Electrochemical sensors for energy and transportation. *The Electrochemical Society Interface,* Summer, 30–35.

Naimushin A.N., Soelberg S.D., Bartholomew D.U., Elkind J.L., and Furlong C.E. (2003) A portable surface plasmon resonance (SPR) sensor system with temperature regulation. *Sens. Actuators B* **96**, 253–260.

Nakagawa H., Okazaki S., Asakura S., Fukuda K., Akimoto H., Takahashi S., and Shigemori S. (2000) An automated car ventilation system. *Sens. Actuators B* **65**, 133–137.

Ni J.Q. and Heber A.J. (2001) Sampling and measurement of ammonia concentration at animal facilities—A review. In: *Proceedings of the 2001 ASAE Annual International Meeting* (The Society for Engineering in Agricultural, Food, and Biological Systems), Sacramento, CA, Paper 01-4090.

Oharal S., Hatakeyama Y., Umetsu M., Sato K., Naka T., and Adschiri T. (2009) Palladium–polyelectrolyte hybrid nanoparticles for hydrogen sensor in fuel cells. *J. Power Sources* **193**(1), 367–379.

Oppelt U. (1999) Auswerteverfahren für die Multisensorik für eine sicherere Meldungsentscheidung bei Bandereignissen. In: *Proceedings of AUBE '99—11th International Conference on Automatic Fire Detection,* Duisburg, Germany, March 16–18, 251–261.

Oto K., Shinobe A., Manabe M., Kakuuchi H., Yoshida Y., and Nakahara T. (2001) New semiconductor type gas sensor for air quality control in automobile cabin. *Sens. Actuators B* **77**, 525–528.

Park Y.H., Song H.K., Lee C.S., and Jee J.G. (2008) Fabrication and its characteristics of metal-loaded TiO_2/SnO_2 thick-film gas sensor for detecting dichloromethane. *J. Ind. Eng. Chem.* **14**(6), 818–823.

Patel S.V., Mlsna T.E., Fruhberger B., Klaassen E., Cemalovic S., and Baselt D.R. (2003) Chemicapacitive microsensors for volatile organic compound detection. *Sens. Actuators B* **96**, 541–553.

Patil P.A., Gorek M., Folberth M., Hartmann A., Forgang S., Fulda C., Hughes B., and Reinicke K.M. (2010) Electrical resistivity and permittivity measurements of oil base muds in the frequency range from 1 MHz to 100 MH. *SPE Drill. Complet.* **25**(3), 380–390.

Persaud K.C., Pisanelli A.M., Szyszko S., Reichl M., Horner G., Krakow W., Keding H.J., and Wessels H. (1999) A smart gas sensor for monitoring environmental changes in closed systems: Results from the MIR space station. *Sens. Actuators B* **55**, 118–126.

Petänen T. and Romantschuk M. (2002) Use of bioluminescent bacterial sensors as an alternative method for measuring heavy metals in soil extracts. *Anal. Chim. Acta* **456**, 55–61.

Petersson H. and Holmberg M. (2004) Initial studies on the possibility to use chemical sensors to monitor and control boilers. In: *Digest of Technical Papers of the 18th European Conference on Solid State Transducers, EUROSENSORS XVIII,* Rome, Italy, P1.54.

Polzius R. and Manns A. (2002) Immunoassays: Highly sensitive and selective analysis. *Dräger Rev.* **88**, 31–35.

Reavell K.S.J., Collings N., Peckham M., and Hands T. (1997) Simultaneous fast response NO and HC measurements from a spark ignition engine. In: *Factors Influencing Emission and Emission Process,* Vol. SP-1275. SAE Publications, Warrendale, PA, 121–128.

Roblin P. and Barrow D.A. (2000) Microsystems technology for remote monitoring and control in sustainable agricultural practices. *J. Environ. Monit.* **2**, 385–392.

Sandia National Laboratories (2002) Biosensors defend at home, on battlefield. *Sandia Technol.* **4**(3), 19–22.

Sberveglieri G., Comini E., Faglia G., Atashbar M.Z., and Wlodarski W. (2000) Titanium dioxide thin films prepared for alcohol microsensor applications. *Sens. Actuators B* **66**, 139–141.

Shankaran D.R., Gobi K.V., Matsumoto K., Imato T., Toko K., and Miura N. (2004) Highly sensitive surface plasmon resonance immunosensor for parts-per-trillion level detection of 2,4,6-trinitrophenol. *Sens. Actuators B* **100**, 450–454.

Schulz H, Schrader B., Quilitzsch R., and Steuer B. (2002) Quantitative analysis of various citrus oils by ATR/FT-IR and NIR-FT Raman spectroscopy. *Appl. Spectrosc.* **56**, 117–124.

Spetz A.L., Tobias P., Uneus L., Svenningstorp H., Ekedahl L.G., and Lundstrom I. (2000) High temperature catalytic metal field effect transistors for industrial applications. *Sens. Actuators B* **70**, 67–76.

Toppozada A.R., Wright J., Eldefrawi A.T., and Eldefrawi M.E. (1997) Evaluation of a fiber optic immunosensor for quantitating cocaine in coca leaf extracts. *Biosens. Bioelectron.* **12**, 113–124.

Utriainen M., Kärpänoja E., and Paakkanen H. (2003) Combining miniaturized ion mobility spectrometer and metal oxide gas sensor for the fast detection of toxic chemical vapors. *Sens. Actuators B* **93**, 17–24.

Wang J.S., Chen W.J., Yang C.S., Tsai Y.H., Wang H.H., Chen R.H., Shen J.L., and Tsai C.D. (2011) Improving stability of photoluminescence of ZnSe thin films grown by molecular beam epitaxy by incorporating Cl dopant. *Appl. Phys. Lett.* **98**, 021908.

Wu M.C. and Micheli A.L. (2004) Calorimetric hydrocarbon sensor for automotive exhaust applications. *Sens. Actuators B* **100**, 291–297.

Young R.C., Buttner W.J., Linnell B.R., and Ramesham R. (2003) Electronic nose for space program applications. *Sens. Actuators B* **93**, 7–16.

Zabler E., Berger J., Herforth A., Harder M., Arnold H., Winner H., Töpler J., Giesel R, and Denner V. (2001) *Sensoren im Kraftfahrzeug.* Bosch, Stuttgart, Germany.

Zhou Q., Kritz D., Bonell L., and Siegel G.H. Jr. (1989) Porous plastic optical fiber sensor for ammonia measurement. *Appl. Opt.* **28**, 2022–2025.

Zhu J., Ren C., Chen G., Yu C., Wu J., and Mu H. (1996) A new automotive air/fuel sensor based on TiO_2-doped Nb_2O_5 thin film by ion-beam-enhanced deposition. *Sens. Actuators B* **32**, 209–213.

Zipser L. and Frane H. (2004) Robust acoustic humidity sensor for industrial drying. In: *Proceedings of the IEEE Sensors Conference 2004*, Vienna, Austria, 107–110.

ZPlus (2008) Oil viscosity. *ZPus™ Tech. Brief #13.* ZPlus LLC (www.zddplus.com).

CHAPTER 7

Chemical Sensor Selection and Operation Guide

G. Korotcenkov

B. K. Cho

1. WHAT IS AN IDEAL CHEMICAL SENSOR?

The main purpose of any chemical sensors is to provide reliable real-time information about the chemical composition of the surrounding environment. As has been shown throughout this series, various devices based on different materials and operating on diverse principles can be used for this task (see Table 7.1).

Ideally, a device designed for chemical sensing should operate continuously and reversibly, without perturbing a sample. An ideal chemical sensor should be low-cost, portable, and fool-proof, working with perfect and instantaneous selectivity with reference to a particular target chemical substance (analyte) which is present in the surrounding environment. The device should produce a measurable signal output at any required analyte concentration (Ho et al. 2001). Chou (2000) believes that the ideal sensor should be enclosed in a small, explosion-proof housing, making it suitable for use in hazardous locations and harsh environments. For portable applications, the ideal sensor should be small and have low energy consumption. In addition, minimal special training should be required for personnel who are to manage the sensor's operation and maintenance. Chemical sensors should provide repetitive measurements of analytes for a long time in multiple or remote locations, supporting environmental monitoring. The ideal sensor should function continuously and reliably without requiring recalibration for more than 30 days at least. Such a sensor should be able to function in an industrial environment for at least 2

Table 7.1. Types of chemical sensors with corresponding sensitive materials and detection principle

Chemical sensor technology		
Type	Sensitive material	Detection principle
Conductometric (chemiresistors)	Semiconductor metal oxides, modified conducting polymers	Resistance change
Field-effect sensors (transistors, diodes)	Semiconductors, catalytic materials	Potential (work-function) change
Piezoelectric sensors (quartz crystal microbalances; SAWs)	Organic or inorganic layers	Frequency change due to mass change
Catalytic sensors (pellistors)	Catalytic materials, metals, metal oxides	Temperature change
Optical sensors	Organic or inorganic layers	Change in optical parameter (reflection, adsorption, fluorescence, etc.)
Surface plasmon	Organic films, thin metal films	Change in the angle of incident light
Fiber optic	Organic layers, metal oxides, semiconductors	Change in reflection, adsorption, refraction index
Infrared sensors		IR adsorption
Electrochemical (potentiometric, amperometric)	Solid or liquid electrolytes	Current or voltage change

years or longer, and it should be replaceable or renewable at a reasonable cost. Also, the sensor should be easy to place into a multipoint system, and either a controller or a controlled distribution system must be able to manage its functioning simply. The needed degree of quantitative reliability, such as precision and accuracy, should be provided as well.

However, no such ideal chemical sensor can be found in the sensor market today, in spite of great advances achieved over the past decades. One will never find a chemical sensor that satisfies all these requirements; there will always be some disadvantages (see Table 7.2).

This situation is a result of the complexity of chemical sensor application, which is related to the technical difficulties associated with analyte determination and with the specific nature (i.e., elemental or molecular) of the chemical substance to be analyzed (Ho et al. 2001). There are a great number of requirements due to the necessity to control and characterize different processes and environments; chemical sensors should be able to detect and recognize a huge number ($>10^6$) of known molecular substances. The sensitivity and selectivity of chemical sensing are affected by the phase, dimensional,

Table 7.2. Comparison of various types of gas sensors

	Type of gas sensor				
Parameter	Semiconductor	Catalytic combustion	Electrochemical	Thermal conductive	Infrared absorption
Sensitivity	e	g	g	b	e
Accuracy	g	g	g	g	e
Selectivity	p	b	g	b	e
Response time	e	g	p	g	p
Stability	g	g	b	g	g
Durability	g	g	p	g	e
Maintenance	e	e	g	g	p
Cost	e	e	g	g	p
Suitability for portable instruments	e	g	p	g	b

e, excellent; g, good; p, poor; b, bad.

Source: Reprinted with permission from Korotcenkov 2007. Copyright 2007 Elsevier.

and temporal aspects of the desired determination. Contaminants may be found in a variety of sample matrixes (e.g., air, water, soil), and their concentrations may vary over a wide range of magnitudes from the percent level to parts per billion or lower. The analyte may also be either persistent or transitory.

2. HOW THE FIELD OF APPLICATION INFLUENCES OUR CONCEPTION OF THE IDEAL SENSOR

No doubt, the field of application creates its own specific requirements for chemical sensors. According to Potyrailo and Mirsky (2008), high reliability, adequate long-term stability, a low false-positive rate, and good sensitivity form the priority list for industrial sensor users, whereas often the size and the cost of sensors are the least important factors. In contrast, medical users focus on the cost of disposable sensors. Specific requirements for medical *in vivo* sensors include compatibility with blood and acceptable size. Resistance to gamma radiation during sterilization, drift-free performance, and cost are the most critical specific requirements for sensors in disposable bioprocess components. The importance of continuous monitoring also differs from application to application. For instance, glucose sensing should be performed two to four times a day using home blood glucose biosensors, whereas blood-gas sensors for use in intensive care should be capable of continuous monitoring with subsecond time resolution.

Environmental monitoring along highways requires controlling specific amounts of toxic components (Yamazoe and Miura 1994). This control should be available for a long time and everywhere under conditions of constantly changing humidity and temperature. However, as a rule, sensors intended for these uses are not required to have extremely high sensitivity. Therefore, chemical sensors different in sensitivity, selectivity, or other characteristics may be applied for environmental monitoring. Stability, reliability, low cost, opportunity for wide-ranging manufacturing, and good matching with measuring chains are the main requirements to sensors to be used for environmental monitoring.

In contrast to environmental monitoring sensor units for chemical weapons treaty verification will likely be custom units produced in very small numbers (e.g., 100 units or less) (Singh 2007). However, the demands for selectivity and sensitivity are very high. For example, limits of detection of the order of 1 ppb are required together with high selectivity in order to avoid incorrect sensor response to common, legitimate chemicals, such as pesticides. This task is particularly difficult because Schedule II compounds (chemicals with limited dual use) are complex organic molecules without distinctive characteristics that might allow easy and rapid identification (Ho et al. 2001). In addition, these compounds can be masked by or mistaken for other organic compounds that may be present in higher concentrations. So, chemical sensors for chemical weapons detection may be unique devices with high cost.

Other scenarios might include stationary stations for monitoring, with permanently installed units of highly sensitive devices. In this case, high-cost sensors needing expensive recalibration may be acceptable. Thus, in real-world applications, the qualities of an ideal sensor are often weighted differently according to the application (Potyrailo and Mirsky 2008).

3. CHEMICAL SENSORS FOR VARIOUS APPLICATIONS: WHAT DETERMINES THE CHOICE?

One of the most frequently asked questions regarding sensors is, "Which sensor is best?" Of course, there is no simple answer to this question (Chou 2001). In terms of the working characteristics of chemical sensors, every type of sensor has specific advantages (see Table 7.3) for specific areas of application, and one type may compete with another type of chemical sensor.

For example, conductometric metal oxide–based gas sensors have low selectivity, noticeable drift of operating characteristics, and high sensitivity to air humidity. Metal oxide sensors can become "dormant" after several days of inactivity due to surface effects. However, in comparison with other sensors, these sensors have excellent sensitivity, very short response time, low cost, low noise, small size, acceptable power consumption for adequate battery life, and very good suitability for portable instruments (see Table 7.3). These advantages compensate for their disadvantages and open up many possibilities for application, such as sensors in alarm systems, portable instruments, electronic nose, and other specific applications, where general-purpose instruments would not be practical. Overlaid on such requirements is the prioritization imposed by the need to monitor different types of hazards in the face of finite economic resources. So, despite many limitations, conductometric gas sensors continue to hold their own in the gas sensor market and appear to have a promising future. Without doubt, in the near future we shall see more of these sensors applied to combustion and environmental monitoring, quality and process control, and safety devices.

The same conclusions can be drawn regarding electrochemical sensors. Electrochemical sensors are suitable only for low-concentration, parts-per-million ranges. In addition, their life expectancy is only

Table 7.3. Comparison of sensing technologies intended for gas control

Type of sensor	Advantages	Disadvantages
Gas chromatographs	Potential analytical accuracy. Reliable with good reproducibility. High sensitivity to vapors (ppb). Wide range of volatile and semivolatile components.	Fixed installation. Sample concentration required. Cannot use *in situ*. Carrier gas required for operation. Fairly expensive per unit. Not portable.
Portable gas chromatographs	Portable. Reliable with good reproducibility. Real-time measurement (in seconds). High sensitivity to vapors (ppb). Remote monitoring capability. No carrier gas required for operation. Wide range of volatile and semivolatile components.	Sample concentration required. Cannot use *in situ*. Fairly expensive per unit.
Ion mobility spectrometers	Potential analytical accuracy. Small packaging. Can be used to detect toxic industrial compounds and chemical warfare agents in ppb range. Integrated alarm for threshold detection. Radioactive ionizing source.	Cannot be used *in situ*. No price information.
Ion-selective FET chemical sensors	Small size. Low power consumption. High sensitivity; Wide range of analyte concentration can be detected. Robust. Fast response. Easy manipulation. Can be stored in dry state. Easy cleaning. The shape can be adapted for various applications.	Need regular calibration. Require reference electrode. Membranes are sensitive and may degrade with time. Drift is possible. Interface electronics. In some cases, miniaturization is difficult.
Catalytic bead sensors	Small size. low cost. Robust. Simple to operate. Easy to install, calibrate, and use. Long-lived with a low life-cycle cost. Fast response. Excellent repeatability. Resistant to temperature and humidity variations. Can detect a variety of flammable gases from 1000 ppm to 100% LEL. Wide operating temperature range. Acceptable for portable instruments.	Commonly not selective. Low sensitivity. Requires oxygen for detection (>10%). Oxygen content must also be monitored. Prolonged exposure to high concentrations of combustible gas may degrade sensor performance. Catalyst can become poisoned or inactive due to contamination. Need regular calibration (1 month is typical) and replacement. Require elevated temperatures for operation. Sensitivity to aromatic and halogenated hydrocarbons is questionable. Passive detection—not fail-safe. Can give false readings in gas-rich atmospheres, i.e., above the upper explosion limit (UEL). Once this sensor is exposed to the upper flammable limit (UFL), it will no longer function properly because the gas will no longer combust in air.
Chemiresistors (metal oxides, MOS)	Low cost. Robust. Small size. Versatile—able to detect a broad range of gases. High sensitivity to combustible gases in wide range of concentrations from ppb to %. Wide operating temperature range. Resistant to corrosive environments and humidity. Long operating life (2–10 years). Compact and durable. Acceptable for portable instruments.	Commonly not selective. Drift of operating characteristics is possible. Nonlinear response. Need oxygen. Sensitivity to aromatic and halogenated hydrocarbons is questionable. Need regular calibration (from 30 days to 6 months). High enough power consumption. Not fail-safe. Certain types of sensors that are not exposed to the target gas for an extended period of time may oxidize and "go to sleep," resulting in lack of response to leaks. Sensitive to water vapor and many other gases that may produce a false reading.

(*Continued on following page*)

Table 7.3. (*Continued*)

TYPE OF SENSOR	ADVANTAGES	DISADVANTAGES
Chemiresistors (polymers)	Small size. Operate at room temperatures. Low consumable power. Good sensitivity to various chemicals. Acceptable for portable instruments.	Commonly not selective. Some polymers react strongly to water vapor. Uncertain durability of polymers in subsurface environments. Need to develop robust packaging. Need regular calibration. May need preconcentrator to detect very low limits. Although reversible, signal may experience hysteresis and a shift in the baseline when exposed to chemicals.
Electrochemical sensors	Low power consumption. Low cost. Response is relatively quick and monitoring is continuous. High sensitivity (ppm range). High accuracy. Can be selective for some gases. Possibility to work in gas and liquid environments. Versatile—responds to wide range of toxic gases. Do not become poisoned. Acceptable for portable and fixed instruments.	Primary sensitivity is for toxic gases and oxygen, not VOCs. Not amenable for *in situ* applications. Membranes are sensitive and may degrade with time. Devices are not very durable and have short shelf lives; average life expectancy is 2–4 years). High concentration of detected gas dramatically reduce useful life. Subject to interfering gases such as hydrogen and vapors of alcohol. Tend to lose sensitivity quickly in a chlorine or hydrogen chloride environment. Limited at low ambient temperatures (<–40°C), Unsuitable for use in dry environments (<15% RH) over several months. Operate in narrow range of pressure (1 ± 0.1 atm). Need calibration (1–2 times per month). Drift is possible. Not fail-safe.
Solid electrolytes (zirconia-type)	Fast response. Acceptable for *in situ* oxygen control in the range from ppb to % (by volume). Suitable for high-temperature environments. Very sensitive. Used in fixed instruments.	Produce a nonlinear response that is also temperature-dependent. Require high operating temperatures (>500°C).
Surface acoustic wave sensors	Small. Low power requirement. Good sensitivity to various chemicals. Can detect chemicals in very low concentrations. Acceptable for portable instruments.	Commonly not selective. May not be able to discriminate among unknown mixtures of chemicals. Need regular calibration. Some polymers react strongly to water vapor. Uncertain durability in subsurface environments. Interface electronics.
Fiber optic sensors	Low power consumption. Can detect various chemicals at very low concentrations.	Limited ability to transmit light through the optical fiber over long distances. Some organic pollutants are not easily differentiated using UV-visible spectroscopy. Concentration range sensitivity may be limited. Sensors that use chemically sensitive coatings may degrade with time.
Optical (colorimetry)	Portable. Simple to use. Visual evidence of gas detection event. Not prone to interferences.	Limited chemical sensitivity to individual VOCs. Need actual water samples (cannot be used *in situ*). Require IR source and detector. Require visual inspection. Not amenable to long-term *in situ* applications.
Optical (infrared sensors)	Immune to contamination, poisoning, and ambiguous above the LEL. Fall- safe operation. No routine calibration. Good durability with minimal maintenance. Ability to operate in the absence of oxygen or in enriched oxygen. Ability to operate in continuous presence of gas. Acceptable for portable and fixed instruments.	Big enough size. The number of detected gases is limited. The gas to be measured must be infrared-active such as a hydrocarbon. Will not work on atomic or diatomic gases that do not absorb infrared radiation (oxygen, chlorine, hydrogen, and nitrogen). Sensors are inherently pressure-sensitive. They can be affected by humidity and water. They can be expensive. Dust

(*Continued on following page*)

Table 7.3. (*Continued*)

Type of sensor	Advantages	Disadvantages
		and dirt can coat the optics and impair response, which is a concern in *in situ* environments. A relatively large volume of gas is required for response testing. Ambient temperature of detector use is limited to 70°C. Do not perform well for multiple gas applications. IR source cannot be replaced the in the field—must be returned to factory for repair.
Optical (UV/vis spectroscopy)	Simple to construct. Require little power. Acceptable for fixed instruments.	Lack sensitivity or selectivity for direct measurements. There are insufficient examples of selective transducers. Need UV source. Fairly expensive per unit.
Optical (luminescence)	Extremely high sensitivity to small quantities of many organic analytes of interest, such as polycyclic aromatic hydrocarbons (PAHs) and other compounds found in fossil fuels.	The broadband character of luminescence spectra and the ubiquitous presence of naturally fluorescent compounds leads luminescence sensors to suffer from a critical lack of selectivity. Sensitive to temperature fluctuations and other environmental factors that quench fluorescence. Need light source. Fairly expensive per unit.
Optical (surface plasmon)		Require laser source. Expensive. Commonly not selective.
Thermal conductivity	Small size. Low cost. Robust. Simple to operate. Easy to install, calibrate, and use. Possibility to control gases with high concentration. Used in the range 0.1–100% v/v. Can be used to complement catalytic detectors when they are used above the LEL. Acceptable for portable and fixed instruments. Do not require the presence of oxygen/air to operate.	Low sensitivity. Commonly not selective. Work well only when the differences in thermal conductivity between the target gas and reference gas (air) are large.
Flame ionization	Fast response. Can be used for most hydrocarbon gases and vapors from ppm to % v/v levels. Acceptable for portable and fixed instruments but mainly fixed.	Low sensitivity. Commonly not selective. Need hydrogen to generate the flame and a clean air supply. Fairly large.
Flame temperature	Fast response. Possibility to control flammable gases from 0 to 100% LEL. Used in fixed instruments.	Low sensitivity. Commonly not selective. Detect only flammable gases. Need hydrogen to generate the flame and a clean air supply. Large size. Halons may cause false readings.
MOSFETs	Small size. Low power consumption. High sensitivity. Wide range of analyte concentration can be detected. Integrated with electronic interface circuits.	Not sensitive to a great number of gases. Odorant reaction product must penetrate gate.
Paper-tape sensors	Low cost. Simple. Easy to install and use. Good selectivity. Operate at room temperatures.	Low sensitivity. Limited range of tested gases. Affected by changes in humidity. Tapes have limited life span and must be stored properly in temperature-controlled rooms. Some tapes must be stored in a freezer. Usually, after a tape package is opened, it should be replaced within 6 weeks.

2–5 years. Moreover, depending on the application, life expectancy may be much shorter. However, the electrochemical sensors have very low power consumption, respond quickly to gas, and are not affected by humidity. Also, these sensors can be exposed to gas periodically, which maximizes sensor life. Therefore, one can conclude that electrochemical sensors are a good choice for portable instruments and alarm/dosimeter systems, including lightweight, personal monitor/alarm devices, rather than for continuous monitors.

Other advantages of electrochemical sensors are conditioned by their mechanism of sensitivity. Generally, the magnitude of the current in electrochemical sensors is controlled by how much of the target gas is oxidized at the working electrode. Therefore, sensors are usually designed so that the gas supply is limited by diffusion and thus the output from the sensor is linearly proportional to the gas concentration. A linear output allows for more precise measurement of low concentrations and much simpler calibration (only baseline and one point are needed). In many other sensor technologies, output must be linearized before it can be used. Diffusion control via changing the diffusion barrier also allows the sensor manufacturer to tailor the sensor to a particular target gas concentration range. In addition, since the diffusion barrier is primarily mechanical, calibration of electrochemical sensors tends to be more stable over time and so electrochemical sensor–based instruments require much less maintenance than some other detection technologies. In principle, the sensitivity can be calculated based on the diffusion properties of the gas path into the sensor (Warburton et al. 1998).

Thus, one can conclude that electrochemical sensors offer many advantages for various applications. However, it is also true that electrochemical sensors are not suitable for every gas. Since the detection mechanism involves the oxidation or reduction of the gas, electrochemical sensors are usually suitable only for gases which are electrochemically active. Inert gases can be detected electrochemically only indirectly, if the gas interacts with another species in the sensor that then produces a response (Pletcher et al. 1991). Sensors for carbon dioxide are an example of this approach.

Information presented in Table 7.4 allows one to evaluate advantages of one or another chemical sensor for portable applications. High sensitivity, selectivity, and robustness are desirable characteristics of any chemical sensor technology in order to meet the specifications of the sensing application. Supplemental to these basic requirements, however, portable instruments also require sensor technologies that consume low power, are compact in size, and incur low electronics overhead (Wilson et al. 2001). Control and measurement electronics overhead can play a major role in increasing the size and power consumption of the overall system beyond what is practical; in many sensing systems, electronics overhead can consume 40% or more of the total space and often contribute at least this much to the overall power consumption as well. Comparisons of CHEMFET, chemiresistor, and surface plasmon resonance (SPR) sensors indicate that, in general, no single sensor technology is ideal by itself for portable instrument design. The chemiresistor is for reasons of its simplicity perhaps the most attractive chemical sensor type for portable applications. However, it also presents significant obstacles in terms of noise, drift, aging, and sensitivity to environmental parameters. The CHEMFET does not exhibit many of the fluctuations and instabilitiers of the chemiresistor, but it also requires more complex signal processing. The SPR sensor holds great promise for miniaturizable optical sensors, since it does not require the use of a laser in a portable system (SPR probes can operate using an LED as the light source). Comparison of other sensors acceptable for portable gas sensors can be found in the review by Brown (2006).

Table 7.4. Comparison of operating parameters of chemical sensors acceptable for portable applications

Sensor technology	Parameter					
	Sensitivity	Selectivity	Robustness	Power	Size	Overhead
Chemoresistor (MOS)	Moderate	Moderate	Moderate	High	Small	Low
Chemoresistor (polymer)	Moderate	Poor	Moderate	Low	Small	Low
CHEMFET (basic)	High	Moderate	High	Moderate	Medium	Moderate
CHEMFET (ISFET)	High	Moderate	Moderate	Moderate	Large	Moderate
CHEMFET (ENFET)	High	High	Poor	Moderate	Large	Moderate
Optical: surface plasmon resonance	High	High	Moderate	Moderate	Moderate	High

Source: Reprinted with permission from Wilson et al. 2001. Copyright 2001 IEEE.

Naturally, in other areas with different technical requirements for the measuring system, preference will be given to other sensors with other sets of advantages (see Table 7.3), which depend not only on the sensor type but also on its version. This facet can be well illustrated using the example of electrochemical sensors, which can be realized either as conductometric, potentiometric, or amperometric (voltammetric) sensors (Brett and Brett 1998; Brett 2001).

In conductometric electrochemical sensors, the concentration of charge is obtained through measurement of solution resistance and is therefore not species-selective. Conductometric detectors can, however, be useful in situations where it is necessary to ascertain, for example, whether the total ion concentration is below a certain permissible maximum level or for use as an online detector after separation of a mixture of ions by ion chromatography. Such situations can arise in electroremediation.

In potentiometric sensors, the equilibrium potential of an indicator electrode is measured against a selected reference electrode using a high-impedance voltmeter, i.e., effectively at zero current. Thus, the current path between the two electrodes can be highly resistive. At an inert redox indicator electrode such as platinum, the potential measured is a mixed potential, a function of all species present in solution and their concentrations. In ion-selective electrodes, careful choice of electrode material can give good selectivity to one particular species, in many cases, with only minimal interference from other ions. Detection limits of the order of 100 nanomoles per liter of the total concentration of the ion present in a particular oxidation state, although down to 10 picomolar differences in concentration can be measured (Brett and Brett 1998). A study has shown that potentiometric electrochemical sensors are very attractive for field operation due to their high selectivity, simplicity, and low cost. Therefore, in the past, potentiometric devices have been more widely used than amperometric electrochemical sensors.

However, potentiometric sensors are less sensitive and often are slower than their amperometric counterparts. Moreover, more information can usually be gained using amperometric sensors. An applied potential in voltammetric sensors can lead to high selectivity and specificity, and thence probing of speciation. Each chemical species as well as each element or oxidation state has an associated potential for oxidation and reduction. Therefore, several species that react at different applied potentials can be determined almost simultaneously in the same experiment without the need for prior separation. As a result, in recent years we have seen an increasing amount of research on amperometric probes, which should gradually shift this balance in the direction of wider application of amperometric electrochemical sensors.

Of course, the selection of an optimal construction (type) for a chemical sensor is also dependent on the sensing material used. For example, Haug et al. (1993) showed that the use of some sensing materials in various devices provides specific advantages and disadvantages. Results of an analysis carried out for polysiloxones are presented in Table 7.5. The polysiloxones are an important class of gas-sensing nonconductive polymers and cage compounds, which can be applied in optical, mass-sensitive, calorimetric, and capacitance devices operated at around room temperature. Analysis of other sensing materials can be found in Volumes 1–3 of this series, *Chemical Sensors: Fundamentals of Sensing Materials*.

So, as we have stated, there are no ideal sensors, and therefore, none can meet all the requirements for workable devices. For example, Ho et al. (2001) analyzed various chemical sensors suitable for chemical weapons detection. Table 7.6 summarizes the results of comparing specification requirements and attributes of candidate sensing technologies using various descriptors. In this table, anticipated problems in meeting specification requirements with available technologies are identified by "P," and particularly desirable features of the candidate systems are labeled "A."

It is seen from Table 7.6 that every analyzed type of sensor has its own specific technical limitations in achieving the required parameters. Therefore, choice of a sensor suitable for chemical weapons detection in many cases will be based not on the sensor's technical characteristics but on available financial resources and current experience in design of specific devices. This means that financial resources and available scientific and technical backlogs can affect the strategy of any sensing task design in many different ways.

More detailed information about the use of various sensors for detection of certain gases is given in Table 7.7. Data presented in this table were extracted from the work of Ho et al. (2001) and many others. One should note that recommendations given in this table do not take into account all existing opportunities for detection of gases, and they do not consider the specificity of gas detection in extremely small concentrations.

Thus, choosing the right chemical sensor is an important part of the activities related to design of an efficient instrument for monitoring a specific analyte. Proper selection is based on a number of inputs: the physical properties of the target analyte, ambient conditions, required sensitivity, maintenance cycle, and method of operation, among others. For example, for H_2 sensing one can conclude that catalytic sensors can be used to detect the presence of higher concentrations of hydrogen; however, catalytic sensors only detect the presence of combustible gas, they are not gas-specific. In areas where other flammable gases are not present, this technology may be appropriate. Electrochemical sensors also can be used to detect hydrogen and offer some selectivity compared to catalytic sensors. Disadvantages

Table 7.5. Advantages and disadvantages of various gas sensors based on polysiloxanes for monitoring organic molecules

	Optical sensor systems	Quartz microbalance sensor systems	Calorimetric sensor systems	Capacitive sensor systems
Advantages	- Low cost of sensor elements - Online registration under continuous flow conditions - High sensitivity - Linear sensor response - Response time ~1 s - Complete reversibility of sensor response - Sensor arrays for determination of different gas components possible with one optical setup and different sensor elements - Application in gas and liquid phases	- Low cost of sensors; medium cost of sensor system (frequency counter) - All sensitive polymers may be used - Online data registration under continuous flow conditions - High sensitivity - Linear sensor response - Response time ~10 s - Complete reversibility of sensor response - Sensor arrays for determination of different gas components possible	- Low cost of sensors; medium cost of accessories (μV meter, chamber with valves and pumps) - Automatic zero-point calibration, complete; reversibility of sensor response - High sensitivity - Linear sensor response - Response time ~10 s - Sensor arrays for determination of different gas components possible with one μV meter and different sensor elements	- Low cost of sensor elements; low cost for sensor systems operating at one frequency - Online data registration under continuous flow conditions - High sensitivity - Linear sensor response - Response time ~20 s - Complete reversibility of sensor response - Sensor arrays for determination of different gas components possible with one impedance meter and different sensor elements
Disadvantages	- Transparent and sensitive polymers are required - Limited frequency range with one type of light source - Expensive accessories (light conductors, xenon light source) - Constant operating temperature is required to avoid cross-sensitivities due to thermal expansion effects of the polymer films.	- Application in the liquid phase is critical	- Sensitive polymers are required with large molar heats ΔH of solution and low response times - Limited online data registration under chopped flow conditions - Application only in the gas phase	- High cost of accessories if impedance meter is required - Sensitive polymers with large relative permittivities required if sensor signals are based on thickness changes - Large cross-sensitivities to variations of relative humidity - Application only in the gas phase

Source: Adapted with permission from Haug et al. 1993. Copyright 1993 Elsevier.

Table 7.6. Candidate sensing technologies for detection of chemical warfare agents

	Candidate sensing technology				
Parameter	GC-MS	IMS	FO	EC	AWA
Response time					
Sensitivity	A	P	P		P
Resolution					
Range					
Linearity					
Detection limit		P			P
Selectivity	A	A		P	P
Accuracy	A			P	
Repeatability	A				
Packaging	P	P	A		
Isolation			A		
Economic					
Development			P		P
Reliability					
Lifetime				P	
Calibration				P	P
Acquisition (data)					
Implementation					
Size				A	
Format					A

Key: P, (potential) problem in meeting specification; A, acceptable attribute. GC-MS, gas chromatography + mass spectroscopy; IMS, ion mobility spectroscopy; FO, fiber optic; EC, electrochemical; AWA, acoustic wave array.

Source: Data extracted from NMAB 1995.

of electrochemical sensors are their required recalibration and periodic maintenance. Solid-state sensors can be used to detect hydrogen at low concentrations; however, they may be subject to interference from various gases such as vapors of alcohol and other reducing gases. Some manufacturers offer activated carbon filters to mitigate alcohol interference. Several companies offer an acoustic sensing technique for H_2 sensing. A special approach to surface acoustic wave (SAW) sensor design offers a specific detection method that is not subject to interference.

For ozone sensing, we draw other conclusions. According to White (2000), there are five basic methods for monitoring ozone:

- Ultraviolet analyzers offer absolute measurements; they are selective, quick, and reliable, but they do respond to interference from some organics and mercury.

- Semiconductor sensors are selective for ozone, inexpensive, and compact, but they require periodic calibration.
- Electrochemical sensors are also inexpensive and compact, but they are slow, unstable, and less selective than other technologies.
- Chemical titration is inexpensive, but it is not selective and it is not a continuous monitoring process.
- Chemiluminescent analyzers are selective, but they need fuel and are expensive.

Regarding sensors designed for detection of other gases, such as CO_2, NH_3, O_2, CH_4, etc., reviews offering comparative analyses are listed in Table 7.8. Reviews that discuss various chemical sensors suitable for specific applications are presented in Table 7.9.

We need to note that most applications require synchronous control of several chemical components. For example, a control system for new gasoline engines, as shown in Figure 7.1, should include sensors for measuring O_2, HC/CO, and NO_x. This situation complicates selection of chemical sensors even more.

Thus, overall, the selection of an optimal sensor is a really difficult, multiple-factor task. However, we hope that this discussion will make this process a bit easier. The given information should help in

Table 7.7. Sensors recommended for various applications

Analyte class	Recommended sensors
Toxic gases	Electrochemical; solid-state (MOS);
Combustible gases	Catalytic; solid-state (MOS); IR; thermal conductivity
Volatile organic compounds	Gas chromatography; mass spectrometry; photoionization; flame ionization; ion mobility spectrometry
Total hydrocarbons (THCs)	Flame ionization
Other gases	Thermal conductivity; colorimetry
Single gas	
O_3	UV photometer; solid-state (MOS)
SO_2	UV photometer; IR; flame ionization
NO_x	Chemoluminescence; solid-state (MOS)
CO_2	IR; ion-selective electrode; solid-state electrolyte
O_2	Electrochemical; paramagnetic; zirconia-based
H_2	Schottky; FET; catalytic; solid-state (MOS)
HCl	IR; mass spectrometry
CO	IR; electrochemical; solid-state (MOS)
H_2S	Solid-state (MOS)
NH_3	Solid-state (MOS)
CH_4	IR; electrochemical; solid-state (MOS)
Multiple gases	UV/IR spectrometry; mass spectrometry Gas chromatography; electronic nose
Ions in water	Electrochemical; ion-selective FET; electronic tongue

Table 7.8. Reviews related to comparative analysis of sensors aimed for detection of specific analytes

Analyte	Type of sensor	Reference
O_2 (high temperatures)	Solid electrolyte	Yamazoe and Miura 1996
	Solid electrolyte	Suzuki et al. 1999
	Solid electrolyte	Riegel et al. 2002
	Solid electrolyte	Docquier and Candel 2002
	Solid electrolyte, semiconductor oxides	Ramamoorthy et al. 2003
	Semiconductor oxides	Aroutiounian 2007
O_2 (RT measurements)	Biosensors, optical, electrochemical	Azevedo et al. 2005
	Optochemical	Ando 2006
O_2 in blood	Biosensors	Ward 2008
O_3	Optical, electrochemical	Parrish and Fehsenfel 2002
	All types of sensors	White 2000
	Optochemical	Ando 2006
H_2	Optochemical	Ando 2006
	Semiconductor oxides	Aroutiounian 2007
	Electrochemical, solid electrolyte	Korotcenkov et al. 2009
VOCs (hydrocarbons)	Luminescence-based	Keefe et al. 2000
	All types of sensors	Pejcic et al. 2007
	Mass spectrometry, optical	Krol et al. 2010a
CH_4	Semiconductor oxides	Basu and Basu 2009
CO	Chromatography, optical	Parrish and Fehsenfel 2002
	Semiconductor oxides	Aroutiounian 2007
CO_2	Solid electrolyte, semiconductor oxides	Yamazoe and Miura 1994
	Solid electrolyte	Yamazoe and Miura 1996
	Optochemical	Ando 2006
	Solid electrolyte	Fergus 2008
	Electrochemical, semiconductor oxides, polymers	Neethirajan et al. 2009
NH_3	All types of sensors	Sazhin et al. 2003
	All types of sensors	Timmer et al. 2005
SO_2	Solid electrolyte, semiconductor oxides	Yamazoe and Miura 1994
	Solid electrolyte, semiconductor oxides	Yamazoe and Miura 1996
	Solid electrolyte	Fergus 2008
H_2S	Optical	Choi and Hawkin 2003
NO	Solid electrolyte, semiconductor oxides	Ménil et al. 2000

(Continued on following page)

Table 7.8. (*Continued*)

Analyte	Type of sensor	Reference
$NO_2(NO_x)$	Solid electrolyte, semiconductor oxides	Yamazoe and Miura 1994
	Solid electrolyte, semiconductor oxides	Yamazoe and Miura 1996
	Optical	Parrish and Fehsenfel 2000
	Solid electrolyte	Fergus 2007
Humidity	Solid state sensors	Greenblatt and Shuk 1996
	All types of sensors	Rittersma 2002
	Solid state sensors	Fontes 2005
	Optochemical	Ando 2006
	Fiber-optic	Yeo et al. 2008
pH	All types of sensors	Midgley 1990
	Optical and fiber optic	Lin 2000
	All types of sensors	Yuqing et al. 2005
	Electrochemical	Zhou 2008
Ethanol	Biosensors	Azevedo et al. 2005
H_2O_2	Optical, biosensors	Parrish and Fehsenfel 2000
	Electrochemical, optical, biosensors	Azevedo et al. 2005
Anions	Luminescence-based	Keefe et al. 2000
	Optochemical, electrochemical	Gale 2003
Metal ions	Luminescence-based	Keefe et al. 2000
	Luminescence-based	Prodi et al. 2000
	Electrochemical	Honeychurch and Hart 2003
Glucose	Different types	Wilkins 1989
	Biosensors	Wilkins and Atanasov 1996
	Fluorescence-based	Pickup et al. 2005
	Various types of sensors	Ferrante do Amaral and Wolf 2008
Biological VOCs	Biosensors	Zhang and Li 2010
Biogenic amines	Chromatography, electrochemical, biosensors	Öna 2007

clarifying the criteria which can be used during selection of an optimal sensor aimed a specific application. Further, in many cases, two or more sensors might fulfill the same requirements. In such cases, the total cost of the device across its life cycle may drive the final selection (Naranjo and Neethling 2007). Of course, advice from company employees and consultants may be part of the process as well.

4. DO SENSORS WITH LOW SELECTIVITY HAVE A FUTURE?

The problem of selectivity of chemical sensors is complicated, and sometimes too much expense is required to resolve it completely. However, chemical sensors with low selectivity may still have applications

Table 7.9. Reviews related to comparative analysis of sensors aimed for several specific applications

ANALYTE	TYPE OF SENSOR	REFERENCE
Environment monitoring	Electrochemical	Fleet and Gunasingham 1992
	Solid electrolyte	Yamazoe and Miura 1994
	All types of sensors	Parrish and Fehsenfel 2000
	All types of sensors	Pejcic et al. 2007
	Gas chromatography, all types of sensors, electronic nose	Muñoz et al. 2010
Clinic and medical applications	Fluorescence-based	Leiner 1991
	All types of sensors	Cutmore and James 2007
	All types of biosensors	Justino et al. 2010
Combustion control	All types of sensors	Docquier and Candel 2002
Exposives detection	All types of sensors	Singh 2007
Food and drink industry	All types of sensors	Pedersen 1991
	All types of biosensors	Mello and Kubota 2002
	All types of biosensors	Patel 2002
Fire alarm systems	All types of sensors	Kohl et al. 2001
	Conventional sensors	NFPA 2003
Agriculture	Various types of sensors	Adamchuk et al. 2004
Car industry, emission control	Solid electrolyte	Igarashi 1986
	Solid electrolyte	Riegel et al. 2002
	Mass spectrometry, spectroscopy, etc.	Maricq 2007
Field application (portable devices)	All types of sensors	Wilson et al. 2001
	All types of sensors	Brown 2006
Electronic nose	All types of sensors	Arshak et al. 2004

in various areas. Even in the absence of high selectivity, one can suggest several different approaches to their utilization (Ho et al. 2001).

- *The use of sensors with limited selectivity in a controlled environment*, such as a manufacturing site, where the nature and range of interferants are known and the normally expected quantities or concentrations of target analyte and interferants can be specified with confidence.
- *Mathematical deconvolution of multiple responses from arrays of imperfectly selective sensors*. This approach was discussed in detail in Chapter 1 of this volume. As shown there, an array of sensors can be employed, with each sensor in the array having a different but known (calibrated) level of response to each analyte or interferant in the sample mixture. Then appropriate mathematical

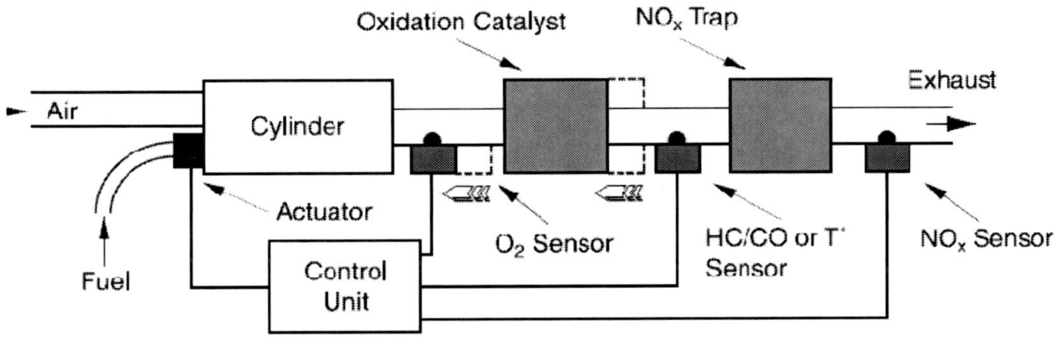

Figure 7.1. Control system for new gasoline direction injection engines. (Reprinted with permission from Docquier and Candel 2002. Copyrigh 2002 Elsevier.)

approaches (e.g., pattern recognition, a topic of "'chemometrics") may permit extraction of the desired analyte response from the fusion of responses from the multiple sensors.

- *Separation of the analytes from the sample matrix, followed by sensing of individual analytes with sensitive but nonselective sensors.* In this case, a large number of different analytes in a sample are separated into pure components. Then a nonselective detector can be used to detect the pure components. Separation by modern techniques of chromatography and electrophoresis is an extremely powerful and enabling technology for analysis of complex chemical mixtures. When combined with a suitable detector and electronic support package, separation-based sensors (i.e., not truly "direct-reading") can potentially provide an analysis with response time similar to the response time of many direct reading sensors but with much better effective selectivity. Miniaturized, total analytical systems are a relatively new area of research in analytical chemistry, but their development could significantly supplement the capabilities of existing chemical sensors.

5. FUTURE TRENDS

Discussions in this volume have confirmed that the capability of chemical sensing technology is substantial and has grown steadily over the past decades. However, achieved progress has been outpaced by the needs and diversity of chemical measurements. Therefore, R&D activity in the field of designing advanced chemical sensors is still very important.

Development trends in sensor technology result from market economics, customer needs, and the requirements for an implementation in the target application. Cost reductions together with improvements in accuracy, reliability, and speed are being attained using measurement methods with better performance, new manufacturing technologies, and sophisticated signal processing methods (Tränkler and Kanoun 2001). Based on the results presented in this series, one can conclude that modern research in the field of chemical sensor elaboration is directed largely toward accomplishing the following tasks.

1. *Sensor miniaturization and incorporation of micromachining technology in sensor fabrication.* Increased efforts in design of device fabrication techniques should result in significant

improvements in the operating parameters of chemical sensors (Simon et al. 2002; Korotcenkov 2007b). While improving fabrication methods, we can expect the appearance of a new generation of devices with advanced repeatability of baseline and sensitivity over previous practice available in very large numbers and at very low cost. A reduction of characteristic dimensions usually results in shorter response times and higher resonance frequencies, so a correspondingly higher speed is achievable in signal generation and processing.

2. *Sensor improvement and adaptation for resolving specific tasks.* Sensors still require improvements in selectivity (chemiresistors, SAWs, etc.) and stability of sensor parameters in corrosive media (polymer-based devices). The MicroChemLab device requires development to detect volatile organic compounds (VOCs) in aqueous environments (Ho et al. 2004). Additional development work is needed to decrease both temporal drift of sensor parameters and sensitivity to fluctuations in environmental parameters (e.g., humidity, temperature). It is necessary to increase the number of analytes which can be analyzed simultaneously. Sampling methods should be developed to introduce aqueous samples to the ion mobility spectrometer (IMS). Additional development is needed to decrease the cost of optical and fiber optic sensors, and so on.
3. *Design of novel methods of assembling and analyzing sensor arrays; design of integrated sensors or "smart" sensors.* Current sensor development is tending toward increased complexity in sensor systems. Wise (2007) believes that greater flexibility and lower production cost, associated with advanced, integrated electronic technology, will allow us to reduce a large and sophisticated computer signal processing system to a microelectronic cheap. For example, smart sensors have transduction, signal amplification, filtering, and other processing on a single substrate. Thereby a big set of available information is managed using sophisticated signal processing techniques, so that the system can achieve better performance. However, from the perspective of the end user, the sensor system now appears simpler, even with its increased functionality and internal complexity.

   Networking of large sensor systems can provide improved spatial and temporal sampling in low-cost, low-maintenance systems. Multisensor systems provide such advantage, so that economical sensors can be used even for the achievement of a high level of precision and reliability. In a network of chemical sensors, the individual outputs from an array of sensitive but only moderately selective transducers can provide a composite indication that is both sensitive and selective with regard to a target chemical species.
4. *Design and application of new advanced materials (1-D materials, nanocomposites, etc.).*
5. *Design of new synthesis technologies; properties stabilization and sensing materials functioning.* Williams (1999) believes that the design of new sensor materials will lead to very significant improvements in stability and rejection of interferences caused by changes in relative humidity or the appearance of additional analytes in the testing surroundings. However, it is necessary to note that the principal technical drivers for sensor development may also come from enabling/supporting technologies that are different from materials technology. Recent advances in many chemical sensors have originated not from the synthesis of new sensing materials but from innovations in low-cost, large-scale manufacturing of interconnections, microelectronics, and micromachining.
6. *Design of new concepts of chemical sensing, including* in situ *control.* The appearance of new concepts and types of devices should promote widening the possible areas of chemical sensor application. New sensor concepts will arise from developments in basic understanding of devices

that operate at room temperature and devices that behave as "self-diagnostic arrays." Advances in the basic understanding of sensor mechanisms will come from systematic study of effects of variations in oxide composition, from studies of surface modification and its consequences, and from application of new tools of computational chemistry.

7. *Development of autonomous sensor systems.* At present the research in this field typically progresses from bench-top prototypes to study systems operated *in situ* and finally to production of operational systems that are accessible to the broad community. The use of wireless systems implies more convenience and leads to a considerable cost reduction. An advantage of wireless sensor systems is that they can be placed without restriction, and can therefore record the measured quantity close to its occurrence and independent of potential harsh circumstances. Energy-autonomous sensors will gain particular importance among wireless sensors (Tränkler and Kanoun 2001), because in this case no wires are necessary, even for electricity supply. Johnson (2003) believes that autonomous observing systems using chemical sensors are becoming well poised for use in the next generation of global atmospheric and biogeochemical studies. Autonomous *in situ* systems can provide efficient monitoring of temporal changes on scales that are cost-prohibitive using shipboard measurements. Mobile platforms, such as gliders, floats, drifters, and volunteer observing ships (VOS) will give sensor systems access to broad geographic regions. In addition to acquiring data at improved spatial and temporal scales, *in situ* analysis can potentially improve data quality by minimizing sample contamination as well as transformations of meta-stable seawater and atmosphere constituents. Autonomous observations cannot currently supply the breadth of specialized measurements that can be made by shipboard scientists, and problems associated with power requirements, endurance, fouling, and sensor response times also weigh against the strengths of *in situ* observations. The long time frames that have been required to bring new sensor systems to an operational status suggests that next-generation, global programs will be influenced primarily by those sensor systems where significant developmental progress has already been made.
8. *More strong efforts for commercialization of achieved results, especially in the field of research and development of sensors destined for dual use.* It is known that sensors are applicable to a wide spectrum of uses. To be effective, they require identification of potential uses and assessment of the degree of suitability. For example, chemical sensors used for detection of chemical warfare agents have numerous possible non-defense applications in areas such as environmental and health monitoring.

In this chapter we will not discuss the potential advantages of the above-mentioned directions in the study of chemical sensors. They are all important, however, because chemical sensors are complicated devices, and their elaboration requires complex approaches.

6. SOME PRACTICAL ADVICE FOR GAS SENSOR SELECTION AND USE

As has been shown in this series, the detection of hazardous gases has always been a complex subject, making the choice of an appropriate gas-monitoring instrument a difficult task. Therefore, we have compiled this guide to provide information about several aspects of sensor selection and use. In preparing this guide we used information presented in numerous references (HSE 1987; Woodfin 1994; UKOOA 1995; BS EN 50241, 1999; COGDEM, 1999; BS EN 61779, 2000; Chou 2000; Kelsey et al. 2000; White 2000; Bradley 2001; Brett 2001; Ho et al. 2001; Walsh et al. 2001; L138 2003; Brown 2006). It should help in making a decision concerning which sensor type is best for a specific application, and how this sensor can be used with maximum efficiency.

One must, of course, take into account that using gas sensors does not prevent the appearance of toxic or flammable gases, and does not indicate what action should be taken. Therefore, implementation of some rules while using gas sensors can provide early problem prevention and help to ensure people's safety.

6.1. SENSOR SELECTION

As has been shown, each sensor has certain capabilities and limitations, and thus the suitability of a given sensor depends largely on the application in which it is to be used. This means that technical characteristics and the chosen sensor's performance should correspond to exploitation conditions. The topology of the land, construction, type of risk, placement, size, and shape of equipment can influence the choice of detection devices sized for particular applications. Some sensors may have physical constraints which limit their location or mounting. For example, gas sensors for detecting air pollutants must be able to operate stably under deleterious conditions, including chemical and/or thermal attack. Some chemical sensors have been designed for intrinsic safety, i.e., for use in potentially explosive atmospheres. Check with the manufacturer to ensure that a specific instrument meets the appropriate intrinsic safety requirements. How strong the requirements for sensors can be depends on the area of application, as one can see from the data presented in Table 7.10.

As mentioned previously, when using any chemical sensors, it is necessary to remember that they will not be 100% specific to the gas of interest. It is important to recognize that no single detection technology can provide the sensitivity and response time required for every gas. Only infrared, oxygen, and electrochemical toxic sensors can honestly be assigned a reasonable degree of specific response to a single gas. Other sensors, including photoionization detectors (PIDs), metal oxide semiconductor (MOS), and hot-wire/catalytic elements are inherently nonspecific and thus cross-sensitive to numerous other compounds. Therefore, depending on the gas to be monitored and the sensor type chosen, one may encounter varying degrees of interference from other gases which are present in the environment.

The composition of the surrounding atmosphere is also very important factor for selection of an optimal sensor. For example, MOS and especially catalytic sensors require oxygen for their operation. Catalytic sensors become less responsive when levels of oxygen fall below 15–16% by volume, which may result in lower than normal readings. Below 10% by volume oxygen, these sensors are virtually useless for lower explosive limit (LEL) detection. At the same time, electrochemical sensors can operate in an oxygen-free atmosphere.

The nature of the target gas also needs to be taken into account. It is necessary to know that the nature of the target gas can influence the lifetime of sensors. For example, the life expectancy of electrochemical sensors designed to detect oxidizing gases such as ozone and chlorine can be less than 12 months, while sensors for reducing gases may last up to 3 years or even more. Electrochemical sensors

Table 7.10. Requirements for ammonia analysis equipment in various application areas

Application	Detection limit	Required response time	Temperature range
	Environmental		
Monitorambient conditions	0.1 ppb to >200 ppm	Minutes	0–40°C
Measure in stables	1 to >25 ppm	~1 min	10–40°C
	Automotive		
Measure NH_3 emission from vehicles	4 to >2000 g/min	Seconds	Up to 300°C
Passenger cabin air control	50 ppm	~1 s	0–40°C
Detect ammonia slip	1–100 ppm	Seconds	Up to 600°C
	Chemical		
Leakage alarm	20 to >1000 ppm	Minutes	Up to 500°C
	Medical		
Breath analysis	50–2000 ppb	~1 min	20–40°C

Source: Reprinted with permission from Timmer et al. 2005. Copyright 2005 Elsevier.

for ammonia also have very short lifetimes. The sensor's life is dramatically reduced from exposure to gas, even in low concentrations. Solid-state sensors are therefore a much better choice for ammonia applications. They perform well over a long period of time and have been used extensively in refrigeration applications. The same is true for hydrogen sulfide. The solid-state sensor performs very well for H_2S, with excellent response and selectivity. Therefore, for H_2S applications, it is generally recommended to use a solid-state sensor, because these applications tend to be in hostile environments, where a solid-state sensor will fare much better than an electrochemical sensor. The life expectancy of photoionization detectors is also sensitive to the surrounding atmosphere. High-electron-volt UV lamps use a salt crystal window that can be damaged by certain corrosive gases.

Additionally, for toxic gases especially, the choice of an appropriate sensor depends strongly on the concentration range to be monitored. For example, for combustion gas monitoring, both catalytic bead sensors and semiconductor (MOS) sensors can be used (see Figure 7.2). Both types of device are cheap, simple, and have long life expectancies. However, solid-state semiconductor sensors tend to be more sensitive than catalytic ones. Experiments have shown that MOS sensors can detect toxic or combustible gases with concentration from the parts-per-billion to the parts-per-million range. At the same time, catalytic bead sensors are sensitive to combustible gases in high concentrations only and will not see any concentration lower than about 1000 ppm. Other technologies are also available for detection of hydrocarbons; some are listed in Figure 7.2 (Miller, 2001). With the exception of infrared (IR) technologies, these techniques are not well suited for LEL-range (explosive atmosphere) monitoring because they do not detect in the proper concentration range and/or are not specific to combustible gases—i.e., they

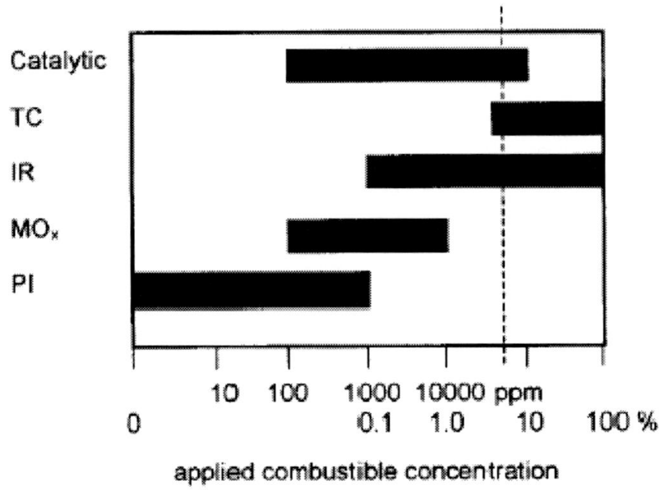

Figure 7.2. Ranges for hydrocarbon analyzers (LEL at dotted line). Sensor type: TC, thermal conductivity; IR, infrared; MOX, semiconductor; PI, photoionization. (Reprinted with permission from Miller 2001. Copyright 2001 IEEE.)

respond to one or more common noncombustible species. (However, parts-per-million-level detection techniques have important applications in monitoring for worker exposure to low-level toxic organics.) Infrared-based sensors are finding increasing application for LEL-range monitoring in permanent installations. However, their size, power requirements, and oftentimes cost, have kept them from widespread use in portable, battery-powered instruments.

Design peculiarities of sensors need to be taken into account as well. For example, for photoionization detectors, the exciting energy of the UV lamp used is of great importance. Compounds with an ionization potential greater than the energy emitted by the lamp cannot be ionized and thus cannot be detected. Most manufacturers typically use 9.6- to 10.6-eV lamps in their PIDs, and therefore these devices cannot detect compounds having ionization energies greater than these values. Among the gases not detected with these common lower-eV lamps are carbon monoxide (14.1 eV), hydrogen cyanide (13.91 eV), methylene chloride (11.35 eV), xylene (11.4 eV), ethane (11.65 eV), and others (Brown 2006). So, a higher excitation energy improves the parameters of PIDs. However, the highest-eV lamp costs more and has the shortest field service life.

The area of monitoring is also important for sensor selection. Devices operate over different monitoring distances. Catalytic, MOS, and electrochemical sensors are single-point devices. These detectors operate best at a monitoring radius of 0.5–2 m. As a result, they are commonly placed near potential sources of gas leaks. Ultrasonic gas leak detectors, as opposed to point detectors, protect larger areas. While monitoring radii varies according to the required sensitivity, the devices have a radial coverage of about 452 m^2 or a 12-m radius. Open-path IR detectors and IR gas cloud imaging cameras offer some of the longest monitoring distances (Hodgkinson and Pride 2005). Open-path IR detectors, for example, typically have sampling paths from 10 to 100 m, while IR gas imaging cameras can cover spans from 100 to 2000 m (McRae and Kulp 1993).

The presence of background gases, which cannot be determined, should toughen our requirements to sensor selectivity. Sensor selectivity (as opposed to cross-sensitivity) relates to the ability of the sensor to differentiate among various gases. A major cause of sensor failure is the presence of background gases that the instrument's manufacturer did not take into consideration. Sensor manufacturers can provide data for each sensor indicating the concentration of an interfering gas that is required to create a sensor response equivalent to the response obtained from a given concentration of the target gas. This is an important criterion for a sensor that is cross-sensitive to other gases that may be found in the same area.

Conditions of exploitation are also important for sensor selection. Some sensors are susceptible to interference from radio-frequency and magnetic fields. Sensors and cables connecting sensors to controllers or other monitoring equipment handle signals of very low magnitude. Especially at the low end of the sensor's sensitivity scale, the signal-to-noise ratio is very low and electromagnetic interference (EMI) or radio-frequency interference (RFI) can cause false alarms if sensors, controllers, and cables are not shielded. Many users report gas alarms triggered by use of hand-held radios. Some process equipment also generates RFI and EMI. Do not forget that in applications requiring the sensors to be constantly exposed to gas, special considerations are required and the sensor specifications and the suppliers of the sensors may need to be consulted.

Stationary and portable gas detectors today are offered worldwide by over 40 firms and may contain three, four, five, or even more of the various types of sensors we have described. However, the majority of sensors are made by only a limited number of independent companies. Therefore, Brown (2006) advises, if you are dissatisfied with one instrument maker's device, do not switch to another supplier's unit until you extract from its maker (not the local sales agent!) the source of the sensor modules inside the instrument. When possible, ask for a 2-week field trial, at no obligation, of a device you are considering purchasing.

The total cost of ownership (TCO) is an important criterion in selecting a sensor system or technology as well. In addition to first purchase cost, it is necessary to consider carefully the cost of expendable parts, maintenance replacements, integration with other systems, expandability, and downtime due to maintenance and false-alarms.

Of course, it is difficult to mention every possible consideration, but a carefully evaluated and studied application can yield savings in both time and money.

6.2. SENSOR PARAMETERS

As we have noted before, selection of an optimal chemical sensor for a specific application is a very difficult task. Some of the criteria which should be considered are shown in Figure 7.3.

Obviously, the sensor should have sensitivity sufficient for detecting gas at the required concentration. The gas detector should be set to alarm at a level low enough to ensure the health and safety of people but high enough to prevent false alarms. Combustible-gas sensors are designed to detect gases at the lower flammable limit (LFL). Other sensors may be designed to detect gas concentrations around the permissible exposure level (PEL), which may be in parts per billion. The PEL for a gas is defined as a safe level to which workers can be exposed for the entire work day. Thus, it will be wise to choose sensors which can operate in a full-scale range of 0.25–10 times the PEL of the gas, which is higher than

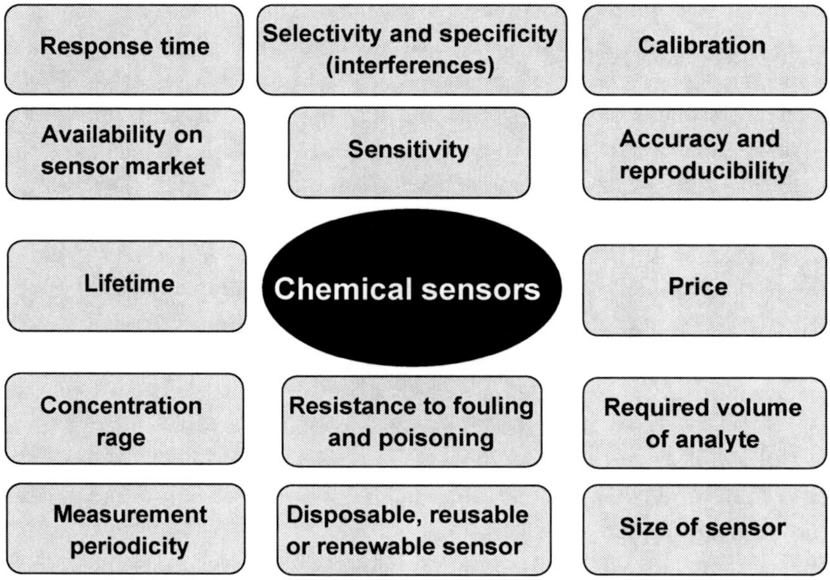

Figure 7.3. Important aspects in choosing chemical sensors.

the PEL to enable setting alarm points at concentration ranges above the PEL. It is necessary to take into account that sensors will not function properly if they are exposed to gas concentrations outside their designed sensitivity range. For example, electrochemical sensors exposed to concentrations more than 10–20 times their designed operating range can become saturated and inoperative (poisoned) for a long time period.

False alarms are most likely to be caused by fluctuations in sensor output due to environmental changes (e.g., ambient temperature, pressure, or humidity), sensitivity to other gases or vapors, or sensor drift. If false alarms are a problem, then one option is to use two detectors—the alarm level must be registered by both detectors before the alarm activates. So, duplication (or triplication) of sensors and control apparatus may be required for continuous monitoring and to prevent false alarms

Response time is another important sensor parameter. The required response time depends on the location, the purpose of the system, and the speed of development of the expected problem. For extractive systems that draw air samples through tubing to a remote sensor, response time also includes the time for air to move through the tube from the sensing location to the sensor. Protection of people from large leaks requires a fast response time, but where the gas concentration is expected to build up slowly, a slower response time may be acceptable. A fast response time is important not only for safety, but also to minimize costly contamination of the workplace.

The lowest detection limit (LDL) is also an important parameter of chemical sensors which must be taken into account during sensor selection. The ability to detect low-level concentrations under operating conditions means earlier detection, before a small leak becomes a significant problem. Several manufacturers provide a published range of detection, but they do not provide this minimum detection limit, due to their inability to detect low concentration levels.

Lifetime, the time between calibrations, and watt consumption are other parameters which one needs to know before purchasing chemical sensors. Just these parameters will determine the expenses related to exploitation of these devices.

6.3. SENSOR CALIBRATION AND TESTING

All types of sensors require periodic calibration and checking to ensure sensor accuracy and system integrity (Chou 2000). For efficiency, liability abatement, and safety, it may be advisable to purchase factory-calibrated replacement parts. The intervals between calibrations can be different from sensor to sensor. Some experts suggest every year, some every half-year, and others, every quarter. Generally, the manufacturer of the sensor recommends a time interval between calibrations. Span calibration, for toxic sensors, is recommended to be performed at least every 6 months. However, some sensor technology requires frequent calibrations, because the calibration is an important factor in the accuracy of an analyzer reading. For example, some companies perform a span calibration every other week. Therefore, they are able to achieve accuracies around ±2%.

Time between recalibrations or maintenance also relates to sensor life. Electrochemical cells exposed to high air flows or high temperatures may tend to dry out faster, leading to more frequent replacement and recalibration. You should do recalibration in accordance with the recommendations of the sensor manufacturer. However, it is good general practice to check the sensor more closely during the first 30 days after installation. During this period, it is possible to observe how well the sensor is adapting to its new environment (Chou 2000). Experience shows that a sensor surviving 30 days after the initial installation will have a good chance of performing its function for the duration expected. In addition, if an analyzer is used for extended periods of time, repeating zero calibration at regular intervals provides the possibility of neutralizing any changes that occur with time. In particular, the zero point will tend to drift to a certain degree with time. This cannot always be compensated electronically, since certain factors are random and a drift that produces a negative result will be automatically treated as zero, since negative concentrations are obviously impossible. Having a rough idea of the calibration value to be expected also allows an experienced service technician to see immediately when a sensor is nearing the end of its useful operating life.

The method and procedure for calibrating sensors should be established at the outset. For example, electrochemical cells exposed to high concentrations of target gas may become saturated (poisoned). Because returning to cell equilibrium takes time and the facility cannot be repopulated until the gas detection system is fully functional, many users will replace the poisoned cell, calibrate the new cell, and return the facility to production. The removed cell can be tested offline, recalibrated, and reused later. Some electrochemical oxidant sensors can "go to sleep" if they are not exposed to gas periodically. The cell electrodes can become coated by oxidation with the electrolyte. Passing target gas over the face of the cell during recalibration reverses the reaction toward the other electrode, effectively cleaning the electrode. Solid-state sensors have a similar problem if they are not exposed periodically to target gas during recalibration. Portable sensors are usually tested and recalibrated more often. Sensing instruments used for confined-space entry are required to be tested before each entry into confined spaces. If the instrument is not within 10% calibration, it must be recalibrated prior to entry into the confined space.

6.3.1. Requirements of a Test System

The construction of a test system is important to obtaining reliable information during chemical sensor calibration and testing. However, we do not offer any actual suggestions for application test systems of different configurations. We offer only one requirement: To ensure qualified performance of gas sensor measurements, a test system should adjust all the conditions that are essential for the sensor's behavior in different applications. Only in this way is it possible to get correct information about sensor properties and sensor behavior under the influence of gases, which is necessary for understanding, development, and optimization of the sensor. Taking this requirement into account, every company and research team will design a test system according to their own notion of the optimal construction, and therefore at present in the literature one can find a lot of different setups used for sensor calibration and testing (see, e.g., Harvey et al. 1989; Demarne et al. 1990; Mousa-Bahia et al. 1993; Nieuwenhuizen and Harteveld 1994; Endres et al. 1995; Magliulo 1996; Tang et al. 1998; Pilling et al. 2003; Sawaguchi et al. 2005; Brahim-Belhouari et al. 2005; Afgan et al. 2006; Korposh et al. 2006; Berry and Brunet 2008; Maziarz and Pisarkiewicz 2008). As is known, different sensors require different measuring methods.

A detailed description of test systems and approaches used for their design was given by Endres et al. (1995). According to Endres et al. (1995), to obtain gas mixtures similar to application conditions, the test system should have a gas mixing system, which should mix many gases in a wide range of combinations and concentrations that cover all desired concentration ranges. Further, the accuracy of the gas combinations has to be maximized. Endres et al. (1995) believe that an optimal gas mixing system will be designed on the basis of a dynamic mixing flow method and provide a constant gas flow through the measurement chamber. The main benefits of this method are that a constant gas flow avoids any problem with gas consumption by the sensor itself and also minimizes problems with adsorption and desorption at the walls of the test system. In addition, a constant gas flow prevents any distortion of the measurement due to changing time constants for gas exchange, forced cooling of heated sensors, and other flow-dependent effects.

Almost all gas sensors show cross-sensitivity to humidity (Sawaguchi et al. 2005; Korotcenkov 2007a). Considering this disturbing influence on the sensor response, it is important to enable the test gas mixture to be humidified. This humidification should satisfy all necessary needs of the natural environment in terms of the absolute humidity and, in combination with keeping a constant temperature in the gas atmosphere, also the relative humidity (RH). Relative humidity is the most familiar measure of moisture in the air. In words, the RH is the ratio of the amount of moisture actually in the air compared to the maximum amount the air can hold at a given temperature and pressure:

$$\mathrm{RH} = \frac{\text{water vapor present}}{\text{max. amount of vapor the air can hold}} \quad \text{or} \quad \mathrm{RH} = \frac{\text{water vapor pressure}}{\text{saturation water vapor pressure}} \tag{7.1}$$

To get a percentage, multiply the above by 100%. The term *relative* in relative humidity reminds us of the fact that the humidity is specified relative to the temperature (and pressure). If the amount of water in the air is a given value but the temperature rises, then the saturation vapor pressure rises and the RH drops. The approach used for air humidification will be described later, in Section 6.3. A relative humidity of 10–80% RH means a water content of 600–4,800 volumes per million (vpm) at 0°C and

4,200–33,500 vpm at 30°C. The inclusion of humidity variation in a test system produces many points that can be considered to ensure good performance of the whole system (Endres et al. 1995).

It is obvious that all thermodynamic processes occurring at the sensor are influenced by the physical conditions in the measurement chamber. Therefore, all important parameters, for example, temperature or pressure, have to be controlled or rather actively directed by the test system to attain stable conditions in all measurements. Undoubtedly, all these requirements for a test gas system necessitate automatic control of all system functions and automatic processing of the sensor data. Also without doubt, only computer-controlled measurement units can guarantee precise control of the environmental conditions and enable the large amount of measurement data that is necessary for gas sensor development and optimization to be received and stored. Thus, the optimal construction of the test system should include three main parts: the gas mixing system, the measurement chamber (cell), and an electronic system based on a personal computer aimed at controlling all parts of the system, including the gas mixing system and the experimental setup used for measuring, controlling, and analyzing the gas sensor parameters (see Figure 7.4.). Of course, the test system may also contain additional instruments both for control of test gas humidity and for analysis of the composition of the exhaust gases.

One possible variant of an electronic control system is shown in Figure 7.5 (Tang et al. 1998). The characterization system is controlled by a personal computer (PC). The PC controls all the instruments, mass flow controllers (MFCs), and valves using multi-IO AD/DA and GPIB cards. The MFCs are controlled by a 12-bit D/A card. A known amount of gas is metered into the chamber through the computer-controlled opening and closing of the MFCs. Temperature of the sensors is controlled by a HP 6626A high-precision programmable power supply. In the sensor dynamic response measurement, the software controls and synchronizes the opening and closing of the MFCs, the sampling of the gas analyzer, and the digital multimeter. The output of the sensor is monitored through a HP 34401A digital multimeter. A gas analyzer (INFICON QX2000), incorporated for real-time evaluation of gas compositions and concentrations, communicates with the PC through the IEEE 488 bus interface.

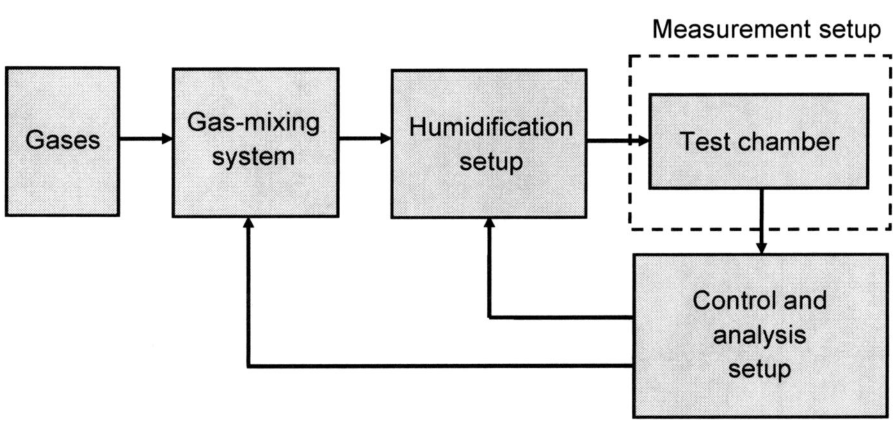

Figure 7.4. Principal scheme of gas sensor measurement setup.

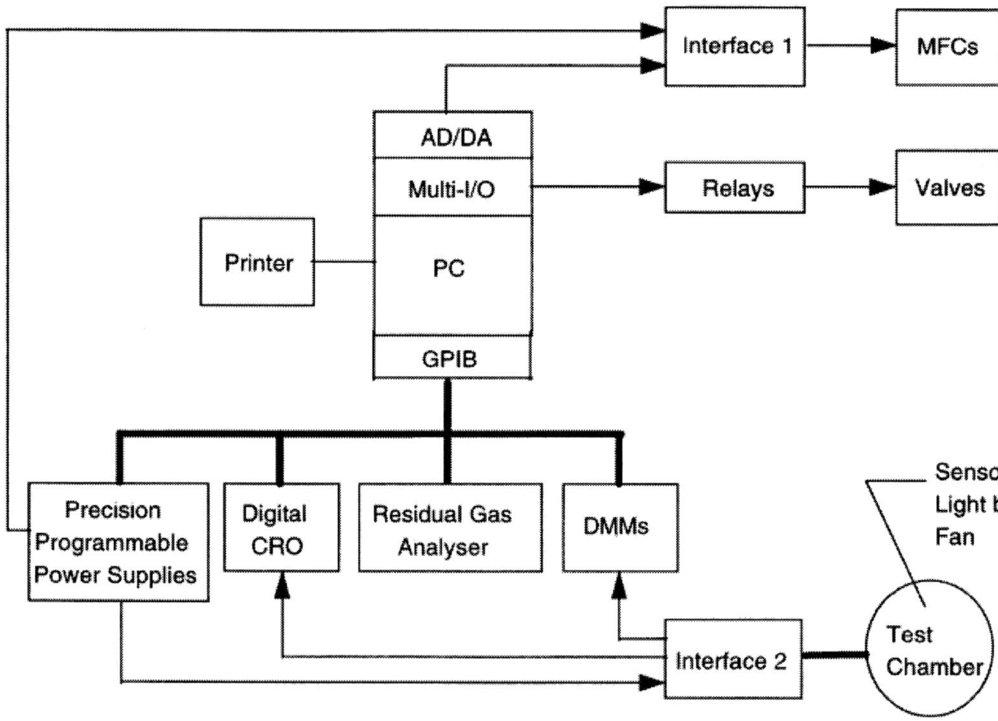

Figure 7.5. Schematic of electronic control system. (Reprinted with permission from Tang et al. 1998. Copyright 1998 Elsevier.)

6.3.2. Some General Rules for Gas Sensor Calibration

Cleaver (2001) analyzed the sources of uncertainty in analytical procedures and found that they can appear at all stages of the measurement process, including sampling, sample preparation, calibration of the instrument, analysis and data acquisition, data processing, and interpretation of results. Factors such as insufficient gas mixture homogeneity, limited number of replicate samples taken, variations in temperature and pressure during sampling, the uncertainty of the certified reference material used in the measuring system, etc., can contribute to mistakes during calibration and testing of chemical sensors. Therefore, to avoid this situation, the process of calibration and testing should be carried out in accordance with all rules established for these processes.

According to Chou (2000), the calibration procedure should be simple, straightforward, and easily executed by regular personnel. It is recommended that, if it is possible, synthetic air (N_2 + O_2) should not be used for calibration. Normal air contains traces of other gases besides nitrogen and oxygen. Also, ambient air normally contains a small percentage of water vapor (see Figure 7.6). At present, in many facilities, shop air is available throughout the plant, making it very accessible and convenient. However, most shop air contains small concentrations of hydrocarbons, carbon monoxide, carbon dioxide, and possibly other interference gases. Also, the air is typically very low in humidity. Of course, the air can be filtered through activated charcoal to remove most of the unwanted gases, and water vapor can be added to the air using a humidifier in the sampling system. After this conditioning, the air can be used to calibrate most types of sensors. However, it is important to note that carbon monoxide is not removed by charcoal filters. Therefore, it is much more realistic and practical to use the air surrounding the sensor for sensor calibration. Of course, this area should be clean. Ideally, the gas to be measured should be used to calibrate the gas detector. If you need air without carbon dioxide, a soda ash filter should be used to remove this gas. This is a very good way to zero carbon dioxide sensors, since placing a soda ash filter in-line with the sampling system will remove all carbon dioxide, thus providing an easily obtainable zero baseline.

Calibration of fixed and portable gas detectors is normally done using a calibrated gas mixture from a cylinder, supplied by a gas company. Usually, all gases are certified to have an accuracy of ±2% of the value indicated on the face of the gas cylinder. The bottles (cylinders) are available in many sizes, but most field calibrators employ smaller, lightweight bottles (Chou 2000). These small, portable bottles come in two different categories, a low-pressure and a high-pressure version. To estimate the volume of a pressurized gas (V_{mix}) in a cylinder, take the total pressure (P) divided by the atmospheric pressure (P_a) and multiply this ratio by the volume of the cylinder (V):

$$V_{mix} = V \cdot \frac{P}{P_a} \tag{7.2}$$

This volume is needed to estimate the time of calibration which can be carried out using this cylinder. Most sensors require approximately 4–8 min of span gas to carry out a single calibration. Consequently, any small transportable cylinder typically should contain sufficient gas for 12–15 calibrations.

Generally, high-vapor-pressure gases with low reactivity, such as methane, carbon monoxide, and carbon dioxide, can be mixed with air and stored under high pressure. Low-vapor-pressure gases, such as liquid hydrocarbon solvents, can only be mixed with air and stored under low pressure. Air for mixing with calibrated gas also can be stored in a cylinder. Compressed air has the advantage that it is easy to regulate and can be carried around in a bottle. To get this highly pressurized gas out of the bottle in

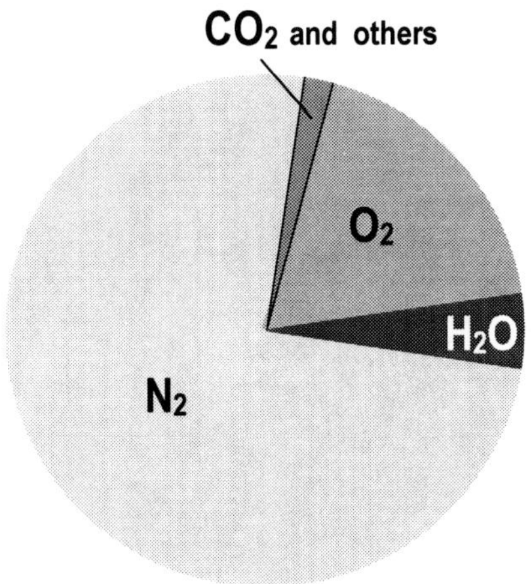

Figure 7.6. Composition of normal air.

order to calibrate the sensor, a regulator assembly is needed (see Figure 7.7). This assembly consists of a pressure regulator, a pressure gauge, and an orifice flow restrictor. The orifice flow restrictor is a fitting with a hairline hole that allows a constant air flow at a given pressure difference.

Many gas sensors are sensitive to pressure, and care must be taken when calibrating instruments that true readings are produced. For example, during calibration of sensors, electrochemical sensors especially, one should make sure that the calibration procedure itself does not pressurize the sensing cell. This is especially important to observe when pressurized cylinders of standard gases are used for calibration. The simplest way to avoid this situation is to fill a bag with gas from the cylinder so that it can be presented to the sensor at atmospheric pressure from the bag.

The environmental conditions (temperature, relative humidity, barometric pressure) of the sensors at the time of calibration should be as near as possible to those that will be encountered during use. Of these three, temperature is most important, because changes in temperature are most often encountered in the field and can cause bias in the readings obtained. Even with the temperature-compensating circuitry employed in most sensors, some time is required for equilibrium to be reached. If it is not possible to calibrate at the working temperature, the user must allow sufficient time for field equilibration of temperature. Changes in barometric pressure are usually less significant than temperature changes and so are of less concern to the user.

Experience shows that for correct calibration the supplied gas mixture should contain the same gases in similar concentrations to those being monitored. The gas should be supplied from a reliable source. To achieve accuracy and consistency, it must be assured that the source is stable and within date code (e.g., for cylinders and permeation tubes). If cylinder gas is used, a "referee" method must be used to ensure that accurate concentrations are being produced. Even certified cylinders may not provide the expected gas concentration, especially after storage.

For most applications, the sample gas should be prepared using humidified dilution air (to approximately 30–60% RH, with 45% RH being optimal). This simulates real-world conditions that are likely to be encountered during an actual event.

When it is not possible to obtain calibration mixtures containing the required gases, another gas mixture should be used which gives a similar detector response to the target gas(es). If more than one gas may be present and there is no obvious calibration mixture, then it is advisable to calibrate for the least sensitive gas.

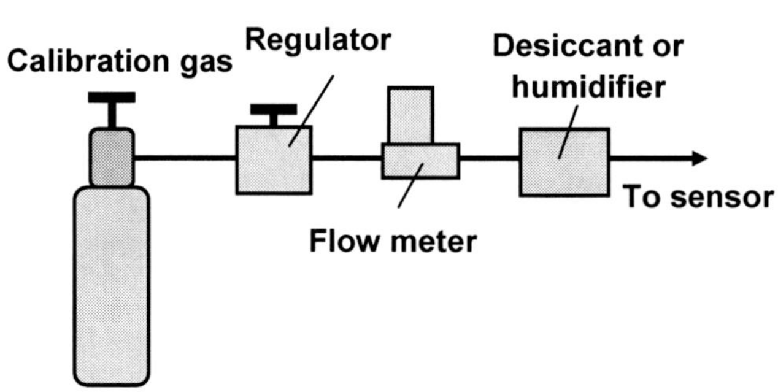

Figure 7.7. Standard scheme for gas sensor calibration.

Taking into account that calibration gases usually are toxic or inflammable, experts advise obeying the following rules (ENERAC 2005).

- Always carry out the calibration procedure in a well-ventilated area, preferably in a fume hood if one is available. Avoid inhaling the gas.
- Handle the gas cylinders carefully. Do not expose them to heat or shock. Do not attempt to fill the cylinders. Do not use them for any other than the intended purpose.
- Make sure the shutoff valves are shut tightly, before disconnecting any cylinders.
- Do not attempt to remove the shutoff valves.
- Do not touch the primary regulator setting.

6.3.3. Flow Meters and Controllers

Keeping the gas flow invariable at the desired level is an important requirement during gas calibration or testing. Many methods can be used for flow rate measurement and control. For a long time, *variable-area (VA) meters* have been the basic devices used for this purpose. The variable-area meter, also commonly called a *rotameter*, consists of a tapered tube, typically made of glass, with a float inside that is pushed up by fluid flow and pulled down by gravity. As flow rate increases, greater viscous and pressure forces on the float cause it to rise until it becomes stationary at a location in the tube that is wide enough for the forces to balance. Floats are made in many different shapes, with spheres and spherical ellipses being the most common. Some are designed to spin visibly in the fluid stream to aid the user in determining whether the float is stuck or not. Rotameters are available for a wide range of liquids but are most commonly used with water or air. They can be made to reliably measure flow down to 1–5% accuracy. Devices such as critical orifices, thermal mass flow controllers, optical flow meters, ultrasonic (Doppler, transit time) flow meters, coriolis flow meters, etc., can also be used for flow rate measurements. In fact, experience shows that critical orifices and thermal mass flow controllers are the most exploitable. It has been found that thermal mass flow controllers are especially suited for controlling gases dynamically, because the mass flow can be changed faster and more easily this way than by other methods, e.g., critical orifices.

Critical orifices are suitable only for gases. In these, the gas flows through a tube at the end of which an orifice is located whose diameter is smaller than the tube diameter; see Figure 7.8. The quantity of gas which flows through the orifice is proportional to the pressure p_1 in front of the orifice and is largely independent of the pressure p_2 behind the orifice. This ratio is valid if the pressure ratio p_2/p_1 is smaller than the critical pressure ratio (ISO 2003). For dosing the air in the calibration gas mixture, different flow rates can be adjusted by changing the primary pressure p_1 or by using orifices with different diameters.

Thermal mass flow controllers are particularly suited for dosing gases but can also be used for liquids. They are composed of a measuring chamber for the determination of the mass flow and a regulation valve which can be controlled electronically. Thermal mass flow meters generally use combinations of heated elements and temperature sensors to measure the difference between static and flowing heat transfer to a fluid and infer its flow with a knowledge of the fluid's specific heat and density. As is known, temperature at the sensors varies depending on the mass flow. The gas transports a certain amount of the

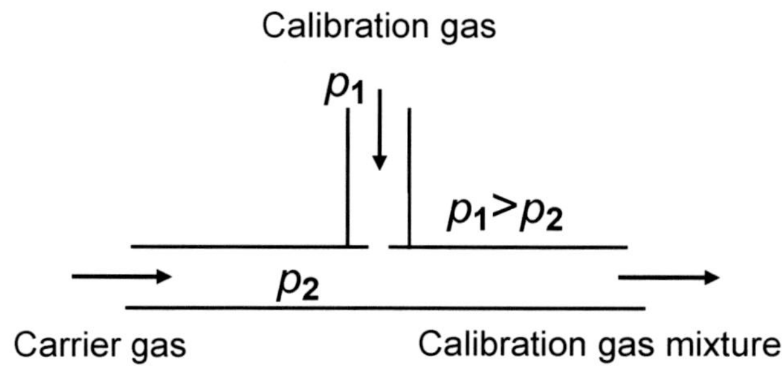

Figure 7.8. Schematic diagram of a critical orifice.

heat away in the direction of the flow. This heat transport depends on the number of molecules passing and can be detected in different ways. One method is called the *constant temperature differential*. Thermal flow meters using this method have two temperature sensors—a heated sensor and another sensor that measures the temperature of the gas. Mass flow rate is computed based on the amount of electrical power required to maintain a constant difference in temperature between the two temperature sensors. A second, less popular, concept is a constant current method. Thermal mass flow meters using this method also have a heated sensor and another one that senses the temperature of the flow stream. The power to the heated sensor is kept constant. Mass flow is measured as a function of the difference between the temperature of the heated sensor and the temperature of the flow stream. Both methods are based on the principle that higher-velocity flows result in a greater cooling effect. And both measure mass flow based on the measured effects of cooling in the flow stream. The fluid temperature can also be measured and compensated for. If the density and specific heat characteristics of the fluid are constant, the meter can provide a direct mass flow readout, and does not need any additional pressure or temperature compensation over its specified range.

A particular advantage of thermal mass flow controllers is that the flow rate can be controlled very easily and flexibly via the regulating valve. This method works best with gas mass flow measurement, because it is difficult to get a strong signal using thermal mass flow meters in liquids, due to considerations relating to heat absorption. Thermal mass flow controllers must be calibrated in accordance with the type of gas. Depending on the type of calibration, the measurements can be directly traceable to mass or volume. Therefore, they are well suited to adjust the air flow for breath alcohol calibration gases. Technological progress has allowed the manufacture of thermal mass flow meters on a microscopic scale as MEMS sensors; these flow devices can be used to measure flow rates in the range of nanoliters or microliters per minute.

Several models of modern flow meters/controllers designed by various companies are shown in Figure 7.9. They are an excellent alternative to the conventional VA meters or rotameters.

Usually, a mass flow controller (MFC) is designed and calibrated to control a specific type of fluid in a particular range of flow rates. The MFC can be given a setpoint from 0 to 100% of its full-scale range but is typically operated at 10–90% of full scale, where the best accuracy is achieved. The device then controls the rate of flow to the given setpoint. In fact, most gases can be measured by a MFC as long as they are fairly clean and noncorrosive. For more aggressive gases, the meter may be made out of special alloys, and predrying the gas also helps to minimize corrosion. MFCs can be either analog or digital; a digital flow controller is usually able to control more than one type of fluid, whereas an analog controller is limited to the fluid for which it was calibrated. All mass flow controllers have an inlet port, an outlet port, a mass flow sensor, and a proportional control valve. The MFC is fitted with a closed-loop control system that is given an input signal by the operator (or an external circuit/computer), which it compares to the value from the mass flow sensor and adjusts the proportional valve accordingly to achieve the required flow. The flow rate is specified as a percentage of its calibrated full-scale flow and is supplied to the MFC as a voltage signal. Mass flow controllers require the supply gas to be within a specific pressure range. Low pressure will starve the MFC of gas, and it may fail to achieve its setpoint. High pressure may cause erratic flow rates.

At present there are a lot of different MFC on the market. Therefore, the basis of good flow meter selection is a clear understanding of the requirements of the particular application. According to expert opinion, some key questions need to answered before selecting a flow meter:

- What is the fluid being measured by the flow meter (air, water, etc)? If the liquid is not water, what is the viscosity of the liquid?
- Do you require rate measurement and/or totalization from the flow meter?
- Do you require a local display on the flow meter, or do you need an electronic signal output?

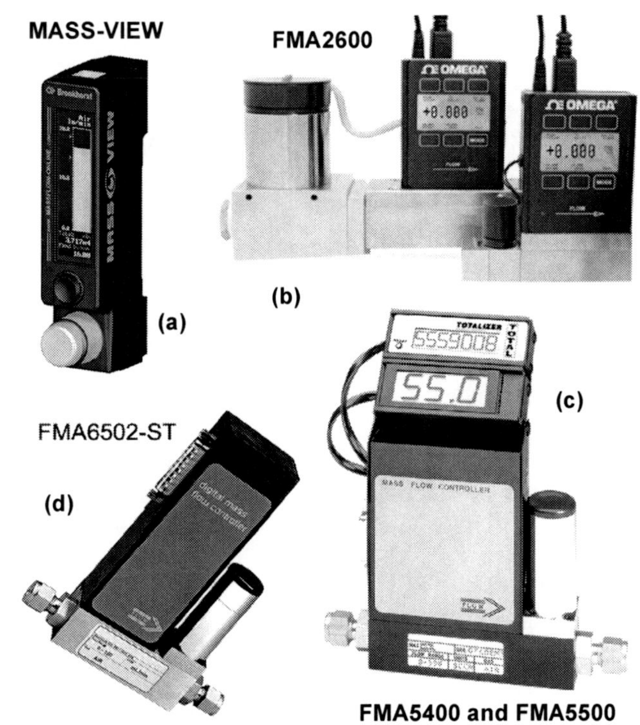

Figure 7.9. Examples of modern gas flow meters/controllers designed by various companies. (Reprinted from www.bronkhost.com and www.omega.com.)

- What are the minimum and maximum flow rates for the flow meter?
- What are the minimum and maximum process pressures?
- What are the minimum and maximum process temperatures?
- Is the fluid clean? Is the fluid chemically compatible with the wetted parts of the flow meter?
- If this is a process application, what is the size of the pipe?

6.3.4. Preparation of Calibration Gas Mixtures

If you do not have accredited cylinders containing calibration gases, gas mixtures for gas sensor calibration can be prepared in the laboratory. This situation is actually prevalent, because usually one does not have cylinders with the required gas concentration. In addition, one needs to know that many gases can be premixed with air and stored under pressure, but some gases can only be mixed in inert gas backgrounds, such as nitrogen. This means that these gases for calibration in many cases can be used only after mixing with air or oxygen. Cross-calibration also usually requires preparing specific gas mixtures which cannot be purchased from gas companies. And, of course, every sensor is subject to interference by other gases.

The techniques of preparation can be divided into those which are (1) static, or low-pressure static dilution, when calibration gas mixtures are prepared statistically, and (2) dynamic, when calibration gas mixtures are prepared dynamically by using various approaches such as permeation or diffusion devices (Dorko and Hughes 1987; Mitchell 1987; Bair 1999; ISO 2001, 2003, 2005, 2009; Gajdošík 2006; Zhou and Gai 2006; Pratzler et al. 2010). A detailed description of the methods used for preparing standard gas mixtures has been given by Barratt (1981). In the case of static methods, calibration gas

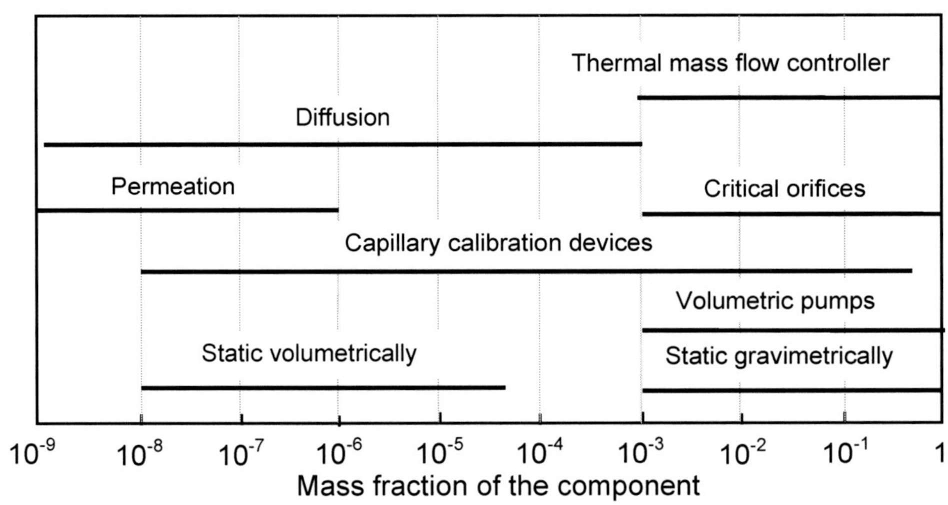

Figure 7.10. Concentration ranges of components for different preparation methods. (Data from ISO 2001, 2003, 2005, 2009; Pratzler et al. 2010.)

mixtures can be prepared gravimetrically or volumetrically, depending on whether the composition is determined by the mass or the volume of the components. In volumetric methods, the prepared gas mixture is saved in a bottle of known volume from which the gas is extracted as necessary. For this technique the partial volume method or the partial pressure method can be used. In the case of the dynamic technique, the gas mixture is prepared continuously and is used immediately for the required purpose. The concentration of the intermediate mixture is at a level where the mixture is known to be stable or is measured just before use. These higher-concentration intermediates are frequently much easier both to stabilize and to analyze by standard chemical techniques. At present the method of mixing flows is the most usable one in many experimental labs during gas sensor design and testing. Dynamic calibration systems are especially useful when working with reactive gas mixtures. If the component of interest is unstable or reactive, the products are swept away and are continually replaced by new and unreacted components. In addition, dynamic systems have a much wider range of applicability than fixed-concentration standard mixtures. With any preparation method, the test mixture is formed in general by combining several test gas substances and a carrying gas for dilution.

For the selection of a suitable preparation method for calibration gas mixtures, various factors must be taken into account. In particular, the following are important (Pratzler et al. 2010).

- *Chemical and physical properties* of the individual components. For example, volumetric pumps are only suitable for mixing gases with each other.
- *Required concentration ranges.* Not every method permits mixing the complementary gas and the components at any desired ratio (see Figure 7.10.). For example, for the saturation method, the achievable concentration range is strongly dependent on the vapor pressure of the components, and general statements cannot be made. Calibration gases with very low concentrations of components are often produced by permeation.
- *Availability of the gas mixture* (pressure, temperatures, dynamic or static use).

6.3.4.1. STATIC TECHNIQUES

6.3.4.1.1. Partial Pressure Method

For a gas mixture at a constant temperature and constant volume, the total pressure p for an ideal gas is equal to the sum of the partial pressures p_i of the individual components:

$$p = \sum_{i=1}^{N} p_i \tag{7.3}$$

where N is the number of gas components and partial pressure, p_i, is given by the equation

$$p_i = \frac{n_i RT}{V} \tag{7.4}$$

Here n_i is the amount of the ith gas component of the mixture, R is the universal gas constant, T is thermodynamic temperature, and V is the total volume of the mixture.

For the partial pressure method it is not necessary to know the volume V of the mixture, but it is necessary to measure the partial pressures p_i of the mixed gases. At temperature T a certain amount of gas is filled to the pressure p_i in an evacuated bottle. Then the bottle is filled by the carrier gas to the total pressure p. Taking into account that in an ideal gas the ratio of partial pressures is the same as the ratio of molecules, the gas component concentration in the gas mixture can be calculated according to the relation

$$X_i = \frac{p_i}{p} \tag{7.5}$$

6.3.4.1.2. Partial Volume Method

The partial volume method is based on the fact that at constant temperature T and constant pressure p, the total volume V is given by the sum of the partial volumes V_i of the partial components of the mixture:

$$V = \sum_{i=1}^{N} V_i \tag{7.6}$$

The partial volume of a particular gas is the volume the gas would have if it alone occupied the volume, with unchanged pressure and temperature. The partial volume can be calculated using Eq. (7.4):

$$V = \frac{n_i RT}{p_i} \tag{7.7}$$

The relative concentration of component X_i in the mixture can be expressed by the relation

$$X_i = \frac{V_i}{V} \tag{7.8}$$

During the preparation of the mixture it is possible to take steps so that the specific volume V_i of the gas is filled into the evacuated bottle, which is then filled up by the carrier gas of the volume, $V - V_i$. The volumes are measured at equal temperatures T and equal pressures p.

6.3.4.2. DYNAMIC TECHNIQUES

At present there are several methods that allow the dynamic production of desired gas calibration mixtures. They include dynamic dilution (single and multistage), diffusion, permeation, electrolytic methods, and chemical reactions (Mitchell 1987).

Electrochemical methods have long been used in analytical chemistry. Using such methods, many gases can be synthesized on a laboratory scale as a by-product or side reaction of electrolytic or coulometric methods. Controlled addition of a gas of interest to the gas flow of a carrying gas (diluent) also may be accomplished using a chemical reaction system. If the material of interest is inordinately reactive, unavailable commercially, or strongly toxic, it may be desirable to produce it continuously as needed. Typical types of reactions that are used for this method of producing calibration mixtures of desired components are recombination, catalytic, thermal decomposition, rearrangements, and photochemical dissociation. The reactions may be carried out in either the gas or liquid phase. However, at present, the methods mentioned above are used only rarely for gas sensor calibration. Permeation/diffusion systems and dynamic dilution systems find much wider application for these purposes. In most situations, either one or a combination of both a permeation/diffusion system and a dilution system will yield a solution to a problem.

6.3.4.2.1. Diffusion Systems

Diffusion systems have been used for a long time in analytical chemistry for preparing calibration gas mixtures (Nakamoto et al. 1991; ISO 2005; Zhou and Gai 2006; Pratzler et al. 2010). The component is filled into a diffusion vessel which is provided with a diffusion tube and is surrounded by the carrier gas flow; see Figure 7.11. Molecules from the liquid phase evaporate into the gaseous phase until the saturated vapor pressure is reached. Then, molecules in the gas phase rise into the diffusion tube and, after diffusion through the diffusion tube, mix with the carrier gas in one step. The quantity of liquid which diffuses into the carrier gas depends on the temperature, the geometry of the diffusion tube, and the vapor pressure of the component. The diffusion rate can be determined by weighing the diffusion vessel and is thus directly traceable to the mass. Zhou and Gai (2006) have shown that the relative uncertainty of the composition of the calibration gas mixture prepared by the use of diffusion tubes is about 5%.

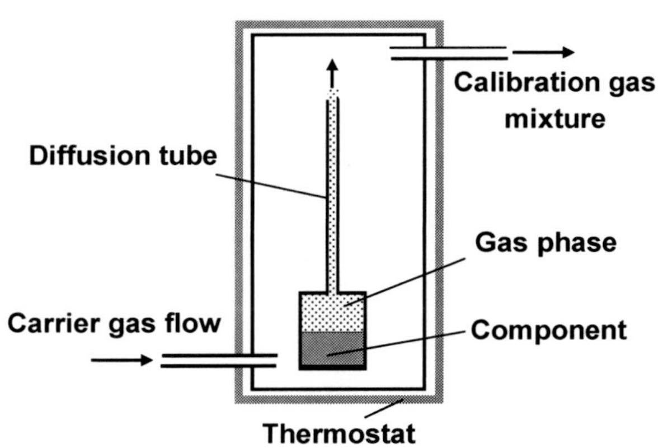

Figure 7.11. Schematic diagram of a diffusion cell.

6.3.4.2.2. Permeation Systems

The operation of permeation systems is very similar to that of diffusion systems (Staude 1992). However, their application is much more handy. A *permeation device* is a sealed small container filled with a pure chemical compound in a two-phase equilibrium between its gas phase and liquid phase. The gas molecules are permeated either through a permeable container wall or through the end cap and mixed with a carrier gas to obtain a known gas mixture used as reference in gas testing equipment. This phenomenon leads to a very versatile method of producing ultralow concentrations of calibration gas mixtures. In preparing a permeation device, the liquid form of the gas is usually sealed inside a polymeric container, typically, polytetrafluoroethylene (Teflon). The rate at which the gas molecules permeate depends on the permeability of the material and the temperature. Therefore, permeation tubes are held in a carrier-gas stream of dry air or nitrogen at a constant temperature. At a known rate of permeation at a given temperature, a constant flow rate of air mixed with the permeated chemicals forms a constant stream of calibration gas. The rate of permeation is constant over long periods of time if the temperature, concentration gradient, and tube geometry remain constants. Thus, a calibrator with constant temperature

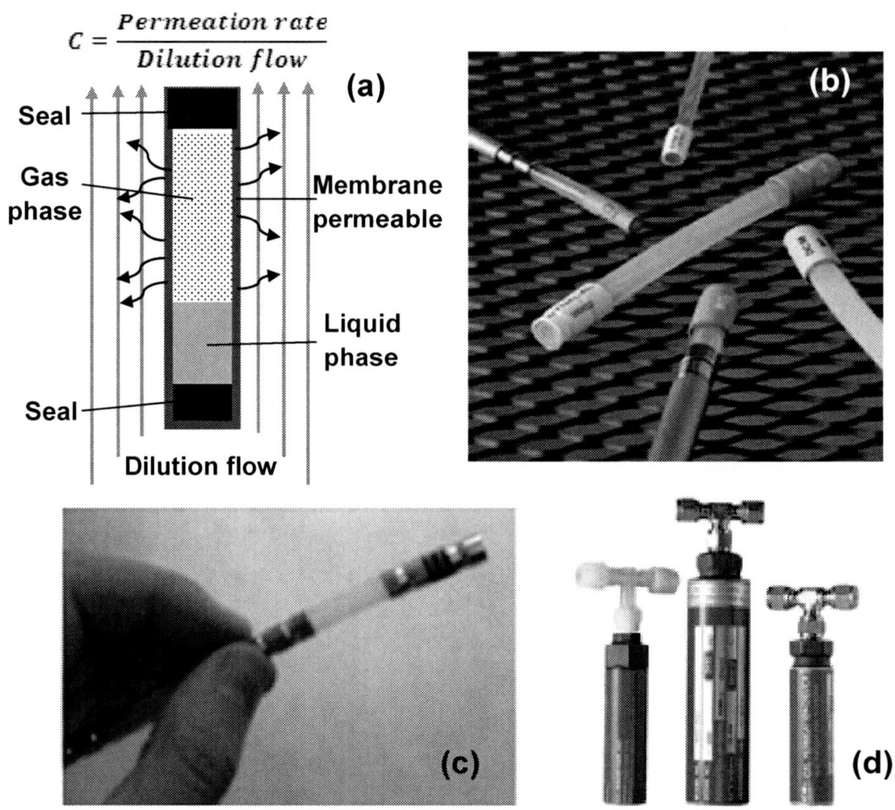

Figure 7.12. (a) Diagram illustrating construction and operation of permeation tube. (b–d) Examples of permeation tubes fabricated by various companies (www.vici.com and ww.owlstonenanotech.com).

and flow regulation is needed for this method. The substances that work best with this method are those that have a critical temperature above 20–25°C. Permeation tubes are available for a large number of substances, including reactive and corrosive gases, such as sulfur dioxide (SO_2), nitrogen dioxide (NO_2), hydrogen sulfide (H_2S), etc. Construction and some typical permeation tubes are shown in Figure 7.12.

Permeation devices have several advantages. They are small, inexpensive, and convenient to use. They are ideal for lab environments. Usually, permeation devices are the size of a pencil (see Figure 7.10). With these devices, there is a low degree of complexity, a wide range of permeation rates can be achieved—normally from 20 ng/min to 500 μg/min—and accurate, stable concentrations range from parts per billion to high parts per million. Permeation devices are ideal for field use, and high precision and excellent accuracy may be obtained for many gases (Mitchell 1987). And, after they are depleted, they can be disposed of easily. However, permeation tubes continuously emit chemicals at a constant rate, thus creating a storage and safety problem (Chou 2000). Also, the rate of permeation for a given gas of interest can be too high or too low for a given application. For example, high-vapor-pressure gases permeate too quickly, while very low-vapor-pressure chemicals have a permeation rate that is too slow to be of any use.

6.3.4.2.3. Dynamic Dilution or the Method of Mixing Flows

A variety of systems have been designed for injection of known, controlled amounts of the component of interest into the flowing diluents stream. One such system is shown in Figure 7.13. Usually a dynamic gas dilution system has sources of calibration and diluent gases (air), and several channels with gas flow

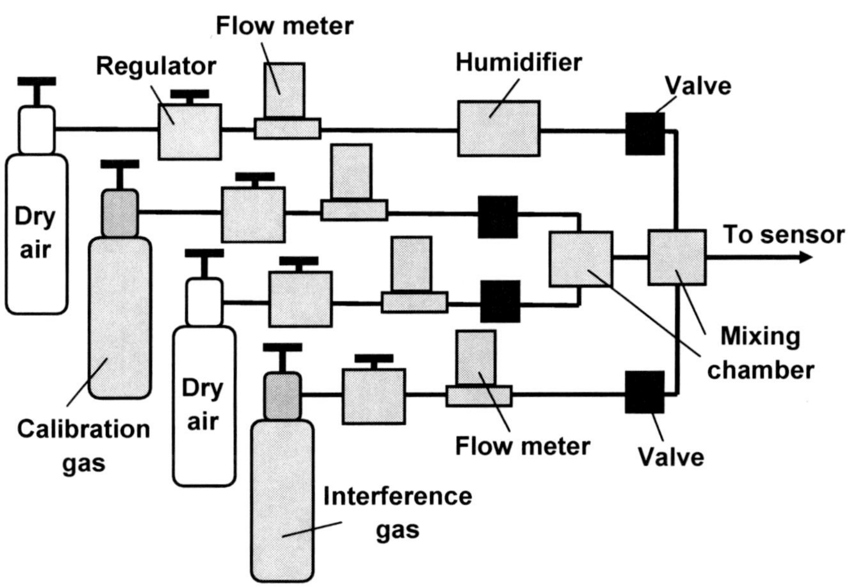

Figure 7.13. Schematic diagram of gas system used for gas sensor calibration.

controllers and mixing chambers. In particular, the system shown in Figure 7.11 has two additional channels intended for calibration-gas humidification and for cross-calibration. At constant pressure p and constant temperature T, the flows q_i of gases are mixed continually.

$$q = \sum_{i=1}^{N} q_i \tag{7.9}$$

The concentration of calibration gas in the output stream of dilution system depends on the initial concentration of calibration gas in the cylinder and the ratio of gas flows in the channels.

If the values of flows q_i are known, the relative concentration X_i can be calculated from

$$X_i = \frac{q_i}{q_i + q_n} = M \tag{7.10}$$

where M is the dilution constant determined by the ratio of the flows q_i and q_n. Systems with modern gas flow controllers can have dilution constants exceeding 100. If greater dilutions are desired, it is possible to connect the dilution units to a cascade. However, multistage dilution techniques that require more than two dilution steps become considerably more difficult (Mitchell 1987). As more dilution steps are added, the system pressure builds up, and the quantity of the gas of interest that is needed becomes the major concern. In such arrangements, as much as 99% of the test gas could be bled off and not used in the output stream. In addition, where multistage systems are used, at the low end of the concentration range consideration must be given to additional problems that may occur from the following (Mitchell 1987):

- *Insufficient flow rate control.*
- *Back diffusion.* Diffusion of the sample outward or contaminant or even a major constituent inward—probably one of the most frequent sources of error in gas analysis. The permeabilities of the calibration system to both the gases of interest and unwanted contaminants have to be determined if reliable results are to be obtained.
- *Surface absorption.* This problem is most critical when considering trace levels of reactive gases. Losses in calibration mixtures may occur either by primary absorption on container walls or connecting gas pipes, or by secondary absorption that occurs through chemical reaction of component of interest with previously absorbed species.

Therefore, for reasons of desired accuracy, it is not recommended to mix the flows with a ratio greater than q_i:$q_n \leq$ 1:100.

The mechanical features of the sampling system can also introduce various errors if they go undetected. Sources of these errors are leaks, faulty seals, corrosion of vital parts, etc. In addition, Jardine et al. (2010) have noted that dynamic methods using diffusion and permeation tubes often suffer from long and tedious gravimetric calibration periods, in which permeation rates are determined by a change in mass over time (weeks for diffusion tubes, months for permeation tubes), the limited commercial availability of compounds (permeation tubes), and their associated high prices (typically \$200–\$1000 for a single permeation tube). Other disadvantages are the need to accurately and precisely control the permeation chamber temperature, degradation of reactive VOCs with time (hydrolysis, hydration, oxidation, etc.), changes in the permeation characteristics of the diffusion membranes over time (permeation tubes), and difficulties preparing complex mixtures of VOCs (each permeation tube is often certified at a different temperature).

6.3.4.3. ORGANIC LIQUID GAS MIXTURE PREPARATION

There are various methods to generate continuously a fixed concentration of volatile analyte from liquid samples instead of preparing gas mixtures in a cylinder with a limited volume. Theoretically, for the preparation of a gas mixture of the vapors of organic liquids, it is possible to use any of the methods mentioned above (Barratt 1981; Vitenberg et al. 1984; Nakamura et al. 1999). For example, with the partial pressure method, a small ampoule containing the necessary volume of organic liquid is placed into the bottle. After the bottle is evacuated, the ampoule is broken. The liquid is evaporated and the partial pressure p_i is created in the bottle. If the volume of organic liquid is sufficient, an equilibrium is created between the gas and liquid phases in the bottle. In a closed systemm a constant, substance-specific vapor pressure—the saturation vapor pressure—is formed above the liquid phase of a pure substance or mixture, which depends on the temperature and the composition in the case of a liquid mixture. In this case, the partial pressure p_i for the closed system at a given temperature can be calculated using the Antoine relation (Boublik et al. 1984):

$$\log p_i = A - \frac{B}{T + C} \tag{7.11}$$

where p_i is the saturated vapor partial pressure, A, B, and C are the experimental constants for the given substance, as tabulated, for example, by Wichterle and Linek (1971), and T is the Celsius temperature of the liquid–gas system. Then the bottle is filled with the carrier gas to the total pressure p. Now the relative concentration is given by the relation

$$X_i' = \frac{p_i}{p} \tag{7.12}$$

where X_i' is the relative concentration of the substance in the gas phase and p is the total pressure of the system.

For most cases it is acceptable from the standpoint of accuracy to use the liquid evaporation in the closed system in the presence of air at atmospheric pressure during saturated vapor preparation. In this case, the real partial pressure is somewhat lower than that which is calculated by the use of the tabulated constants. To prepare a calibration gas mixture, it is possible to dilute the obtained saturated vapor of the concentration X_i' using dry clean air to the desired concentration X_i. For the partial volume preparation method it is possible to use the equation

$$X_i = X_i' \frac{V_i'}{V} \tag{7.13}$$

where V_i' is the volume of saturated vapor volume used, V is the total volume of the prepared final concentration mixture X_i, and $V - V_i'$ is the volume of dilution carrier gas.

However, as a rule, the saturation method in dynamic mode is used for the preparation of calibration gases containing vapors of organic liquids (Mackay-Sim et al. 1993; Bodenhofer et al. 1996; Smith et al. 2004; Pratzler et al. 2010). A major disadvantage of using static methods for the preparation of VOC standards is the absorptive losses on surfaces such as bags and gas cylinder walls (Jardine et al. 2010). This loss is difficult to account for, especially in the parts-per-billion range and is particularly significant for VOCs with low vapor pressures, such as semivolatile organic compounds and other compounds with strong surface interactions such as oxygenated VOCs. Pure dynamic methods minimize wall effects, allow rapid variations in concentration, and enable the generation of standards when needed.

If a gas is passed through the liquid, it enriches itself with the vapor of the liquid components. At thermodynamic equilibrium, the concentration in the vapor phase corresponds to that of the saturation state (Boublik et al. 1984). The bubbler method, in particular, has been utilized extensively in chemical vapor deposition systems as well as in sensor measurement systems because of its simple setup. An important point in such bubblers is to achieve a steady, controllable flux of an analyte sample. Common restrictions are the formation of aerosols leading to a higher analyte concentration in the gas phase, and the decrease in pickup efficiency due to sample depletion around the bubble-passing channels (Kim et al. 2007). However, it is possible to minimize these problems through optimized bubbler design (Smith et al. 2004) and correcting via a system calibration (Mackay-Sim et al. 1993). To reach the state of equilibrium, several steps of enriching the concentration in the vapor phase may have to be realized successively. A conventional realization of a calibration gas generator working according to a three-step saturation principle is shown in Figure 7.14. The purified pressurized air flows successively through three bottles filled with the same organic liquid solution of known concentration, of which the temperature is controlled (see Figure 7.15). Of course, one- or two-step saturation can be used as well (Kim et al. 2007).

The setup in Figure 7.14 can be used as an air humidifier as well. Data related to water saturated vapor pressure and density of water vapor are presented in Table 7.11.

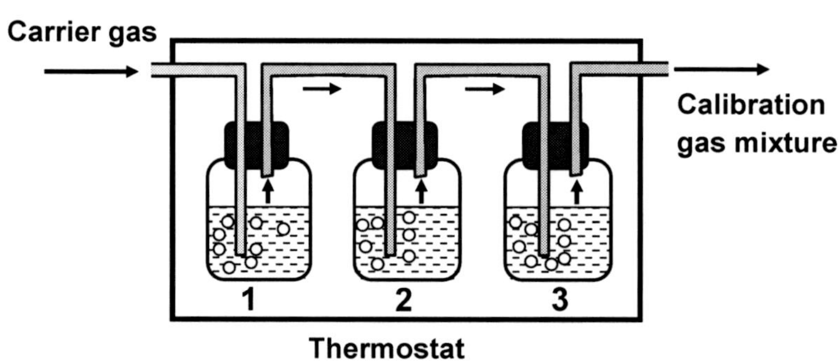

Figure 7.14. Setup of a three-step calibration gas generator according to the saturation method.

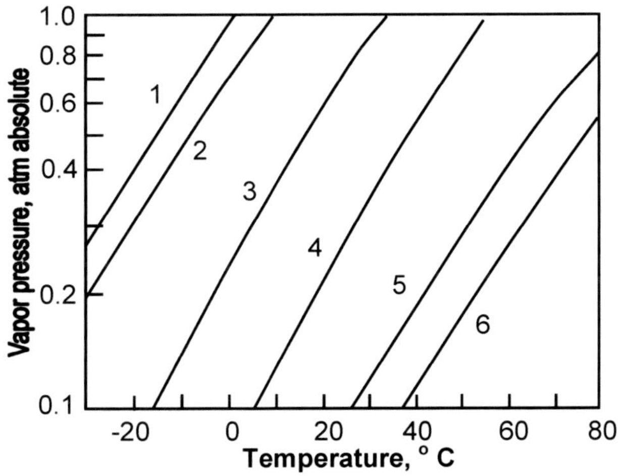

Figure 7.15. Temperature influence on the saturated vapor pressure of several organic substances: 1, butane; 2, neo-pentane; 3, diethyl ether; 4, methyl acetate; 5, fluorobenzene; 6, 2-heptene (http://en.wikipedia.org).

The vapor pressure of water may be approximated by the following relation:

$$p = \exp\left(20.386 - \frac{5132}{T}\right) \tag{7.14}$$

where p is the vapor pressure (mm Hg) and T is the temperature in kelvin.

Table 7.11. Water saturated vapor pressure and saturated vapor density

Temperature (°C)	Saturated vapor pressure (10³ Pa)	Saturated vapor pressure (mm Hg)	Saturated vapor density (g/m³)
0	0.6105	4.58	4.85
5	0.8722	6.54	6.8
10	1.228	9.21	9.4
20	2.338	17.54	17.3
30	4.243	31.8	30.4
40	7.376	55.3	51.1
50	12.33	92.5	83.1
60	19.92	149.4	130.5
70	31.16	233.7	198.6
80	47.34	355.1	293.8
90	70.10	525.8	424.1
100	101.3	760.0	598.0

Source: Data from www.vivoscuola.it/US and www.engineeringToolbox.com.

If one uses additional dilution of vapor flow applying the method of mixing flows, the final concentration of the required substance, X_i, in the gas phase can be calculated from

$$X_i = X_i' \frac{q_i'}{q_i' + q_n} = X_i' M \tag{7.15}$$

where q_i' is the flow of the air saturated by the liquid vapor and q_n is the flow of diluting clean air.

Of course, the methods described above can be used for gas sensor testing during gas sensor design and study. More detailed descriptions of methods used for preparing calibration gas mixtures could be found in the literature (Barratt 1981; Chou 2000; ISO 2001, 2003, 2005, 2009; Gajdošík 2006; Pratzler et al. 2010).

6.3.5. Measurement (Test) Chamber

There are no strong general requirements for a test chamber aimed at calibration or study of chemical sensors. Every chamber is designed for a specific task and therefore configuration of the chamber first of all has to simulate or reproduce the conditions of exploitation and provide control of specific functional sensor parameters in the proper way. Therefore, depending on the task, the outward appearance of the chamber can change considerably. As a result, in practice one can find a lot of different configurations of those chambers. Several of them are shown in Figure 7.16.

It can be seen that the chambers shown in Figure7.16 differ in size, component parts, and end uses. For example, the chamber shown in Figure 7.16b was designed for field applications, the chamber in Figure 7.16d for testing room-temperature gas sensors, and the environmental sensing chamber (ESC) in Figure 7.16c was designed for testing wireless chemical sensor networks (WCSNs). The ESC was constructed with multiple inlet/outlet ports to allow for seeding of the environment at a number of different locations. The spatial arrangement of the five chemical sensor nodes used to form the WCSN in the ESC was for acetic acid plume tracking through the chamber. Sensors 1–4 were arranged in ascending order in the channel at increasing distances from the acetic acid input point. Sensor 5 was positioned outside and above the channel close to the extraction point. The chamber shown in Figure 7.16a has a volume equal to 2 m^3 (2 m × 1 m × 1 m) and 5 installed sensors, while the chamber shown in Figure 7.16d has a volume of only 30 cm^3 and can maintain 16 sensors. The flow channel in this chamber has a gap of 3 mm. However, we cannot say that some of these chambers are better than others: There is too big a difference in the end uses of these chambers. For example, during design of the chamber for an infrared gas sensor based on a pyroelectric detector (see Figure 7.16a), besides observing requirements related to the gas stream, the developers also had to observe requirements related to optical beam passing (Han et al. 2010). In particular, (1) to improve detection accuracy, the chamber had to provide multiple passes of the optical beam; (2) to reflect more infrared light to the pyroelectric detector, the internal wall of the optical air chamber had to have small surface roughness, high reflectivity to infrared light, stable chemical properties, and no adhesion to the detected gas; (3) the optical air chamber had to have an airtight structure and at the same time have several holes for gas; and so on.

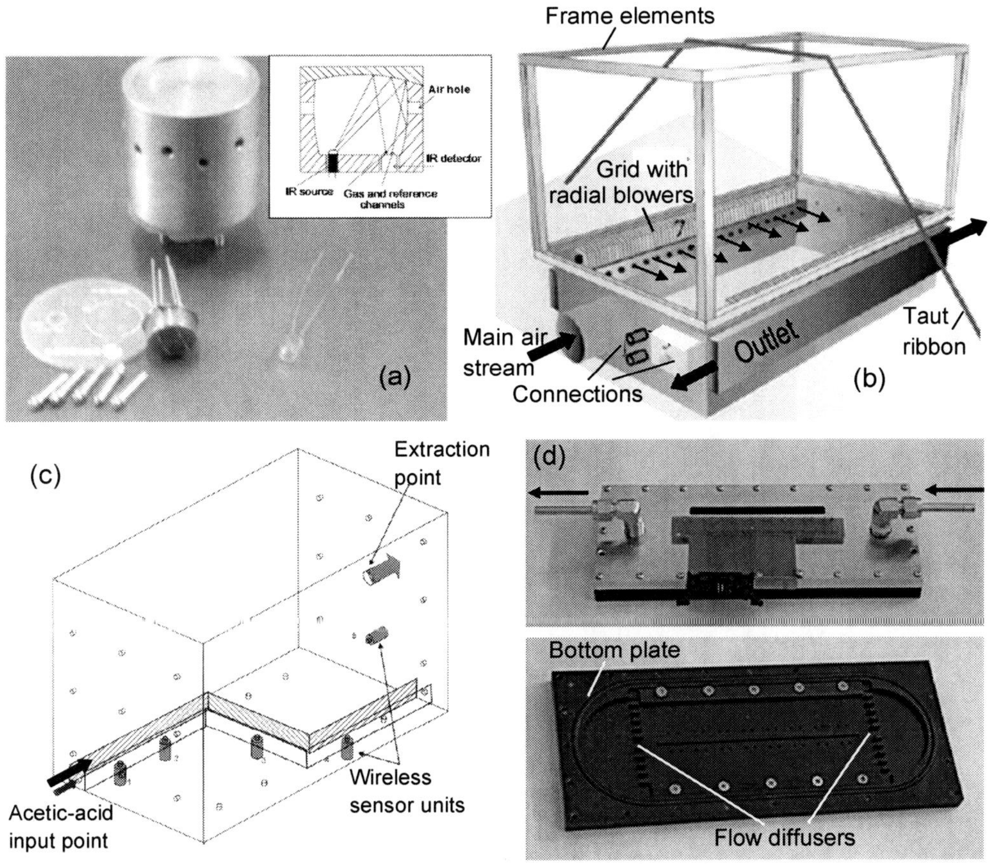

Figure 7.16. Views and schematic diagrams of various test chambers for chemical sensors. (a) Side-opening chamber for IR optical gas sensor. (Reprinted with permission from Han et al. 2010. Copyright 2010 IEEE.) (b) An automated, portable, and non–steady-state through-flow canopy–chamber assembly for measurements of CO_2 and water vapor fluxes on field crops and for simultaneous recording of environmental variables within the chambers and in the field. (Reprinted with permission from Muller et al. 2009. Copyright 2009 Elsevier.) (c) Environmental sensing chamber for plume delivery and extraction study. (Reprinted with permission from Shepherd et al. 2007. Copyright 2007 Elsevier.) (d) A detection chamber for testing 16 polymer-based gas sensors. (Reprinted with permission from Kim et al. 2007. Copyright 2007 Elsevier.)

Taking into account the huge number of applications where specific test chambers are required, we will consider further only chambers designed for testing solid-state gas sensors. Normally, the gas sensor is placed in a chamber, and the gas flow through the chamber provides the sensor ambient. Through many experiments it has been established that the measurement chamber, which houses the sensors, is really an important component of the test system, because chamber geometry greatly affects sample concentration inside the chamber (Falcitelli et al. 2002). It is known that solid-state gas sensors respond to the local sample concentration on the surface of the sensitive layer rather than to the average concentration inside the chamber. This means that data collected during characterization of these sensors will be affected by sensor position and concentration transients. Rapid establishment of a uniform concentration is therefore desirable to avoid poor measurement reproducibility due to random fluctuations, as happens when recirculating regions are present. Di Francesco et al. (2005) note correctly that quite often the exposure of the sensor to the sample is not considered a particularly critical phase of the measurement process, hence resources devoted to the design of these devices are limited. As a result, in many cases the chamber design and sensor arrangement do not guarantee identical exposure conditions. Different distances from sensors to the inlet cause a time shift in concentration profiles over the sensitive layers, while sudden changes in the direction of the flow, sharp variations of the hydraulic section, and dead volumes produce recirculating zones that lead to an uneven distribution of the sample concentration.

Experience has shown that for better reproducibility the gas mixture in the test chamber first of all has to be as homogeneous as possible. This effect is mainly influenced by the geometry of the measurement chamber. This means that the sensor chamber should be optimally designed for the gas flow. The chamber shown in Figure 7.17a is one of the first successful attempts to resolve this problem. The chamber designed by Endres et al. (1995) has a linear configuration. The linear chamber with small volume was developed for a suitable gas stream to prevent poorly purged locations. Parts responsible for the homogeneity and gas flow optimization of the gas mixture can also be added to a test chamber. In particular, two flow diffusers are furnished in the chamber shown in Figure 7.16d, near both the inlet and the outlet along the perpendicular direction to the flow, for the purpose of improving flow uniformity over the sensors and minimizing the inefficient stagnation zone (Kim et al. 2007). However,

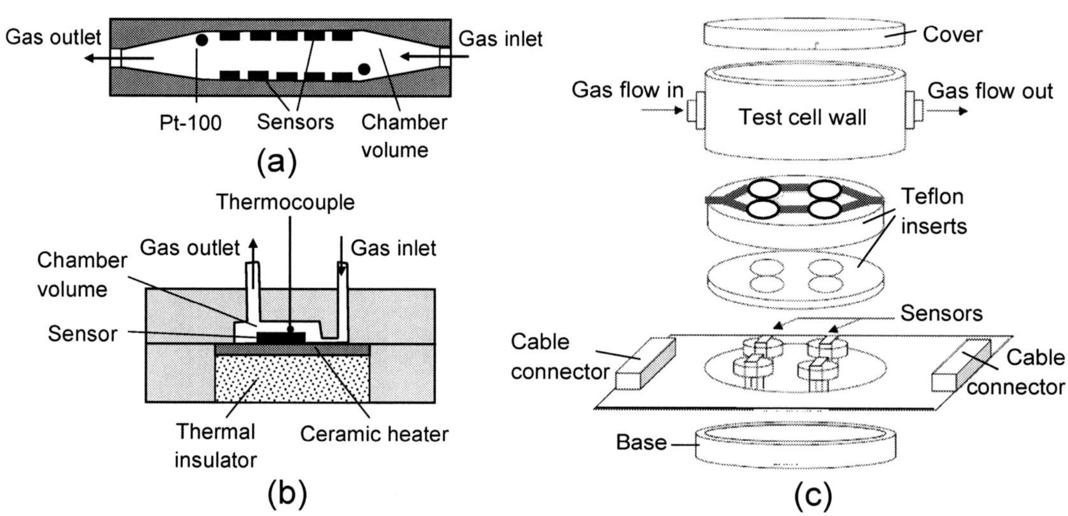

Figure 7.17. Schematic diagrams of test chambers with small volumes. (a) Chamber designed by Endres et al. (1995). (Reprinted with permission from Endres et al. 1995. Copyright 1995 Elsevier.). (b) Chamber designed by Kida et al. (2008). (Reprinted with permission from Kida et al. 2008. Copyright 2008 Elsevier.) (c) Chamber designed by by Pilling et al. (2003). (Reprinted with permission from Pilling et al. 2003. Copyright 2003 Elsevier.)

usually these special units for gas mixing in small test chambers are not necessary. Most test gases have low molecular weights and therefore have large enough diffusion constants in air. Additionally, the gases are mixed by turbulent flows, caused by valves, etc.

As is known, sensitivity and response and recovery times are important parameters of gas sensors. However, it is necessary to admit that most researchers direct their efforts toward the measuring magnitude of sensor response only and therefore they usually use test chambers with large volumes. In the literature one can find descriptions of measurement chambers with volumes up to 1–5 × 10^3 cm^3 and higher. However, as modeling and experiment carried out by Lezzi et al. (2001) for test chambers with volumes of 1000 cm^3 and 233 cm^3 and for flow from 100 to 500 cm^3/min have shown, the use of a large-volume chamber is a faulty approach even for experiments in steady-state conditions. Numerical results confirmed that geometric parameters such as the measurement chamber volume, the chamber shape, and the inlet and outlet positions play important roles for the fluid flow and the concentration fields inside the chamber as well. With a poorly designed test chamber, characterized by large volumes of almost stagnant fluid and/or by a peripheral position of the sensor with respect to the gas stream, the concentration of calibration gas near the sensor surface does not correspond to the concentration of the calibration gas after preparation, and the calibration gas concentration on the sensor surface may continue to increase for quite a long time. It has been established that in many cases the time taken for replacement of the gas atmosphere in a gas chamber from base air to sample gases is much longer than the time required for the sensor to reach steady state. In this case, the device response time is more a measure of the change of tested gas concentration inside the chamber than a characteristic feature of the device. This means that these chambers cannot be used for measuring operating characteristics of gas sensors in transient mode and determining their response and recovery times. Even chambers with volumes of 100 cm^3 are too large to quickly introduce gases into the chamber within a sufficiently short time.

There are two ways to exchange the ambient quickly. Either the chamber has a normal volume and the gas has extremely high flow, or the gas flow is normal and the chamber is very small. As is known, the time constant of gas exchange in a measurement cell depends on its volume, the construction of the gas tubes, and the flow rate. Approximately, the purging time (t_{purge}) of the measurement chamber can be calculated according to the relation (Endres et al. 1995)

$$t_{purge} = 3\frac{V_{chamber}}{V_{tot}} \qquad (7.16)$$

where $V_{chamber}$ is the volume of the chamber, V_{tot} is the gas flow, and 3 is a security factor. The first principle was realized by Gerblinger and Meixner (1991), who mounted nozzles of 10 injection valves into a wall of a sensor chamber. However, this approach is not optimal because a gas stream moving at high speed strongly affects the temperature of the heated sensor.

There are variants of gas sensor position relative to the gas stream which contribute to considerable decrease of gas stream influence on the temperature parameters of sensor operation. The most commonly utilized variant is shown in Figure 7.18b. However, due to the additional volume in the measurement cell and the presence of an additional membrane between the gas flow and the gas sensor (see Figure 7.18b), the time to establish steady state in such a system after exchanging the gas atmosphere is longer than in the flow system shown in Figure 7.18a. In addition, the higher the speed of gas flow, the more

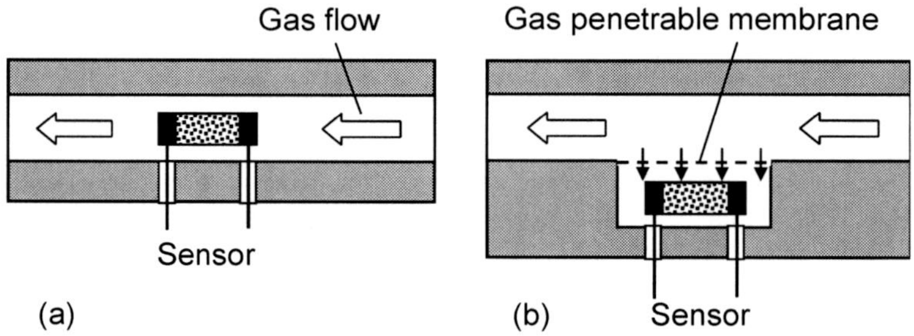

Figure 7.18. Variants of gas sensor position relative to gas stream.

calibration gas is used. This last factor considerably decreases the time during which measurements can be carried out using standard sources of calibration gases. This limitation is very important for devices designed for field applications Therefore, reducing the volume of the measurement chamber is a better choice than increasing the speed of the gas stream. Such an approach, aimed at decreasing the volume of the test chamber, was used by Pilling et al. (2003), Korotcenkov et al. (2004, 2007), Kim et al. (2007), and Kida et al. (2008), among others. Schematic diagrams of several chambers designed for gas sensor study are shown in Figure 7.16. For example, the chamber designed by Kim et al. (2007) for 16 sensors mounted in TO-8 packages (see Figure 7.17d) had a volume of 30 cm^3, the chamber designed by Pilling et al. (2003) for 4 sensors (see Figure 7.17c) had a volume 18 cm^3, while the chambers designed by Korotcenkov et al. (2004, 2007) and by Kida et al. (2008) (see Figure 7.17b) had volumes smaller than 0.4–0.6 cm^3. Further decrease of the test chamber size is not expedient, because at some point the diameter of pipes used in the test system starts to limit the time of gas atmosphere exchange.

The measurement chamber shown in Figure 7.17b allows exchanging the gas atmosphere in 1–2 s. To achieve fast exchange of the gas atmosphere in the chamber, the gas flow apparatus was equipped with a high-speed switching valve. Of course, to achieve this level of result, other parameters of the test system, such as the diameter of pipes and the distance between the mixing point and the target gas cell, should also be minimized. In particular, in the chamber shown in Figure 7.17b, the diameters of the gas inlet and exhaust tubes were set to 0.5 and 2 mm, respectively. This difference in size can reduce the pressure drop in the chamber by facilitating gas exhaust, thus achieving a maximum gas flow rate of 120 cm^3/min. A ceramic heater (MS-1; Sakaguchi E. H. VOC, Tokyo, www.sakaguchi.com) was placed at the bottom of the chamber, while three thermal insulating sheets (mullite) were placed directly below the heater so as to allow heating at 450°C at the maximum. According to Pilling et al. (2003), the estimated time constant of a chamber with a volume of 18 cm^3 was ~4 s.

It is necessary to note that test chambers developed to study fast processes in gas sensor response as a rule are designed to use unpackaged sensors, which do not have additional filters and membranes inhibiting penetration of the test gas to the active element of the gas sensor.

With respect to other approaches used for measuring fast response and recovery times, it is necessary to note first of all "the moving gas outlet technique" developed by Tobias et al. (1997, 1999) and first applied to SiC-based catalytic-gate field-effect gas sensors. This measurement technique allows producing an almost instantaneous change in the gas atmosphere at a sensor surface (Tobias et al. 1999). The measurement setup, designed to measure the short response times of gas sensors in the laboratory while using moderate flows of about 100 ml/min, is shown schematically in Figure 7.19. The equipment is placed in a box, which is flushed with nitrogen, to further decrease diffusion of active species from the surroundings to the sensor.

The small aluminum block shown in Figure 7.19 contains two gas chambers, one in each half. Gas tubes lead reducing gas into the first chamber and oxidizing gas into the second. Each chamber has an opening at the top where the gases flow out. The chambers are filled with glass wool to make the outflow steadier and independent from the chamber geometry. Four thick wires hold the aluminum block about 5 cm above the bottom plate. They are hooked at one end into the bottom plate and at the other end into the aluminum block. Another thick wire connects the aluminum block to a wheel. When the wheel rotates, the aluminum block moves like a windscreen wiper under the sensor and exposes the sensor alternately to gas from the first outlet and to gas from the second outlet. The amplitude of the periodic movement is 4 mm—i.e., there are 8 mm between the left turning point and the right turning point. The frequency can be chosen between 1.8 and 52 Hz. At a frequency of 20 Hz, the maximal lateral speed of the outlets is 0.5 mm/ms. The distance from the outlets to the sensor is less than 1 mm to keep small the amount of gas which has to be exchanged. The outlets are formed like triangles, as shown in the lower part of Figure 7.19, to meet two requirements: First, the exchange of the sensor ambient at the border between the two gases should be fast; and second, the gas flows in the periphery should be high enough to prevent surrounding gas from diffusing to the sensor. To monitor the actual position of the

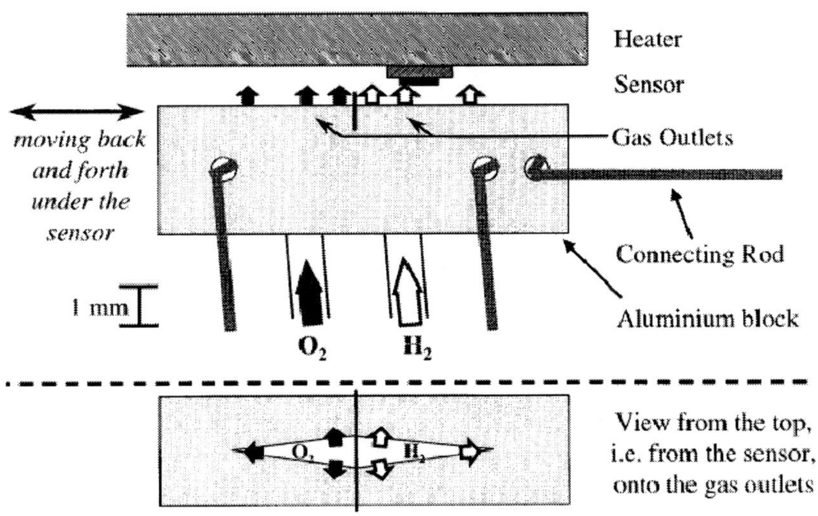

Figure 7.19. Schematics of the moving gas outlets, which can exchange the ambient around the sensor within milliseconds. In the position shown, the hydrogen-containing gas flows around the sensor. The connecting rod pulls the outlets to the right and, from a certain point on, the oxygen-containing gas replaces the old ambient of the sensor. The other end of the connecting rod is fixed to a rotating wheel. The gas outlets move back and forth and expose the sensor half of the time to oxygen and half of the time to hydrogen. At the border between the gas outlets, each outlet is wider, and the gas flow is stronger than farther away, as indicated by the density of the flow arrows. (Reprinted with permission from Tobias et al. 1999. Copyright 1999 Elsevier.)

gas outlets, the rotating wheel interrupts the beam of a light barrier during one half of the rotation. This position signal indicates when the transition between the two gases occurs.

This technique in various modifications has been realized with good results (Schalwig et al. 2002; Ghosh et al. 2005; Helwig et al. 2007, 2008). Schematic diagrams of these chambers are shown in Figure 7.20. The principle of operation of these apparatuses is understandable from the title of the technique and the schemes shown in Figure 7.20. Using a moving gas outlet, it was possible to change the gas around the sensor from one component to the next within several milliseconds in a stepwise manner.

In the chamber designed by Ghosh et al. (2005) (Figure 7.20a), the planar sensor is mounted face down. The gas outlets move sideways below the fixed sensor. In the "moving part," the two gas jets are separated by a thin barrier, extending 0.3 mm above the surface. Using a set of calibrated shims, the separation between the sensor active surface and the moving gas outlet can be varied between 0.85 and 1.15 mm. During measurements the gas environment around the sensor is determined by the upward-flowing gas stream impinging on the active sensor surface. Typical flow rates are in the range of 100–500 ml/min. The electrical actuator can rotate the moving part between two stable positions. The actuation translates the gas outlets by 3 mm in 2 ms, thereby switching the gas under the sensor.

In the apparatus shown in Figure 7.20b, reactive gases are supplied in the form of forced gas flows directly to the heated sensor surface. Rotating around its axis, the rotator wheel alternately admits a gas stream of either gas 1 (synthetic air) or gas 2 (synthetic air with an admixture of a reactive gas) into jet nozzles blowing onto the sensor surface. So, a rotator wheel with a slot acts as switching element. As a result, there are step-function changes in the gas concentration directly at the sensor surface. The sensor itself is mounted on a ceramic heater substrate to control the sensor temperature. Experiments carried out by Helwig et al. (2007, 2008) showed that in this mode of operation the intrinsic—i.e., the surface reaction–limited—response and recovery time constants of a gas sensor can be determined.

Of course, to determine the response time of sensors, one can use methods such as modulation of the hydrostatic pressure while keeping the composition constant (Tragut and Härdtl 1991), the rapid

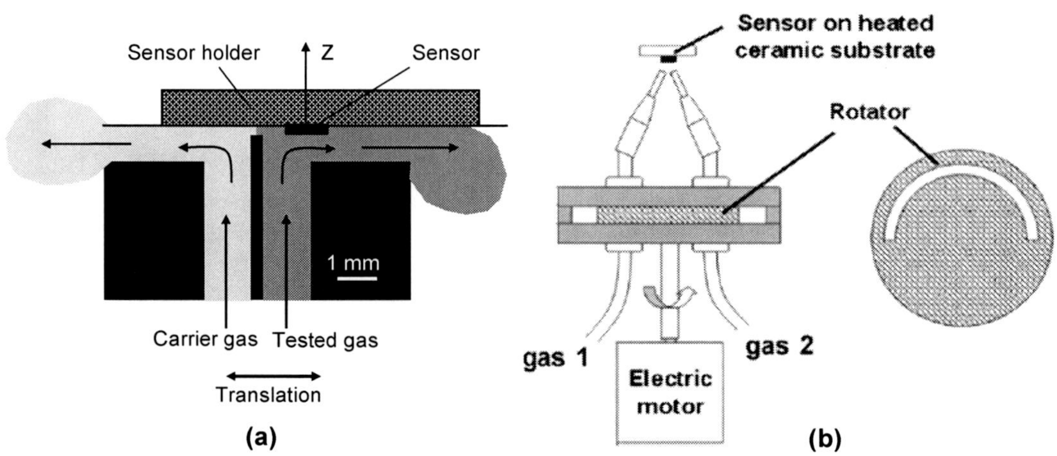

Figure 7.20. Schematic diagrams illustrating the configuration and operation of moving-gas outlet setups. [(a) Reprinted with permission from Ghosh et al. 2005. Copyright 2005 IEEE; (b) reprinted with permission from Helwig et al. 2007. Copyright 2007 Elsevier.]

gas composition changes resulting from combustion in an engine (Lampe et al. 1992; Tobias et al. 1997), or using vacuum for fast transport of gas analytes to a surface (Kelleter et al. 1992). However, none of these approaches is optimal due to the complexity of the test system and difficulties in arranging the measurement process.

Frequently, test systems for gas sensors should produce measurement data for pattern-recognition tasks. For that reason it is necessary to merge different sensors into one sensor array, where all the individual sensors are measured simultaneously. Additionally, higher accuracy of sensor response applies as well to the comparison of different sensor types, attainable by putting several sensors together into one gas combination, instead of measuring them in succession (Endres et al. 1995). According to Di Francesco et al. (2005), an ideal chamber for these purposes should have the following features: (1) allow all the sensors to be exposed simultaneously under the same conditions and make it possible to reproduce accurately the shape of the input concentration signal at each sensor position; (2) yield the same concentration profile in repeated measurements and short analyte concentration rise/fall times; (3) avoid memory effects and sample dilution, and (4) use a small amount of sample. Similar statements can be made about electronic noses and tongues (Falcitelli et al. 2002; Lezzi et al. 2001). We need to note that many of chambers we have described provide the ability to measure several sensors simultaneously, from 4 sensors in the chamber shown in Figure 7.17c to 16 sensors in the chamber shown in Figure 7.16d. Various approaches can be used for design of such chambers (Endres et al. 1995; Di Francesco et al. 2005). In particular, chambers can have either linear (Figures 7.16d and 7.17a) or circular (Figure 7.21) configurations.

In contrast to the linear geometry, where sensors at the gas inlet heat the test gas and the succeeding gas sensors are exposed to the disturbed gas stream, every sensor in the circular chamber receives the same gas flow directly from the inlet. As shown in Figure 7.21, the sensors are arranged around the flow splitter and face the channels conveying the sample. Thus, the radially symmetric measurement chamber shown in Figure 7.21 guarantees homogeneous flow conditions, low velocity gradients, and absence of stagnant volumes. The small internal volume of the chambers and their geometry allow the simultaneous exposure of all the sensors to gas samples under the same conditions. The circular chamber is especially promising for high-temperature and gas-consuming sensors. The chamber can be made using either polymeric materials such as polytetrafluoroethylene (PTFE) and polyether ether ketone (PEEK), or stainless steel, metals, quartz, or fused silica (Di Francesco et al. 2005).

6.4. SENSOR LOCATION

Safety codes typically do not dictate specific locations for sensors. Sensor locations should be reviewed with the local authority having jurisdiction as well as with the facility safety and risk managers. However, as a rule, sensors are typically provided in the following locations:

- Gas cabinets (usually in the exhaust)
- Any place where fittings are not welded, such as valve manifold boxes (VMBs) and tool gas jungles or pump enclosures
- Annular space in double-containment gas lines

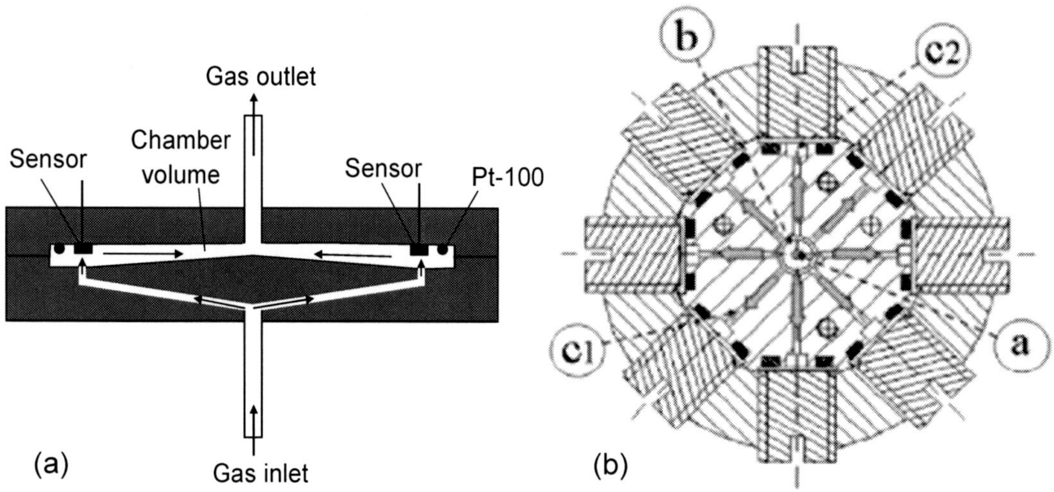

Figure 7.21. (a) Plan of the circular measurement chambers designed by (a) Endres et al. (1995) and (b) Di Francesco et al. (2005). Sample path in (b) is marked by arrows, while the tracts are indicated by circled letters: a, input connection; b, inlet collector; c1 and c3, radial channels; c2, sensor cylindrical chambers; d, outlet collector; e, output connection. [(a) Reprinted with permission from Endres et al. 1995. Copyright 1995 Elsevier; (b) reprinted with permission from Di Francesco et al. 2005. Copyright 2005 Elsevier.)

- Tool exhausts
- Operator breathing zones
- Chemical delivery/service corridors
- Gas purifier cabinets and ozonators
- Exhaust treatment system discharge areas (problematic due to sensor clogging)
- Gas storage rooms
- Vacuum pump enclosures

6.5. SENSOR INSTALLATION

There are no set rules regarding where sensors must be installed, but the judgment of trained personnel and good common sense should always be followed. During sensor installation it is necessary to take into account the following:

- Locations of sensors should provide easy access for maintenance and calibration. Sensors which are properly installed can save hours of maintenance and provide trouble-free operation.
- Reliable application of chemical sensors depends on a system that pairs the detectors and sampling techniques to the target gas and ambient conditions of the monitored area. Do not install sensors near sources of steam; steam will damage the sensors. Do not install sensors, especially

electrochemical ones, next to sources which are constantly leaking gas. Limit moisture condensation on the sensor.

- Avoid installing sensors too close to walls or floors. These surfaces can absorb and emit gases with changes in temperature, affecting the reading of the sensor.
- The temperature range in which the sensor is to be installed should be within the sensor specifications and should be suitable for the gases to be monitored.
- Solid-state gas sensors operate best at a monitoring radius of 0.5–2 m. Therefore, sensors should be located at or near points where there is the possibility of a gas release.
- A suitable safety margin should also be incorporated to account for ventilation dead spots, where vapors could accumulate. Therefore sensors should be positioned to detect any gas accumulation before it creates a serious hazard.
- The path of the gas or its dispersion characteristics depend on the density of the gas and the ventilation patterns. Therefore, the density of the target gas should be used to determine at what height sensors should be positioned relative to the potential source. For example:
  - Methane (natural gas) and H_2 are less dense than air and so will rise in an air atmosphere.
  - Vaporized liquefied petroleum gas (LPG) is heavier than air, so it will fall or remain at a low level.
  - Landfill gas (a mixture of methane, carbon dioxide, hydrogen, and trace gases) will have a density that depends on its composition. For example, mature landfill sites generate gas that is slightly less dense than air.

  However, this does not mean that sensors should be installed on the floor or ceiling to monitor these gases.
- The sensors or sample points should not be positioned where they may be susceptible to excess vibration or heat, contamination, mechanical damage, or water damage. This will reduce the operational life of the sensor.
- In areas of new construction, sensors should be installed after operations such as sandblasting, painting, welding, etc., are completed. During such operations, sensors present should be protected.

6.6. SAMPLING

Successful quantitative analysis starts with sample collection and ends with the interpretation of the data collected. Therefore, sensor stability over time and reduction of variability between different sensors of the same kind are the main issues, though the development of standard procedures for sample collection, preparation, and reproducible sampling systems has similar importance. Sampling systems transfer the sample from the source to the sensors and then to the waste, keeping all factors influencing sensor responses under control, so that the odor composition is the only variable present (Di Francesco et al. 2005).

In environmentally related gas analysis, continuous emissions monitoring, vehicle emissions testing, or LEL warning indication, and in process applications such as combustion efficiency monitoring, boiler trim, ethylene plants or petrochemical facilities, sample gas needs to be extracted from the gas stream. It is obvious that for reliable analysis the sample gas must be extracted free of ambient gas leaking into the system while at the same time the dew point of the sample must remain stable. Therefore, sampling systems in gas analyzers are often complex installations, which require careful design (Giles 2010). For example, many instruments applying extractive gas analysis or dry-basis measurement systems need dry gas, and therefore in many sampling systems a dryer is included in the system. Sample drying in fixed gas detection systems can be achieved by cooling the gas down below its dew point. Consequently, any moisture will condense and be removed from the system. In other applications, such as engine emission tests or combustion control, the gas tested is at a high temperature and therefore specific cooling equipment also can be incorporated in the sampling system. These cooling systems can be designed, for example, using thermoelectric chillers or water-cooled heat exchangers. Thus, the key requirements of a sampling system are that the system design includes all of the components required for safe and proper operation, including probes, valves, filters, coolers, pressure regulators and relief valves, pumps, piping, etc., thus ensuring that a sample which is representative of the overall process is admitted to the sensor for analysis (Cleaver 2001; Giles 2010). Additional information about the components of a sampling system can be found on the websites of various companies, such as www.Buhler-technologies.com; www.jiskoot.com; www.isotechlabs.com; www.analyticaltechnology.com; etc.

Filters, for instance, can be produced from various materials, depending on the analyte being tested for. In this regard, it is necessary to know that filters with larger surfaces have longer lifetimes.

The materials used for the sampling line should be selected carefully to prevent dilution or contamination of the sample by leakage, diffusion or sorption of the vapors, corrosion, or the ingress of air. For example, plastics are not gas-tight, since gases are able to diffuse through the walls of tubing and containers made of plastic (Kjeldsen 1993). Calculations carried out by Kjeldsen (1993) showed that diffusion of oxygen through plastic tubing and reactors into anoxic water can also be a serious problem for a number of plastic materials. Therefore, the use of some plastic materials, particularly silicone rubber, in experimental and sampling equipment should be prohibited by oxygen diffusion. Care should also be taken to avoid blockages in the sampling line. These can be caused by particulates, water condensation, or liquids with high boiling points. To minimize this effect, a filter may be needed to remove particulates and a trap to remove entrained liquid. It may also be necessary to heat the sampling line.

Various approaches are used for gas sampling (see Figure 7.22). In many fixed gas detection systems, the sensor units are designed to use natural diffusion as the sampling method. However, natural diffusion as a sampling method can be slow. In addition, the concentration profile at a particular instant during sampling is a function of many variables, including temperature, fluid velocity at the sampler face, sampler response time to changes in concentration, etc. Therefore, in many cases a faster response is needed, and the sample is transported to the sensor using a sampling pump. For pumped (aspirated) systems it is normal to flow the gas to waste and allow the instrument to draw the mixture from a tee-piece or reservoir in the line. We need to note that the transportation of sample gas requires very reliable sample pumps. Of course, the sample gas should have an equal and steady flow, but very often the corrosive nature of the gas and the presence of condensate form the real task for the pump. In addition, care must be taken not to set the flow too high or low, as similar problems will occur as for diffusion instruments. Usually, the sampling system has a configuration as shown in Figure 7.23. The configuration shown in Figure 7.23a is recommended for systems in which there is no significant pressure drop between the gas trap and the sensor, while the configuration shown in Figure 7.23b is recommended for systems with significant pressure drop.

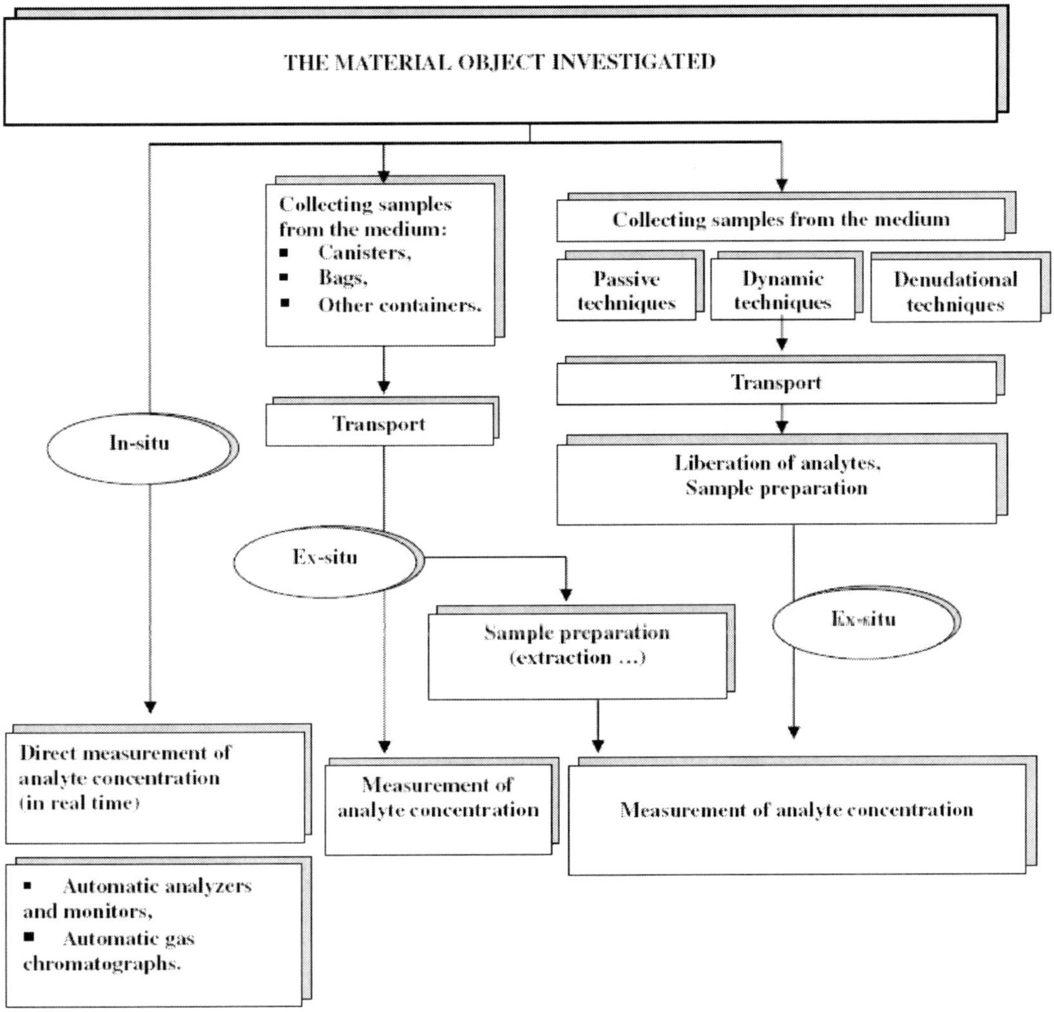

Figure 7.22. Sampling techniques and monitoring equipment used during atmospheric air monitoring. (Reprinted with permission from Krol et al. 2010. Copyright 2010 Elsevier.)

Selection of sampling locations requires special attention during sampling system design. According to Cleaver (2001), the sample point should be chosen to: (1) provide the most accurate information on the composition of the process stream; (2) minimize time lags, particularly important when analyzer outputs are used for process control; (3) ensure that temperature, pressure, dryness, particulate loading, etc., are analyzer-compatible wherever possible, to minimize the need for coolers, relief valves, filters, etc.; and (4) ensure that sample take-off points are readily accessible. In continuous emission monitoring (CEM) systems, the distance between the extraction point of sample gas and the analyzer cabinet is normally bridged by tubes or hoses. Depending on the method of analysis, these gas lines can be unheated or must be heated to prevent unwanted condensate formation.

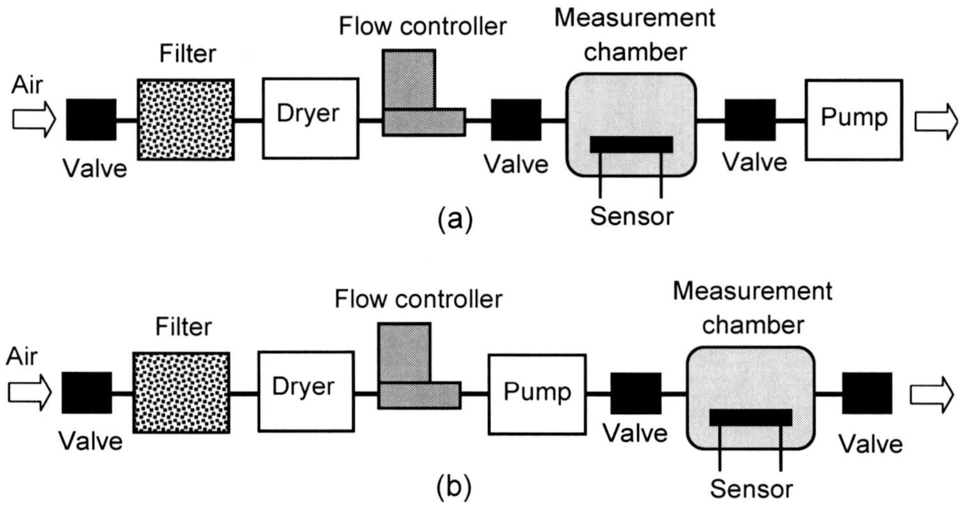

Figure 7.23. Configurations of sampling systems used in gas analyzers.

Gas sampling while monitoring gas composition in high-pressure gas mains (e.g., for natural gas) requires special conditions, which can be found in various references (GPA 1986; API 2001; Kelner et al. 2002; George et al. 2004).

In addition to the techniques mentioned above, various other sample collection techniques can be employed as well (Vrana et al. 2005; Seethapathy et al. 2008; Krol et al. 2010b). These techniques depend on the nature of the matrix, which can be gas, solid, or liquid. The type of analytical instrumentation used also influences the sample preparation technique required. For example, volatile samples can be pumped into an empty specific sampling container, which is then sent to the laboratory for analysis. Usually these canisters are specially treated containers which are evacuated before deployment. Sample collection in the field is based on the differential pressure existing between the inside and the outside of the canister. This method can provide a very accurate measurement of contaminant concentrations in air, because in this case it is possible to use high-precision gas analysis. Sampling tools used to collect different types of media samples for laboratory analysis should be constructed of materials that are compatible with the media and the constituents that are being tested for. In other words, the sampler should not be constructed of a material that could cause a loss or gain in contaminant concentrations measured due to sorption, desorption, degradation, or corrosion. Because stainless steel, Teflon, and glass are inert substances, sampling equipment made of these materials should be preferentially selected over equipment made of other materials (Byrnes 2009). Polyvinyl fluoride (Tedlar), a fluorinated ethylene and propylene copolymer (Teflon), or polyterephthalic ester (Nalofan) can also be used for preparing air bags (Krol et al. 2010b). Air bags made from synthetic materials are used usually for collecting samples of gases, especially those with a noxious odor. The advantages of Tedlar bags are that they are cheap to use and that a wide range of volumes (0.5–100 L) can be collected. The downsides of Tedlar bags are that they are easily punctured, analytes can be lost through leakage, and some VOCs become unstable if they are stored for more than 48 h. To eliminate the effect of air humidity on sample composition, double-film and triple-film bags have been introduced.

Sample preconcentration steps such as cryogenic freeze-out, solid-adsorbent trapping, or derivatization onto cartridges containing a chemical reactant can also be used in the sampling of gases. Conventional preconcentrators, so-called microtraps, comprise a stainless steel tube or glass capillary tube packed with one or more granular absorbent materials. Silica gel, polyurethane foam (PUF), and other sorbents can be used for this purpose. Sorbents usually used for atmospheric air monitoring are shown in Table 7.12.

For desorption, the current is passed through the stainless steel tube or through a metal wire coiled around the glass capillary tube. Desirable features of the preconcentrator device include operation at

Table 7.12. Sorbents usually used for atmospheric air control

Type of sorbent	Specific area (m^2/g)	Maximum temperature (°C)	VOC analytes retained	Preferred analyte liberation technique
Porous organic polymers				
Tenax TA and GR	35	<350	Aromatic compounds, nonpolar VOCs	Thermal desorption
Chromosorb 102 (styrene-divinylbenzene copolymer)	750	20–250	Alcohols, ketones, aliphatic hydrocarbons	Thermal desorption
Chromosorb 106 (polystyrene)	350	20–250	C_5–C_{12} hydrocarbons, oxygenated VOCs	Thermal desorption
Porapak N (polyvinylpyrrolidone)	300	20–190	Acetylene, aliphatic hydrocarbons	Thermal desorption
Porapak Q (ethylvinyl-benzene-divinylbenzene copolymer)	550	20–190	Acetylene, aliphatic hydrocarbons, and oxygenated VOCs	Solvent extraction
Graphitized carbon blacks (specific surface depends on the degree of graphitization)				
Carbotrap	12	>400	Hydrocarbons less than C_{20}, alkylbenzenes	Thermal desorption
Carbograph	100	>400	Ketones, alcohols, and	
Carbopack	12	>400	aldehydes	
Molecular sieves (formed during the pyrolysis of organic polymers, e.g., polyvinyl chloride)				
Spherocarb	1200	>400	Very volatile organic compounds,	Thermal desorption
Unicarb	>1000	>400	methanol, acetone	
Carbosieve III	800	400		
Carboxen	>1200	400		
Molecular Sieve 13X, 5A		350–400	1,3-butadiene	
Activated carbon (formed by the low-temperature oxidation of vegetable carbon)				
Activated carbon	>1000	400	Aliphatic and aromatic hydrocarbons, e.g., C_2–C_4	Solvent extraction or thermal desorption

Source: Reprinted with permission from Krol et al. 2010b. Copyright 2010 Elsevier.

high flow rates, thermal heating with short time constants, and selective sampling of the analyte(s) of interest. The gas or vapor preconcentrator can be used with a variety of analytical systems, including chemical sensors (Shen et al. 2004; Voiculescu et al. 2008). The preconcentrator device serves the important function of concentrating and purifying a chemical sample on a sorptive material at the inlet of an analytical system. Concentrating analyte in a stage prior to the detection system can improve detection limits for the analyte of interest. The preconcentrator device requires time for analyte collection and therefore is used usually in situations when analyte detection is performed over a period of time.

A potential problem encountered in these techniques is reactions occurring during the sampling between the concentrated organic gases of interest and other reactive air constituents such as ozone, halogens, the hydroxyl radical, nitrogen oxides, water or hydrogen peroxide. Reactions of this kind may alter the quantities of the trace gases of interest and may also contribute to the formation of artifact products which may mistakenly be interpreted as atmospheric constituents (Helmig 1997). Helmig (1997) has also noted that in VOC sampling onto solid adsorbents and onto derivatization cartridges, artifacts may be formed not only by reactions of reactive air constituents with the previously adsorbed analyte compounds but also from reactions with the solid adsorbent bed or the derivatizing agent, which in many cases is based on an organic polymer, such as poly(oxy-*m*-terphenyl-2′,5′-ylene) (Tenax) or polyurethane foam. A similar problem is encountered in derivatization methods, for example, in the sampling of aldehydes on cartridges coated with 2,4-dinitrophenylhydrazine (DNPH), where either the formed products or the derivatizing agent itself may be depleted during the sampling. The investigation of interferences from reactions during sampling is an important part of method development and improvement strategies for atmospheric monitoring.

Solid samples can be dissolved and diluted. This approach is quite common for the analysis of drug substances. The solid or liquid sample can also be placed into a glass vial, which is heated, with enough empty space in the vial above the sample to allow the analyte to reach equilibrium between the gas and solid (or liquid) phase. Then the gas phase can be sampled and analyzed by standard methods. This type of sampling is generally used for analyzing trace amounts of volatile samples (Ahuja and Diehl 2006). A number of different diffusion tubes and passive membranes are also available for sampling of various chemical species from water or other liquids (Vrana et al. 2005). The principles of this approach to sampling were discussed in detail in previous sections related to the preparation of calibration gas mixtures.

One also needs to take into account that every test protocol discussed here requires that the gas samples be evaluated in terms of both repeatability and reproducibility. Therefore, when using any of the collection techniques mentioned above, it is necessary to take into account that samples collected for environmental investigations can be of three types: (1) grab samples, (2) composite samples, and (3) integrated samples (Peirce et al. 1998; Byrnes (2009).

A *grab sample* refers to the physical collection of a sample from a single location for analysis. When collecting a grab, the sample is transferred directly from the sampling tool into the sample jar or bottle. No mixing or compositing is performed when collecting this type of sample. Grab samples should be collected when one would like to know the range of concentration levels (minimum and maximum) for a contaminant present at a site. Grab sampling is effective in collecting samples for site characterization, waste characterization, risk assessment, feasibility study, remedial design, etc.

A *composite sample* is collected either by taking multiple grab samples from different locations and homogenizing them together, or by collecting samples from multiple depths from the same sampling location and homogenizing these depth intervals together for analysis. Composite sampling is most frequently used to help reduce the cost of site characterization, and in some cases is used for waste characterization activities. However, it is important to recognize that the results provide only the mean concentration for the composited intervals, and will not provide a reliable estimate of the range of concentration levels present at a site. Composite sampling is less frequently used to collect data for human health or environmental risk assessment, because it is important to know the full range of contaminant levels to which a receptor may be exposed. Composite sampling is generally not recommended for site closeout sampling, because it tends to "dilute" the analytical results. When collecting composite samples, it is important that the composite interval not be so large that the resulting data are diluted beyond the point of providing meaningful information.

Integrated sampling involves the collection of a sample from one location over an extended period of time (e.g., weeks or months). Integrated sampling is most commonly performed when assessing surface water, groundwater, or air quality over time. It is different from composite sampling in that samples are collected from the same location over an extended period. For example, one integrated surface water sample may be collected from a location downstream from a chemical plant, where a small volume of water is collected every week throughout the 3-month wet season. At the end of the 3-month period, the integrated sample is analyzed to identify the mean concentration of contaminants in the surface water over that time period. Integrated sampling should not be used when the contaminants of concern are volatile organic compounds.

6.7. TRAINING

Technology marches forward: Sensors get better each year, smaller/lighter batteries appear with higher energy density, microprocessor circuit design enhancements are offered, etc. However, no improved instrument design feature can offset poor or inadequate training of the instrument operator and the possible consequences to your special tactical unit from misunderstanding an instrument's limitations. Therefore, in your activity you need to follow the advice of V. R. Brown, president of ENMET Corporation (Brown 2006): When buying 6 units or more, why not make "on-site/no charge" training by factory personnel a specification in the bid set? A reputable instrument supplier who really deserves your business will be happy to accommodate your request.

7. SUMMARY

In summarizing the results presented and discussed in this series, one should note that in spite of the many problems that have been discussed, the development of chemical sensors as well as materials aimed for sensors design is going to continue, since the market remains promising. However, the huge potential of research will be realized in new chemically sensitive devices only if both expansion of the spectrum of materials tested and extension of research targeted toward understanding the nature and mechanism of chemical sensor functioning takes place. According to Gopel and Reinhard (1998), deep understanding of chemical sensor's operation can be achieved by conducting spectroscopic and microscopic control of sensing materials and identification of their electronic and ionic conduction mechanisms through a detailed characterization of geometric arrangements of atoms, elemental compositions, electronic structures, dynamic structures, and coverage of adsorbed species. The influence of systematic changes in the sensing materials and operation parameters on sensing and catalytic properties should be studied as well. That understanding would allow optimizing either existing or new chemical sensor materials to attain good long-term stability, high sensitivity, and pronounced selectivity, which are very important for real applications.

On the other hand, we need to recognize that it is very hard to fulfill all these requirements. For example, in the case of solid-state gas sensors, we cannot design selective sensors. For fundamental reasons, cross-sensitivity can only be reduced, not eliminated. The impossibility of producing universal gas sensors which can precisely detect or measure the target gas under any conditions and for any gas composition is another important conclusion of the present discussions.

Therefore, we have to agree with Lample et al. (1996) that the elaboration of sensors assigned for more specific applications, rather than for general applications, is a more optimal approach to chemical sensor design, because even such sensors might satisfy most market needs. This last aspect is important for increasing the level of chemical sensor application and for promoting further development of specific sensors, able to operate in various environments. As is known, sensors usually operate in basically known surroundings with fixed gas compositions.

8. ACKNOWLEDGMENT

This work was supported by the World Class University (WCU) Program at the Gwangju Institute of Science and Technology (GIST) through a grant (Project No. R31-20008-000-10026-0) provided by the Ministry of Education, Science and Technology (MEST) of Korea, for the project titled "Development of Maritime Environmental Sensor Using Nano and Photonic Technology," funded by the Ministry of Land, Transport and Maritime Affairs, Korea, and by a Korean Science and Engineering Foundation (KOSEF) NCRC grant funded by the Korean government (MEST) (No. R15-2008-006-01002-0). G. Korotcenkov and B. K. Cho are also grateful to the Korean BK21 Program for support of their scientific research.

REFERENCES

Adamchuk V.I., Hummel I.W., Morgan M.T., and Upadhyaya S.K. (2004) On-the-go soil sensors for precision agriculture. *Comput. Electron. Agric.* **44**(1), 71–91.

Afgan N.H., Carvalho M.G., Pilavachi P.A., Tourlidakis A., Olkhonski G.G., and Martins N. (2006) An expert system concept for diagnosis and monitoring of gas turbine combustion chambers. *Appl. Thermal Eng.* **26**, 766–771.

Ahuja S. and Diehl D. (2006) Sampling and sample preparation. In: Ahuja S. and Jespersen N. (eds.), *Comprehensive Analytical Chemistry*, Vol. 47. Elsevier, Amsterdam, 15–40.

Ando M. (2006) Recent advances in optochemical sensors for the detection of H_2, O_2, O_3, CO, CO_2 and H_2O in air. *TrAC Trends Anal. Chem.* **25**(10), 937–948.

API (2001) Natural gas fluids measurement, Section 1—Collecting and handling of natural gas samples for custody transfer. In: *Manual of Petroleum Measurement Standards*, 5th ed. American Petroleum Institute, Washington, DC, chap. 14.

Aroutiounian V. (2007) Metal oxide hydrogen, oxygen, and carbon monoxide sensors for hydrogen setups and cells. *Int. J. Hydrogen Energy* **32**(9), 1145–1158.

Arshak K., Moore E., Lyons G.M., Harris J., and Clifford S. (2004) A review of gas sensors employed in electronic nose applications. *Sensor Rev.* **24**(2), 181–198.

Azevedo A.M., Prazeres D.M.F., Cabral J.M.S., and Fonseca L.P. (2005) Ethanol biosensors based on alcohol oxidase. *Biosens. Bioelectron.* **21**(2), 235–247.

Bair M. (1999) *The Dissemination of Gravimetric Gas Flow Measurements Through an LFE Calibration Chain*. DH Instruments, Inc., Phoenix, AZ.

Barratt R.S. (1981) The preparation of standard gas mixtures: A review. *Analyst* **106**, 817–849.

Basu S. and Basu P.K. (2009) Nanocrystalline metal oxides for methane sensors: Role of noble metals. *J. Sensors* **2009**, 861968.

Berry L. and Brunet J. (2008) Oxygen influence on the interaction mechanisms of ozone on SnO_2 sensors. *Sens. Actuators B* **129**, 450–458.

Bodenhofer K., Hierlemann A., Noetzel G., Weimar U., and Gopel W. (1996) Performances of mass-sensitive devices for gas sensing: Thickness shear mode and surface acoustic wave transducers. *Anal. Chem.* **68**, 2210–2218.

Boon-Brett L., Bousek J., Black G., Moretto P., Castello P., Hubert T., and Banach U. (2010) Identifying performance gaps in hydrogen safety sensor technology for automotive and stationary applications. *Int. J. Hydrogen Energy* **35**(1), 373–384.

Boublik T., Fried V., and Hála E. (1984) *The Vapour Pressures of Pure Substances*. Elsevier, Amsterdam.

Bradley P. (2001) *Fixed Flammable Gas Detectors*, TD5/035, www.hse.gov.uk.

Brahim-Belhouari S., Bermak A., Shi M., and Chan P.C.H. (2005) Fast and robust gas identification system using an integrated gas sensor technology and Gaussian mixture models. *IEEE Sensors J.* **5**(6), 1433–1444.

Brett C.M.A. and Brett A.M.O. (1998) *Electroanalysis*. Oxford University Press, Oxford UK.

Brett C.M.A. (2001) Electrochemical sensors for environmental monitoring. Strategy and examples. *Pure Appl. Chem.* **73**(12), 1969–1977.

Brown V.R. (2006) Sensor selection for hand-held portable gas detection. In: *The Grey House Safety and Security Directory*. Grey House Publishing, Millerton, NY, 291–293 (www.enmet.com).

BS EN 50073 (1999) *Guide for the Selection, Installation, Use and Maintenance of Apparatus for the Detection and Measurement of Combustible Gases or Oxygen*. British Standards Institution, London, UK (www.bsi-global.com).

BS EN 50241 (1999) *Specification for Open Path Apparatus for the Detection of Combustible or Toxic Gases and Vapours. Part 1—General Requirements and Test Methods. Part 2—Performance Requirements for Apparatus for the Detection of Combustible Gases*. British Standards Institution (www.bsi-global.com).

BS EN 61779 (2000) *Electrical Apparatus for the Detection and Measurement of Flammable Gases* (Parts 1–5). British Standards Institution (www.bsi-global.com).

Byrnes M.E. (2009) *Field Sampling Methods for Remedial Investigations*. CRC Press, Taylor & Francis Group, Boca Raton, FL.

Choi M.M.F. and Hawkin P. (2003) Development of an optical hydrogen sulphide sensor. *Sens. Actuators B* **90**, 211–215.

Chou J. (2000) *Hazardous Gas Monitors: A Practical Guide to Selection, Operation and Application*. McGraw-Hill, New York.

Cleaver K.D. (2001) The analysis of process gases: A review. *Accred. Qual. Assur.* **6**, 8–15.

COGDEM (1999) *Gas Detection and Calibration Guide*. The Council of Gas Detection and Environmental Monitoring, Hitchin, UK (www.cogdem.org.uk).

Cutmore T.R.H. and James D.A. (2007) Sensors and sensor systems for psychophysiological monitoring: A review of current trends. *J. Psychophysiol.* **21**(1), 51–71.

Demarne V., Grisel A., Sanjines R., and Lévy F. (1990) Integrated semiconductor gas sensors evaluation with an automatic test system. *Sens. Actuators B* **1**, 87–92.

Di Francesco F., Falcitelli M., Marano L., and Pioggia G. (2005) A radially symmetric measurement chamber for electronic noses. *Sens. Actuators B* **105**, 295–303.

Docquier N. and Candel S. (2002) Combustion control and sensor: A review. *Prog. Energy Combustion Sci.* **28**, 107–150.

Dorko W.D. and Hughes E.E. (1987) Special calibration systems for reactive gases and other difficult measurements. In: Taylor J.K. (ed.), *Sampling and Calibration for Atmospheric Measurements*, ASTM STP 957. American Society for Testing and Materials, Philadelphia, 132–137.

Endres H.E., Jander H.D., and Gottler W. (1995) A test system for gas sensors. *Sens. Actuators B* **23**, 163–172.

ENERAC (2005) *Span Gas Calibration KIT: Instruction Manual*. ENERAC, Inc., Westbury, NY.

Falcitelli M., Benassi A., Di Francesco F., Domenici C., Marano L., and Pioggia G. (2002) Fluid dynamic simulation of a measurement chamber for electronic noses. *Sens. Actuators B* **85**, 166–174.

Fergus J.W. (2007) Materials for high temperature electrochemical NO_x gas sensors. *Sens. Actuators B* **121**, 652–663.

Fergus J.W. (2008) A review of electrolyte and electrode materials for high temperature electrochemical CO_2 and SO_2 gas sensors. *Sens. Actuators B* **134**, 1034–1041.

Ferrante do Amaral C.E. and Wolf B. (2008) Current development in non-invasive glucose monitoring. *Med. Eng. Phys.* **30**, 541–549.

Fleet B. and Gunasingham H. (1992) Electrochemical sensors for monitoring environmental pollutants. *Talanta* **39**, 1449–1457.

Fontes J. (2005) Humidity sensors. In: Wilson J.S. (ed.), *Sensor Technology Handbook*. Elsevier, Oxford, UK, 271–284.

Gajdošík L. (2006) Tin dioxide gas sensors. In: Grimes C.A., Dickey E.C., and Pishko M.V. (eds.), *Encyclopedia of Sensors*, Vol. 10. American Scientific Publishers, Valencia, CA, 361–375.

Gale P.A. (2003) Anion and ion-pair receptor chemistry: Highlights from 2000 and 2001. *Coord. Chem. Rev.* **240**(1–2), 191–221.

George D.L., Hart R.A., and Nored M. (2004) Evaluation of a proposed gas sampling method performance verification test protocol. Final report. Southwest Research Institute, San Antonio, TX.

Gerblinger J. and Meixner H. (1991) Fast oxygen sensor on sputtered strontium titanate. *Sens. Actuators B* **4**, 99–102.

Ghosh R.N., Tobias P., Hu H., and Koochesfahani M. (2005) Fast solid state gas sensor characterization technique. In: *Proceedings of IEEE Sensors Conference, IEEE Sensors,* Oct. 31–Nov. 3, Irvine, CA, 1411–1413.

Giles J.G. (2010) Sampling. In: Boyes W. (ed.), *Instrumentation Reference Book*, 4th ed. Elsevier, Burlington, MA, 661–676.

Gopel W. and Reinhard G. (1998) Metal oxide sensors: New devices through tailoring interfaces on the atomic scale. In: *Sensor Update.* Part 3. Wiley-VCH, Weinheim, Germany, 47–121.

GPA (1986) *Obtaining Natural Gas Samples for Analysis by Gas Chromatography*, GPA Standard 2166-86. Gas Processors Association, Tulsa, OK.

Greenblatt M. and Shuk P. (1996) Solid-state humidity sensors. *Solid State Ionics* **86–88**, 995–1000.

Han Y., Liang T., Yang X.-J., Ren X.-L., and Yin Y.-F. (2010) Research on optical air chamber of infrared gas sensor. In: *Proceedings of First International Conference on Pervasive Computing, Signal Processing and Applications. IEEE,* Sept. 17–19, Harbin, China, 33–36.

Harvey I., Coles G., and Watson J. (1989) The development of an environmental chamber for the characterization of gas sensors. *Sens. Actuators* **16**(4), 393–405.

Haug M., Schierbaum K.D., Gauglitz G., and Göpel W. (1993) Chemical sensors based upon polysiloxanes: Comparison between optical, quartz microbalance, calorimetric, and capacitance sensors. *Sens. Actuators B* **11**, 383–391.

Helmig D. (1997) Ozone removal techniques in the sampling of atmospheric volatile organic trace gases. *Atmos. Environ.* **31**, 3635–3651.

Helwig A., Muller G., Sberveglieri G., and Faglia G. (2007) Gas response times of nano-scale SnO_2 gas sensors as determined by the moving gas outlet technique. *Sens. Actuators B* **126**, 174–180.

Helwig A., Muller G., Sberveglieri G., and Faglia G. (2008) Catalytic enhancement of SnO_2 gas sensors as seen by the moving gas outlet method. *Sens. Actuators B* **130**, 193–199.

Ho C.K., Itamura M.T., Kelley M., and Hughes R.C. (2001) Review of chemical sensors for in-situ monitoring of volatile contaminants. Sandia Report SAND2001-0643, Unlimited release, Sandia National Laboratories, Albuquerque, NM (www.sandia.gov/sensor).

Ho C.K., Robinson A., Miller D.R., and Davis M.J. (2004) Sensors for environmental monitoring and long-term environmental stewardship. Sandia Report SAND2004-4596, Unlimited release, Sandia National Laboratories, Albuquerque, NM (http://prod.sandia.gov/techlib).

Hodgkinson J. and Pride R.D. (2005) Gas leak imaging. In: *Exploration & Production: The Oil & Gas Review,* Issue 2. Touch Briefings, www.touchbriefings.com/pdf.

Honeychurch K.C. and Hart J.P. (2003) Screen-printed electrochemical sensors for monitoring metal pollutants. *TrAC Trends Anal. Chem.* **22**(7), 456–469.

HSE (1987) Industrial use of flammable gas detectors. Guidance Note CS1 (withdrawn). www.hse.gov.uk.

HSG193 (2003) *COSHH Essentials: Easy Steps to Control Chemicals—Control of Substances Hazardous to Health Regulations*, 2nd ed. HSE Books, ISBN 0 7176 2737 3 (www.hse.gov.uk).

Igarashi I. (1986) New technology of sensors for automotive applications. *Sens. Actuators* **10**, 181–193.

ISO 6142 (2001) *Gas Analysis—Preparation of Calibration Gas Mixtures—Gravimetric Method.* www.iso.org.

ISO 6145-2 (2001) *Gas Analysis—Preparation of Calibration Gas Mixtures Using Dynamic Volumetric Methods: Part 2. Volumetric Pumps.* www.iso.org.

ISO 6144 (2003) *Gas Analysis—Preparation of Calibration Gas Mixtures—Static Volumetric Method.* www.iso.org.

ISO 6145-6 (2003) *Gas Analysis—Preparation of Calibration Gas Mixtures Using Dynamic Volumetric Methods: Part 6. Critical Orifices.* www.iso.org.

ISO 6145-8 (2005) *Gas Analysis—Preparation of Calibration Gas Mixtures Using Dynamic Volumetric Methods: Part 8. Diffusion Method.* www.iso.org.

ISO 6145-5 (2009) *Gas Analysis—Preparation of Calibration Gas Mixtures Using Dynamic Volumetric Methods: Part 5. Capillary Calibration Devices.* www.iso.org.

ISO 6145-7 (2009) *Gas Analysis—Preparation of Calibration Gas Mixtures Using Dynamic Volumetric Methods: Part 7. Thermal Mass-Flow Controllers.* www.iso.org.

ISO 6145-9 (2009) *Gas Analysis—Preparation of Calibration Gas Mixtures Using Dynamic Volumetric Methods: Part 9. Saturation Method.* www.iso.org.

Jardine K.J., Henderson W.M., Huxman T.E., Abrell L., and Shartsis T. (2010) Dynamic solution injection: A new method for preparing pptv-ppbv standard atmospheres of volatile organic compounds. *Atmos. Meas. Tech. Discuss.* **3**, 3047–3066.

Johnson K.J. (2003) Chemical sensors for autonomous and Lagrangian platforms. In: *Proceedings of the Autonomous and Lagrangian Platforms Workshop*, March 31 to April 2, Sea Lodge, La Jolla, CA.

Justino C.I.L., Rocha-Santos T.A., Duarte A.C., and Rocha-Santos T.A. (2010) Review of analytical figures of merit of sensors and biosensors in clinical applications. *TrAC Trends Anal. Chem.* **29**(10), 1172–1183.

Keefe M.K., Benkstein K.D., and Hup J.T. (2000) Luminescent sensor molecules based on coordinated metals: A review of recent developments. *Coord. Chem. Rev.* **205**(1), 201–228.

Kelleter J., Schüll G., Kohl C.D., Grüner U., Kraft A., and Wollnik H. (1992) Novel method for studying the kinetics of surface reactions at submolecular exposures: Decomposition of acetic acid on palladium foil. *Surf. Interface Anal.* **19**, 581–585.

Kelner E., Sparks C.R., and Behring K.A. (2002) *Metering Research Facility Program: Natural Gas Sample Collection and Handling—Phase III: Experimental Investigation of Gas Sampling Techniques and Equipment.* Gas Research Institute, Des Plaines, IL.

Kelsey A., Hemingway M.A., Walsh P.T., and Connolly S. (2000) Evaluation of the effectiveness of the location of flammable gas detectors from dispersion trial data. In: *Proceedings of Major Hazards Offshore Conference,* London. ERA, Leatherhead, UK, 5.4.1–5.4.12.

Kida T., Kuroiwa T., Yuasa M., Shimanoe K., and Yamazoe N. (2008) Study on the response and recovery properties of semiconductor gas sensors using a high-speed gas-switching system. *Sens. Actuators B* **134**, 928–933.

Kim Y.S., Ha S.-C., Yang H., and Kim Y.T. (2007) Gas sensor measurement system capable of sampling volatile organic compounds (VOCs) in wide concentration range. *Sens. Actuators B* **122**, 211–218.

Kjeldsen P. (1993) Evaluation of gas diffusion through plastic materials used in experimental and sampling equipment. *Water Res.* **27**(1), 121–131.

Kohl D., Kelleter J., and Petig H. (2001) Detection of fires by gas sensors. In: Baltes H., Gopel W., and Hesse J. (eds.), *Sensors Update*, Vol. **9**(1). Wiley-VCH, Weinheim, Germany, 161–223.

Korotcenkov G. (2007a) Metal oxides for solid state gas sensors. What determines our choice? *Mater. Sci. Eng. B* **139**, 1–23.

Korotcenkov G. (2007b) Practical aspects in design of one-electrode semiconductor gas sensors: Status report. *Sens. Actuators B* **121**, 664–678.

Korotcenkov G., Brinzari V., Cerneavschi A., Golovanov V., Matolin V., and Tadd A. (2004) Acceptor like behavior of reducing gases on the surface of *n*-type In_2O_3. *Appl. Surf. Sci.* **227**, 122–131.

Korotcenkov G., Brinzari V., Stetter J.R., Blinov I., and Blaja V. (2007) The nature of processes controlling the kinetics of indium oxide-based thin film gas sensor response. *Sens. Actuators B* **128**, 51–63.

Korotcenkov G., Han S.-D., and Stetter J.R. (2009) Review of electrochemical hydrogen sensors. *Chem. Rev.* **109**(3), 1402–1433.

Korposh S.O., Takahara N., Ramsden J.J., Lee S.-W., and Kunitake T. (2006) Nano-assembled thin film gas sensors. I. Ammonia detection by a porphyrin-based multilayer film. *J. Bio. Phys. Chem.* **6**, 125–132.

Krol S., Zabiegata B., and Namiesnik J. (2010a) Monitoring VOCs in atmospheric air. I. On-line gas analyzers. *TrAC Trends Anal. Chem.* **29**(9), 1092–1100.

Krol S., Zabiegata B., and Namiesnik J. (2010b) Monitoring VOCs in atmospheric air II. Sample collection and preparation. *TrAC Trends Anal. Chem.* **29**(9), 1101–1112.

L138 (2003) Dangerous substances and explosive atmospheres. In*: Dangerous, Substances and Explosive Atmospheres Regulations.* HSE Books, ISBN 0 7176 2203 7 (www.hse.gov.uk).

Lampe U., Gerblinger J., and Meixner H. (1992) Comparison of transient response of exhaust-gas sensors based on thin films of selected metal oxides. *Sens. Actuators B* **7**, 787–791.

Lampe U., Fleischer M., Reitmeier N., Meixner H., McMonagle J.B. and Marsh A. (1996) New materials for metal oxide sensors. In: Baltes H., Gopel W., and Hesse J. (eds.), *Sensors Update*, Part 2. Wiley-VCH, Weinheim, Germany, 1–36.

Lee D.-D. and Lee D.-S. (2001) Environmental gas sensors. *IEEE Sensors J.* **1**, 214–224.

Leiner M.J.P. (1991) Luminescence chemical sensors for biomedical applications: Scope and limitations. *Anal. Chim. Acta* **255**, 209–222.

Lezzi A.M., Beretta G.P., Comini E., Faglia G., Galli G., and Sberveglieri G. (2001) Influence of gaseous species transport on the response solid state gas sensors within enclosures. *Sens. Actuators B* **78**, 144–150.

Lin J. (2000) Recent development and applications of optical and fiber-optic pH sensors. *TrAC Trends Anal. Chem.* **19**(9), 541–552.

Mackay-Sim A., Kennedy T.R., Bushell G.R., and Thiel D.V. (1993) Sources of variability arising in piezoelectric odorant sensors. *Analyst* **118**, 1393–1398.

Magliulo V. (1996) Automatic scanning of multiple leaf chambers for gas exchange measurements. *Comput. Electron. Agric.* **15**, 149–160.

Maricq M.M. (2007) Chemical characterization of particulate emissions from diesel engines: A review. *J. Aerosol Sci.* **38**(11), 1079–1118.

Maziarz W. and Pisarkiewicz T. (2008) Gas sensors in a dynamic operation mode. *Meas. Sci. Technol.* **19**, 055205.

McRae T. G. and Kulp T. G. (1993) Backscatter absorption gas imaging: A new technique for gas visualization. *Appl. Opt.* **32**, 4037–4050.

Mello L.D, and Kubota L.T. (2002) Review of the use of biosensors as analytical tools in the food and drink industries. *Food Chem.* **77**(2), 237–256.

Ménil F., Coillard V., and Lucat C. (2000) Critical review of nitrogen monoxide sensors for exhaust gases of lean burn engines. *Sens. Actuators B* **67**, 1–23.

Midgley D. (1990) A review of pH measurement at high temperatures. *Talanta* **37**, 767–781.

Miller J.B. (2001) Catalytic sensors for monitoring explosive atmospheres. *IEEE Sensors J.* **1**(1), 88–3.

Mitchell G.D. (1987) Trace gas calibration systems using permeation devices. In: Taylor J.K. (ed.), *Sampling and Calibration for Atmospheric Measurements*, ASTM STP 957. American Society for Testing and Materials, Philadelphia, 110–120.

Mousa-Bahia A.A., Coles G.S.V., and Watson J. (1993) A gas injector for an automatic environmental test chamber for the characterization of gas sensors. *Sens. Actuators B* **2**, 141–145.

Muller J., Eschenroder A., and Diepenbrock W. (2009) Through-flow chamber CO_2/H_2O canopy gas exchange system—Construction, microclimate, errors, and measurements in a barley (*Hordeum vulgare L.*) field. *Agric. Forest Meteorol.* **149**, 214–229.

Muñoz R., Sivret E.C., Parcsi G., Lebrero R., Wang X., Suffet I.H., and Stuetz R.M. (2010) Monitoring techniques for odour abatement assessment. *Water Res.* **44**(18), 5129–5149.

Nakamoto T., Fukuda T., and Moriizumi T. (1991) Gas identification system using plural sensors with characteristics of plasticity. *Sens. Actuators B* **3**, 1–6.

Nakamura K., Nakamoto T., and Moriizumi T. (1999) Prediction of quartz crystal microbalance gas sensor responses using a computational chemistry method. *Sens. Actuators B* **61**, 6–11.

Naranjo E. and Neethling G.A. (2007) Integrated solutions for fixed gas detection. General Monitors Report, 2–20 (www.generalmonitors.com).

Neethirajan S., Jayas D.S., and Sadistap S. (2009) Carbon dioxide (CO_2) sensors for the agri-food industry—A review. *Food Bioprocess. Technol.* **2**, 115–121.

NFPA (2003) *Operation of Fire Protection Systems.* National Fire Protection Association, Quincy, MA.

Nieuwenhuizen M.S. and Harteveld J.L.N. (1994) An automated SAW gas sensor testing system. *Sens. Actuators A* **44**, 219–229.

NMAB (1995) Expanding the vision of sensor materials. National Materials Advisory Board, National Academy Press, Washington, DC (http://books.nap.edu).

Öna A. (2007) A review: Current analytical methods for the determination of biogenic amines in foods. *Food Chem.* **103**(4), 1475–1486.

Parrish D.D. and Fehsenfel F.C. (2000) Methods for gas-phase measurements of ozone, ozone precursors and aerosol precursors. *Atmos. Environ.* **34**(12–14), 1921–1957.

Patel P.D. (2002) Biosensors for measurement of analytes implicated in food safety: A review. *TrAC Trends Anal. Chem.* **21**(2), 96–115.

Pedersen L.D. (1991) Assessment of sensors used in the food industry. *Food Control* **2**(2), 87–98.

Pejcic D., Eadington P., and Ross A. (2007) Environmental monitoring of hydrocarbons: A chemical sensor perspective. *Environ. Sci. Technol.* **41**(18), 6333–6342.

Pickup J.C., Hussain F., Evans N.D., Rolinski O.J., and Birch D.J.S. (2005) Fluorescence-based glucose sensors. *Biosens. Bioelectron.* **20**, 2555–2565.

Pilling R.S., Bernhardt G., Kim C.S., Duncan J., Crothers C.B.H., Kleinschmidt D., Frankel D.J., Lad R.J., and Frederick B.G. (2003) Quantifying gas sensor and delivery system response time using GC/MS. *Sens. Actuators B* **96**, 200–214.

Peirce J.J., Weiner R.E., and Vesilind E.A. (1998) *Environmental Pollution and Control.* Butterworth-Heinemann, Boston.

Pletcher D., Evans J., Warburton P.R., and Gibbs T.K. (1991) Acidic Gas Sensors and Method of Using the Same. U.S. Patent 5,071,526.

Potyrailo R.A. and V.M. Mirsky V.M. (2008) Combinatorial and high-throughput development of sensing materials: The First 10 Years. *Chem. Rev. B* **108**, 770–813.

Pratzler S., Knopf D., Ulbig P., and Scholl S. (2010) Preparation of calibration gas mixtures for the measurement of breath alcohol concentration. *J. Breath Res.* **4**, 036004.

Prodi L., Bolletta F., Montalti M., and Zaccheron N. (2000) Luminescent chemosensors for transition metal ions. *Coord. Chem. Rev.* **205**(1), 59–83.

Ramamoorthy R., Dutta P.K., and Akbar S.A. (2003) Oxygen sensors: Materials, methods, designs and applications. *J. Mater. Sci.* **38**, 4271–4282.

Reichel P. (2005) Development of a Chemical Gas Sensor System. Ph..D thesis, University of Tubingen, Germany.

Riegel J., Neumann H., and Wiedenmann N.-M. (2002) Exhaust gas sensors for automotive emission control. *Solid State Ionics* **152–153**, 783–800.

Rittersma Z.M. (2002) Recent achievements in miniaturised humidity sensors—A review of transduction techniques. *Sens. Actuators A* **96**, 196–210.

Sawaguchi N., Shin W., Izu N., Matsubara I., and Murayama N. (2005) Effect of humidity on the sensing property of thermoelectric hydrogen sensor. *Sens. Actuators B* **108**, 461–466.

Sazhin S.G., Soborover E.I., and Tokarev S.V. (2003) Sensor methods of ammonia inspection. *Russ. J. Nondestruct. Test.* **39**(10), 791–806.

Schalwig J., Bosch-v. Braunmuhl C., Muller G., Lee A. (2002) Semiconductor Gas Sensors as Control Monitors for NO*x* Storage Catalytic Converters, SAE Technical Paper 2002-01-1095.

Seethapathy S., Gorecki T., and Li X. (2008) Passive sampling in environmental analysis. *J. Chromatogr. A* **1184**, 234–253.

Shen C.Y., Huang C.P., and Huang W.T. (2004) Gas-detecting properties of surface acoustic wave ammonia sensors. *Sens. Actuators A* **101**, 1–7.

Shepherd R., Beirne S., Lau K.T., Corcoran B., and Diamond D. (2007) Monitoring chemical plumes in an environmental sensing chamber with a wireless chemical sensor network. *Sens. Actuators B* **121**, 142–149.

Simon I., Bârsan N., Bauer M., and Weimar U. (2001) Micromachined metal oxide gas sensors: Opportunities to improve sensor performance. *Sens. Actuators B* **73**, 1–26.

Singh S. (2007) Sensors—An effective approach for the detection of explosives. *J. Hazard. Mater.* **144**, 15–28.

Smith L.M., Odedra R., Kingsley A.J., Coward K.M., Rushworth S.A., Williams G., Leese T.A., Purdie A.J., and Kanjolia R.K. (2004) Trimethylindium transport studies: The effect of different bubbler designs. *J. Crystal Growth* **272**, 37–41.

Staude E. (1992) *Membranen und Membranprozesse.* Verlag Chemie,Weinheim, Germany.

Suzuki S., Noda Y., and Sawaki N. (1999) Market overview: Oxygen sensors. In: Baltes H., Gopel W., and Hesse J. (eds.), *Sensors Update.* Vol. **6**(1). Wiley-VCH, Weinheim, Germany, 381–396.

Tang Z., Fung S.K.H., Wong D.T.W., Chan P.C.H., Sin J.K.O., and Cheung P.W. (1998) An integrated gas sensor based on tin oxide thin-film and improved micro-hotplate. *Sens. Actuators B* **46**, 174–179.

Timmer B., Olthuis W., and van den Berg A. (2005) Ammonia sensors and their applications—A review. *Sens. Actuators B* **107**, 666–677.

Tobias P., Baranzahi A., Lloyd Spetz A., Kordina O., Janzen E., and Lundstrom I. (1997) Fast chemical sensing with metal-insulator silicon carbide structures. *IEEE Electron. Dev. Lett.* **18**(6), 287–289.

Tobias P., Mårtensson P., Göras A., Lundström I., and Spetz A.L. (1999) Moving gas outlets for the evaluation of fast gas sensors. *Sens. Actuators B* **58**, 389–393.

Tragut C. and Härdtl K.H. (1991) Kinetic behaviour of resistive oxygen sensors. *Sens. Actuators B* **4**, 425–429.

Tränkler H.-R. and Kanoun O. (2001) Recent advances in sensor technology. In: *Proceedings of IEEE Instrumentation and Measurement Technology Conference,* Budapest, Hungary, May 21–23.

UKOOA (1995) Guidelines for Fire and Explosion Hazard Management. Issue 1 (www.ukooa.co.uk).

Voiculescu I., Zaghloul M., and Narasimhan N. (2008) Microfabricated chemical preconcentrators for gas-phase microanalytical detection systems. *TrAC Trends Anal. Chem.* **27**(4), 327–343.

Vrana B., Mills G.A., Allan I.J., Dominiak E., Svensson K., Knutsson J., Morrison G., and Greenwood R. (2005) Passive sampling techniques for monitoring pollutants in water. *TrAC Trends Anal. Chem.* **24**(10), 845–868.

Vitenberg A.G., Kostkina M.I., and Ioffe B.V. (1984) Preparation of standard vapor-gas mixtures for gas chromatography: Continuous gas extraction. *Anal. Chem.* **56**, 2496–2500.

Walsh P T, Hedley D., and Pritchard D. K. (2001) Framework for HSE guidance on gas detectors. (On-line chekking of flammability monitoring equipment–Final report) HSL Internal Report EC/01/10, FM/01/01.

Warburton P.R., Pagano M.P., Hoover R., Logman M., and Crytzer K., Warburton Yi.J. (1998) Amperometric gas sensor response times. *Anal. Chem.* **70**(5), 998–1006.

Ward J.P.T. (2008) Oxygen sensors in context. *Biochim. Biophys. Acta* **1777**(1), 1–14.

Wichterle I. and Linek J. (1971) *Antoine Constants of Pure Compounds.* Academia, Prague.

Wilkins E.S. (1989) Towards implantable glucose sensors: A review. *J. Biomed. Eng.* **11**(5), 354–361.

Wilkins E. and Atanasov P. (1996) Glucose monitoring: State of the art and future possibilities. *Med. Eng. Phys.* **18**(4), 273–288.

Williams D. (1999) Semiconducting oxides as gas-sensitive resistors. *Sens. Actuators B* **57**, 1–16.

Wilson D.M., Hoyt S., Janata J., Booksh K., and Obando L. (2001) Chemical sensors for portable, handheld field instruments. *IEEE Sensors J.* **1**, 256–274.

Wise K.D. (2007) Integrated sensors, MEMS, and microsystems: Reflections on a fantastic voyage. *Sens. Actuators A* **136**, 39–50.

White L.T. (2000) *Hazardous Gas Monitoring: A Guide for Semiconductor and Other Hazardous Occupancies.* Noyes Publications/William Andrew Publishing, Norwich, NY.

Woodfin W.J. (1994) Portable electrochemical sensor methods. In: Eller P.M. and Cossinelli M.E. (eds.), *NIOSH Manual Analytical Methods*, 4th ed. U.S. Department of Health and Human Services, Cincinnati, OH, 70–73.

Yamazoe N. and Miura N. (1994) Environmental gas sensing. *Sens. Actuators B* **20**, 95–102.

Yamazoe N. and Miura N. (1996) Prospect and problems of solid electrolyte-based oxygenic gas sensors. *Solid State Ionics* **86–88**, 987–993.

Yeo T.L., Sun T., and Grattan K.T.V. (2008) Fibre-optic sensor technologies for humidity and moisture measurement. *Sens. Actuators A* **144**, 280–295.

Yuqing M., Jianrong C., and Keming F. (2005) New technology for the detection of pH. *J. Biochem. Biophys. Meth.* **63**(1), 1–9.

Zhang Z. and Li G. (2010) A review of advances and new developments in the analysis of biological volatile organic compounds. *Microchem. J.* **95**(2), 127–139.

Zhou Z. and Gai L. (2006) Preparation of calibration gas mixtures of water and nitrogen by using diffusion tubes. *Accred. Qual. Assur.* **11**, 205–207.

Zhou D.D. (2008) Microelectrodes for *in-vivo* determination of pH. In: Zhang X., Ju H., and Wang J. (eds.), *Electrochemical Sensors, Biosensors and Their Biomedical Applications.* Academic Press, Boston, 261–305.

APPENDIX

Books Related to the Field of Chemical Sensors

Before 1970

Dole M. (1941) *The Glass Electrode.* John Wiley, New York.

Ives D.J.G. and Janz G.J. (eds.) (1961) *Reference Electrodes: Theory and Practice.* Academic Press, New York.

Bard A.J. (ed.) (1966) *Electroanalytical Chemistry.* Marcel Dekker, New York.

Eisenman G., Bates G., Mottock G., and Friedman S.M. (1966) *The Glass Electrode.* Wiley-Interscience, New York.

Eisenman G. (1967) *Glass Electrodes for Hydrogen and Other Cations.* Marcel Dekker, New York.

Durst R.A. (ed.) (1969) *Ion-Selective Electrodes.* Special Publication 314, National Bureau of Standards, Washington, DC.

1970–1979

Moody G.J. and Thomas J.D.R. (1971) *Selective Ion Sensitive Electrodes.* Merrow, Watford, UK.

Bates R.G. (1973) *Determination of pH, Theory and Practice*, 2nd ed. John Wiley, New York.

Lakshminarayanaiah N. (1976) *Membrane Electrodes.* Academic Press, New York.

Wehry E.L. (1976) *Modern Fluorescence Spectroscopy*, Vols. 1 and 2. Plenum Press, New York.

Baiulescu G.E. and Cosofret V.V. (1977) *Applications of Ion Selective Membrane Electrodes in Organic Analysis.* Ellis Horwood, Chichester, UK.

Vesely J., Weiss D., and Stulik K. (1977) *Analysis with Ion-Selective Electrodes.* John Wiley, New York.

Long D.A. (1977) *Raman Spectroscopy.* McGraw-Hill, London.

Schulman S.G. (1977) *Fluorescence and Phosphorescence Spectroscopy: Physicochemical Principles and Practice.* Pergamon Press, Oxford, UK.

Freiser H. (ed.) (1978) *Ion-Selective Electrodes in Analytical Chemistry*, Vols. 1 and 2. Plenum Press, New York.

Hitchman M.L. (1978) *Measurements of Dissolved Oxygen.* John Wiley, New York.

Thomas R.C. (1978) *Ion-Sensitive Intracellular Microelectrodes: How to Make and Use Them.* Academic Press, London, New York.

Thomas J.D.R. (ed.) (1979–1992) *Ion-Selective Electrode Reviews*, Vols. 1–14. Pergamon Press, Oxford, UK.

1980–1985

Bailey P.L. (1980) *Analysis with Ion-Selective Electrodes*, 2nd ed. Heyden & Son, London.

Freiser H. (ed.) (1980) *Ion-Selective Electrodes in Analytical Chemistry II.* Plenum Press, New York.

Svehla C. (1981) *Comprehensive Analytical Chemistry.* Elsevier, New York.

Wehry E.L. (1981) *Modern Fluorescence Spectroscopy*, Vols. 3 and 4. Plenum Press, New York.

Koryta J. (1982) *Ions, Electrodes, and Membranes.* Wiley, New York.

Koryta J. and Stulik K. (1983) *Ion-Selective Electrodes.* Cambridge University Press, London.

Culshaw B. (1984) *Optical Fibre Sensing and Signal Processing.* Peter Peregrinus, London.

Czanderna A.W. and Lu C. (eds.) (1984) *Applications of Piezoelectric Quartz Crystal Microbalances.* Elsevier, New York.

Desiderio D.M. (1984) *Analysis of Neuropeptides by Liquid Chromatography and Mass Spectrometry.* Elsevier. Amsterdam.

Fung C.D., Cheung P.W., Ko W.H., and Fleming D.G. (eds.) (1985) *Micromachining and Micropackaging of Transducers.* Elsevier, Amsterdam.

Martellucci S. and Chester A.N. (eds) (1985) *Analytical Laser Spectroscopy.* Plenum Press, New York.

Morgan D.P. (1985) *Surface-Wave Devices for Signal Processing.* Elsevier, Amsterdam.

1986

Ammann D. (1986) *Ion-Selective Microelectrodes.* Springer-Verlag, Berlin.

Cantle J.E. (ed.) (1986) *Atomic Absorption Spectrometry.* Elsevier, Amsterdam.

Culshaw B. (1986) *Optical Fibre Sensing and Signal Processing.* Peter Peregrinus, London.

Dahmen E.A.M.F. (1986) *Electroanalysis.* Elsevier, Amsterdam.

Howe R.T., Muller R.S., Pisano A.P., and White R.M. (1986) *Microsensors and Microactuators.* IEEE Press, Piscateway, NJ.

Schuetzle D. and Hammerle R. (1986) *Fundamentals and Applications of Chemical Sensors.* American Chemical Society, Washington, DC.

1987

Jones B.E. (ed.) (1987) *Current Advances in Sensors.* Adam Hilger, Bristol, UK.

Madou M.J. (ed.) (1987) *Chemical Sensors for Extreme Environments.* SRI Microscience Program Volume on Sensors. NIST Publishing, Washington, DC.

Madou M.J. and Morrison S.R. (1987) *Chemical Sensing with Solid State Devices.* Academic Press/Harcourt Brace Jovanovich, Boston, New York.

Madou M.J., Gaisford G.S., and Gentile F. (1987) *Chemical Sensors, Business Intelligence Program.* SRI International, NIST Publishing, Washington, DC.

Moseley P.T. and Tofield B.C. (eds.) (1987) *Solid State Gas Sensors.* Adam Hilger, Bristol, UK, and Philadelphia.

Riley T. and Tomlinson C. (1987) *Principles of Electroanalytical Methods.* John Wiley, New York.

Turner A.P.F., Karube I., and Wilson G.S. (eds.) (1987) *Biosensors. Fundamentals and Applications.* Oxford University Press, Oxford, UK.

1988

Cesareo R. (ed.) (1988) *Nuclear Analytical Techniques in Medicine.* Elsevier, Amsterdam.

Dakin J. and Culshaw B. (eds.) (1988) *Optical Fiber Sensors: Principles and Components,* Vol. 1. Artech House, Boston.

Edmonds T.E. (1988) *Chemical Sensors.* Chapman & Hall, New York.

Florido A. (1988) *Dynamic Characteristics of Ion-Selective Electrodes.* CRC Press, Boca Raton, FL.

Hurst G.S. and Payne M.G. (1988) *Principles and Applications of Resonance Ionisation Spectroscopy.* Hilger Publications, Bristol, UK.

Massart D.L., Vandeginste B.G.M., Deming S.N., Michotte Y., and Kaufman L. (eds.) (1988) *Chemometrics: A Textbook.* Elsevier, Amsterdam.

Rosenbaum J.F. (1988) *Bulk Acoustic Wave Theory and Devices.* Artech House, Boston.

Ruzicka J. and Hansen E.H. (1988) *Flow Injection Analysis.* John Wiley, New York.

Schulman S.G. (1988) *Molecular Luminescence Spectroscopy: Methods and Applications.* John Wiley, New York.

Seiyama T. (ed.) (1988) *Chemical Sensor Technology,* Vol. 1. Elsevier, Amsterdam.

Valcárcel M. and Luque de Castro M.D. (1988) *Automatic Methods of Analysis.* Elsevier, Amsterdam.

1989

Culshaw B. and Dakin J. (eds.) (1989) *Optical Fiber Sensors: Systems and Applications,* Vol. 2. Artech House, Boston.

Grandke T. and Ko W.H. (eds.) (1989) *Sensors. A Comprehensive Survey,* Vol. 1: *Fundamentals and General Aspects.* Wiley-VCH, Weinheim, Germany.

Janata J. (1989) *Principles of Chemical Sensors.* Kluwer Academic Publishing, Dordrecht, The Netherlands/Plenum Press, New York–London.

Karlberg B. and Pacey G.E. (1989) *Flow Injection Analysis.* Elsevier, Amsterdam.

Madou M.J. and Morrison R.S. (1989) *Chemical Sensing with Solid State Devices.* Academic Press, London.

Middelhoek S. and Audet S.A. (1989) *Silicon Sensors.* Academic Press, London.

Murray R.W., Dessy R.E., Heineman W.R., Janata J., and Seitz W.R. (eds.) (1989) *Chemical Sensors and Microinstrumentation.* American Chemical Society, Washington, DC.

Scheller F. and Schubert F. (1989) *Biosensoren.* Birkhäuser Verlag, Basel.

1990

Cass A.E.G. (ed.) (1990) *Biosensors—A Practical Approach.* IRL Press, Oxford University, Oxford, UK.

Dakin J.P. (ed.) (1990) *The Distributed Fibre Optic Sensing Handbook.* IFS Publications, London.

Hall E.A.H. (1990) *Biosensors.* Open University Press, Cambridge, UK.

Seiyama T. (ed.) (1990) *Chemical Sensor Technology,* Vol. 2. Elsevier, Amsterdam.

Tölgyessy J., Havránek E., and Dejmková E. (1990) *Radionuclide X-Ray Fluorescence Analysis with Environmental Application.* Elsevier, Amsterdam.

Umezawa Y. (1990) *Handbook of Ion Selective Electrodes: Selectivity Coefficients.* CRC Press, Boca Raton, FL.

Bard A.J. (ed.) (1991) *Electroanalytical Chemistry*, Vol. 17. Marcel Dekker, New York.

Gardner J.W. and Bartlett Ph.N. (eds.) (1991*) Sensors and Sensory Systems for an Electronic Nose.* NATO ASI Series E: Applied Sciences—Vol. 212. Kluwer Academic Publishing, Dordrecht, The Netherlands.

Gopel W., Hesse J., and Zemel J.N. (eds.) (1991) Sensors: *A Comprehensive Survey,* Vol. 2. *Part I. Chemical and Biochemical Sensors.* VCH, Weinheim, Germany.

Lakowicz J.R. (1991) *Topics in Fluorescence Spectroscopy: Principles.* Plenum Press, New York.

Madou M. and Joseph J. (eds.) (1991) *Opportunities for Innovation: Chemical and Biological Sensors.* NIST Publication GCR 91-593-1, Washington, DC.

Moseley P.T., Norris J.O.W., and Williams D.E. (eds.) (1991) *Techniques and Mechanisms in Gas Sensing.* Adam Hilger, Bristol, UK.

Muller R.S., Howe R.T., Senturia S.D., Smith R.L., and White R.M. (1991) *Microsensors.* IEEE Press, Piscataway, NJ.

Scheller F.W. and Schubert F. (1992) *Biosensors.* Elsevier, Amsterdam.

Smolkova-Keulemansova E. and Feltl L. (1991) *Analysis of Substances in the Gaseous Phase.* Elsevier, Amsterdam.

Udd E. (ed.) (1991) *Fiber Optic Sensors: An Introduction for Engineers and Scientists.* John Wiley, New York.

Weeks I. (1991) *Chemiluminescence Immunoassay.* Elsevier, Amsterdam.

Wise D. (ed.) (1991) *Bioinstrumentation and Biosensors.* Marcel Dekker, New York.

Wise D. and Wingard L.B. (eds.) (1991) *Biosensors with Fiberoptics.* Humana Press, Clifton, NJ.

Wolfbeis O.S. (ed.) (1991) *Fiber Optic Chemical Sensors and Biosensors*, Vols. 1 and 2. CRC Press, Boca Raton, FL.

Yamazoe N. (ed.) (1991) *Chemical Sensor Technology*, Vol. 3. Elsevier, Amsterdam.

1992

Ciureanu P. and Middelhoek S. (eds.) (1992) *Thin Film Resistive Sensors.* IOP Publishing, Bristol, UK.

Colomban Ph. (ed.) (1992) *Proton Conductors.* Cambridge University Press, Cambridge, UK.

Gardner J.W. and Bartlett P.N. (eds.) (1992) *Sensors and Sensory Systems for an Electronic Nose.* Kluwer Academic Publishers, Dordrecht, The Netherlands.

Gopel W., Hesse J., and Zemel J.N. (eds.) (1992) *Sensors: A Comprehensive Survey*, Vol. 3. *Part II. Chemical and Biochemical Sensors.* VCH, Weinheim, Germany.

Krohn D.A. (1992) *Fiber Optic Sensors: Fundamentals and Applications.* Instrument Society of America, Research Triangle Park, NC.

Lambrechts M. and Sansen W. (1992) *Biosensors: Microelectrochemical Devices.* IOP Publishers, Bristol, UK.

Rosen C.Z., Hiremath B.V., and Newnham R. (eds.) (1992) *Piezoelectricity.* Springer-Verlag, New York.

Sberveglieri G. (ed.) (1992) *Gas Sensors—Principles Operation and Developments.* Kluwer Academic Publishers, Dordrecht, The Netherlands.

Scheller F. and Schmid R.D. (eds) (1992) *Biosensors: Fundamentals, Technologies and Applications.* VCH, Weinheim, Germany.

Smyth M.R. and Vos J.G. (eds.) (1992) *Analytical Voltammetry.* Elsevier, Amsterdam.

Stoeppler M. (ed.) (1992) *Hazardous Metals in the Environment.* Elsevier, Amsterdam.

Wagner E., Danliker R., and Spenner K. (eds.) (1992) *Sensors: A Comprehensive Survey, Vol. 6, Optical Sensors.* (Gopel W., Hesse J., and Zemel J.N., series eds.). John Wiley, New York.

Wolfbeis O.S. (ed.) (1992) *Fluorescence Spectroscopy: New Methods and Applications.* Springer-Verlag, Heidelberg.

Yamauchi S. (ed.) (1992) *Chemical Sensor Technology*, Vol. 4. Elsevier, Amsterdam.

1993

Barceló D. (ed.) (1993) *Environmental Analysis*. Elsevier, Amsterdam.

Brett C.M.A. and Brett A.M.O. (1993) *Electrochemistry: Principles, Methods, and Applications*. Oxford University Press, Oxford, UK.

Hauptmann P. and Pownall T. (1993) *Sensors Principles and Applications*. Prentice-Hall, New York.

Mandelis A. and Christofides C. (eds.) (1993) *Physics, Chemistry and Technology of Solid State Gas Sensor Devices*. John Wiley, New York.

Ohba Y.J. (1992) *Intelligent Sensor Technology*. John Wiley, New York.

Wolfbeis O.S. (ed.) (1993) *Fluorescence Spectroscopy: New Methods and Applications*. Springer-Verlag, Berlin–Heidelberg.

Yinon J. and Zitrin S. (1993) *Modern Methods and Applications in Analysis of Explosives*. John Wiley, Chichester, UK.

1994

Bau H.H., de Rooij N.F., and Kloeck B. (eds.) 1994) *Sensors: A Comprehensive Survey*, Vol. 7. VCH Verlag, Weinheim, Germany.

Crow D.R. (1994) *Principles and Applications of Electrochemistry*, 4th ed. Blackie Academic and Professional, Glasgow, UK.

Ihokura K. and Watson J. (1994) *The Stannic Oxide Gas Sensor: Principle and Applications*. CRC Press, Boca Raton, FL.

Gardner J.W. (1994) *Microsensors: Principles and Applications*. John Wiley, Chichester, UK.

Grattan K.T.V. and Grattan K.T. (eds.) (1994) *Optical Fibre Sensor Technology*. Chapman & Hall, London.

Herber R.F.M. and Stoeppler M. (eds.) (1994) *Trace Element Analysis in Biological Specimens*. Elsevier, Amsterdam.

Lakowicz J.R. (ed.) (1994) *Topics in Fluorescence Spectroscopy, Vol. 4: Probe Design and Chemical Sensing*. Plenum, New York.

Mallouk T.E. and Harrison D.J. (eds.) (1994) *Interfacial Design and Chemical Sensing*. American Chemical Society, Washington, DC.

Prudenziati M. (1994) *Thick-Film Sensors*. Elsevier. Amsterdam.

Ristic L. (ed.) (1994) *Sensor Technology and Devices*. Artech House, Norwood, MA.

Sze S.M. (1994) *Semiconductor Sensors*. John Wiley, New York.

Valcárcel M. and Luque de Castro M.D. (1994) *Flow-Through (Bio)chemical Sensors*. Elsevier, Amsterdam.

Westbrook M.H. and Turner J.D. (1994) *Automotive Sensors*. IOF Publishing, Philadelphia.

1995

D'Amico A. and Sberveglieri G. (eds.) (1995) *Sensors for Domestic Applications*. World Scientific, Singapore.

Donald G. and Buerk D.G. (1995) *Biosensors*. CRC Press, Boca Raton, FL.

Harsányi G. (1995) *Polymer Films in Sensor Applications*. Technomic Publishing, Lancaster, PA/Basel.

Frank R. (1995) *Understanding Smart Sensors*. Artech House, Boston.

Göpel W., Hesse J., Zemel N., Meixner H., and Jones R. (eds.) (1995) *Sensors, A Comprehensive Survey*, Vol. 8. Wiley-VCH, Weinheim, Germany.

Quevauviller Ph., Maier E.A., and Griepink B. (eds.) (1995) *Quality Assurance for Environmental Analysis.* Elsevier, Amsterdam.

1996

Ballantine D.S., White R.M., Martin S.J., Ricco A.J., Zellers E.T., Frye G.C., and Wohltjen H. (1996) *Acoustic Wave Sensors: Theory, Design, Physico-Chemical Applications.* Academic Press, London.

Baltes H., Gopel W., and Hesse J. (eds.) (1996) *Sensors Update,* Vols. 1 and 2. Wiley-VCH Verlag, Weinheim, Germany.

Boisde G. and Harmer A. (1996) *Chemical and Biochemical Sensing with Optical Fibers and Waveguides.* Artech House, Boston–London.

Brignell J.E. and White N.M. (1996) *Intelligent Sensor Systems,* 2nd ed. IOP Publishing, Bristol, UK.

Campbell M. (ed.) (1996) *Sensor Systems for Environmental Monitoring: Sensor Technologies,* Vols. 1 and 2. Kluwer Academic Publishing, Dordrecht, The Netherlands.

Culshaw B. and Dakin J. (eds.) (1996) *Optical Fiber Sensors: Components and Subsystems,* Vols. 3 and 4. Artech House Books, Boston.

Di Natale C. and D'Amico A. (eds.) (1996) *Sensors and Microsystems.* World Scientific, Singapore.

Fraden J. (1996) *Handbook of Modern Sensors: Physics, Design and Applications.* Springer-Verlag, Berlin and Heidelberg.

Göpel W., Boll R., Ricolfi T., Grandke T., Bau H.H., Wagner E., and Meixner H. (1995) *Sensors, A Comprehensive Survey.* Vol. 9: *Cumulative Index.* Wiley-VCH, Weinheim, Germany.

Kissinger P.T. and Heineman W.R. (eds.) (1996) *Laboratory Techniques in Electroanalytical Chemistry,* 2nd ed. Marcel Dekker, New York.

Kress-Rogers E. and Nicklin S. (eds.) (1996) *Handbook of Biosensors: Medicine, Food and the Environment.* Taylor & Francis Group, CRC Press, Boca Raton, FL.

Kupriyanov L.Yu. (ed.) (1996) *Semiconductor Sensors in Physico-Chemical Studies.* Elsevier, Amsterdam.

Lobinski R. and Marczenko Z. (1996) *Spectrochemical Trace Analysis for Metals and Metalloids.* Elsevier, Amsterdam.

Masel R.I. (1996) *Principles of Adsorption and Reaction on Solid Surfaces.* Wiley International, New York.

Meixner H. and Jones R. (eds.) (1996) *Sensors: A Comprehensive Survey,* Vol. 8. VCH Verlag, Weinheim, Germany.

Moseley P.T. and Crocker A.J. (1996) *Sensor Materials (Sensors Series).* IOP, Bristol, UK.

Nenov T.G. and Yordanov S.P. (1996) *Ceramic Sensors: Technology and Applications.* Technomic Publishing, Lancaster, PA.

Taylor R.F. and Schultz J.S. (eds.) (1996) *Handbook of Chemical and Biological Sensors.* Institute of Physics Publishing, Bristol, UK–Philadelphia.

Usher M.J. and Keating D.A. (1996) *Sensors and Transducers, Characteristics, Applications, Instrumentation, Interfacing,* 2nd ed. Macmillan, London.

Vanýsek P. (ed.) (1996) *Modern Techniques in Electroanalysis.* Wiley-Interscience, New York.

Verga Scheggi A.M., Martellucci S., Chester A.N., and R.Pratesi, (1996) *Biomedical Optical Instrumentation and Laser-Assisted Biotechnology* (NATO A.S.I. Series Vol. 325). Kluwer Academic Publishers, Dordrecht, The Netherlands.

Yamasaki H. (ed.) (1996) *Intelligent Sensors (Handbook of Sensors and Actuators).* Elsevier, Amsterdam.

1997

Asher R.C. (1997) *Ultrasonic Sensors for Chemical and Process Plant (Sensors Series).* Taylor & Francis, New York.

Ballantine D.S., White R.M., Martin S.J., Ricco A.J., Zellers E.T., Frye G.C., and Wohltjen H. (1997) *Acoustic Wave Sensors: Theory, Design, and Physico-Chemical Applications.* Academic Press, San Diego, CA.

Blum L.G. (1997) *Bio- and Chemi-Luminescent Sensors*. World Scientific, Singapore.

Cattrall R.W. (1997) *Chemical Sensors*. University Press, Oxford–New York–Melbourne.

Dakin J. and Culshaw B. (1997) *Optical Fiber Sensors, Vol. IV: Applications, Analysis, and Future Trends.* Artech House, London.

Hennion M.-C. and Barcelo D. (eds.) (1997) *Trace Determination of Pesticides and Their Degradation Products in Water*. Elsevier, Amsterdam.

Hogan J.D. (ed.) (1997) *Specialty Gas Analysis: A Practical Guidebook.* Wiley VCH, New York.

Huijsing, J.H. and Meijer G.C.M. (eds.) (1997) *Smart Sensor Interfaces.* Kluwer Academic Publishing, Dordrecht, The Netherlands.

Fraden J. (1997) *Handbook of Modern Sensors*, 2nd ed. Springer-Verlag, Berlin.

Kress-Rogers E. (ed.) (1997) *Handbook of Biosensors and Electronic Noses: Medicine, Food, and the Environment.* CRC Press, Boca Raton, FL.

Madou M.J. (1997) *Fundamentals of Microfabrication.* CRC Press, Boca Raton, FL.

Paré J.R.J. and Bélanger J.M.R. (eds.) (1997) *Instrumental Methods in Food Analysis.* Elsevier, Amsterdam.

Tabib-Azar M. (1997) *Microactuators: Electrical, Magnetic, Thermal, Optical, Mechanical, Chemical, and Smart Structures.* Kluwer Academic Publishing, Dordrecht, The Netherlands.

Thompson M. and Stone D.C. (1997) *Surface-Launched Acoustic Wave Sensors: Chemical Sensing and Thin-Film Characterization.* John Wiley, New York.

Winefordner J.D. (ed.) (1997) *Chemical Analysis: A Series of Monographs on Analytical Chemistry and Its Applications.* John Wiley, New York.

1998

Akmal N. and Usmani A.M. (eds.) (1998) *Polymers in Sensors: Theory and Practice.* American Chemical Society, Washington, DC.

Baltes H., Gopel W., and Hesse J. (eds.) (1998) *Sensors Update*, Vols. 3 and 4. Wiley-VCH Verlag, Weinheim, Germany.

Busch-Vishniac I. (1998) *Electromechanical Sensors and Actuators.* Springer-Verlag, New York.

Campbell C.K. (1998) *Surface Acoustic Wave Devices for Mobile and Wireless Communications.* Academic Press, San Diego, CA.

Colin F.P. and Quevauviller P. (eds.) (1998) *Monitoring of Water Quality: The Contribution of Advanced Technologies.* International Water Centre, Elsevier, New York.

Cunningham A.J. (1998) *An Introducion to Bioanalytical Sensors.* John Wiley, New York.

Diamond D. (ed.) (1998) *Principles of Chemical and Biological Sensors.* John Wiley, New York.

Elwenspoek M. and Jansen H. (1998) *Silicon Micromachining.* Cambridge University Press, Cambridge, UK.

Grattan K.T.V. and Meggitt B.T. (eds.) (1998). *Optical Fiber Sensor Technology*, Vols. 1, 2, and 3. Kluwer Academic Publishing, Dordrecht, The Netherlands.

Kovacs G.T.A. (1998) *Micromachined Transducers Source-book.* WCB and McGraw-Hill, Boston.

Naji B.O. and Wang A. (1998) *Sensors and Controls for Advanced Manufacturing.* SPIE, Bellingham, WA.

Pérez-Bendito D. and Rubio S. (1998) *Environmental Analytical Chemistry.* Elsevier, Amsterdam.

Scott A.O. (1998) *Biosensors for Food Analysis.* Woodhead Publishing, Cambridge, UK.

Spichiger-Keller U.E. (1998) *Chemical Sensors and Biosensors for Medical and Biological Applications.* Wiley-VCH, Weinheim, Germany.

Solomon S. (ed.) (1998) *Sensors Handbook.* McGraw-Hill, New York.

Webster J.G. (ed.) (1998) *The Measurement, Instrumentation and Sensors Handbook.* CRC Press, Boca Raton, FL.

1999

Anderson G. and Hadden D. (1999) *The Gas Monitoring Handbook.* Avocet Press, New York.

Baltes H., Gopel W., and Hesse J. (eds.) (1999) *Sensors Update*, Vols. 5 and 6. Wiley-VCH Verlag, Weinheim, Germany.

Gardner J.W. and Bartlett P.N. (1999) *Electronic Noses: Principles and Application.* Oxford University Press, Oxford, UK.

Grattan K.T.V., Meggitt B.T., and Grattan L.S. (eds.) (1999) *Optical Fiber Sensor Technology, Vol. 4. Chemical and Environmental Sensing.* Kluwer Academic Publishing, Dordrecht, The Netherlands.

Hurst W.J. (ed.) (1999) *Electronic Noses and Sensor Array Based Systems: Design and Applications.* Technomic Publications, Lancaster, PA.

Hornung M.R. and Brand O. (1999) *Micromachined Ultrasound-Based Proximity Sensors.* Kluwer Academic Publishing, Dordrecht, The Netherlands.

Lakowicz J.R. (1999) *Principles of Fluorescence Spectroscopy*, 2nd ed. Kluwer Academic/Plenum Publishers, New York.

Maier EA. and Quevauviller Ph. (1999) *Interlaboratory Studies and Certified Reference Materials for Environmental Analysis.* Elsevier, Amsterdam.

Sharma A. and Schulman S.G. (1999) *Introduction to Fluorescence Spectroscopy.* John Wiley, New York.

Sorenson T.S. (ed.) (1999) *Surface Chemistry and Electrochemistry of Membranes.* CRC Press, Boca Raton, FL.

Yinon J. (1999) *Forensic and Environmental Detection of Explosives.* John Wiley, Chichester, UK.

2000

Baltes H., Gopel W., and Hesse J. (eds.) (2000) *Sensors Update*, Vols. 7 and 8. Wiley-VCH Verlag, Weinheim, Germany.

Barceló D. (ed.) (2000) *Sample Handling and Trace Analysis of Pollutants.* Elsevier, Amsterdam.

Chou J. (2000) *Hazardous Gas Monitors: A Practical Guide to Selection, Operation and Application.* McGraw-Hill Professional, New York.

Harsányi G. (2000) *Sensors in Biomedical Applications: Fundamentals, Technology and Applications.* CRC Press, Boca Raton, Fl.

Grattan L.S. and Meggitt B.T. (eds.) (2000) *Optical Fiber Sensor Technology: Advanced Applications—Bragg Gratings and Distributed Sensors.* Springer-Verlag, New York.

Martellucci S., Chester A.N., and Mignani A.G. (eds.) (2000) *Optical Sensors and Microsystems: New Concepts, Materials, Technologies.* Kluwer Academic/Plenum Press, New York.

Mulchandani A. and Sadik O.A. (eds.) (2000) *Chemical and Biological Sensors for Environmental Monitoring.* American Chemical Society, Washington, DC.

Osada Y. and De Rossi D.E. (2000) *Polymer Sensors and Actuators (Macromolecular Systems—Materials Approach).* Springer-Verlag, Berlin.

Soloman S. (ed.) (2000) *Sensors Handbook.* McGraw-Hill, New York.

Vaezi-Nejad S.M. (ed.) *Advanced Topics in Solid State and Fibre-Optic Sensors.* The Institution of Engineering and Technology, London, UK.

Vaezi-Nejad S.M. Haigh D., and Soin R. (eds.) (2000) *Selected Topics in Advanced Solid-State and Fibre-Optic Sensors.* The Institution of Engineering and Technology, London, UK.

Varney M.S. (ed.) (2000) *Chemical Sensors in Oceanography.* Gordon & Breach, Amsterdam.

Wang J. (2000) *Analytical Electrochemistry*, 2nd ed. Wiley-VCH, New York.

Wright J.D. and Sommerdijk N.A.J.M. (2000) *Sol-Gel Materials: Chemistry and Applications (Advanced Chemistry Texts).* CRC Press, Boca Raton, FL.

Yang V.C.-M. and Ngo T.T. (2000) *Biosensors and Their Applications.* Kluwer Academic, New York.

2001

Baltes H., Gopel W., and Hesse J. (eds.) (2001) *Sensors Update*, Vol. 9. Wiley-VCH Verlag, Weinheim.

Bard A.J. and Faulkner L.R. (2001) *Electrochemical Methods: Fundamentals and Applications*, 2nd ed. John Wiley, New York.

Gardner J.W., Varadan V.K., and Awadelkarim O.O. (2001) *Microsensors, MEMS, and Smart Sensors.* John Wiley, Chichester, UK.

Grattan K.T.V., Meggitt B.T., and Grattan L.S. (2001) *Optical Fiber Sensor Technology: Advanced Applications—Bragg Gratings and Distributed Sensors.* Springer-Verlag, Dordrecht, The Netherlands.

Fry C. and Langley S.E.M. (2001) *Ion-Selective Electrodes for Biological Systems.* Taylor & Francis, Amsterdam.

Kress-Rogers E. and Brimelow C.J.B. (eds.) (2001) *Instrumentation and Sensors for the Food Industry.* Woodhead Publishing, Cambridge, UK.

Monk P.M.S. (2001) *Fundamentals of Electroanalytical Chemistry.* John Wiley, New York.

Ruppel C.C.W. and Fjeldly T.A. (eds.) (2001) *Advances in Surface Acoustic Wave Technology, Systems and Applications.* World Scientific, Singapore.

Sellergren B. (ed.) (2001) *Molecularly Imprinted Polymers: Man-Made Mimics of Antibodies and Their Applications in Analytical Chemistry.* (Techniques and Instrumentation in Analytical Chemistry, Vol. 23). Elsevier, Amsterdam.

Senturia S. (2001) *Microsystem Design*. Kluwer Academic, Dordrecht, The Netherlands.

Stefan R.I. and Van Stead J.F. (2001) *Electrochemical Sensors in Bioanalysis.* Marcel Dekker, New York.

Tomita A. (ed.) (2001) *Emissions Reduction.* NO_x/SO_x *Suppression*. Elsevier, Amsterdam.

Tonshoff H.K. and Inasaki I. (eds.) (2001) *Sensors in Manufacturing* (Hesse J., Gardner J.W., and Gopel W., series eds.). Wiley-VCH, Weinheim, Germany.

Valeur B. (2001) *Molecular Fluorescence: Principles and Applications*, Wiley-VCH, Weinheim, Germany.

Valeur B. and Brochon J.-C. (eds.) (2001) *New Trends in Fluorescence Spectroscopy. Applications to Chemical and Life Sciences*. Springer-Verlag, Berlin–Heidelberg.

Van Stade J.F.and Stefan R.-I. (2001) *Electrochemical Sensors in Bioanalysis.* CRC Press, Boca Raton, Fl.

2002

Baltes H., Gopel W., and Hesse J. (eds.) (2002) *Sensors Update*, Vols. 10 and 11. Wiley-VCH Verlag, Weinheim, Germany.

Eggins B.R. (ed.) (2002) *Chemical Sensors and Biosensors*. Wiley-VCH, Weinheim, Germany.

Gassmann O. (ed.) (2002) *Sensors in Intelligent Buildings* (Hesse J., Gardner J.W., and Gopel W., series eds.). Wiley-VCH, Weinheim, Germany.

Gautschi G. (ed.) (2002) *Piezoelectric Sensorics*. Springer-Verlag, Berlin.

Gizeli E. and Lowe C.R. (eds.) (2002) *Biomolecular Sensors*. Taylor & Francis, London, UK.

Kordal R.J., Usmani A.M., and Law W.T. (eds.) *Microfabricated Sensors: Application of Optical Technology for DNA Analysis.* American Chemical Society, Washington, DC.

Law W.T., Usmani A.M., and Akmal N. (eds.) (2002) *Biomedical Diagnostic Science and Technology.* Informa Healthcare, London, UK.

Lopez-Higuera J.M. (ed.) (2002) *Handbook of Optical Fibre Sensing Technology*. John Wiley, Chichester, UK.

Lange D., Brand O., and Baltes H. (2002) *CMOS Cantilever Sensor Systems: Atomic Force Microscopy and Gas Sensing Applications.* Springer-Verlag, Berlin.

Pawliszyn J. (2002) *Sampling and Sample Preparation in Field and Laboratory*. Elsevier, Amsterdam.

Sadana A. (2002) *Engineering Biosensors: Kinetics and Design Applications*. Academic Press, San Diego, CA.

Scholz F. (ed.) (2002) *Electroanalytical Methods*. Springer-Verlag, New York.

Silverstein A., Silverstein V., and Silverstein Nun L. (2002) *Smelling and Tasting* (Senses and Sensors). 21st Century, Brookfield, CT.

Zolotov Yu.A., Ivanov V.M., and Amelin V.G. (eds.) (2002) *Chemical Test Methods of Analysis*. Elsevier, Amsterdam.

2003

Alegret S. (ed.) (2003) *Integrated Analytical Systems*. Elsevier, Amsterdam.

Baltes H., Fedder G., and Korvink J. (eds.) (2003*) Sensors Update*, Vols. 12 and 13. Wiley-VCH, Weinhein, Germany.

Barth F.G., Humphrey J.A.C., and Seecomb T.W. (eds.) (2003) *Sensors and Sensing in Biology and Engineering*. Springer-Verlag, New York.

Callaway E.H.J. and Callaway E.H. (2003) *Wireless Sensor Networks: Architectures and Protocols*. CRC Press, Boca Raton, FL.

Doll T. (ed.) (2003) *Advanced Gas Sensing: The Electroadsorptive Effect and Related Techniques*. Kluwer Academic, Boston.

Knepper T., De Voogt P., and Barcelo (2003) *Analysis and Fate of Surfactants in the Aquatic Environment*. Elsevier, Amsterdam.

Marek J., Trah H.-P., Suzuki Y., and Yokomori I. (eds.) (2003) *Sensors Applications, Vol. 4, Sensors for Automotive Applications* (Hesse J., Gardner J.W., and Gopel W., series eds.). Wiley-VCH, Weinheim, Germany.

Mester Z. and Sturgeon R. (2003) *Sample Preparation for Trace Element Analysis*. Elsevier, Amsterdam.

Mitra S. (ed.) (2003) *Sample Preparation Techniques in Analytical Chemistry*. John Wiley, Hoboken, NJ.

Oosterbroek R.E. and Van der Berg A. (eds.) (2003) *Lab-on-a-Chip: Miniaturized Systems for (Bio) Chemical Analysis and Synthesis*. Elsevier, Amsterdam.

Pearce T.C., Schiffman S.S., Nagle H.T., and Gardner J.W. (eds.) (2003) *Handbook of Machine Olfaction: Electronic Nose Technology*. Wiley-VCH, Weinheim, Germany/New York.

Prosser S.J. and Lewis E. (2003) *Sensors and Their Applications XII (Series in Sensors)*. Taylor & Francis, Boca Raton, FL.

Tschulena G.R. and Lahrmann A. (eds.) (2003) *Sensors in Household Appliances*, Vol. 5 (Hesse J., Gardner J.W., and Gopel W., series eds.). Wiley-VCH, Weinheim, Germany.

Wallace G.G., Spinks G.M., Kane-Maguire L.A.P., and Teasdale P.R. (2003) *Conductive Electroactive Polymers: Intelligent Materials Systems*, 2nd ed. CRC Press, New York.

2004

Bard A.J. (ed.) (2004) *Electrogenerated Chemiluminescence*. Marcel Dekker, New York.

Cannons A.C., Harwood V.J., and Cannons A.C. (2004) *Sensor Technology for Water Quality Monitoring: Fiber-Optic Biosensor*. IWA Publishing, London, UK.

Fernández Alba A.R. (2004) *Chromatographic-Mass Spectrometric Food Analysis for Trace Determination of Pesticide Residues*. Elsevier, Amsterdam.

Fraden J. (ed.) (2004) *Handbook of Modern Sensors: Physics, Designs, and Applications*. Springer-Verlag, New York–Berlin.

Frymier P., Ren S., and Lajoie C. (2004) *Sensor Technology for Water Quality Monitoring: Bioluminescent Microorganisms*. Water Environment Research Foundation, Alexandria, VA.

Gardner J. and Yinon J. (eds.) (2007) *Electronic Noses and Sensors for the Detection of Explosives* (NATO Science Series II: Mathematics, Physics and Chemistry. Vol. 159). Kluwer Academic Publishers, Dordrecht, The Netherlands.

Geschke O., Klank H., and Telleman P. (eds.) (2004) *Microsystem Engineering of Lab-on-a-Chip Devices.* Wiley-VCH, Weinheim, Germany.

Girault H.B. (2004) *Analytical and Physical Electrochemistry.* Marcel Dekker, New York.

Hac A. (2004) *Wireless Sensor Network Designs.* John Wiley, New York.

Jackson R.G. (2004) *Novel Sensors and Sensing.* CRC Press, Boca Raton, FL.

Mirsky V.M. (ed.) (2004) *Thin Film Biosensors* (Springer Series on Chemical Sensors and Biosensors, Urban G., series ed.). Springer-Verlag, Berlin.

Mirsky V.M. (ed.) (2004) *Ultrathin Electrochemical Chemo- and Biosensors. Technology and Performance* (Springer Series on Chemical Sensors and Biosensors, Urban G., series ed.). Springer-Verlag, Berlin.

Narayanaswamy R. and Wolfbeis O.S. (eds.) (2004) *Optical Sensors. Industrial, Environmental and Diagnostic Applications* (Springer Series on Chemical Sensors and Biosensors, G. Urban, series ed.). Springer-Verlag, Berlin.

Sen C.K. and Semenza G. (eds.) (2004) *Oxygen Sensing.* Elsevier, New York.

Sun Y. and Ong K.Y. (2004) *Detection Technologies for Chemical Warfare Agents and Toxic Vapors.* CRC Press, Boca Raton, FL.

Togawa T. and Spelman F.A. (eds.) (2004) *Sensors in Medicine and Health Care.* Wiley-VCH, Weinheim, Germany.

Toko K. (2004) *Biomimetic Sensor Technology.* Cambridge University Press, Cambridge, UK.

2005

Connolly D. (2005) *Sensor Technology for Water Quality Monitoring.* IWA Publishing, London, UK.

Hierlemann A. (2005) *Integrated Chemical Microsensor Systems in CMOS Technology (Microtechnology and MEMS).* Springer-Verlag, Berlin.

Eren H. (2005) *Wireless Sensors and Instruments: Networks, Design, and Applications.* CRC Press, Boca Raton, FL.

Fierro J.L.G. (ed.) (2005) *Metal Oxides: Chemistry and Applications (Chemical Industries).* CRC Press, Boca Raton, FL.

Gorton L. (ed.) (2005) *Biosensors and Modern Biospecific Analytical Techniques.* Elsevier, Amsterdam.

Khandpur R.S. (2005) *Biomedical Instrumentation: Technology and Application.* McGraw-Hill, New York.

Marina M.L., Ríos A., and Miguel Valcárcel M. (eds.) (2005) *Analysis and Detection by Capillary Electrophoresis.* Elsevier, Amsterdam.

Orellana G. and Moreno-Bondi M.C. (eds.) (2005) *Frontiers in Chemical Sensors. Novel Principles and Techniques* (Springer Series on Chemical Sensors and Biosensors, G. Urban, series ed.). Springer-Verlag, Berlin.

Raghavendra C.S., Krishna M., Sivalingam K.M., and Taieb Znati T. (eds.) (2005) *Wireless Sensor Networks.* Springer-Verlag, New York.

Thompson R.B. (ed.) (2005) *Fluorescence Sensors and Biosensors.* CRC Press, Boca Raton, FL.

Wilson J.S. (ed.) (2005) *Sensor Technology Handbook.* Elsevier, Amsterdam.

Worsfold P., Townshend A., and Poole C. (eds.) (2005) *Encyclopedia of Analytical Science,* 2nd ed. Elsevier, Oxford, UK.

Yurish S. and Gomes M. (eds.) (2005) *Smart Sensors and MEMS. NATO Science Series III: Mathematics, Physics and Chemistry,* Vol. 181. Kluwer Academic Publishers, Dordrecht, The Netherlands.

2006

Ahuja S. and Jespersen N. (eds.) (2006) *Modern Instrumental Analysis.* Elsevier, Amsterdam.

Baldini F., Chester A.N., Homola J., and Martellucci S. (eds.) (2006) *Optical Chemical Sensors* (NATO Series). Springer-Verlag, Amsterdam.

Cao G. and Fryxell G.E. (eds.) (2006) *Environmental Applications of Nanomaterials: Synthesis, Sorbents and Sensors.* Imperial College Press, London, UK.

Chakrabarty K. and Zeng J. (eds.) (2006) *Design Automation Methods and Tools for Microfluidics-Based Biochips.* Springer-Verlag, Dordrecht, The Netherlands.

Dorf R.C. (ed.) (2006) *Sensors, Nanoscience, Biomedical Engineering, and Instruments: Sensors Nanoscience Biomedical Engineering (The Electrical Engineering Handbook*, 3rd ed.). CRC Press, Boca Raton, FL.

Grimes C.A., Dickey E.C., and Pishko M.V. (eds.) (2006) *Encyclopedia of Sensors*, Vols. 1–10. American Scientific Publishers, Stevenson Ranch, CA.

Gründler P. (2007) *Chemical Sensors: An Introduction for Scientists and Engineers.* Springer-Verlag, Berlin.

Henry C.S. (ed.) (2006) *Microchip Capillary Electrophoresis. Methods and Protocols.* Humana Press, Totowa, NJ.

Homola J. (ed.) (2006) *Surface Plasmon Resonance Based Sensors* (Springer Series on Chemical Sensors and Biosensors, Urban G., series ed.). Springer-Verlag, Berlin.

Kneipp K., Moskovits M., and Kneipp H. (eds.) (2006) *Surface-Enhanced Raman Scattering: Physics and Applications.* Springer-Verlag, Berlin.

Knopf G.K. and Bassi A.S. (eds.) (2006) *Smart Biosensor Technology.* CRC Press, Boca Raton, FL.

Korvink J. and Haber O. (eds.) (2006) *MEMS: A Practical Guide to Design, Analysis and Applications.* William Andrew Publishing, Norwich, NY/Springer-Verlag, Heidelberg.

Laconte J., Fkandre D., and Raskin J.-P. (2006) *Micromachined Thin-Film Sensors for SOI-CMOS Co-integration.* Springer-Verlag, Dordrecht, The Netherlands.

Nalwa H.S. (eds.) (2006) *Polymeric Nanostructures and Their Applications.* American Scientific Publishers, Valencia, CA.

Priego Capote F. and Luque de Castro M.D. (2006) *Analytical Applications of Ultrasound.* Elsevier, Amsterdam.

Raghavendra C.S. (2006) *Wireless Sensor Networks.* Lavoisier, Paris.

Srinivasan S. (2006) *Fuel Cells: From Fundamentals to Applications.* Springer-Verlag, New York.

Thompson R.B. (200) (ed.) *Fluorescence Sensors and Biosensors.* CRC Press, Boca Raton, FL.

Wang J. (2006) *Analytical Electrochemistry*, 3rd ed. Wiley-VCH, Hoboken, NJ.

2007

Albani J.R. (2007) *Principles and Applications of Fluorescence Spectroscopy.* Blackwell Science, Oxford, UK.

Alegret S. and Merkoci A. (eds.) (2007) *Electrochemical Sensor Analysis (Wilson and Wilson's Comprehensive Analytical Chemistry*, Vol. 49). Elsevier, Amsterdam.

Arruda M.A.Z. (2007) *Trends in Sample Preparation.* Nova Science, New York.

Aswal D.K. and Gupta S.K. (eds.) (2007) *Science and Technology of Chemiresistor Gas Sensors.* Nova Science, New York.

Baltes H., Lange D., and Brand O. (2007) *CMOS Cantilever Sensor Systems.* Springer-Verlag, Berlin.

Bhushan B. (ed.) (2007) *Handbook of Nanotechnology*, 2nd ed. Springer-Verlag, Berlin.

Ellis B. and Smith R. (eds.) (2007) *Polymers: A Property Database*, 2nd ed. CRC Press, Boca Raton, FL.

Freund M.S. and Deore B.A. (2007) *Self-Doped Conducting Polymers.* JohnWiley, New York.

Graf M., Barrettino D., Baltes H.P., and Hierlemann A. (2007) *CMOS Hotplate Chemical Microsensors (Microtechnology and MEMS).* Springer-Verlag, Berlin.

Greenwood R., Mills G., and Vrana B. (eds.) (2007) *Passive Sampling Techniques in Environmental Monitoring.* Elsevier, Amsterdam.

Lee H., Ham D., and Westervelt R.M. (eds.) (2007) *CMOS Biotechnology.* Springer Science & Business Media, New York.

Marks R.S., Lowe C.R., Cullen D.C., Weetall H.H., and Karube I. (eds.) (2007) *Handbook of Biosensors and Biochips.* Wiley-VCH, Weinheim, Germany.

Millner P. (ed.) (2007) *Biosensors: Electrochemical and Mechanical Biosensors.* Scion Publishing, Banbury, UK.

Perez D.M. (ed.) (2007) *Pattern Recognition in Nanoscience, Environmental Engineering and Archeology.* Nova Science, New York.

Petrovic M. and Barcelo D. (eds.) (2007) *Analysis, Fate and Removal of Pharmaceuticals in the Water Cycle.* Elsevier, Amsterdam.

Ripka P. and Tipek A. (eds.) (2007) *Modern Sensors Handbook.* ISTE, London, UK.

Rodriguez J.A. and Fernández-García M. (eds.) (2007) *Synthesis, Properties, and Applications of Oxide Nanomaterials.* Wiley-Interscience, New York.

Schrader T. (ed.) (2007) *Creative Chemical Sensor Systems.* Springer-Verlag, Berlin.

Skotheim T.A. and Reynolds J.R. (eds.) (2007) *Handbook of Conducting Polymers,* 3rd ed. CRC Press, Boca Raton, FL.

Steinem C. and Janshoff A. (eds.) (2007) *Piezoelectric Sensors* (Springer Series on Chemical Sensors and Biosensors, Urban G., series ed.). Springer-Verlag, Berlin.

Thomas O. and Burgess C. (eds.) (2007) *UV-Visible Spectrophotometry of Water and Wastewater.* Elsevier, Amsterdam.

Woodfin R.L. (ed.) (2007) *Trace Chemical Sensing of Explosives.* Wiley-Interscience, Hoboken, NJ.

Zhang X., Ju H., and Wang J. (eds.) (2007) *Electrochemical Sensors, Biosensors and Their Biomedical Applications.* Academic Press, Boston.

2008

Arregui F. (ed.) (2008) *Sensors Based on Nanostructured Materials.* Springer-Verlag, New York.

Bernards D.A., Owens R.M., and Malliaras G.G. (eds.) (2008) *Organic Semiconductors in Sensor Applications.* Springer-Verlag, Berlin.

Daya I. Sooryadas (2008) *Biomimetric Sensor for In-vivo Applications.* VDM Verlag, Berlin.

Demchenko A.P. (2008) *Introduction to Fluorescence Sensing.* Springer Science + Business Media, The Netherlands.

Eftekhari A. (ed.) (2008) *Nanostructured Materials in Electrochemistry.* Wiley-VCH, Weinheim, Germany.

Gasik M. (ed.) (2008) *Materials for Fuel Cells.* Woodhead, Cambridge, UK/CRC Press, Boca Raton, FL.

Inzelt G. (2008) *Conducting Polymers: A New Era in Electrochemistry.* Springer-Verlag, Berlin.

Kolev S. and McKelvie I. (eds.) (2008) *Advances in Flow Injection Analysis and Related Techniques.* Elsevier, Amsterdam.

Ligler F. (ed.) (2008) *Optical Biosensors: Today and Tomorrow,* 2nd ed. Elsevier, Amsterdam.

Meijer G. (ed.) (2008) *Smart Sensor Systems.* Wiley-Interscience, Weinheim, Germany.

Mirsky V.M. and Potyrailo R.A. (eds.) (2008) *Combinatorial Methods for Chemical and Biological Sensors.* Springer-Verlag, Berlin.

Mukhopadhyay S.C. and Gupta G.S. (eds.) (2008) *Smart Sensors and Sensing Technology.* Springer-Verlag, Berlin.

Pabby A.K., Rizvi S.S.H., and Sastre A.M. (eds.) (2008) *Handbook of Membrane Separations: Chemical, Pharmaceutical, Food, and Biotechnological Applications.* CRC Press, Boca Raton, FL.

Picó Y. (ed.) (2008) *Food Contaminants and Residue Analysis.* Elsevier, Oxford, UK.

Tok J.B.H. (ed.) (2008) *Nano and Microsensors for Chemical and Biological Terrorism Surveillance.* Royal Society of Chemistry, London.

Vives A.A. (ed.) (2008) *Piezoelectric Transducers and Applications.* Springer-Verlag, Berlin.

Wallace G.G., Spinks G.M., Kane-Maguire L.A.P., and Teasdale P.R. (2008) *Conductive Electroactive Polymers: Intelligent Polymer Systems,* 3rd ed. CRC Press, New York.

Yin S., Ruffin P.B., and Yu F.T.S. (2008) (eds.) *Fiber Optic Sensors*, 2nd ed. CRC Press, Boca Raton, FL.
Zhuiykov S. (2008) *Electrochemistry of Zirconia Gas Sensors*. CRC Press, Boca Raton, FL.

2009

Alkire R.C., Kolb D.M., Lipkowski J., and Ross P. (eds.) (2009) *Advances in Electrochemical Science and Engineering*, Vol. 11. *Chemically Modified Electrodes*. Wiley-VCH, Weinheim, Germany.
Arregui F.J. (ed.) (2009) *Sensors Based on Nanostructured Materials*. Springer-Verlag, Berlin.
Comeaux R. and Novotny P. (eds.) (2009) *Biosensors: Properties, Materials and Applications*. Nova Science, New York.
Comini E., Faglia G., and Sberveglieri G. (eds.) (2009) *Solid State Gas Sensing*. Springer-Verlag, Berlin.
Decoster D. and Harari J. (eds.) (2009) *Optoelectronic Sensors*. Wiley-ISTE, London, UK.
Deric J. (ed.) (2009) *Biomedical Sensors*. Momentum Press, New York.
Dill K., Liu R.H., and Grodzinsky P. (eds.) (2009) *Microarrays: Preparation, Microfluidics, Detection Methods, and Biological Applications*. Springer-Verlag, New York.
Gerlach G. and Arndt K.F. (eds.) (2009) *Hydrogel Sensors and Actuators* (Springer Series on Chemical Sensors and Biosensors, Urban G., series ed. Springer-Verlag, Heidelberg.
Goldys E.M. (ed.) (2009) *Fluorescence Applications in Biotechnology and Life Sciences*. John Wiley, Hoboken, NJ.
Gouma P. (2009) *Nanomaterials for Chemical Sensors and Biotechnology*. Pan Stanford Publishing, Singapore.
Hayashi K. (ed.) (2009) *Electroanalytical Chemistry Research Trends*. Nova Science, New York.
He N. and Nalwa H.S. (eds.) (2009 *Biochips Nanotechnology*. American Scientific, Valencia, CA.
Huang K. and Goodenough J.B. (2009) *Solid Oxide Fuel Cell Technology: Principles, Performance and Operations*. Woodhead, Cambridge, UK/CRC Press, Boca Raton, FL.
Fan X. (ed.) (2009) *Advanced Photonic Structures for Biological and Chemical Detection (Integrated Analytical Systems)*. Springer-Verlag, New York.
Islam S.K. and Haider M.R. (2009) *Sensors and Low Power Signal Processing*. Springer-Verlag, New York.
Janata J. (2009) *Principles of Chemical Sensors*, 2nd ed. Springer-Verlag, New York.
Kharton V.V. (ed.) (2009) *Solid State Electrochemistry*: Vol. 1, *Fundamentals, Materials and Their Applications*; and Vol. 2, *Electrodes, Interfaces and Ceramic Membranes*. Wiley-VCH, Weinheim, Germany.
Lee J.W. (ed.) (2009) *Methods in Molecular Biology*. Humana Press, Totowa, NJ.
Musgrave G.E., Larsen A.M., and Sgobba T. (eds.) (2009) *Safety Design for Space Systems*. Elsevier, Oxford, UK.
Merkoçi A. (ed.) (2009) *Biosensing Using Nanomaterials*. John Wiley, Hoboken, NJ.
Mukhopadhyay S.C., Gupta G.S., and Huang Y.-M. (eds.) (2009) *Recent Advances in Sensing Technology* (Lecture Notes in Electrical Engineering). Springer-Verlag, Berlin.
Naidi S.M.J. and Matsuura T. (eds.) (2009) *Polymer Membranes for Fuel Cells*. Springer-Verlag, New York.
Orellana G., Moreno-Bondi M.C., and Wolfbeis O.S. (eds.) (2009) *Frontiers in Chemical Sensors: Novel Principles and Techniques*. Springer-Verlag, Berlin.
Potyrailo R.A. and Mirsky V.M. (eds.) (2009) *Combinatorial Methods for Chemical and Biological Sensors (Integrated Analytical Systems)*. Springer-Verlag, New York.
Rao G. (ed.) (2009) *Optical Sensor Systems in Biotechnology*. Springer-Verlag, Berlin.
Raol J.R. (2009) *Multi-sensor Data Fusion with MATLAB*. CRC Press, Boca Raton, FL.
Righini G.C., Tajani A., and Cutolo A. (2009) *An Introduction to Optoelectronic Sensors*. World Scientific, London, UK.
Ríos A., Escarpa A., and Simonet B.T. (2009) *Miniaturization of Analytical Systems: Principles, Designs and Applications*. John Wiley, Chichester, UK.

Ryan M.A., Shevade A.V., Taylor C.J., Homer M.L., Blanco M., and Stetter J.R. (eds.) (2009) *Computational Methods for Sensor Material Selection*. Springer-Verlag, New York.

Shinar R. and Shinar J. (eds.) (2009) *Organic Electronics in Sensors and Biotechnology.* McGraw-Hill, New York.

Soloman S. (2009*) Sensors Handbook*, 2nd ed.). McGraw-Hill, New York.

Tian W.-C. and Finehout E. (eds.) (2009) *Microfluidics for Biological Applications.* Springer-Verlag, New York.

Tinoco A.V. (2009) *Improving the Performance of Micro-machined Metal Oxide Gas Sensors.* VDM Verlag, Saarbrücken, Germany.

Tseng T.-Y. and Nalwa H.S. (eds.) (2009) *Handbook of Nanoceramics and Their Based Nanodevices*. American Scientific, Valencia, CA.

Turner J. (ed.) (2009) *Automotive Sensors.* Momentum Press, New York.

Umasankar Y., Kumar S.A., and Chen S.-M. (eds.) (2009) *Nanostructured Materials for Electrochemical Biosensors.* Nova Science, New York.

Vielstich W., Gasteiger H.A., and Yokokawa H. (eds.) (2009) *Handbook of Fuel Cells: Fundamentals, Technology, Applications*, Vols. 5 and 6: *Advances in Electrocatalysis, Materials, Diagnostics and Durability*. John Wiley, New York.

Zribi A. and Fortin J. (eds.) (2009) *Functional Thin Films and Nanostructures for Sensors.* Springer-Verlag, New York.

2010

Bakeev K. (ed.) (2010) *Process Analytical Technology.* Wiley-Blackwell, Chichester, UK.

Baronas R., Ivanauskas F., and Kulys J. (2010) *Mathematical Modeling of Biosensors: An Introduction for Chemists and Mathematicians.* Springer-Verlag, Berlin.

Cejka J., Corma A., and Zones S. (eds.) (2010) *Zeolites and Catalysis. Synthesis, Reactions and Applications.* Wiley-VCH, Weinheim, Germany.

Harrison R.V. (ed.) (2010) *Chemical Sensors: Properties, Performance and Applications.* Nova Science, New York.

Haus J. (2010) *Optical Sensors. Basics and Applications.* Wiley-VCH, Berlin.

Hierlemann A., Graf M., Barretino D., and Baltes H.P. (2010) *CMOS Hotplate Chemical Microsensors.* Springer-Verlag, Berlin.

Fraden J. (ed.) (2010) *Handbook of Modern Sensors: Physics, Designs, and Applications*, 4th ed. Springer-Verlag, New York/Berlin.

Geddes C.D. (ed.) (2010) *Metal-Enhanced Fluorescence.* John Wiley, Hoboken, NJ.

Ghafar-Zadeh E. and Sawan M. (2010) *CMOS Capacitive Sensors for Lab-on-Chip Applications.* Springer-Verlag, Dordrecht, The Netherlands.

Grattan L.S. and Meggitt B.T. (eds.) (2010) *Optical Fiber Sensor Technology.* Vol. 4: *Chemical and Environmental Sensing.* Springer-Verlag, Dordrecht, The Netherlands.

Jamal M. (2010) *Biosensor for Androgens.* Lamp Lambert Academic Publishing, Saarbrücken, Germany.

Kilcast D. (ed.) (2010) *Sensory Analysis for Food and Beverage Quality Control: A Practical Guide.* Woodhead Publishing, Cambridge, UK.

Korotcenkov G. (ed.) (2010) *Chemical Sensors: Fundamentals of Sensing Materials.* Vol. 1: *General Approaches.* Momentum Press, New York.

Korotcenkov G. (ed.) (2010) *Chemical Sensors: Fundamentals of Sensing Materials.* Vol. 2: *Nanostructured Materials.* Momentum Press, New York.

Lange D., Brand O., and Baltes H. (2010) *CMOS Cantilever Sensor Systems.* Springer-Verlag, Berlin.

Lobnik A. and Mizaikoff B. (2010) *Handbook of Optical Chemical Sensors.* Springer-Verlag, Berlin.

Mirsky V.M. and Yatsimirsky A.K. (2010) *Artificial Receptors for Chemical Sensors*. Wiley-VCH, Weinheim, Germany.

Mukhopadhyay S.C. and Huang Y.-M. (eds.) (2010) *Sensors: Advancements in Modeling, Design Issues, Fabrication and Practical Applications* (Lecture Notes in Electrical Engineering, Vol. 21). Springer-Verlag, Berlin.

Narayanaswamy R. and Wolfbeis O.S. (eds.) (2010) *Optical Sensors. Industrial, Environmental and Diagnostic Applications* (Springer Series on Chemical Sensors and Biosensors, Wolfbeis O.S., series ed.). Springer-Verlag, Berlin/Heidelberg.

Pawliszyn J. and Lord H.L. (eds.) (2010) *Handbook of Sample Preparation*. John Wiley, Hoboken, NJ.

Pierce D.T. and Zhao J.X. (eds.) (2010) *Trace Analysis with Nanomaterials*. Wiley-VCH, Weinheim, Germany.

Prasada Rao T., James D., and Elias M.T. (2010) Electrochemical Sensing of Deadly Toxin-Atrazine: An Overview. Nova Science, New York.

Ram M.K. and Bhethanabotla V.R. (eds.) (2010) *Sensors for Chemical and Biological Applications*. CRC Press, Boca Raton, FL.

Rivas G.A., Luque G., Rubianes M.R., Pedano M.L., Ferreyara N.F., and Miscoria S.A. (2010) *Carbon Nanotubes: A New Alternative for Electrochemical Sensors*. Nova Science, New York.

Schrader T. (ed.) (2010) *Creative Chemical Sensor Systems*. Springer-Verlag, Berlin.

Torriero A.A.J. and Shiddiky M.J.A. (eds.) (2010) *Electrochemical Properties and Applications of Ionic Liquids*. Nova Science, New York.

Vetelino J. and Reghu A. (2010) *Introduction to Sensors*. CRC Press, Boca Raton, FL.

Zourob M. and Lakhtakia A. (eds.) (2010) *Optical Guided-Wave Chemical and Biosensors II* (Springer Series on Chemical Sensors and Biosensors). Springer-Verlag, Berlin.

2011

Baldrich E. and Garcia-Aljaro C. (2011) *Pathogen Detection Methods: Biosensor Development*. Nova Science, New York.

Castillo J.M. (2011) *Relative Humidity: Sensors, Management and Environmental Effects*. Nova Science, Hauppauge, NY.

Cusano A., Cutolo A., and Albert J. (eds.) (2011) *Fiber Bragg Grating Sensors: Recent Advancements, Industrial Applications and Market Exploitation*. Bentham Science, Oak Park, IL.

Du W. (2011) *Resistive, Capacitive, and Inductive Based Sensing Technologies*. CRC Press, Boca Raton, FL.

Eranna G. (2011) *Metal Oxide Nanostructures as Gas Sensing Devices*. CRC Press, Boca Raton, FL.

Khanna V.K. (2011) *Nanosensors: Physical, Chemical and Biological*. CRC Press, Boca Raton, FL.

Korotcenkov G. (ed.) (2011) *Chemical Sensors: Fundamentals of Sensing Materials*. Vol. 3: *Polymers and Other Materials*. Momentum Press, New York.

Korotcenkov G. (ed.) (2011) *Chemical Sensors: Comprehensive Sensor Technologies*. Vol. 4: *Solid State Devices*. Momentum Press, New York.

Korotcenkov G. (ed.) (2011) *Chemical Sensors: Comprehensive Sensor Technologies*. Vol. 5: *Electrochemical and Optical Sensors*. Momentum Press, New York.

Korotcenkov G. (ed.) (2011) *Chemical Sensors: Comprehensive Sensor Technologies*. Vol. 6: *Chemical Sensors Applications*. Momentum Press, New York.

Lai D.T.H. (ed.) *Healthcare Sensor Networks: Challenges Toward Practical Implementation*. CRC Press, Boca Raton, FL.

Lange U. (2011) *Electrochemical Transistor and Chemoresistor Based Sensors: Measurement Technique, Materials and Applications*. SVH Verlag, Saarbrücken, Germany.

Li S., Singh J., Li H., and Banerjee I.A. (eds.) (2011) *Biosensor Nanomaterials*. Wiley-VCH, Weinheim, Germany.

Okada C.T. (ed.) (2011) Humidity Sensors: Types, Nanomaterials and Environmental Monitoring. Nova Science, New York.

Reithmaier J.P., Paunovic P., Kulisch W., Popov C., and Petkov P. (eds.) (2011) *Nanotechnological Basis for Advanced Sensors* (NATO Science for Peace and Security Series B: Physics and Biophysics). Springer-Verlag, Berlin.

Ren F. and Pearton S.J. (eds.) (2011) *Semiconductor Device-Based Sensors for Gas, Chemical, and Biomedical Applications.* CRC Press, Boca Raton, FL.

Schlücker S. (ed.) (2011) *Surface Enhanced Raman Spectroscopy: Analytical, Biophysical and Life Science Applications.* Wiley-CVH, Weinheim, Germany.

Shin W., Nishibori M., and Matsubara I. (2011) *Integration Technologies for Sensors. Ceramic Integration and Joining Technologies: From Macro- to Nanoscales.* John Wiley, New York.

Silverman S.K. (ed.) (2011) *Nucleic Acid Switches and Sensors.* Springer-Verlag, New York.

Stone D.C. (2011) *Chemical and Biological Sensors.* Royal Society of Chemistry, London.

Tagawa T., Tamura T., and Oberg P.A. (2011) *Biomedical Sensors and Instruments*, 2nd ed. CRC Press, Boca Raton, FL.

Weng Q. (ed.) (2011) *Advances in Environmental Remote Sensing: Sensors, Algorithms, and Applications* (Remote Sensing Applications Series). CRC Press, Boca Raton, FL.

Wide P. (ed.) (2011) *Artificial Human Sensors: Science and Applications.* CRC Press, Boca Raton, FL.

Woo H.-G. and Li H. (eds.) (2011) *Advances Functional Materials.* Springer-Verlag, Berlin.

2012

Arregui F.J., Cusano A., Cutolo A., and Giordano M. (2012) *Optical Chemical Nanosensors.* CRC Press, Boca Raton, FL.

Herold K.E. and Rasooly A. (eds.) (2012) *Biosensors for Cancer Diagnostics.* CRC Press, Boca Raton, FL.

Higson S. (ed.) (2012) Biosensors for Medical Applications. Woodhead, Cambridge, UK.

Iniewski K. (ed.) (2012) *Sensors for Chemical, Optical, Biological, and Medical Applications.* CRC Press, Boca Raton, FL.

Korotcenkov G. (ed.) (2012) *Chemical Sensors: Simulation and Modeling.* Vol. 1: *Microstructural Characterization and Modeling of Metal Oxides.* Momentum Press, New York.

Korotcenkov G. (ed.) (2012) *Chemical Sensors: Simulation and Modeling.* Vol. 2: *Conductometric-Type Sensors.* Momentum Press, New York.

Korotcenkov G. (ed.) (2012) *Chemical Sensors: Simulation and Modeling.* Vol. 3: *Solid-State Sensors.* Momentum Press, New York.

Korotcenkov G. (ed.) (2012) *Chemical Sensors: Simulation and Modeling.* Vol. 4: *Optical Sensors.* Momentum Press, New York.

INDEX